COMPREHENSIVE ANALYTICAL CHEMISTRY

Elsevier
Radarweg 29, PO Box 211, 1000 AE Amsterdam, The Netherlands
Linacre House, Jordan Hill, Oxford OX2 8DP, UK

First edition 2007

Library of Congress Cataloging-in-Publication Data
A catalog record for this book is available from the Library of Congress

British Library Cataloguing in Publication Data
A catalogue record for this book is available from the British Library

ISBN: 978-0-444-53052-3
ISSN: 0166-526X

Printed and bound in The Netherlands

07 08 09 10 11 10 9 8 7 6 5 4 3 2 1

COMPREHENSIVE ANALYTICAL CHEMISTRY

ADVISORY BOARD

Wilson & Wilson's

COMPREHENSIVE ANALYTICAL CHEMISTRY

Edited by

D. BARCELÓ

Research Professor
Department of Environmental Chemistry
IIQAB-CSIC
Jordi Girona 18-26
08034 Barcelona
Spain

Wilson & Wilson's

COMPREHENSIVE ANALYTICAL CHEMISTRY

VOLUME 50

ANALYSIS, FATE AND REMOVAL OF PHARMACEUTICALS IN THE WATER CYCLE

Edited by

M. PETROVIĆ

Department of Environment Chemistry, IIQAB-CSIC, c/Jordi Girona 18-26, 08034 Barcelona, Spain
and
Institució Catalana de Recerca i Estudis Avançats (ICREA), Passeig Lluis Companys 23, 80010 Barcelona, Spain

D. BARCELÓ

Department of Environmental Chemistry, IIQAB-CSIC, Jordi Girona 18-26, 08034 Barcelona, Spain

ELSEVIER

AMSTERDAM – BOSTON – HEIDELBERG – LONDON – NEW YORK – OXFORD – PARIS
SAN DIEGO – SAN FRANCISCO – SINGAPORE – SYDNEY – TOKYO

CONTRIBUTORS TO VOLUME 50

Javier Adrian
Applied Molecular Receptors Group (AMRg), Department of Biological Organic Chemistry, IIQAB-CSIC, Jordi Girona, 18-26, 08034-Barcelona, Spain

William A. Arnold
Department of Civil Engineering, University of Minnesota, 500 Pillsbury Dr. SE, Minneapolis, MN 55455, USA

Damià Barceló
Department of Environmental Chemistry, IIQAB-CSIC, Jordi Girona 18-26, 08034 Barcelona, Spain

Kenneth Carlson
Department of Civil and Environmental Engineering, Colorado State University, Fort Collins, CO 80523-1372, USA

Josefina Carpinteiro
Department of Analytical Chemistry, Nutrition and Food Sciences, Institute of Food Research and Analysis, University of Santiago de Compostela, 15782 Santiago de Compostela, Spain

Christian G. Daughton
Environmental Chemistry Branch, Environmental Sciences Division, National Exposure Research Laboratory, U.S. Environmental Protection Agency, 944 East Harmon Ave., Las Vegas, NV, USA

M. Silvia Díaz-Cruz
Department of Environmental Chemistry, IIQAB-CSIC, Jordi Girona 18-26, 08034 Barcelona, Spain

Jörg E. Drewes
Environmental Science & Engineering, Colorado School of Mines, Golden, CO 80401-1887, USA

Amadeo Rodríguez Fernández-Alba
Department of Analytical Chemistry, University of Almería, 04120 Almería, Spain

Roger Galve
Applied Molecular Receptors Group (AMRg), Department of Biological Organic Chemistry, IIQAB-CSIC, Jordi Girona, 18-26, 08034-Barcelona, Spain

Meritxell Gros
Department of Environmental Chemistry, IIQAB-CSIC, Jordi Girona 18-26, 08034 Barcelona, Spain

Thomas Heberer
Federal Institute for Risk Assessment, Department for Residues of Medicinal Products (55), Dledersdorfer Weg 1, 12277 Berlin, Germany

Maria Dolores Hernando Guil
Department of Analytical Chemistry, University of Almería, Ctra. Sacramento, 04120 Almería, Spain

Martin Jekel
Department of Water Quality Control, Technical University Berlin, Sekr. KF 4, Strasse des 17. Juni 135, 10623 Berlin, Germany

Oliver A.H. Jones
Department of Biochemistry, University of Cambridge, Tennis Court Road, Cambridge, CB2 1GA, UK

Sung-Chul Kim
Department of Civil and Environmental Engineering, Colorado State University, Fort Collins, CO 80523-1372, USA

Mark Kreuzer
Applied Molecular Receptors Group (AMRg), Department of Biological Organic Chemistry, IIQAB-CSIC, Jordi Girona, 18-26, 08034-Barcelona, Spain

Marina Kuster
Department of Environmental Chemistry, IIQAB-CSIC, Jordi Girona 18-26, 08034 Barcelona, Spain

Katherine Langford
Norwegian Institute for Water Research (NIVA), Gaustadalléen 21, NO-0349 Oslo, Norway

John N. Lester
Centre for Water Science, Cranfield University, Cranfield, Bedfordshire, MK43 OAL, UK

Maria J. Lopez de Alda
Department of Environmental Chemistry, IIQAB-CSIC, Jordi Girona 18-26, 08034 Barcelona, Spain

M.-Pilar Marco
Applied Molecular Receptors Group (AMRg), Department of Biological Organic Chemistry, IIQAB-CSIC, Jordi Girona, 18-26, 08034-Barcelona, Spain

Kristopher McNeill
Department of Chemistry, University of Minnesota, 207 Pleasant St. SE, Minneapolis, MN 55455, USA

Chris D. Metcalfe
Environmental and Resource Studies, Trent University, 1600 West Bank Drive, Peterborough, Ontario, K9J 7B8, Canada

Xiu-Sheng Miao
Environmental and Resource Studies, Trent University, 1600 West Bank Drive, Peterborough, Ontario, K9J 7B8, Canada

Mira Petrović
Department of Environmental Chemistry, IIQAB-CSIC, Jordi Girona 18-26, 08034 Barcelona, Spain

Daniel G. Pinacho
Applied Molecular Receptors Group (AMRg), Department of Biological Organic Chemistry, IIQAB-CSIC, Jordi Girona, 18-26, 08034-Barcelona, Spain

Anke Putschew
Department of Water Quality Control, Technical University Berlin, Sekr. KF 4, Strasse des 17. Juni 135, 10623 Berlin, Germany

José Benito Quintana
Department of Analytical Chemistry, Nutrition and Food Sciences, Institute of Food Research and Analysis, University of Santiago de Compostela, 15782 Santiago de Compostela, Spain

Jelena Radjenovic
Department of Environmental Chemistry, IIQAB-CSIC, Jordi Girona 18-26, 08034 Barcelona, Spain

Antonio Rodríguez Fernández -Alba
Department of Chemical Engineering and Analytical Chemistry, University of Alcalá de Henares, 28871 Alcalá de Henares, Spain

Isaac Rodríguez
Department of Analytical Chemistry, Nutrition and Food Sciences, Institute of Food Research and Analysis, University of Santiago de Compostela, 15782 Santiago de Compostela, Spain

Sara Rodríguez-Mozaz
Department of Environmental Chemistry, IIQAB-CSIC, Jordi Girona 18-26, 08034 Barcelona, Spain

J.-Pablo Salvador
Applied Molecular Receptors Group (AMRg), Department of Biological Organic Chemistry, IIQAB-CSIC, Jordi Girona, 18-26, 08034-Barcelona, Spain

Francisco Sánchez-Baeza
Applied Molecular Receptors Group (AMRg), Department of Biological Organic Chemistry, IIQAB-CSIC, Jordi Girona, 18-26, 08034-Barcelona, Spain

Kevin V. Thomas
Norwegian Institute for Water Research (NIVA), Gaustadallèen 21, NO-0349 Oslo, Norway
Nikolaos Voulvoulis
Centre for Environmental Policy, Imperial College London, London SW7 2AZ, UK

WILSON AND WILSON'S

COMPREHENSIVE ANALYTICAL CHEMISTRY

VOLUMES IN THE SERIES

Vol. 1A	Analytical Processes
	Gas Analysis
	Inorganic Qualitative Analysis
	Organic Qualitative Analysis
	Inorganic Gravimetric Analysis
Vol. 1B	Inorganic Titrimetric Analysis
	Organic Quantitative Analysis
Vol. 1C	Analytical Chemistry of the Elements
Vol. 2A	Electrochemical Analysis
	Electrodeposition
	Potentiometric Titrations
	Conductometric Titrations
	High-Frequency Titrations
Vol. 2B	Liquid Chromatography in Columns
	Gas Chromatography
	Ion Exchangers
	Distillation
Vol. 2C	Paper and Thin Layer Chromatography
	Radiochemical Methods
	Nuclear Magnetic Resonance and Electron Spin Resonance Methods
	X-Ray Spectrometry
Vol. 2D	Coulometric Analysis
Vol. 3	Elemental Analysis with Minute Sample
	Standards and Standardization
	Separation by Liquid Amalgams
	Vacuum Fusion Analysis of Gases in Metals
	Electroanalysis in Molten Salts
Vol. 4	Instrumentation for Spectroscopy
	Atomic Absorption and Fluorescence Spectroscopy
	Diffuse Reflectane Spectroscopy
Vol. 5	Emission Spectroscopy
	Analytical Microwave Spectroscopy
	Analytical Applications of Electron Microscopy
Vol. 6	Analytical Infrared Spectroscopy
Vol. 7	Thermal Methods in Analytical Chemistry
	Substoichiometric Analytical Methods
Vol. 8	Enzyme Electrodes in Analytical Chemistry
	Molecular Fluorescence Spectroscopy
	Photometric Titrations
	Analytical Applications of Interferometry

Volumes in the series

Volumes in the series

Vol. 39	Integrated Analytical Systems
Vol. 40	Analysis and Fate of Surfactants in the Aquatic Environment
Vol. 41	Sample Preparation for Trace Element Analysis
Vol. 42	Non-destructive Microanalysis of Cultural Heritage Materials
Vol. 43	Chromatographic-mass spectrometric food analysis for trace determination of pesticide residues
Vol. 44	Biosensors and Modern Biospecific Analytical Techniques
Vol. 45	Analysis and Detection by Capillary Electrophoresis
Vol. 46	Proteomics and Peptidomics New Technology Platforms Elucidating Biology
Vol. 47	Modern Instrumental Analysis
Vol. 48	Passive Sampling Techniques in Environmental Monitoring
Vol. 49	Electrochemical (Bio) Sensor Analysis

Contents

Chapter 2. Analysis of pharmaceuticals as environmental contaminants

Editors' preface

Pharmaceuticals are a diverse group of chemicals used in veterinary medicine, agricultural practices, human health and cosmetic care. Many are highly bioactive, most are polar, many are optically active, and all (when present in the environment) occur usually in not more than trace concentrations.

Pharmaceuticals are a new class of so-called emerging contaminants that have raised great concern in the last years. Pharmaceuticals deserve attention, firstly, because of their continuous introduction via effluents from sewage treatment facilities and from septic systems. They are referred to as "pseudo" persistent contaminants (i.e. high transformation/removal rates are compensated by their continuous introduction into environment). Secondly, they are developed with the intention of performing a biological effect. Thirdly, they often have the same type of physico-chemical behaviour as other harmful xenobiotics (persistence that avoid the substance to be inactive before having a curing effect, and lipophilicity in order to be able to pass through membranes) and fourthly they are used by man in rather large quantities (i.e. similar to those of many pesticides).

The continuous introduction of pharmaceuticals and their bioactive metabolites into the environment may lead to high long-term concentrations and promote continual, but unnoticed adverse effects on aquatic and terrestrial organisms. During the last few years attention is being given to develop a better understanding of the toxicology issues including low-dose multi-generational exposure to multiple chemical stressors and how human and ecological risks might be affected by these chemical cocktails.

The main objective of this book is to provide the reader with a well-founded overview of the state of the art of the analytical methods for trace determination of pharmaceuticals in the environmental samples, and to give a review of the fate and occurrence of pharmaceuticals in the water cycle (elimination in waste- and drinking-water treatment), including latest developments in the treatment technologies, such as membrane bioreactors, advance oxidation and natural attenuation processes.

To reach these objectives, the book includes a concise and critical compilation of the information published in the past years regarding all analytical aspects of the trace determination of pharmaceutical residues in the environmental samples, including advanced technologies for sample preparation of aqueous and solid samples, clean-up protocols and analysis. The analytical methodology for the determination of trace pharmaceuticals in complex environmental matrices is still evolving and the number of methods described in the literature has grown considerably. Ten years ago, GC-MS with derivatisation was the method of choice due to the possibility to go as low as low nanogram per litre levels. Nowadays, LC-MS-MS and hybrid MS systems involving time-of-flight (TOF) and other analysers are the methods used due to their increased availability, high sensitivity and the fact that there is no need of derivatisation of the samples, as it is in GC-MS.

The *first chapter* deals with the general introduction of the problem of pharmaceuticals as environmental contaminants indicating their sources and management options.

The *second chapter* comprises the largest part of the book and it is devoted to the analysis of pharmaceuticals and consists of eight sub-chapters dealing with modern analytical techniques for the unequivocal detection of all main classes of pharmaceuticals (antibiotics, anti-inflammatory drugs, β-blockers, lipid regulating agents, sex hormones, X-ray contrast agents and psychiatric drugs) in liquid (wastewater, surface, ground- and drinking water) and solid matrices (soil, sediment and sludge). The chapter is mainly devoted to highly sophisticated and established hyphenated mass spectrometric methods such as LC-MS and LC-MS-MS and GC-MS. In addition, sample preparation methods are thoroughly evaluated for all groups of pharmaceuticals including their major metabolites. Finally, one of the sub-chapters addresses the application of bioassays and biosensors for the analysis of pharmaceuticals in the environment.

The *third chapter* gives an overview on the occurrence data in all environmental compartments including sewage sludge, as well as transformation processes of pharmaceuticals in the environment, including photolysis and other processes, and a final sub-chapter provides an overview of toxicological data.

The *fourth chapter* deals with the removal of pharmaceuticals in waste- and drinking-water treatment, including discussions on biotic and abiotic removal mechanisms. Of the treatment techniques discussed, not only conventional wastewater treatment (activated sludge) is evaluated, but also innovative treatment technologies such as membrane reactors and advanced oxidation processes are evaluated.

Finally, the *fifth chapter* summarizes the current state of the art in the field and outlines future trends and research needs.

Overall the book is certainly timely since the interest and the developments on the analysis, fate and removal of pharmaceuticals in the environment have grown considerably during the last few years. This book will be of interest to a broader audience of analytical chemists and environmental scientists who are working in the field of pharmaceuticals in the water cycle as well as newcomers who want to learn more about this emerging contamination problem.

Finally, we would like to thank all the contributing authors of this book for their time and efforts in preparing their chapters. Without their cooperation and engagement this volume would certainly not have been possible.

M. Petrović and D. Barceló
Barcelona, January 4th 2007

Series editor's preface

As Series Editor of CAC I feel that I have certain duties. The first one is to be able to acquire new titles for this successful series in the field of analytical chemistry. As a second duty I feel that I need to bring also titles from my own field of expertise. In this respect in 2003 I was the co-editor of Volume 40 of the series on *Analysis and Fate of Surfactants in the Aquatic Environment* together with my two old friends, Thomas Knepper and Pim de Voogt. In 2007, and ten volumes later, Volume 50 is being published, and again I am the co-editor together with my colleague Mira Petrović of this book entitled *Analysis, Fate and Removal of Pharmaceuticals in the Water Cycle.*

Pharmaceuticals in the aquatic environment has been a topic of interest in conferences and in the literature for the past ten years. One of the reasons for the increasing concern on pharmaceuticals has certainly been the improvement in analytical techniques.

Nowadays, after sampling and conventional solid phase extraction of a surface river water sample followed by liquid chromatography–tandem mass spectrometry it is possible to easily detect nanogram per litre levels of common pharmaceutical residues in natural water samples. After analysis, environmental analytical chemists start performing monitoring programmes and this is the reason why every month several papers on the topic of occurrence of pharmaceutical residues in the water cycle are published, mainly surface waters and wastewaters.

The fate of pharmaceuticals during sewage treatment is a key issue, since wastewater treatment processes represent point source pollution of human pharmaceuticals. Investigation of removal technologies is also of high interest to the scientific community, always linked to analytical chemistry, because engineers need the high level of expertise provided by analytical chemists to determine pharmaceuticals residues at low nanogram per litre levels in very complex wastewaters and sludge matrices which is the challenging task. Finally, the growing occurrence of human and veterinary pharmaceuticals in the environment is leading to toxicological studies and publications on ecological and risk assessment.

All the above-mentioned topics have been included in the book. So, analytical chemists will find a detailed chapter containing eight subchapters on analysis but at the same time the other chapters will bring the necessary information to understand the problems related to the sources, fate, toxicity and removal of pharmaceuticals. The book brings a comprehensive view on the problems associated with this new and emerging problem of pharmaceutical residues in the environment and it is addressed to a broad audience, from experts in the field to newcomers.

The book is also well balanced with regard to the geographical location of the contributing authors, with US/Canada and European scientists, indicating a very similar problem to be tackled on both sides of the Atlantic Ocean. Finally, I would like to thank all the authors, many of them friends and colleagues for their efforts in compiling the literature references and writing their book chapters. My special thanks go to my co-worker and colleague at the Department, Mira Petrović, for her efforts and time spent communicating with the different contributors and correcting and harmonizing the different chapters of this comprehensive book on pharmaceuticals in the environment.

D. Barceló
Department of Environmental Chemistry, IIQAB-CSIC
Barcelona, Spain

Glossary of acronyms and abbreviations

ACN	Acetonitrile
ADME	Absorption, distribution, metabolism, and elimination
AMDOPH	1-acetyl-1-methyl-2-dimethyl-oxamoyl-2-phenylhydrazide
ANDR	Androstenedione
AOI	Adsorbable organic bound iodine
AOP	Advanced oxidation processes
AP	Alkaline phosphatase
AP	Alkylphenol
APCI	Atmospheric pressure chemical ionisation
APEC	Alkylphenol ethoxy carboxylates
APE_nO	Alkylphenol ethoxylates (*n* is number of ethoxy groups)
API	Active pharmaceutical ingredient
API	Atmospheric pressure ionisation
APPI	Atmospheric pressure photoionization
ASE	Accelerated solvent extraction
BAC	Benzalkonium chloride
BOD_5	Biochemical oxygen demand
BPA	Bisphenol
BRET	Bioluminescence resonance energy transfer
BSTFA	N,O-bis(trimethylsilyl)fluoroacetamide
CAAP	Concentrated aquatic animal production
CAFO	Confined (concentrated) animal feeding operations
CAS	Conventional activated sludge
CBZ	Carbamazepine
COD	Chemical oxygen demand
CSA	Controlled Substances Act
DAD	Diode-array detector
DEA	Drug Enforcement Administration
DHT	Dihydrotestosterone

DMF	Dimethylformamide
DOC	Dissolved organic carbon
DOM	Dissolved organic matter
DTC	Direct-to-consumer advertising
DTE	4-dithioerythritol
E1	Estrone
E1-3G	Estrone 3-(β-D-glucuronide)
E1-3S	Estrone 3-sulfate
E2	Estradiol
E2-17-acet.	Estradiol-17-acetate
E2-3,17diS	17β-estradiol 3,17-disulfate
E2-3G	17β-estradiol 3-(β-D-glucuronide)
E2-17G	17β-estradiol 17-(β-D-glucuronide)
E2-17-valer.	Estradiol-17-valerate
E2-3S	17β-estradiol 3-sulfate
E2-3S	17G, 17β-estradiol 3-sulfate 17-glucuronide
E3	Estriol
E3-3G	Estriol 3-(β-D-glucuronide)
E3-16G	Estriol 16α-(β-D-glucuronide)
E3-3S	Estriol 3-sulfate
EA	Environmental Assessments
EC_{50}	Median effect concentration
ECD	Electrochemical detection
EDCs	Endocrine disrupting compounds
EE	Ethynyl estradiol
FBCl	Pentafluorobenzylbenzene
EDC	Endocrine disrupting compund
EI	Electron ionization
EIA	Enzyme immunoassay
ELBRA	Enzyme-linked bioluminescent receptor assay
ELISA	Enzyme-linked immunosorbent assay
ELRA	Enzyme linked receptor assay
EPA	Environmental Protection Agency
ER	Estrogen receptor
ERA	Environmental risk assessment
ESI	Electrospray ionisation
FDA	Federal Drug Administration
FIA	Fluoroimmunoassays
FID	Flame ionisation detector (used as detection system in GC)
FL	Fluorescence (detection)

FMAT	Fluorometric microvolume assay technology
FP	Fluorescent polarization
FRET	Fluorescence resonance energy transfer
GAC	Granular activated carbon
GC	Gas chromatography
GCxGC	Two-dimensional gas chromatography
GC-MS	Gas chromatography-Mass spectrometry
GMO	Genetically modified organisms
GOx	Glucose oxidase
GPC	Gel permeation chromatography
GWR	Groundwater replenishment
HHRA	Human health risk assessment
HR	High resolution
HRMS	High resolution mass spectrometry
HRP	Horseradish peroxidase
HRT	Hydraulic residence time
IA	Immunoaffinity
ICM	Iodinated X-ray contrast media
IISF	Induced in-source fragmentation
IP	Ionophores
IS	Ionspray or pneumatically assisted electrospray
ISE	Ion-selective electrode
IT	Ion trap
LC	Liquid chromatography
LC-MS	Liquid chromatography-Mass spectrometry
LEV	Levonorgestrel
LIT	Linear ion trap
LLE	Liquid-liquid extraction
LOEC	Lowest observed effect concentration
LOD	Limit of detection
LOQ	Limit of quantification
LPME	Liquid-phase microextraction
LSE	Liquid solid extraction
LRMS	Low resolution mass spectrometry
LTCF	Long-term care facilities
LVI	Large volume injection
MAE	Microwave assisted extraction
MBR	Membrane bioreactors
MDWPU	Mobile drinking water purification units
MES	Mestranol
MF	Microfiltration

MIP	Molecularly imprinted polymers
ML	Macrolides
MLSS	Mixed liquor suspended solids
MRM	Multiple reaction monitoring
MS	Mass spectrometry
MS-MS	Tandem mass spectrometry
MSTFA	N-methyl-N-(trimethylsilyl)-trifluoroacetamide
MTBSTFA	*N*-methyl-*N*-(*tert*.-butyldimethylsilyltrifluoroacetamide
MWCO	Molecular weight cutoff
NBBS	Butyl benzenesulfonamides
NCI	Negative chemical ionisation
NF	Nanofiltration
NOEC	No observed effect concentration
NOR	Norethindrone
NOR-acetate	Norethindrone acetate
NP	4-nonylphenol
NSAID	Non-steroidal anti-inflammatory drugs
OA	Oxolinic acid
oaTOF	Orthogonal-acceleration time-of-flight
OP	Octylphenol
OTC	Over-the-counter (drugs)
OTC	Oxytetracycline
PB	Particle beam
PBT	Persistence, bioaccumulation and toxic
PCB	Polychlorinated biphenyls
PCP	Personal care products
PEC	Predicted environmental concentrations
PEN	Penicillin
PFBBR	Pentafluorobenzyl bromide
PFBCl	Pentafluorobenzoyl chloride
PFIA	Polarization Fluorescent Immunoassay
PFPA	Pentafluropropionic anhydride
PhAC	Pharmaceutically Active Compounds
PLE	Pressurized liquid extraction
PLOT	Porous layer open tubular
PMP	Plant-made pharmaceuticals
PNEC	Predicted no-effect concentration
POP	Persistent organic pollutant
POTW	Publicly owned treatment works
PPCP	Personal care products
PPRI	Photochemically produced reactive intermediate

PREG	Pregnenolone
PROG	Progesterone
QA/QC	Quality assurance/quality control
QqQ	Triple quadrupole (MS)
Qq-LIT	Quadrupole-linear ion trap
Qq-TOF	Quadrupole-time-of-flight
RIA	Radioimmunoassay
RAM	Restricted access material
RO	Reverse osmosis
ROC	Receiver Operation Characteristic
RRA	Radio receptor assay
SA	Sulfonamide
SAT	Soil-aquifer treatment
SBSE	Stir bar sorptive extraction
SCP	Sulfachloropyridazine
SFE	Supercritical fluid extraction
SIM	Selected ion monitoring
SM-MIPs	Surface modified molecularly imprinted polymers
SOX	Soxhlet (extraction)
SPA	Scintillation proximity assay
SPE	Solid phase extraction
SPME	Solid phase microextraction
SPR	Surface Plasmon Resonance
SRM	Selected reaction monitoring
SRT	Solids retention time
SS	Surrogate standard
SS	Suspended solids
STP	Sewage treatment plant
TBDMCS	*tert*-butyldimethylchlorosilane
TBT	Tributyltin
TC	Tetracycline
TCPP	Tris-propylphosphate
TEA	Triethylamine
TEST	Testosterone
TFA	Trifluoroacetic
TIE	Toxicity identification evaluation
TMCS	Trimethylchlorosilane
TMD	Therapeutic drug monitoring
TMSI	Trimethylsilylimidazole
TOC	Total organic carbon
TOC/DOC	Total/dissolved organic carbon

ToF	Time of flight
TRF	Time-resolved fluorescence
TSP	Termospray
TYL	Tylosin
UF	Ultrafiltration
ULPRO	ultra-low-pressure reverse osmosis
USE	Ultrasonic extraction
US-EPA	United States Environmental Protection Agency (http://www.epa.gov/)
UV	Ultraviolet
WTW	Wastewater treatment works
WW	Wastewater
WWTP	Wastewater treatment plant
WFD	Water Framework Directive (Directive 2000/60/EC; info at: http://europa.eu.int/comm/environment/water/water-framework/index_en.html
WHO	World Health Organisation
YAS	Yeast androgen screen
YES	Yeast estrogen screen

Chapter 1

Pharmaceuticals in the environment: sources and their management

Christian G. Daughton

1.1 INTRODUCTION

An issue that began to receive more attention by environmental scientists in the late 1990s was the conveyance of pharmaceuticals to the environment by way of their use in human and veterinary medical practices and personal care. Pharmaceuticals and personal care products (PPCPs) comprise a remarkably diverse array of thousands of unique chemical substances, most of which are purchased for use directly by, or for, consumers and medical and agricultural practices. Of the so-called "emerging" pollutants, PPCPs perhaps serve as the prototypical examples because they amply illustrate the many dimensions and new questions associated with non-regulated pollutants. Of the many aspects that set them apart from conventional pollutants, a defining one is their diffuse, dispersed origins from the combined and varied actions, behaviors, and activities of multitudes of individuals. At the same time, many of these compounds experience significant parallel uses in agriculture. As consumer items, the minuscule contributions from each individual, while meaningless by themselves, combine to yield measurable environmental residues. Although, pollutants from dispersed sources and origins are not fully amenable to engineered solutions for controlling their entry to the environment, the potential for significant reductions is possible through comprehensive environmental stewardship strategies such as pollution prevention.

Of the many aspects of PPCPs as environmental contaminants, this chapter focuses only on two—sources and their management. Numerous sources and pathways serve to convey PPCPs to the environment after their use by humans and in animals. Some have origins from both natural and anthropogenic sources. After all, the design of many

Comprehensive Analytical Chemistry 50
Petrović and Barceló (Eds)
Volume 50 ISSN: 0166-526X DOI: 10.1016/S0166-526X(07)50001-2

synthetic drugs was patterned or inspired from naturally occurring substances that possess extraordinary pharmacologic activity. Careful consideration of these sources can help guide the development of strategies for reducing or minimizing the introduction of PPCPs to the environment. These strategies include a wide spectrum of possibilities that span drug design, commercial distribution, end usage, disposal, and treatment for waste and drinking water. The last-mentioned approach—treatment technologies—is the subject of other chapters. This chapter's focus is on the pollution prevention aspects of environmental stewardship; for a general discussion of the principals of environmental stewardship, see EPA [1]. An important perspective to keep in mind regarding pollution prevention relates to the many unknowns regarding the significance of PPCPs in the environment—why should effort be devoted to reducing the disposition of PPCPs to the environment if the potential for adverse effects on the environment or humans is largely unknown. The answer resides partly in the precautionary principal [2] and partly in the fact that other, unanticipated benefits can derive from implementing the principals of environmental stewardship. These collateral benefits could include improvement of healthcare effectiveness, reduction in health care costs, and reduction in human and animal accidental (and purposeful) poisonings—all potentially deriving directly from reduced usage rather than from reduced exposure to environmental residues.

1.1.1 Scope: the universe of pharmaceuticals

By itself, the word "pharmaceutical" refers to a chemical prepared or dispensed in pharmacies and which treats or prevents or alleviates the symptoms of disease or physiologic function. Technically, this limits the scope solely to prescription drugs. A narrow definition of "pharmaceutical" therefore excludes over-the-counter (OTC) drugs, diagnostics (e.g., X-ray contrast media, including certain radiologicals), nutritional and dietary supplements (ephedra is an example), illicit drugs, cosmetic or lifestyle drugs not essential for medical purposes, and the broad range of personal care product (PCP) ingredients.

For the purposes of this chapter, however, the universe of chemicals encompassed in the scope of PPCPs will be defined to include all chemicals used for humans, domestic animals, or agricultural crops that (i) treat disease, (ii) alter or improve physiological, cosmetic, or emotional function, appearance, or status, (iii) prevent disease (prophylaxis) or maintain health, (iv) help in the diagnosis or monitoring of health or disease, or (v) serve to formulate the active ingredient into a commercial product (e.g., excipients and delivery vehicles). The scope includes all preparations intended for topical, pulmonary, or parenteral (injection) administration or ingestion, as well as suppositories and enemas. The obvious galaxies of chemicals in this universe are the diverse arrays of human and veterinary prescription and OTC medications. But others include diagnostic agents (e.g., X-ray contrast media, radiopharmaceuticals), vaccines, and "nutraceuticals" (bioactive dietary supplements such as huperzine A and functional foods), and food supplements (including vitamins). Drug consumption originates not just from approved usages, but also from unapproved (e.g., extra-label) and illegal usage. Illicit drugs, in particular, comprise an unknown but possibly significant fraction of total drug usage, and consequently contribute to individual environmental residues and to the overall environmental loading of PPCPs.

As nutraceuticals (alternative spelling "nutriceuticals") and functional foods become more sophisticated, the demarcation between these exclusively naturally derived substances (e.g., phytochemicals) and the predominantly synthetic pharmaceuticals will become less distinct. According to this delineation of scope, also included would be materials resulting from genetically modified organisms (GMOs) and other biotechnology products, as well as radiopharmaceuticals and nanoscale materials used in medicine. The scope includes both licit and illicit drugs. PPCPs comprise both anthropogenic and naturally occurring substances, derived from such sources as microorganisms (e.g., antibiotics and toxins), tissues, plants, animals, petroleum products, and nanoscale materials. Traditionally included in the scope of PPCPs from natural sources are the toxicologically important group of endogenously synthesized and excreted steroid hormones (e.g., the estrogens, androgens, progestagens, gluoco/mineralcorticoids, and thyroid hormones); these naturally produced substances are often included in the scope of PPCPs because their mechanisms or modes of action are so similar to their synthetic relatives ("artificial hormones" and "mimics") and because they are sometimes directly formulated as their unaltered or prodrug forms in medications (e.g., estradiol, testosterone). Of importance to note is that estrogenic and androgenic activity together with other forms of endocrine modulation (e.g., "endocrine disruption") in the environment can originate from both synthetic and natural sources, and only an unknown portion originates from PPCPs. Synthetic and endogenous relatives sharing the same mechanism or mode of action, therefore, jointly contribute to cumulative and aggregate exposures.

Although the topic of endocrine disruption, especially with respect to the effects of sewage effluents on aquatic life, is far beyond the scope of this chapter, the major unknowns and complexities surrounding the subject are articulated by Sumpter and Johnson [3].

For the purposes of this document, the universe of PPCPs includes not just the parent form of the chemicals (whether the active ingredients or prodrugs (a ‘‘prodrug’’ is an inactive precursor that is converted to the active form by normal metabolic processes; the fibrates are an example)), but also their bioactive metabolites and transformation products (including conjugates). Note that an intersection exists between pharmaceuticals and pesticides, where several are registered for both uses (lindane, triclosan, and triclocarban are but three examples). Another intersection occurs with endocrine disrupting chemicals (EDCs) and PPCPs.

Some miscellaneous statistics supply ample perspective for the breadth and size of the pharmaceutical market. As of March 2005, the FDA lists in its Orange Book [4], whose updating frequency ranges from daily to monthly, over 11,000 distinct prescription drug formulations and dosages, comprising pharmaceutical equivalents, pharmaceutical alternatives, and therapeutic equivalents, as well as those that have been discontinued from marketing [5]; it must be noted that these numbers are not distinct drug entities, which are a subset of the total numbers of drug preparations.

The number of drugs cataloged in 2005 from Germany and the EU [6] include roughly 9,000 preparations in over 11,000 different dosage forms and 35,000 products. The total number of medicinal preparations worldwide probably exceeds tens of thousands, 60,000 being a figure cited by Tropsha [7]. Each of these products is marketed for one or more of over 80 major therapeutic categories, but so-called off-label prescribing for non-approved conditions is also widespread. Of significance with respect to the occurrence of PPCP residues in the environment is that the relative usage rates for individual drugs can vary dramatically not just between countries (whose approved drugs are not necessarily the same), but also among geographic locales within a country, as a function of prescribing practices and preferences [8]. Many of the excipients used in drug formulation are also used in the processing of foods, but derivation of most excipient residues from drugs is probably minor compared with their sources from foods.

From 1999 to 2000, 44% of all Americans were taking at least one or two prescription drugs during a prior month, and nearly one in five were taking three or more. Nearly, 84% of all Americans aged 65 and older were taking at least one or two prescription drugs, and nearly half were taking three or more. Other trends in increasing medical drug usage are presented in the report prepared by Department of Health and Human Services for Congress [9].

The types of drugs most commonly prescribed or used, together with their usage rates, vary over time, as distinct therapeutic classes and their individual members change in popularity and as new active pharmaceutical ingredients (APIs or "drug substances") are introduced to commerce; an API is any chemical constituent having pharmacological activity or other useful effect in the diagnosis, cure, mitigation, treatment or prevention of disease, or that affects the structure or a function of the body. Types of drugs also vary with geographic locale (from country to country and even within adjacent regions) because of varying prescribing practices and customs, differing health needs, and continually evolving patient desires and expectations. The broad spectrum of the types of human pharmaceuticals (i.e., therapeutic categories and drug classes and subdivisions) that are most frequently prescribed can be perused by referring to any "preferred drug list" (also known as a drug "formulary") maintained in the U.S. by the healthcare plan industry; example formularies, under Medicare, can be accessed for all U.S. states (see database maintained by Medicare [10]; a specific example selected at random can be seen at [11]). Other countries have similar formularies (e.g., for Britain, see [12]). Analogous formularies exist for veterinary medicine [13]. The most commonly prescribed individual drugs in each class or subdivision can be obtained from various pharmaceutical data providers; for example, the most frequently prescribed pharmaceuticals in the U.S. are compiled by [14]. The publicly available list currently comprises the top 300 most frequently dispensed medications in the U.S. in 2004 [14]; each of these medications can be assigned to one or more therapeutic classes/subdivisions mentioned earlier. Those pharmaceuticals from each class that have been repeatedly identified in various monitoring studies are reported in various overviews [15,16].

A compilation of the top 200 prescriptions for 2004, according to U.S. sales, is also provided by [17]. Worldwide sales in 2004 for the top selling 500 drugs were nearly \$300US billion [18]. Per capita sales in the US (\$552) were nearly 40% higher than those of the next highest country, Japan [19]. Global audited sales of all pharmaceuticals reached \$518US billion in 2004 [20]. Coincident with these measures of absolute consumption is the attendant annual growth in usage as reflected by annual sales and dispensing; for example, IMS Health

[21] reported U.S. prescription drug sales increased 5.4% from $238.9US billion in 2004 to $251.8US billion in 2005, generic sales increased nearly 21%, and dispensed prescriptions in the U.S. increased 4.7% over the same period. IMS Health predicts continued five-year growth of 5–8%.

Although the number of existing drugs is quite large, the rate at which new drug entities are introduced to the U.S. is very small. For the 10-year period 1993–2003, the number of new molecular entities (NMEs) approved yearly by the U.S. FDA ranged from 9 to 35 [22]; an NME is a medication whose active ingredient has never before been approved in the U.S. for marketing in any form. While thousands of distinct drug entities exist, and hundreds are used routinely throughout the world, roughly only 100 or so PPCPs have been routinely identified so far in various environmental samples. This discrepancy results largely from the fact that (1) not all drugs are used in quantities sufficient to be detected as environmental residues, (2) many drugs are either extensively metabolized, lessening the excretion of the parent chemical, or rapidly transformed by engineered or natural processes, and (3) not all pharmaceuticals are easily detectable at low concentrations in complex environmental matrices using current chemical analysis methodologies. But also important to keep in mind is that those drugs that have been detected in the environment have resulted from targeted monitoring; and therefore, those not targeted escape detection—just as with any unregulated pollutant [23]. Many additional PPCPs undoubtedly occur as contaminants but they have yet to be targeted for monitoring. Surprisingly, a comprehensive compilation of all PPCPs reported in the literature as environmental contaminants does not yet exist in a publicly available database.

It is important to note that this very brief presentation of some facts regarding human drug use is pertinent neither to veterinary drugs (and those used in animal husbandry) nor to PCPs. Entirely different databases need to be consulted to glean this information. A major lesson is that information regarding drug usage (whether human or animal) is extremely difficult to obtain easily or without cost. Subscription services are required for accessing human prescription drug usage (e.g., see references cited by Daughton [8]). It is also important to recognize that prescription drugs represent but a small portion of overall drug usage compared with OTC drug sales. Illicit drugs and those obtained on the black market represent additional major origins for drugs [24], but their contribution to overall residues in the environment is unknown, with investigations just beginning [25,26].

Without access to proprietary data, it is difficult if not impossible to gain a sound perspective regarding the total numbers and quantities of distinct drug entities used in the U.S., if not worldwide. This has major ramifications with regard to assessing those PPCPs that might play the most important roles as environmental pollutants. Information concerning the PPCPs that are in use commercially, coupled with their quantities (active ingredients), comprise one of several factors that determine the types and quantities of the parent chemical and transformation products that can eventually become pollutants as a result of their intended end use. Neither total numbers of distinct chemicals in use as PPCPs, nor their sales data, nor prescription data are by themselves useful for even predicting the potential occurrence of these chemicals in the environment. Even more data are required for predicting the flux of distinct entities (including parent drugs and transformation products, sometimes referred to as "degradates") to the environment.

Regardless, this brief sketch of the commercial significance of pharmaceuticals provides the backdrop reflecting the extent of widespread usage of pharmaceuticals, at least in Western countries, and superficially explains the ubiquitous occurrence of at least trace residues of these consumer products in certain environmental compartments. With this discussion aside, it is also critical to keep in mind that the absolute quantities of individual PPCP residues is only one factor required in assessing their significance in the environment. Other key attributes include biological potency. A simple example would compare the overall significance of a weakly estrogenic xenobiotic (e.g., nonylphenol, a breakdown product of the ubiquitous nonylphenolethoxylate surfactants) with a potent estrogenic drug (e.g., ethynylestradiol). The former is manufactured and introduced to the environment in quantities orders of magnitude greater than the latter, but their relative potencies, which also differ by many orders of magnitude, serve to place them on comparative footings with regard to impacts in the environment.

1.1.2 Background regarding the acronym "PPCPs"

The acronym "PPCPs" was coined in a review article by Daughton and Ternes [15]. Its original intent was merely to serve as a shorthand in that article to refer to "pharmaceuticals and personal care products." The term was subsequently assimilated into the environmental science literature, presumably for convenience. This broad collection of substances includes any product consumed by individuals or domestic animals for any number of countless reasons pertinent to health, performance, cognitive and physical function, or appearance. Note that the similar, truncated acronym "PCPs" is sometimes used when reference is made solely to "PCPs," exclusive of pharmaceuticals. This sometimes creates confusion if PPCPs and PCPs are interspersed frequently in the same discussion.

While the use of acronyms is often a bane of science—at best unnecessarily confusing or at worst an obfuscation to communication — acronyms are very useful when researching the published literature. This is especially true for the topic of "PPCPs" as environmental pollutants as none of these "key" words has any specific meaning useful for literature searches. Searching the literature relevant to PPCPs in the environment is made difficult because there are no search terms specific to the topic other than several rather unique acronyms, such as PPCPs. The words "drug," "pharmaceutical," "medicine," "medication," "medicament," "medicinal," "therapeutant," "diagnostic agent," "active ingredient," or "PCP" are all much too broad by themselves, encompassing a vast literature, largely irrelevant to environmental science. General searches can be better focused, however, by coupling any combination of these terms (or names of specific PPCPs) with others that are used more specifically in the environmental literature but much less so in the traditional medical literature: "aquatic," "sewage," "sludge," "manure," "pollutant," "pollution," "contaminant," etc. Even though PPCPs as a pollution concern derive primarily from the practice of human and veterinary medicine, the topic has been rarely discussed in the medical literature itself (early instances first being Zucatto et al. [27] and then Daughton [28] but few since, e.g., Sherer [29]). Another way to better target literature searches when using subscription databases is to limit keyword searches to the environmental literature.

The advantage of a distinctive acronym is that it allows for more focused literature searches—providing results that are almost always directly relevant to the topic. But a proliferation of yet other acronyms adds to the already difficult task of performing targeted literature searches. In addition to the acronym "PPCPs," some other acronyms that have appeared in the literature include "PhACs" (pharmaceutically active compounds), which was coined by Sedlak et al. [30]; sometimes the shorter acronym "PACs" is used in its place. While encompassing therapeutically active drugs, PhACs would not include non-therapeutic pharmaceuticals (e.g., diagnostic agents, X-ray contrast media being one example), nor would it include PCPs (such as synthetic musks or parabens). Yet another acronym appeared in 2003—"PCPIs" (personal care product ingredients); PCPIs is analogous to

PhACs (referring to the actual "active" ingredients), but specific for PCPs [31]. The term "PhPCPs" has also been used recently. It is worth noting that both PhACs and PCPIs (both of which are subsets of PPCPs) exclude the so-called inert or "inactive" ingredients used in product formulation (e.g., excipients); but even the "inert" ingredients can have biological effects (examples include alkylphenolic surfactants, parabens, and phthalate esters, used in various PCPs) or alter the absorption or metabolism of the API; the role of "inert ingredients" might become more significant as nanomaterials become more widely used in medicine. Another expression that aptly captures the pollution aspect of PPCPs is "feral pharmaceuticals," a term coined by Fisher and Borland [32].

1.1.3 An historical perspective regarding the published literature and PPCPs

The annual rate of published articles directly relevant to PPCPs has grown exponentially since the mid-1990s. The published English literature had been rather scarce up until the mid-1990s. An informal assessment of one compilation of the published literature [33] reveals that by 1998, the yearly publication rate was merely several dozen. In 2000, the yearly rate multiplied but was still less than 100. Beginning in 2001 and continuing through 2003, the yearly rate climbed past 100, and in 2004 it exceeded 200. As of 2005, it has become increasingly difficult to locate and digest all of the citations that are relevant to PPCPs because the breadth of journals covering the issue has greatly expanded, because these journals often carry multiple PPCP articles in each issue, and because the number of topics encompassed by the field continues to grow.

Discussions of the environmental ramifications of PPCPs originally focused on their occurrence and monitoring, primarily in surface/ground waters and untreated/treated sewage. This work was driven primarily by environmental analytical chemists, as new instrument technologies expanded the types of unknowns that could be easily identified, as instrument sensitivity increased, and as detection limits of analytical methods were lowered. This focus continued until the late 1990s, when it began to expand to waste treatment and fate/transport. In the last couple of years, more attention is beginning to be devoted to exotoxicology, pollution prevention, and environmental stewardship. Likewise, the scope of environmental compartments under investigation has expanded from primarily waters to now include sediments (and suspended particulates), sewage sludge (and biosolids), air (e.g., PPCPs sorbed to suspended particulates), and biota.

As the literature on the many aspects of PPCPs continues to grow, it is only possible to cite a select few articles that cover some of the facets summarized in this chapter. The vast majority of pertinent references must necessarily be omitted because there are simply too many; this is not to be interpreted in any way as a reflection of the quality of these many works. But by referring to the literature cited in a limited number of key articles, the reader can readily gain access to a more expansive literature. Some useful articles that offer broad perspectives on either human or veterinary pharmaceuticals, especially regarding sources and origins, include: Boxall et al. [34], Daughton and Jones-Lepp [35], Daughton and Ternes [15], Díaz-Cruz et al. [36], Halling-Sørensen et al. [37], Heberer [38], Jorgensen and Halling-Sørensen [39], Kolpin et al. [40], Kümmerer [41,42], Petrović et al. [43], and Ternes [44].

1.2 SOURCES AND PATHWAYS FOR PHARMACEUTICALS TO THE ENVIRONMENT

1.2.1 Importance of understanding sources and origins

The following discussion of the sources and origins of PPCPs occurring as residues in the environment is necessary for gaining an appreciation of the scope and magnitude of the entire issue. An understanding of origins and sources is required so that knowledge gaps can be assessed and so that future research or actions can be most effectively targeted. A key concern regarding sources is whether they lead to immediate or delayed, direct exposures of biota or humans without additional transport being required (e.g., via the trophic food chain); the recently discovered link between diclofenac-treated cattle and die-offs of scavenging vultures in Southeast Asia is but one example [45]. Thorough understanding of sources and origins is also essential for implementing not just engineered control measures, but also for designing effective pollution reduction measures, a topic that will be discussed in the second part of this chapter. A comprehensive inventory of the types of sources for each PPCP ingredient does not exist; such a system overlain with geographic information would be extremely useful.

Although the significance of pharmaceuticals as trace environmental pollutants in waterways, and on land to which treated sewage sludge or wastewater has been applied, is largely unknown, the fact that certain PPCPs with short environmental half-lives can nonetheless have continual persistence (noted by Daughton and Ternes [15], and later referred to as "pseudo-persistence" by Daughton [28], because of their continual introduction via effluents from sewage treatment facilities and from septic systems, poses two immediate concerns. First, with respect to ecological integrity, the potential for adverse effects on biota is largely unknown, especially for aquatic life, and secondarily for those organisms that are part of the food chain involving sewage-amended land. Second, drug residues that make their way to drinking water sources could pose the potential for significant problems with regard to public acceptance of, and trust in, their water supplies. This second concern is not widely appreciated and results from the complex ways in which risk is perceived [46]. The overarching environmental concerns associated with PPCPs as pollutants have been summarized by Daughton [8]; the potential for subtle effects, in contrast to overt acute effects, was identified as the primary concern by Daughton and Ternes [15].

One of the major attributes that distinguishes PPCPs from other chemicals that become pollutants is the fact that they are primarily marketed as products for use by the public. As such, they do not fit into the conventional mold of pollutants that result from commercial activities, such as manufacturing, or from waste treatment practices (e.g., incineration). While most of these conventional sources of pollution are well-defined point sources, PPCPs instead emanate from the confluence of individually minuscule contributions from each of multitudes of individuals or animals. The private individual as polluter, as a result of direct use of chemicals for personal purposes, is only a recently recognized phenomenon. At the same time, it is important to recognize that PPCPs are not the only galaxy of chemicals that the public directly uses (or creates). Other galaxies contributing to pollution include household products used for cleaning and maintenance, wastes from electronics, fuel combustion, and even food; some of these, however, can also serve as sources for certain PPCPs (e.g., caffeine from foods, and broad-spectrum biocides such as triclosan and triclocarban, which are used in many consumer products). There is also an intersection between pharmaceuticals and pesticides, a small select number of which serve double duty as both registered pesticides and as PPCPs; some examples are presented later (see section "Multiple Aggregate Sources"). Many of the issues discussed here are relevant not just to PPCPs, but also to other unregulated pollutants, including the so-called "emerging" pollutants and chemical stressors in general (see discussions at [23,47]).

Drugs can enter the environment by a number of distinct and varied routes. The two general means by which they gain entry are indirectly (involuntarily) by excretion and bathing, and directly (purposefully) by disposal. Disposal of drugs that are no longer needed or wanted occurs by discarding to trash (which in turn usually goes to landfills) or by directly discarding to sewage systems (usually via the toilet). Although the long-accepted means of disposing to sewerage by flushing down toilets is now known to maximize the ability of a drug to enter the environment, the rationale behind this approach is to minimize the chances of consumption by others for whom the drug was not intended. Poisoning of adults and children by medications discarded by others is a problem of increasing concern to healthcare professionals. To date, however, there are no widely available alternative means, at least in the U.S., for drug disposal that are inherently protective of human safety. Drug disposal is a deceivingly complex topic. The many issues and dimensions surrounding drug disposal (and other approaches to pollution reduction) were covered in a 2-part monograph [8,48] and will be further addressed in the second part of this chapter.

It is also important to note that while most drugs enter the environment from individually dispersed sources, primarily via sewerage (much of which is recombined into flows leading to publicly owned treatment works—POTWs—which in turn yield point-source discharges into receiving waters), some sources are extremely localized (e.g., privately owned septic leach fields, straight piping, cemeteries, etc.); straight piping is the practice where untreated, raw sewage is illegally discharged without treatment directly to the environment immediately from the point of origin, often private residences. Some origins therefore may have broad significance for the environment while others might have significance only in certain special, local circumstances.

1.2.2 Sources/origins

Before discussing sources and origins of PPCPs in the environment, we must recognize that such a discussion can get confusing without defining some rather arbitrary boundaries and artificial definitions. The distinction between source, origin, and fate is often vague. At any point along a pollutant's environmental transport chain, a variety of different exposure and effects scenarios can come into play. Any point in the chain can be considered to be a source (but not necessarily an origin). For example, sediments or edible plants become an environmental compartment in terms of transport and fate, but as a result, they also become a source in terms of being a reservoir for subsequent rerelease (e.g., to another compartment such as water) or exposure (e.g., as food for humans or wildlife). "Sources" include routes to and from the end-user.

The topic of PPCPs as pollutants is intimately tied to a bewildering array of phenomena that transport and transform these chemicals from one place to another via a multitude of distinct "routes" by which the chemical is emitted, dispersed, or otherwise introduced to the next "compartment" or ultimately to a biological receptor. As an example, any point in the water "cycle" or in a waste treatment process chain can be considered a "source" for the downstream connecting points, which in turn then become sources themselves. What constitutes an actual "source" is often difficult to define as a PPCP leaves a manufacturer and progresses until it leaves the supply-consumption cycle and is released to the environment, where it can reside in any number of environmental compartments and exchange among them. If we limited our discussion to those "sources" that serve as the points where PPCPs leave the consumption cycle and enter the environment, the discussion would be rather simple. Any discussion of "sources" will therefore necessarily intermingle with discussions about fate and transport, which are covered in more detail in other chapters of this book. A hint of this complexity can be seen in Fig. 1.1. So we will recognize from the outset that some overlap with other chapters will be inevitable but discussion of processes traditionally considered to constitute "fate and transport" will be minimized. This is perhaps made clearer by distinguishing "origin" (as the point at which something comes into existence) from "source" (as the point from which something is derived or obtained).

Anthropogenic pollutants gain entry to air, surface and ground waters, land, and biota as a result of manufacturing emissions, power generation, waste disposal (e.g., incineration, landfills), accidental releases (e.g., spills), purposeful introduction (e.g., pesticide application, groundwater recharge, sewage sludge application to land, illegal discharge and dumping), and consumer activity (which includes both the excretion and purposeful disposal of a wide range of naturally occurring and anthropogenic chemicals, PPCPs being but one expansive galaxy of such chemicals). All of these sources but the last have long been recognized as major potential routes of pollutant release. Once released to the environment, PPCPs (like other pollutants) can take up residence in "storage reservoirs," which can be viewed as secondary sources for further releases; examples are residues that have been concentrated by sorption to sediment, biosolids, or biota. Consumer activities, however, have only recently been recognized as a potentially major, long-standing source of uncontrolled non-point pollution.

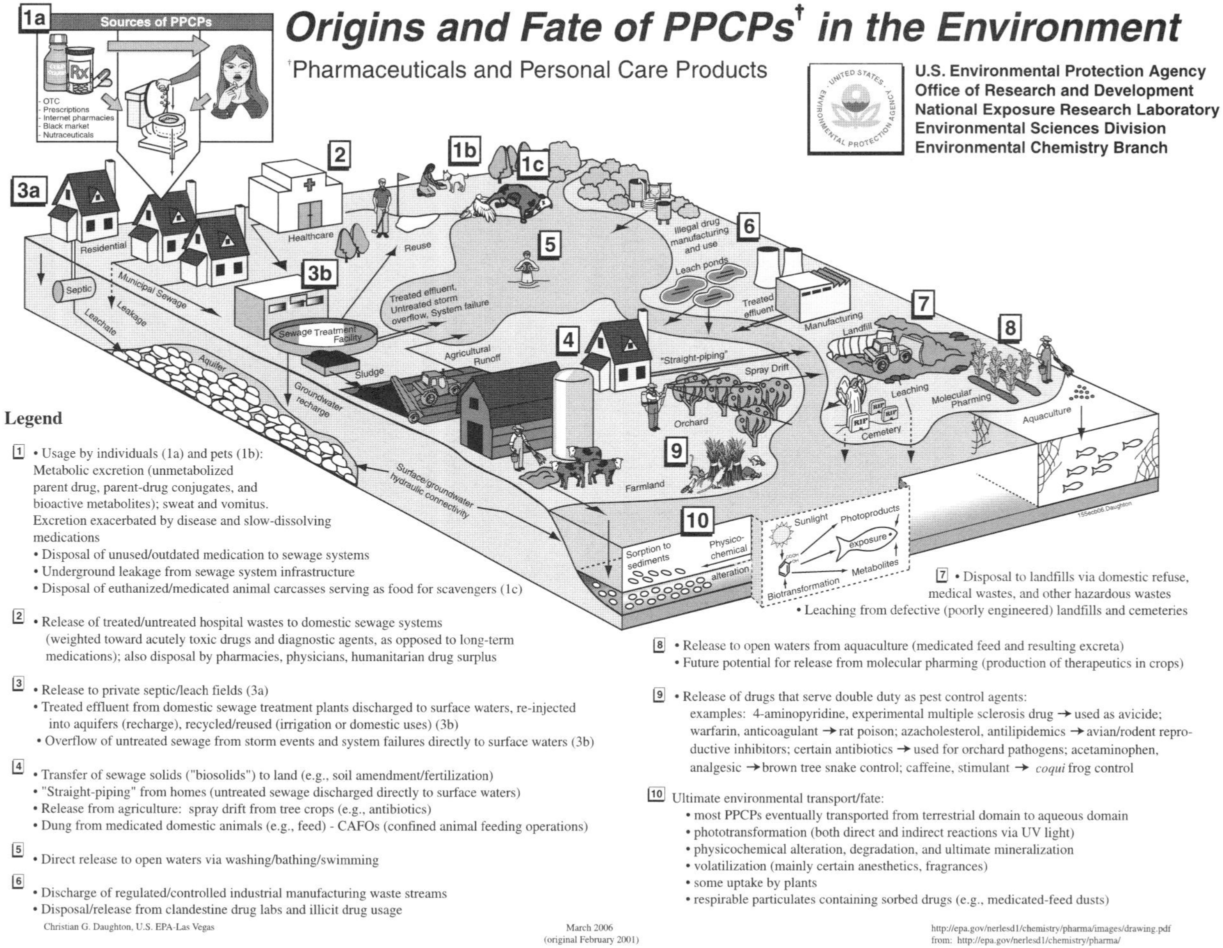

Fig. 1.1. Origins and fate of PPCPs in the environment. NOTE: Whatever format and resolution is best suited for publication will be provided on request. This figure was revised in March 2006. Copy of figure currently available here: http://epa.gov/nerlesd1/chemistry/pharma/images/drawing.pdf

An obvious source for PPCPs as environmental pollutants includes residues from their manufacturing. But since the discharge of pharmaceuticals and synthesis materials and by-products from manufacturing are already well defined and controlled in the U.S., they are not part of the scope of this chapter. The loss of the API during manufacturing is very small because APIs represent a significant monetary investment; in the U.S., the chemicals in manufacturing waste streams are primarily left-over intermediates and by-products from synthesis (none of which occurs in the final commercial product), not the drugs themselves. These emissions are all subject to existing regulations [49,50]. For more information regarding manufacturing discharges, see EPA [51].

It is worth noting, however, that of the major chemical synthesis industries, the pharmaceutical industry produces the most waste (from by-products, catalysts, solvents, salts, and intermediates) per unit of actual product. The ratio of waste mass produced per unit of API produced ranges roughly from 25 to 100. In comparison, for example, the petrochemical sector operates at a waste-per-product ratio of about 0.01. For proper perspective, however, the annual production volumes of final product (i.e., the API itself) are many orders of magnitude lower in the pharmaceutical sector, somewhere between 1 and 1000 tons [52], using the E-factor approach adapted from Sheldon. The industry is also making continual progress in adopting green chemistry approaches in developing alternative synthesis routes for new and existing APIs.

When discussing quantities of drugs manufactured or disposed, it is important to recognize that the actual API represents only a portion of the overall mass of the finished formulated drug (which includes the other ingredients composing the finished drug, namely the excipients [53]). While manufacturing is indeed a potential source for APIs, albeit minor in the U.S., this chapter focuses on the importance of the activities, actions, and behaviors of the individual consumer and other end users of healthcare and veterinary medicines. The significance of the individual in directly contributing to the combined load of chemicals in the environment has been largely overlooked. PPCPs in the environment illustrate the immediate, intimate, and inseparable connection of the individual with the environment. These diffuse sources include the excretion of ingested drugs and bioactive metabolites, the washing of externally applied drugs and PCPs [54,55], and the direct disposal of PPCPs to terrestrial sites and domestic sewage [8,48]. The importance of dispersed, diffuse, minute "discharges" of anthropogenic chemicals to the environment has been overshadowed for decades by the more obvious point sources.

Several factors are driving increases in usage of pharmaceuticals, at least in the U.S. These include: direct-to-consumer advertising (DTC); generic switches (reduced cost of medication previously available only by prescription); ease of access (Internet, black market); aging of the population (growing popularity of "anti-aging" pharmacy) and the consequent growing incidence of polypharmacy (e.g., as a necessity in geriatric medicine or as a result of patients retaining multiple providers, sometimes known as "doctor shopping" or "double doctoring"); new uses for existing drugs (e.g., lifestyle drugs and cosmetic pharmacology); increasing off-label prescribing (partly a result of the growing numbers of drug targets as revealed by genomics; one estimate gives at least 10 times as many potential molecular targets for future therapy than have been exploited to date [56]); distribution of medicines free of charge (e.g., to elderly patients as disease preventatives); and the continued growing use of illicit drugs and abuse of legal drugs. Trends for reducing usage include the advent of individualized (personalized) therapy (e.g., ability to test for polymorphisms, obviating the ineffective use of medication for certain sub-populations) and advanced technology for delivery of smaller doses directly to target sites.

Over the last decade, environmental scientists have established a large, diverse, and sometimes unexpected variety of pathways that serve to convey pharmaceuticals originating from the practice of human and veterinary medicine to various environmental compartments (e.g., see Fig. 1.1). Despite the analogies and actual connections between human health and the integrity of our environment [8], it is worth noting that little exists in the formal medical literature that explores the significance that the practice of medicine can have on the environment [28]. Indeed, involvement of the medical community in the issues surrounding PPCPs should be considered a major objective for environmental scientists.

Tracing the sources of PPCP residues in the environment necessarily involves consideration of a broad spectrum of possible routes that connect their origination (at time of manufacture) to their deposition (or creation of transformation products) in the various environmental compartments, at which time they then acquire the potential to become involved with exposure of humans or the environment. These routes range from obvious ones, such as excretion of parent drugs and their metabolites, bathing, and purposeful disposal to sewerage, to more obscure ones, such as burial of heavily medicated bodies to the feeding of wildlife scavengers on discarded carcasses of euthanized or medicated domestic animals.

An examination of the sources of PPCPs in the environment poses a number of unresolved questions, many of which are highlighted in this chapter. These represent unmet needs defining future research. A compilation of research needs relevant to all the aspects of PPCPs as pollutants is available [57].

1.2.2.1 Sources: general considerations

Several dimensions define the scope of sources of PPCPs in the environment. As with most chemicals, PPCPs can find their way to the major compartments, including: (i) water (both ground and surface waters—lakes, rivers, streams, marine), (ii) solids (sediments and soils, including agricultural lands), (iii) air, and (iv) biota. With respect to air, the vast majority of drugs (in contrast to PCPs) have insufficient vapor pressures but they can gain entry to the air by dispersal while sorbed to fine particulates (e.g., medicated feed dusts used in confined animal feeding operations). Important to note is that both spatial and temporal dimensions exist for all sources. Some sources release transient, discontinuous spikes or pulses (especially disposal events), while others provide more continuous releases (e.g., excretion of a particular drug by significant portions of the local population), depending on the time scale of measurement, but nonetheless influenced by diurnal and seasonal patterns; an example of the occurrence of transient concentrations is shown by Lissemore et al. [58]. This consideration plays a critical role in the design of sampling protocols and can dictate whether small, discrete samples (e.g., grab samples) will be representative or whether time-and-space "integrative" sampling is required [59]. These considerations determine whether the resulting data can be used to calculate accurate environmental fluxes or loads (e.g., "predicted environmental concentrations," PECs) required for environmental or risk assessments. Other dimensions include the levels, amounts, doses, or concentrations of PPCPs that different sources can contribute for an exposure event. Discussion of the risk assessment process (with a focus on veterinary drugs) is provided by Montforts [60,61].

The best-documented outcome from exposures is the possibility of chronic, low-level exposures (especially in the aquatic environment) that hold the potential primarily for subtle effects [15]. Others, however, can result in acute, high-level exposures, such as those resulting from the improper disposal or storage of PPCPs, or from the exposure of scavenging animals to drug-tainted carcasses or medications improperly disposed in trash. Finally, some sources are best characterized as dispersed (e.g., release of residues via private residence sewerage) while others more resemble conventional point sources, such as private sewage leach fields, outfalls from sewage treatment plants, or confined (concentrated) animal feeding operations—CAFOs (see [62] for more information).

The distribution/supply chain for PPCPs, especially that for drugs, can be considered as well controlled and, therefore, an insignificant source for drugs in the environment. For drugs that remain unsold in pharmacies, the system of reverse distributors in the U.S. serves to ensure that unwanted pharmaceuticals are either returned to manufacturers or properly disposed according to regulations; for those countries without a reverse distribution system (Korea, for example see [63]), pharmacies must figure out how to dispose of expired pharmaceuticals themselves. For this reason, the fountainhead of sources for PPCPs as environmental pollutants in the U.S. begins with the consumer and other end-users.

The major route by which human-use PPCPs gain entry to the environment is from their intended, direct end-use. After systemic absorption due to topical, pulmonary, or parenteral administration or most commonly by ingestion, residues of the parent PPCP (as well as sometimes a complex array of metabolites) are either excreted or are dislodged from skin by sweating, bathing, or swimming (e.g., dermally applied drugs, such as topical antibiotics and hormones); even systemic drugs can be excreted through the skin, an example being the appearance of loratadine on the skin 40 min after ingesting a 10-mg oral dose of the antihistamine [64]. With respect to excretion, these residues are associated primarily with the feces and urine (the relative partitioning between which depends on the pharmacokinetics of the individual drug), and less so via sweat, vomitus, and saliva. These residues include unmetabolized parent drug, bioactive metabolites (responsible for either intended therapeutic effects or adverse side effects), and inactive metabolites (including parent-drug metabolic conjugates, which can be subsequently hydrolyzed after excretion to release the parent drug; e.g., via microbial deglucuronidation via β-glucuronidase [65]). Conjugates and parent PPCPs sorbed to sediments/particulates can essentially serve as secondary sources or hidden reservoirs.

The relative ratios among the different routes and of excreted forms can be dramatically altered by the health/disease status of the individual, as dictated by numerous factors including genetics, gender, age, and individual metabolic idiosyncracies, as well as by the formulation of the drug (e.g., some slow-dissolving tablet forms can lead to poor absorption and therefore enhance excretion of the unaltered parent drug). It is important to understand that the pharmacokinetics for a drug (as described by the absorption, distribution, metabolism, and elimination [ADME]) as documented in the literature can differ profoundly from reality as a result of the health status, diet, and genetics of the individual. Some of the many factors that dictate the absorption of a drug have been summarized by Surian [66]. Also of significance is that the extent of metabolism of a drug (and therefore the extent of excretion of the parent form) is not necessarily related to the frequency with which it is detected in the environment. Some extensively metabolized drugs (those for which only a very small percentage of the parent form is excreted, such as carbamazepine) can nonetheless establish widespread environmental occurrence [67].

While consideration of sources tends to focus on parent, unaltered PPCPs, it is important to not disregard that many of the sources of parent PPCPs also serve as sources of transformation products—not just excreted metabolites, but also environmental transformation products such as from microbial metabolism and phototransformation ("degradates"). Consider carbamazepine (CBZ) as one example. Pharmacologically, CBZ is an extremely "dirty" or "promiscuous" drug, capable of eliciting numerous side effects as a result of its action on multiple receptors, partly as a result of a plethora of metabolites. Although carbamazepine-10,11-epoxide is the major initial (and bioactive) metabolite of CBZ, it is rather efficiently converted to the diol and a host of thirty-some other metabolites [68], some of which undoubtedly are responsible for the multitude of human side effects. While CBZ is extensively metabolized by humans (roughly only 3% of the parent drug is excreted unchanged), its introduction to the environment would likely be accompanied by numerous metabolites. Indeed, Miao and Metcalfe [69] revealed the occurrence of five CBZ metabolites in the influent to sewage treatment plants: 10,11-dihydro-10,11-epoxycarbamazepine; 10,11-dihydro-10,11-dihydroxycarbamazepine; 2-hydroxycarbamazepine; 3-hydroxycarbamazepine; and 10,11-dihydro-10-hydroxycarbamazepine. The 10,11-dihydro-10,11-dihydroxycarbamazepine was also detected in surface water, at a three-fold higher level than the ubiquitous CBZ.

A secondary route by which PPCPs gain entry to the environment is by direct disposal. The primary routes for disposal are via flushing to sewerage from toilets (and other drains) and from discard to domestic trash, which is then usually buried in landfills. Note that while the majority of excreted urine and fecal material passes into sewage collection systems, a smaller but potentially significant portion is disposed to landfills by way of baby diapers and adult incontinence products; this constitutes another route of disposal to landfills (albeit of excreted drugs in contrast to unused drugs). Even properly engineered landfills can serve as delayed sources of drug residues, especially if leachates seep into the ground or are actively pumped out for disposal at water treatment facilities. Landfills and PPCPs have been discussed by Bound and Vouvoulis [70]. Disposal to sewerage occurs not just in domestic residences but also in certain healthcare facilities such as those used for long-term care. The driving forces behind the necessity of disposal include the expiration of medication, cessation of therapeutic need, and patients' "non-adherence" (non-compliance) such as discontinuation of medication because of adverse effects, failure to treat, or lack of motivation to continue therapy.

Little appreciated in the many aspects of work regarding PPCPs over the last decade or so is the critical importance of recognizing that the alternative to drug disposal from private residences is on-site long-term storage (e.g., in medicine cabinets and kitchens). Because of storage, PPCPs are responsible for a preponderance of poisonings in the U.S. for both children and adults. Detailed therapeutic-class and substance-specific data are maintained by the American Association of Poison Control Centers (AAPCC) [71]. Since storage of PPCPs within domestic residences is a major contributor to accidental and purposeful poisonings for humans (adults, children, infants) and pets [72], it is critical to minimize the quantities that are stored, and confine storage to proper areas; for example, conventional vials and bottles will not ensure pet safety, as dogs for example, can chew through plastic containers. Accumulation of multiple containers of drugs can become confusing, especially for older patients and minors, and increases the risk that the wrong medication could be consumed, especially for those practicing polypharmacy (see example presented later below).

Storage of PPCPs in healthcare facilities can also result in diversion to the black market. Storage of drugs in excess of those needed for immediate use also encourages abuse, a current example of which among teens is illustrated by the popularity of "pharming" [73]. Also worth noting is that prescription drugs are not necessarily the most hazardous PPCPs for infants and toddlers; for example, high-potency iron supplements and widely used OTC products (e.g., those containing acetaminophen and stimulants) are major causes of poisoning.

One of the most important aspects of the controversies surrounding drug disposal in the U.S. is the lack of recognition for the direct connection between ways to minimize the introduction of drugs to the environment and the ways in which properly designed storage and disposal programs could protect human health and reduce poisonings—environmental concerns aside. Strategies to facilitate the collection, inventory, and destruction of unnecessarily prescribed/purchased pharmaceuticals is a very important risk management tool, especially given its potential to mine information critical to continually reduce future medication errors, reduce accidental and purposeful poisonings, reduce abuse, reduce controlled substance diversion, and reduce inappropriate and dangerous drug therapy resulting from the lack of appropriate diagnosis or prescribing by care givers.

In considering disposal as a source of environmental residues, while the major aspect probably involves discard of the unused medication (regardless of formulation), consideration must also be given to the concentrated residues contained in used dispensers and delivery devices (e.g., dermal patches, gel packs, bottles, pumps, inhalers, syringes); this is particularly relevant to hormonal preparations (e.g., testosterone, estrogens, progestins, and illegal anabolic steroids) and analgesic controlled substances (e.g., fentanyl). Toxicologically significant residues can remain in the used devices. Another minor consideration is the unintentional, incidental direct release of drugs and excipients simply as a result of correct usage of a product; examples are the release of propellants and volatile active ingredients to the air during use of inhalers and anaesthetic gases.

A tangentially related issue regarding sources, but one not covered here, is the fate of the packaging materials used for PPCPs, such as the materials used for plastic vials, IVs, and syringes, including the drug residues contained therein. Incineration and weathering of these materials are processes perhaps leading to a number of additional unknown products.

1.2.2.2 The role of source in the perception of risk

The significance of the real and perceived connections between our waste products with sources of drinking water and food can be greatly amplified by the presence of drug residues—regardless of how minute—as they can profoundly impact the perception of risk. In this respect, understanding the origins of these chemicals in the environment is extremely important and has ramifications with respect to our understanding of the water cycle [46]. Drinking water as a source of PPCPs, no matter how minuscule the concentrations, could be a key issue with regard to public acceptance (or rejection) of water recycled from wastewater [46].

One of the ways in which risk is subconsciously framed or valued during its perception derives from a form of "logic" or valuation based on what are known as the "common laws of magic" [74]. One of these laws is the Law of Association, which in turn comprises the sub-laws of Similarity and of Contact or Contagion. The "magical law of contagion" constitutes one of the sympathetic laws of magic as introduced over a century ago by anthropologists. These "laws" partly originated with the alchemists. Of particular relevance to drinking water as a source of PPCPs is the Law of Contagion, which holds that once contaminated, always contaminated: "Things that have once been in contact with each other continue to act on each other at a distance even after physical contact has been severed." Once objects come into contact with each other, they will continue to influence each other, even after separation. The presence of PPCPs essentially serves as a reminder that the drinking water was at one time in "contact" with human waste. This can lead to rejection by the consumer of recycled water for drinking [46,75].

1.2.2.3 Specific sources

The following provides a summary of the major sources for PPCP residues in the environment as well as some of those that are less discussed. The summary, however, should by no means be considered comprehensive. Previously unexpected sources are at times revealed. It is also important to note that the significance or magnitude of many of these sources is difficult to document, as they fall outside the normal domain of information addressed in the peer-reviewed literature; some are simply "common knowledge."

Beginning with the end-user (or end use) as the ultimate source, the principal groups are consumers, healthcare providers, hospitals, veterinarians, and those working in agriculture (including farming, CAFOs, and aquaculture). Additional but less obvious sources exist, most of which have localized impacts. Examples are the accumulation of drugs donated during humanitarian relief efforts and the cemetery burial of bodies that have received large doses of drugs (including radionuclides).

Consumers

Already discussed above are the major routes from the consumer to the environment, primarily from excretion and direct disposal. With regard to the significance of disposal as a source, it cannot be overemphasized that current knowledge cannot establish the portion of PPCPs in the environment that originate from disposal versus excretion. This is a major unanswered question deserving some concerted investigation. Distinguishing flushed drugs from excreted drugs is currently not possible by chemical monitoring. Instead, this is currently achievable only by consumer surveys. But only rudimentary data from limited questionnaire surveys are available to offer some insights. These have been summarized by Daughton [48]. These surveys have addressed the manner in which unwanted drugs are disposed by consumers but not the absolute or relative quantities that are actually disposed. The latter data are very scarce. For example, Berckmans et al. [76] cite work claiming that about 40% of the drugs marketed in France annually remain unused, but with no reference as to what their disposition is.

The determination of the significance of disposal with regard to environmental loads is a major unmet research need—one that should be addressed if environmental residues are going to continue to be used as a justification for the need for comprehensive drug "take-back" programs in order to preclude disposal to sewerage or trash. It can probably be assumed that disposal might very well represent a significant source for a limited number of widely and heavily used, inexpensive drugs that can be purchased in large quantities (leading to expiration before they can be consumed); disposal probably does not represent a significant source for expensive drugs or for those that are unit packaged, but this is merely speculation.

Many factors lead to the storage of unwanted PPCPs in domestic residences. The level of adherence (compliance) by patients to prescribed medication regimes is one of the major factors that determines the accumulation and eventual expiration of unused drugs in the household, although the purchase of unnecessarily large quantities of OTC drugs is another reason (e.g., bulk containers). Adherence to medication is an extremely complex issue with a wide spectrum of causes. This important topic is discussed later.

Using non-compliance statistics as a starting point, the rate of disposal could eventually be indirectly inferred by determining the reasons for the non-compliance, as not all result in disposal. Only a portion of the reasons for non-compliance would lead to leftover medications that might eventually require disposal. For example, non-compliance results from the failure to take medications at the correct time or frequency, but neither of these failures on the part of the patient necessarily leads to leftover drugs. Another action classified as non-compliance is failure to have a prescription filled; this clearly would never result in leftover drugs.

Household surveys are another way to determine the portion of drugs that are disposed. There have been few such surveys conducted [48]. A recent survey of 400 households in England [77] found that nearly one half did not finish their course of medication. For those portions that were disposed and not stored indefinitely, nearly two-thirds were disposed to trash, about a fifth were returned to the pharmacy, and about one-tenth were disposed to sewerage. Disposal to sewerage, however, did vary depending on the type of medication; for example, some classes (e.g., hormones) were either disposed to trash or returned to the pharmacy, with none being discarded to sewerage.

Theoretically, one possible way to directly determine by chemical monitoring the contributions to environmental loads of a particular drug originating from disposal versus excretion would entail analyzing for skewness in the relative ratios of optical isomers (the "enantiomeric fraction") from chiral drugs that are racemates. Such an approach would follow from the example of Sedlak and Fono [78] and Fono and Sedlak [79], who used enantiomer ratios from racemic metoprolol to distinguish sewage originating from waste treatment (the equi-enantiomer ratio is changed by selective action of biodegradation) from that of sewage having experienced no treatment (e.g., overflow events or straight-piping). Using this approach, sewage-influent drug residues from racemic drugs originating from disposal might be distinguishable from those that were excreted by having insufficiently enriched ratios of optical isomers (because of a lack of metabolic transformation). Another approach would work only on selected drugs—those that are extensively metabolized (where little of the parent drug is excreted). The discovery of parent drug residues in sewage would then be a likely indicator of direct disposal.

Also worth noting is that the private individual is the major contributor of illicit drugs to the environment. Illicit drugs have received surprisingly little attention from environmental scientists, especially given their unknown effects on aquatic biota. To date, only two publications have focused on illicit drug residues in an environmental context [24,25], but additional investigations are underway [26].

Once a drug is disposed or excreted (along with its metabolites), it passes dissolved or suspended in sewage to engineered sewage treatment facilities, to septic facilities, leach fields, or directly into receiving waters (e.g., via illegal privies or "straight-piping"); straight-piping serves to maximize the availability to the environment of any PPCPs that are present since no treatment is used to remove residues. Raw, untreated sewage can also enter the environment from sewage distribution and treatment systems as a result of storm events (overflows), system failures, and overcapacity; this is a common problem in those locales with aging infrastructures or rapidly expanding populations. Sewage distribution systems are all prone to underground leakage, especially from decaying sewage distribution infrastructure. Together with private septic systems and leach fields, these serve as potential sources for groundwater contamination. A particular approach that some municipalities are designing to deal with sewage overflows involves large-diameter subsurface deep-rock storage tunnels for accepting diverted flows until the treatment capacity for the wastewater treatment facility is restored. These tunnels, like any subsurface sewage conveyance infrastructure, provide an opportunity for seepage of untreated waste into aquifers. Another source for introduction of residues to groundwater is active recharge (groundwater reinjection) using treated sewage (reclaimed water). Some of the problems associated with groundwater contamination and the perception of risk are discussed by Daughton [46].

At sewage treatment facilities, the residues are subjected to various treatment regimes (depending on the size and sophistication of the treatment plant), resulting in "removals" that range from nearly complete to nearly zero. The removal efficiencies are a function of the individual PPCP as well as the treatment process(es). Removal of PPCPs is essentially a collateral or incidental function of a sewage treatment plant, as these facilities were never specifically designed to remove exotic, bioactive xenobiotics.

Two principal effluents result from sewage treatment—one consisting of the liquid effluent and the other sludge. While the liquid effluent is usually discharged to surface waters, it is sometimes used for irrigation; PPCPs are known to occur in the reclaimed water and to accumulate in and migrate through irrigated soils at concentrations in the nanogram-per-gram range [80]. The sludges are usually disposed to land, sometimes after being upgraded to "biosolids," a process intended to effect logarithmic removal of microorganisms but which also coincidentally reduces PPCPs (but to unknown degrees); the need for sludge disposal can be avoided only by incineration. Sludge disposal to land (even in the form of soil amendments or fertilizer) can result in leaching of residues into the ground or lead to contaminated wet weather runoff to receiving waters. Both the irrigation waters and the sludges derived from sewage can lead to direct exposure of organisms (e.g., worms and insects). Plants can also systemically absorb PPCP residues [81,82]. A third waste stream that can result from water treatment (primarily for drinking water) is the brine rejection stream from membrane filtration; the concentrations of PPCPs as well as large numbers of other pollutants will be enriched in these streams and therefore require special attention.

The many questions associated with the disposal of drugs often raise some of the very same questions for PCPs. PCPs are analogous to pharmaceuticals in that their intended end-use can distribute to the environment the ingredients that compose their formulations (e.g., see [55]), as well as the chemicals contained by their packaging. But unlike most drugs, this occurs primarily not by excretion, but rather by washing the product from skin, hair, and mouth, where its ingredients (both active and inactive ingredients) can then enter the environment via the sewerage pathway. Packaging used for PCPs, which is usually much more elaborate and substantial than for drugs, is discarded to trash where it can eventually weather, with the resulting release of additional chemicals from the combined actions of microbial degradation (especially fungal), UV photolysis, physical deterioration (e.g., action of heat), and chemical processes (e.g., leaching by water). PCPs also contrast with pharmaceuticals in that the latter are produced in relatively small quantities (sometimes as low as the kg/year range) and are largely designed to be biologically active. PCPs, in contrast, are more similar to high-volume chemicals, are not purposefully designed with bioactivity, and comparatively less is known regarding their interactions with organisms (in part because this is not always registration requirement).

With regard to environmental ramifications, PCPs (such as cosmetics) can differ from pharmaceuticals in three major respects: (1) the design of the packaging discourages disposal of the contents to sewage (because of the added difficulty of emptying package contents), (2) the "active" ingredients in PCPs are generally not engineered or designed to interact with biological receptors that regulate essential cellular functions, and (3) PCPs are used predominately external to the body. When applied to skin or the mouth, however, those chemicals in PCPs that are lipophilic are subject to absorption through the skin or mucosal membranes (e.g., parabens, phthalates, UV screens, and synthetic musks). Indeed, one of the paradoxes of consumer risk perception relates to the relatively high concentrations and plethora of types of chemicals formulated in PCPs that are applied directly to the skin, versus the concentrations of some of these same chemicals that might be found in drinking water, but at many orders of magnitude lower concentrations; the former is often deemed risk-free by the consumer but the latter not.

Unlike pharmaceuticals, mainly as a result of the packaging design and the way in which they are used, the disposal of unused or unwanted PCPs (e.g., cosmetics, shampoos) to sewerage has not been a concern with regard to the potential for environmental pollution. With this said, the ingredients comprising PCPs (both the active ingredients and the so-called "inactive" ingredients) are used in much larger quantities in end-user commercial products than are pharmaceuticals. These active ingredients and even some of the inactive ingredients pose the potential for exposure to aquatic organisms (from residues discharged with sewage) and to terrestrial animals (by scavengers foraging in municipal refuse). The UV-filters used in sunscreens serve as one example [83,84]. Substantial, sustained exposures pose unknown risks (e.g., subtle effects such as behavioral change) for certain organisms, especially those that are subject to continual exposure, such as aquatic organisms; this is especially true for lipophilic compounds that can bioconcentrate. An important perspective to maintain, however, relates to those ingredients that share the same mechanism or mode of action. For example, the alkylphenolethoxylates used extensively in PCPs have extremely weak estrogenic activity compared with the estrogenic drugs, but their environmental residue levels are also far greater. To determine relative exposure risks, both potency and concentration need to be considered in tandem.

With regard to the disposition of PCPs in the environment, the principles that could guide the creation of products that are most environmentally friendly would be those that fall under the stewardship concept of "cradle-to-cradle design" [8,48]. This ecologically intelligent design paradigm, as formulated by McDonough and Braungart [85], can be applied not just to the ingredients used in formulating these products, but also to the design and composition of the packaging and to the way the final product is distributed and consumed in the distribution commerce chain. For example, the types and amounts of materials used in manufacturing the packaging itself could be selected for minimal environmental impact (whether that be the sheer volume of packaging added to landfills, or the chemicals released by weathering, or by combustion of refuse packaging). At the same time, the packaging can be designed to maximize the consumer's ability to use the contents to the fullest extent possible, ensuring that the ingredients are directed to sewage treatment facilities (e.g., during bathing) where they can at least be degraded, rather than discarding partially empty containers in the trash.

Consumer PCPs also illustrate the potential importance of the so-called "inactive" or "inert" ingredients (such as the solvents/carriers) with regard to unanticipated exposure routes. PCPs are becoming established as a source of previously unrecognized air pollution. Although the active ingredients in PPCPs (with the exception of certain anaesthetic gases and synthetic musk fragrances) are probably without impact on air, the more prevalent "inert" ingredients can contribute to general indoor air pollution and serve as precursors to smog. California regulators, for example, are becoming more cognizant of the individually minuscule but significant combined effects of the chemicals released by consumerism [86].

Finally, with regard to consumer use, PCPs can also serve as significant sources for conventional pollutants. Obvious examples include: phthalates (especially diethyl and dibutyl), solvents, dyes, and parabens (4-hydroxybenzoic acid alkyl esters), all of which are commonly used in dermal products; alkylphenolic surfactants (major ingredients in shampoos and soaps); pesticides (some of which are used as PPCPs); lead (Pb) and other metals, which can comprise significant percentages by weight of various Ayurveda and folk remedies. Lead (Pb) in particular is used in litargirio (or litharge), sometimes at upwards of 80% by weight [87,88]; likewise, mercury is used in certain (banned) skin-lightening creams and disinfectant soaps (upwards of 3% mercuric iodine, wt/wt, in soaps and 10% ammoniated mercury in skin lightening creams) [89,90]. Metals and organometallics are also used in pharmaceuticals, one of the more notable instances being ethylmercury (as ethylmercurithiosalicylate-sodium, Thimerosal), added as a preservative to certain vaccines; others include barium and lithium. Extractables and leachables in dispensing devices and containers can also be a significant source of certain conventional chemicals (e.g., plasticizers, nitrosamines, and acrylonitrile, deriving from plastics adhesives, antioxidants, coatings, vulcanizers, accelerants, adhesives). Worth noting is the significant distinction between EU and U.S. policy in the regulation of cosmetics, as reflected by the hundreds of ingredients in U.S. cosmetics that are not permitted in EU products as a result of purported linkages with genetic or reproductive effects.

Healthcare providers

The major sector of the healthcare community that contributes to the environmental load of PPCPs is probably long-term care facilities (LTCFs) and hospices, a topic discussed by Daughton [48]. Patients at LTCFs are often under the care of multiple physicians and receive multiple medications (polypharmacy). Their prescriptions are also subjected to frequent change, resulting in unusually large amounts of unused medications. LTCFs often dispose of unused medication to sewerage (in some states this is a legislated requirement), especially if the drug is a controlled substance. Physicians (and dentists) also sometimes dispose of out-dated manufacturers' samples and pharmaceuticals used in-practice to sewerage and to trash; pharmacies are minor sources, as they can use the reverse-distribution system and must also abide by laws (e.g., RCRA [91]) regulating the disposal of hazardous waste [48]. Improved efficiency in the way drugs are dispensed at LTCFs, namely with computerized unit-dose dispensing, could greatly reduce the quantities needing disposal. LTCFs are an example of a point source that could have ramifications at the local level.

Hospitals

The medications used in hospitals differ with respect to their types, doses, per-capita consumption, and relative quantities consumed compared with those used by the consumer. These drugs are weighted toward those with higher acute toxicity and genotoxicity (e.g., cytotoxics, oncolytics) and which are used for short-term therapy and diagnostics (e.g., radionuclides), rather than toward long-term maintenance. For this reason, the suite of drugs that occur in waste streams from hospitals can differ in both classes and quantities from those emanating from private residences. Locales having a confluence of hospitals may pose unique circumstances for municipal waste treatment plants and their effluents, depending on whether the hospitals practice waste pretreatment and how sophisticated the pretreatment might be.

Veterinarians

The complete list of drugs available for use with animals in the U.S. is captured in the *Green Book*, which is published by the FDA's Center for Veterinary Medicine [92]. Key information regarding the environmental assessment of veterinary pharmaceuticals (i.e., formal environmental assessments (EAs), findings of no significant impact (FONSIs), and environmental impact statements (EISs)) can be found at FDA [93]. While there is significant overlap among the drugs used in veterinary and human medicine, some are unique. Veterinary use of drugs leads also to some unique sources and routes of exposure. Veterinary use of drugs for domestic animals, such as pets, leads to the direct deposition of residues on land via excrement; any drug residues are then subject to entrainment in wet-weather run-off to storm drains or receiving waters.

While the primary significance of veterinary drugs in the environment derives from their routine usage with raising domestic animals for commerce (and the resulting issues concerning CAFOs, grazing livestock, and aquaculture), the consequences for some veterinary uses have involved significant but little recognized instances of acute poisonings of wildlife. Two examples illustrate the profound ecological consequences that can result from these sources. One is the improper discarding of carcasses from animals that have been euthanized or heavily medicated. The principle drug used for animal euthanasia is pentobarbital. High doses are used, and most of the body-burden residue escapes excretion and persists indefinitely in the body. If not disposed properly, the carcasses can be consumed by scavenger wildlife. But determined wildlife can even uncover well-buried carcasses. Wildlife pentobarbital poisonings had been recorded in at least 14 states since the mid-1980s, the U.S. Fish and Wildlife Service at one point having documented more than 130 bald and golden eagle casualties. Wildlife vulnerable to accidental pentobarbital poisoning (or to any other drug used for euthanasia) include a wide range of birds (especially eagles), foxes, bears, martens, fishers, coyotes, lynx, bobcats, cougars, and otters. Domestic dogs can be poisoned, and zoos have documented the deaths of tigers, cougars, and lions that were accidentally fed tainted meat. As a result, in July 2003, the FDA's Center for Veterinary Medicine required an environmental warning to be added to animal euthanasia products [94].

A second example is the massive poisonings of vultures in Southeast Asia by their feeding on carcasses of cattle that had been treated with diclofenac. Beginning in the early 1990s, vultures (especially white-backed vultures such as *Gyps bengalensis*) experienced dramatic population declines (as great as 95%) in southern Asia. The causative agent had led to acute renal failure (manifested as visceral gout from accumulation of uric acid), leading to death of the breeding population. At least some of these die-offs were eventually linked to poisoning with diclofenac [45]. Although primarily a human anti-inflammatory in the U.S., diclofenac was used in veterinary medicine in other countries. In India, diclofenac was used for cattle, whose carcasses are a major food source for *Gyps*. Diclofenac seemed to be selectively toxic to *Gyps* spp. versus other carrion-eating raptors. As of 2005, India committed to phasing out the veterinary use of diclofenac.

These two examples show some of the unexpected routes by which PPCPs can gain access to the environment. They also show the types of unanticipated, acute ecological effects that can occur from seemingly innocuous drugs and their routine usage.

Agriculture and aquaculture

Drugs are widely and heavily used in a spectrum of agriculture practices. But unlike with human use, the numbers of targeted biological endpoints are limited. Although there are human drugs from a wide spectrum of therapeutic classes, agricultural use tends to focus on antibiotics and steroidal hormones. Discussions in the literature of environmental aspects of drugs used in agriculture are usually separate from those used in human medicine. It is important to recognize, however, that many of the drugs used in agriculture and human medicine are identical or belong to the same chemical classes; some, however, have exclusive uses. The residues that get introduced to the environment also hold the potential for effecting ecological and human exposures. What sets agricultural uses apart from others are the quantities of drugs that can be released and the localized manner in which they are released (e.g., with CAFOs) and during open-range grazing (e.g., impacting run-off to local water bodies). Especially unique for agricultural use is that drugs can be introduced directly to the environment as a direct result of their use, similar to pesticides (e.g., crop spraying and aquaculture). Aquaculture can release drugs directly to open waters (from excess medicated feed and from excreta). In contrast to human use, agricultural introduction via CAFOs tends to be localized, more resembling point sources. Aquaculture also experiences off-label and illegal usage of certain drugs [95], especially highly toxic antibiotics such as chloramphenicol, furazolidone, and nitrofurazone, all of whose use is banned in many countries but continues nonetheless; this source can lead to direct human exposure via consumption of contaminated fin and shell fish). For thorough background on the environmental aspects of concentrated aquatic animal production (CAAP) and the role of pharmaceuticals, refer to the materials available from EPA [96]. In contrast to human use, agricultural use also poses concerns with regard to occupational exposure, an example being the inhalation of medication sorbed to dust particles generated by the handling of medicated feeds [97].

Agricultural use of drugs ranges from crops (e.g., use of antibiotics for plant disease control), CAFOs (antibiotics and estrogenic and androgenic steroids, for both therapeutic treatment and growth promotion), and aquaculture (e.g., antibiotics for disease prevention and treatment). A major unknown is the relative portions of residues (especially antibiotics) in the environment that emanate from agricultural versus human use. As with determining overall drug usage rates (in terms of quantities), unequivocal statistics are not even available for the usage rates (e.g., for antibiotics) by agriculture versus others. A major issue with regard to CAFOs is the integrity of lagoons and other storage areas that detain or treat wastes from CAFOs. These wastes can contain high levels of PPCPs as well as endogenous hormones (especially estrogens and their conjugates). The holding areas (lagoons) are vulnerable to overflow during extreme wet-weather events as well as leaching to groundwater. Manure and sludges are also widely dispersed as amendments or fertilizer to land. An approach for prioritizing the veterinary medications deserving concerted attention with respect to assessing human exposure has been presented by Capelton et al. [98].

Potential sources for human exposure include not just drinking water and the well-known but less publicized routes such as domestic livestock and fish treated with veterinary drugs, but also the less-known route of edible plants. When excrement from domestic animals treated with veterinary medicines is used on arable lands (e.g., as soil amendment or fertilizer), plants have the potential to remove the drug residues that partition to the soil pore water. These residues can accumulate in shoots and roots. For a limited number of targeted drugs evaluated under controlled conditions, the residues found to accumulate in certain plants were calculated to hold the potential for yielding intakes that approached 10% of the accepted daily intake [81].

A related issue regarding agriculture as a source involves "plant-made pharmaceuticals" (PMPs) derived from the crop-based transgenic production of proteinaceous therapeutics by genetically altered plants ("molecular farming"—"biopharming"). Current transgenic biotechnology has the potential for using food crop species (primarily corn, soybeans, rice) for producing hundreds of distinct proteinaceous therapeutics (especially enzymes, hormones, vaccines, monoclonal antibodies). PMPs raise a host of questions regarding risk, primarily centered around allergenicity and consumer toxicity in the form of direct endocrine disruption or other mechanisms. Less-recognized concerns include possible hazards to non-target organisms (e.g., foragers and insects), whose interactions with crops are extremely difficult to prevent. Drugs based on peptides and proteins would ordinarily not be expected to persist in the environment because they are easily degraded. A possible exception is the cyclic peptides and circular proteins. Natural products of the former include Cyclosporin (an immunosuppressant) and Gramicidin S (an antibiotic); these are distinguished from the circular proteins in being synthesized by enzymatic pathways as opposed to being synthesized ribosomally. Synthetic versions of these chemicals (which cross over into the domain of self-assembling nanostructures) can be designed with broad-ranging biological activities, especially antimicrobial. The significant aspects of this class of drugs are that they resist chemical, thermal, and enzymatic alternation and therefore have the potential to persist in the environment.

Miscellaneous sources
There are probably numerous miscellaneous sources for drugs in the environment, such as from the discharge of sewage (both treated and raw) from cruise ships. Such sources are characterized by being insignificant with respect to contributing to overall environmental loads but in certain localized situations could prove significant with respect to the ecology. Two examples serve to illustrate the range of ways in which miscellaneous sources can be unforeseen. First is the accumulation of drugs donated during humanitarian relief efforts. Drug donations have long proved problematic to humanitarian efforts because of the sometimes-massive quantities of inappropriate or outdated medications, or simply because of large surpluses that cannot be used before expiration. Thousands of tons of drugs are sometimes received and necessitate storage and eventual disposal; this is discussed further in the second section of this chapter. A second example, but one that could only have a possible effect on local groundwater, relates to the burial of bodies in cemeteries. Bodies can sometimes serve as reservoirs of large quantities of multiple drugs if heroic life-saving measures had been attempted. More detail on these examples (and others) is provided in [8,48].

Another example pertains to "manufacturing." An exception to manufacturing being an insignificant environmental source of drugs is the release of certain highly potent drugs and chemical synthesis agents from illegal, clandestine drug laboratories ("clan labs"). The growing problem of clan labs, especially for methamphetamine, continues to reveal a wealth of previously unrecognized sources, including those labs that are mobile or easily hidden. While clan labs can release hazardous amounts of synthesis ingredients to the environment, the amount of the active ingredient itself that is accidentally or purposefully discarded is unknown. Buildings that are used for meth labs, however, can pose acute risks to first responders, clean-up crews, and even to those who subsequently occupy the structures after remediation has been attempted, because of the large quantities of methamphetamine that have been absorbed by porous building materials (e.g., concrete and masonry) and which slowly migrate back to the surfaces.

Related to clandestine drugs is the escalating occurrence of counterfeit drugs, a problem also partly related to drug importation and drug diversion. Although counterfeit drugs often comprise non-pharmacologic ingredients (but sometimes harmful), they sometimes contain active ingredients (sometimes sub-potent, sometimes a different drug) [99]. The analogous problem exists with nutraceuticals and food supplements, which are sometimes adulterated with drugs; one example was the marketing in 2003 of an OTC dietary supplement (Viga) that actually contained sildenafil.

Finally, new technologies introduced to medicine hold the potential to serve as previously unanticipated sources of new types of chemicals. A totally new class of chemicals being introduced to medicine are those comprising nanoscale materials (nanomaterials). These materials are touted as presenting unprecedented, revolutionary opportunities for medicine. In contrast to "conventional" chemicals, the properties of these materials are dictated more by their molecular or particle size and shape than by their chemical structures or compositions. Nanomaterials comprise particles with diameters ranging roughly from 1 nm (10 Å, about the size of 10 hydrogen atoms) to 100 nm. The advent of "nanomedicine" holds the potential as another source of medically related materials in the environment. Current applications include vastly improved delivery of drugs to target organs and tissues, thereby improving therapeutic outcomes and minimizing side effects and adverse reactions—all with greatly lower doses. Futuristic uses are vast, including the use of "nanobots" that can roam and diagnose disease, monitor health status, correct cellular defects, repair damaged tissue, or enhance biological performance, or that can be used in the fabrication of biocompatible materials that substitute for biological tissues. While nanotechnology holds the potential to reduce the introduction of conventional drugs to the environment, the environmental ramifications of these materials themselves include release of totally new types of pollutants derived from the manufacture, use, and weathering of nanomedicines and nanodevices [47].

Multiple aggregate sources

Some drugs can have multiple origins, which pose opportunities for aggregate exposure. A special case includes those chemicals that have dual uses as therapeutants and as pesticides. Examples include: triclosan/triclocarban (broad spectrum antimicrobials used as general biocides; triclosan is also used as a gingivitis agent used in toothpaste); 4-aminopyridine (an experimental multiple sclerosis drug and avicide); warfarin (an anti-coagulant and rat poison); azasteroids (antilipidemics and avian/rodent/insect reproductive inhibitors); certain antibiotics (control of orchard pathogens); acetaminophen (analgesic and control of the Brown Tree snake [100]); caffeine (stimulant and used experimentally for control of the *Coqui* frog in Hawaii [101]; also repels and kills snails and slugs at concentrations exceeding 0.5% [102]); lindane and permethrins/pyrethrins (insecticide and control of ticks, fleas, and body and head lice as a shampoo ingredient); and nicotine (a broad spectrum insecticide).

1.2.2.4 Data needs

Comprehensive data on PPCP sources is key to understanding and predicting the occurrence of PPCPs in the environment (which can be done via modeling and by directed target-based monitoring). It is also important for understanding the best approaches for reducing accidental poisonings from stored medications or from those being improperly disposed. In this regard, nation-wide databases, based on geographic information systems, would be invaluable. An ideal system would provide real-time prescription and OTC sales/usage and disposal data. In the U.S., neither the absolute usage rates for PPCPs nor their geographic variations are available in public databases. Geographic drug usage patterns are partly a function of local prescribing customs, patient preferences and fads, and distribution of disease and illness. A real-time GIS database showing drug usage by geographic locale would greatly aid modeling and monitoring efforts; but the proprietary nature of the pharmaceutical industry, widespread OTC availability of veterinary and agricultural drugs (especially antibiotics), and the availability of drugs from outside manufacturer distribution networks (e.g., via the Internet, foreign countries, and black markets) are major barriers to gaining accurate information. Understanding geographic anomalies in prescribing and usage is important as it could result in localized residue levels that are higher than the PECs predicted on the basis of geographically unbiased usages [8].

1.3 MEANS FOR MINIMIZING THESE SOURCES (e.g., POLLUTION PREVENTION)

With better understanding of the sources or origins of PPCPs as pollutants, those sources most amenable to lessening or minimizing their connections with the environment can be identified. Pollution reduction (or minimization) encompasses a wide range of actions, including reduced dosage, waste treatment, waste containment/storage (which often is simply a form of pollution "postponement"), recycling/reuse, disposal, and pollution prevention (also known as source reduction). Note that the formal definition of pollution prevention itself does not include any of the aforementioned activities, but it is distinguished by serving to eliminate or reduce the need for those activities.

A wide spectrum of actions and activities could be designed and implemented to reduce the environmental residues contributed from many of the major sources of PPCPs. These pollution reduction approaches fall into all of the primary categories just listed. Among these categories, a wide spectrum of approaches for pollution prevention, aimed at all aspects of the regulated distribution/sales chain (which spans drug discovery, manufacturing, packaging, distribution, dispensing, and retailing) as well as how a drug is formulated and administered have been summarized by Daughton [8,48,103]. As for the other categories (especially waste treatment), they have been covered in many other publications, including this book and several prior books [35,42]. As but one of numerous possible examples, alternative delivery mechanisms, such as intranasal (which bypasses first-pass hepatic metabolism) can be used for better targeting the dose and thereby reducing dosages and minimizing undesirable metabolic products.

The remainder of this chapter will therefore focus on the two categories that have generated the most attention in the U.S.—disposal and reuse—both of which come into play once a drug exits the regulated sales chain and enters the largely unregulated realms of the consumer and other end-users. Although these two topics have also been covered in Daughton [8,48], various aspects will be developed in more detail here.

1.3.1 Drug disposal

The disposal of drugs by consumers has been a controversial and confusing topic in the U.S. for two major reasons. First, a number of federal and state regulations limit the options available for disposal of unwanted drugs. Analogous regulations do not exist in many other countries. Second, because of these imposed constraints, selecting "prudent" options for drug disposal forces a mutually exclusive choice between ensuring public safety and protecting ecologic integrity. No widely available, cost-effective mechanism or procedure is currently available to do both. In distinct contrast, note that drug disposal in certain other countries is handled in a straightforward manner with "take-back" or "returns" programs, where consumers simply return their unwanted PPCPs to drop-off points such as local pharmacies.

Consequently, much of the following discussion pertains primarily to the U.S.

The need for a simple, universal option for disposing of unused medications is driven by the following considerations and scenarios, all of which are all known to occur as a result of either storing leftover medications or of improper disposal: (i) accidental (and sometimes purposeful) poisoning of infants, children, adults, pets, and wildlife, (ii) increased risk of medications being used past expiry (at which time their efficacy can decrease and/or toxicity can increase), (iii) accumulation of multiple drugs (even if they have not expired) increases the chances of adverse drug interactions, especially if polypharmacy and self-medication are practiced past the date when the original prescription was intended, (iv) accumulation of multiple drugs (even if they have not expired) increases the chances of improper self-medication simply as a result of confusion (this is a long-standing problem for the aging population, especially for those practicing polypharmacy), (v) stored drugs encourage self-medication by those for whom they were not prescribed, increasing the risk of adverse events, (vi) accumulation of stored drugs increases the risk of burglary (by those seeking drugs) and of diversion (e.g., "pharming" parties), and (vii) leftover drugs are a symptom of inefficiencies and/or errors in physician prescribing or patient compliance, and as such, represent increased costs for the healthcare community and consumers as well as reduced or jeopardized therapeutic outcomes.

Prior to any discussion of drug disposal, it is critical to understand the motivation and perceived need driving "environmentally sound" practices for drug disposal. Key to this is recognizing that the portion of environmental drug residues originating from direct disposal by consumers and other end users compared with the portion originating from excretion and bathing is simply not known. The relative contributions to environmental loadings from direct (controllable) disposal of unwanted PPCPs (to sewage and trash) versus indirect (involuntary or inadvertent) excretion and washing to sewerage are known neither for the total environmental burden of PPCPs nor for specific, individual drugs. Does this fraction vary from drug to drug, or among packaging types (e.g., bulk bottles versus blister packs)? The relative significance of direct disposal versus excretion is therefore a major question whose answer is important with regard to justifying drug disposal or take-back programs. This consideration has been overlooked by all assessments made to date of drug disposal, and it represents one of the numerous research needs for the many facets of PPCPs as environmental pollutants [57] and as highlighted by Daughton [48]. In the absence of this data, the inability to predict the outcome (if any) that might result from successfully implementing a nationwide, environmentally sound disposal program is problematic. This would be the case even if the disposal of drugs to sewerage or trash were completely eliminated.

It is possible that direct disposal may indeed be a significant source of environmental residues for a limited number of drugs, such as for OTC medicines (especially those that are bulk purchased in such large quantities that they expire before being completely used); in contrast, disposal is probably not a source for those drugs provided by unit dispensing and for those that are costly or prescribed in short courses. *It is quite possible, therefore, that even if environmentally sound drug disposal*drug disposal *could be implemented, the resulting reduction in overall environmental loads of PPCPs*PPCPs *might be negligible (at least for most drugs)*. This prompts the obvious question of why options such as take-back programs are needed or desired, especially if they are perceived as adding further cost to health care. The answer is several-fold.

The desire to minimize ecological exposure to PPCPs is not the only driving force behind the need for prudent drug disposal. Two other drivers are: (1) the need to protect human safety (e.g., accidental and purposeful poisonings made possible by unwanted drugs that are stored and not disposed), and (2) the public's fundamental desire to be proactive in removing as many possible xenobiotics from the environment (especially from drinking water sources), regardless of any known adverse toxicology. This latter point is important and reveals a fundamental relationship of society with chemicals in general—namely that aversion to involuntary or inadvertent chemical exposure to certain chemicals, even in the stark absence of any known hazard, can result *solely because the chemicals occur where they are not expected or desired*. Such substances have been termed ''chemical weeds'' [23].

It is also important to keep in mind that prudent drug disposal is but one of many possible facets of a larger, holistic environmental stewardship program [8,48,103]. A multitude of pollution prevention approaches can be applied to the many facets of the existing production-distribution-consumption chain for PPCPs. These facets include everything from drug design, drug manufacturing (e.g., green chemistry approaches), drug delivery, package design, distribution, prescribing (e.g., individualization of therapy), dispensing, marketing/advertising, patient compliance, education for health care practitioners, disposal, to data mining (e.g., from unused medications) and others. Implementation of one or more of these stewardship measures (in addition to proper drug disposal) affords two more possible advantages:

(a) Improvement in therapeutic outcomes and patient health, as well as reducing healthcare costs; these are part of the philosophy behind cradle-to-cradle stewardship [8]. A holistic stewardship program also could yield collateral benefits for consumer/public health, such as by improving the awareness of the consumer and the medical community of environmental ramifications and increasing the prudent use of drugs.
(b) Possible achievement of even greater reductions in PPCP loadings to the environment than by drug disposal alone. In fact, it is worth asking if the resulting reduction in human and ecological exposure from such a stewardship program could be accomplished with far less investment of resources than required for further research (e.g., environmental toxicology) and development of end-of-pipe control technologies. A stewardship program designed for minimizing the introduction of PPCP residues to the environment might be particularly advantageous for dealing with the foreseeable increase and expansion in drug usage (e.g., as the population ages and as new therapies continue to be developed).

A suitably designed drug take-back program would be capable of improving overall health care and lowering health costs. This would be accomplished by inventorying returned drugs and the reasons for their return. Every medication that goes unused, eventually needing disposal, represents a prescription or purchase that was either not needed or not complied with. Either represents wasted health care resources and the possibility of adverse or suboptimal therapeutic outcomes. By mining the information that could be obtained from drug returns, knowledge could be developed for continually adjusting and improving prescribing practices and for lessening health care expenditures. The data that can be obtained from drug returns can also be used in prioritization models for selecting those drugs being used (or disposed of) most frequently in particular geographic locales and which might therefore have a significant environmental presence. This can then better guide and tailor the selection of drugs targeted for environmental monitoring. An example of the type of information that can be obtained just from a small, local take-back event is available from NERC [104].

Currently, the only aspect of PPCPs known to directly impact human morbidity and mortality is their major contribution to accidental and purposeful poisonings [71]. One of the factors determining or encouraging inappropriate or undesired access to drugs is the prevalence of improper storage or misguided attempts at disposal, which is in turn caused by the accumulation of left-over drugs. Many factors lead to the unnecessary storage of unwanted PPCPs in domestic residences. The level of adherence (compliance) by patients to prescribed medication regimes is one of the major factors that determines the accumulation and eventual expiration of unused drugs in the household. Adherence to medication is an issue of great importance to health care. Its causes are many and complex (e.g., see [105]). A variety of ways to improve compliance, ranging from simple to technologically sophisticated, currently exist or are under development (e.g., see [106]).

The critical importance of medication adherence is shown by the fact that one- to two-thirds of all hospital admissions in the U.S. related to medicine result from poor medication adherence, leading to medical costs of about $100 billion per year [107]. An unknown portion of non-adherence, which undoubtedly varies wildly among classes of drugs (e.g., being roughly 50% for long-term medications prescribed for chronic conditions), is one of the contributing factors to the accumulation of unused drugs in the household and therefore contributes to their direct disposal. The information that is completely lacking is the percentage of medications (once purchased) that are never used.

Once the consumer has accumulated a certain number of unuseable or unwanted medications in the home, the question of disposal is confronted. Conflicting needs and motivations make disposal of PPCPs a confusing issue. Water treatment facilities increasingly no longer want drugs unnecessarily discharged via sewers, while at the same time poison control centers have long-advised against discarding them to trash and have always recommended discarding to sewerage (since this is historically the easiest means available for protecting humans and pets from accidental and purposeful poisonings). Drugs discarded to municipal trash/landfills pose not just future environmental exposure risks but also ongoing risks with regard to reuse by those who scavenge for them (e.g., human "gleaners" or animal scavengers). Discard to sewerage, in contrast, is also the surest simple means for preventing drug diversion.

Solutions to the drug disposal quandary might seem to be easily addressable. However, an array of local, state, and federal regulations—promulgated to ensure occupational and consumer health, safety, and privacy—make any solution much more challenging. Statutes that must be considered include: (i) Federal and State hazardous wastes regulations, (ii) Controlled Substances Act (CSA, see below) as administered by the Drug Enforcement Administration (DEA); (iii) State regulations for long-term care facilities (where disposal to sewerage is sometimes required by law [8]; and (iv) HIPAA, the Health Insurance Portability and Accountability Act; for HIPAA, it is unclear whether any recommendations that might be made to consumers regarding disposal must also inform them of the privacy protections afforded by HIPAA; for example, should patients be encouraged to remove their personal information, but not the prescription information, from drug labels prior to disposal.

The DEA regulates certain drugs under the CSA of 1970 [108], which classifies these drugs within five "Schedules" (I–V); note that Schedule I is reserved for those drugs having no recognized medical use and which are therefore deemed to be the most dangerous. The CSA through a series of amendments also regulates a list of chemicals that are used in the illicit synthesis of controlled substances. Several of these "listed chemicals" also happen to be non-controlled active ingredients of licit OTC drug products; these include ephedrine, pseudoephedrine, and phenylpropanolamine. Once prescribed, a controlled substance cannot be transferred to any other entity (including the original prescriber, a pharmacy, reverse distributor, or even a hazardous waste facility) other than DEA-"exempted" law enforcement. "The CSA also creates a closed system of distribution for those authorized to handle controlled substances. The cornerstone of this system is the registration of all those authorized by the DEA to handle controlled substances. All individuals and firms that are registered are required to maintain complete and accurate inventories and records of all transactions involving controlled substances, as well as security for the storage of controlled substances." [108].

"The overall goal of the CSA and of DEA's regulations in Title 21, Code of Federal Regulations (CFR), Parts 1300–1316 is to provide a closed distribution system so that a controlled substance is at all times under the legal control of a person registered, or specifically exempted from registration, by the DEA until it reaches the ultimate user or is destroyed. DEA achieves this goal by registering manufacturers, distributors, importers, exporters, and dispensers of controlled substances as well as analytical laboratories and researchers. Thus, any movement of controlled substances between these registered persons is covered by DEA regulations, which ensure that all controlled substances are accounted for from their creation until their dispensing or destruction. When a controlled substance has become outdated or otherwise unusable, the registrant who possesses the substance must dispose of it.

However, over the past decade, environmental concerns and regulatory changes have caused drug manufacturers and government agencies (including the DEA and State authorities) to become increasingly reluctant to be involved in the disposal process. Thus, some disposal options are no longer available." [109].

With this as background to the CSA, the DEA also provides specific answers to two key questions: (1) "Can an individual return their controlled substance prescription medication to a pharmacy?" [110] and (2) "Can a long-term care facility (LTCF) return a resident's unused controlled substance medication to a pharmacy?" [111]. The answers are: (1) Quoting from the DEA [110], "An individual patient may not return their unused controlled substance prescription medication to the pharmacy. Federal laws and regulations make no provisions for an individual to return their controlled substance prescription medication to a pharmacy for further dispensing or for disposal. There are no provisions in the CSA or Code of Federal Regulations (CFR) for a DEA registrant (i.e., retail pharmacy) to acquire controlled substances from a non-registrant (i.e., individual patient). "An individual may dispose of their own controlled substance medication without approval from the DEA. Medications should be disposed of in such a manner that does not allow for the controlled substances to be easily retrieved." (2) Quoting from the DEA [111], "There are no provisions in the CSA for a DEA registrant (i.e., retail pharmacy) to acquire controlled substances from a non-registrant (i.e., resident of a LTCF). Most [LTCFs] are not licensed by their respective state to handle controlled substances and therefore are not registered with the DEA. The [LTCFs] act in a custodial capacity, holding controlled substances that, pursuant to a prescription, have been dispensed to and belong to the resident of the LTCF. Federal laws and regulations make no provisions for controlled substances that have already been dispensed to patients, regardless of the packaging method, to be returned to a pharmacy for further dispensing or disposal."

One of the ways the CSA impacts drug returns programs results from the fact that once a controlled substance is dispensed, the prescription label has no marking or indication that the drug is a controlled substance; such markings (e.g., "CII") are only on the manufacturer's original packaging. This makes it difficult for anyone other than a licensed pharmacist to determine the status of the medication. Consequently, in the absence of any legislated change in labeling standards, for any program designed to accept the return of unwanted drugs, a pharmacist must be present to physically separate controlled from non-controlled medications.

Note that drugs used by consumers are often treated differently than those administered in the same household by a licensed health care provider. Consumer household hazardous waste, including pharmaceuticals, is exempted from RCRA. Unwanted pharmaceuticals are considered waste materials only when declared as wastes. Consumer discharge of drugs to sewerage does not violate water regulations as drug residues are not covered by regulations for water quality. Pharmacies, however, cannot dispose of those medications containing ingredients that are considered hazardous under RCRA (e.g., P- or U-listed drugs; see Table 1, page 782, Daughton [48]; also Smith [91]). A wide spectrum of state laws (many of which have conflicting ramifications with respect to drug disposal) govern the handling and disposition of unused drugs; these are compiled in the database "Current Substance Abuse Legislation" [112].

With all of this as background, there are really only four current options for consumers to dispose of unwanted drugs: (1) discard to sewerage after removing from all packaging, (2) pick-up (of non-controlled substances) by community hazardous waste handler, (3) discarding with municipal trash, and (4) drop off at local sites that host drug take-back events overseen by DEA-exempted law enforcement. The second and fourth options, however, are available only in certain locales.

The second option (disposal with community hazardous waste) is a rather confusing area. Transfer of controlled substances would violate the CSA (although most hazardous waste facilities may not be aware of this). A further complication is the difficulty of expecting the consumer to understand which drugs are controlled substances. Absent any eventual take-back program, drug labeling could be used to provide advice on environmental disposition and possible environmental ramifications of improperly disposed materials. As an example, an "environmental labeling" classification system is being developed in Sweden in a collaborative project between Sweden's Department of the Environment and the Stockholm County Council Pharmaceutical Unit [113,114]. Another example is from the European Medicines Agency [115]: "Appropriate disposal of unused pharmaceuticals, e.g., when shelf life is expired, is considered important to reduce the exposure of the environment. In order to enhance environmental protection, it is therefore recommended that—even for medicinal products that do not require special disposal measures—package leaflets (patient information leaflets) should include the following general statement: 'Medicines no longer required should not be disposed of via wastewater or the municipal sewage system.''' But note that Europe has the option of returning medications to pharmacies.

Absent any imminent nationwide system for drug returns, and with the growing emphasis of local agencies emphasizing the importance of avoiding disposal to sewerage, the most straightforward solution for most consumers is to dispose of their unwanted medications to domestic trash. But note that trash is usually stored in landfills, which can be considered a form of "pollution postponement" [23]. Moreover, the single most important aspect of disposal to trash to keep in mind is that it poses imminent risks for both children and "gleaners" (those who rummage through trash), as well as for domestic, feral, and wild animal scavengers (e.g., coyotes, racoons, bears, dogs). Medications improperly stored or disposed with domestic refuse can be accidentally ingested, especially by infants and children. This is the major impetus behind the recommendations of poison control centers to dispose to sewerage. Disposal to trash can also be limited by obstacles posed by little-recognized transportation rules.

With the potential for future pollution aside, a safe and effective protocol for disposal to trash would surprisingly require considerable explanation, as it would involve attention on the part of the consumer to what would probably prove to be too many details; for example, special attention would need to be devoted to medical patches, some of which still contain very toxic levels of residual drug (e.g., fentanyl). Some of the details to consider for safe disposal to trash include the following. Medicine containers should have the name of the patient obliterated (but not the name of the medicine—in case a poisoning should later occur). To minimize the chances of others gaining access to the disposed medications, the medicine should be placed in leak-proof, double (nested), opaque containers and tightly sealed (e.g., with heavy-duty packing tape); the containers should not have originally contained food (to discourage their opening by others). This will also prevent casual inspection by others who might then be enticed to consume or sell the medication. To further minimize access by others, the packaged medications should be placed at curb-side as close as possible to the actual time of pick up. One particular note of caution. There has been considerable discussion about the need to render unwanted medications unsuitable for consumption (e.g., by adding reactive chemicals, by heating, or by disassembly of capsules or crushing tablets). Such procedures could be hazardous because they promote the unnecessary handling of active ingredients and can lead to dermal or pulmonary exposure (e.g., by hand contact or inhalation of dusts) or the generation of highly hazardous vapors (e.g., if denaturing chemicals, such as bleach, or heat are used).

Becoming more widely accepted or recognized is the fact that none of these options embodies the dual objectives of protecting human health and safety together with ecological integrity. This is what leads to the need for creating take-back programs, preferably those that are statewide or nationwide in scope.

In 2004, State of Maine legislation enacted the nation's first statewide program for take-back of unused drugs: "An Act to Encourage the Proper Disposal of Unused Pharmaceuticals" [116]. As of early 2007, this program had not yet been implemented, but it would allow individuals to safely dispose of their unused medications by mailing unused pharmaceuticals in a prepaid mailer to the Mainc Drug Enforcement Agency for destruction (via incineration). This particular approach could be amenable to templating across the U.S. A summary of recommendations regarding the Maine program is available [117].

Attention to the need for take-back programs is also developing within medical associations and pharmaceutical organizations. First steps include: (i) a position adopted by the American Society of Consultant Pharmacists [118] where unused non-controlled substances dispensed by the LTCF pharmacies may be returned to the pharmacy for reuse, (ii) a resolution adopted by the U.S. Pharmacopeia Convention [119] aimed at working with "appropriate constituencies to continue developing programs to promote safe medication use and disposal," (iii) the May 2005 Assembly of the American Psychiatric Association endorsed a paper encouraging state and federal legislation for programs aimed at the proper disposal of unused pharmaceuticals, and (iv) In April 2006, the National Association of Boards of Pharmacy adopted a resolution to "Develop Legal and Environmentally Safe Programs for the Disposal of Unwanted Medications."

The prospects for future advances in designing more effective take-back programs hinge largely on whether current regulatory practices can be modified, especially with respect to improved means for handling the disposition of controlled substances between non-registrants (e.g., the consumer) and other entities; the most likely target for legislative change would be via modification of 21 CFR part 1307.21 [120].

1.3.1.1 Example of the hazards associated with storage of drugs at the home

The following illustrates the hazards associated with maintaining easy access to multiple medications (prescription and OTC alike) for both children and adults by on-site storage in the home. As an extreme example, even medications formulated specifically for infants can prove toxic when consumed by adults. Consider the case of NSAIDs (non-steroidal anti-inflammatories) and specifically acetaminophen; note that acetaminophen is not an NSAID, as it is not an anti-inflammatory, but it is often loosely lumped under the NSAID category. Every NSAID (like any drug) has a maximum: safe unit dose, cumulative daily dose, and duration of dosing. Safe, recommended doses of any particular NSAID can be easily exceeded even when following prudent/safe-use instructions for the individual product. This situation results from aggregate exposure (ingesting another medication containing the same active ingredient); acetaminophen is used in antihistamines, cough and cold preparations, flu medications, and analgesics. Many individual formulations contain the maximum safe dosage for a particular NSAID. Unwitting toxic cumulative exposure can result from the consumption of multiple formulations of the same drug. Even different formulations from same manufacturer can contain different amounts of the same active ingredient, leading to confusion regarding total amount ingested. For example, different products from a particular, single manufacturer can contain acetaminophen in dosage amounts of 80, 160, 325, 500, 650, or 1000 mg. These dosages are provided in 11 distinct adult formulations and 13 distinct child formulations. If an adult used 2 tablespoons of concentrated drops formulated for infants, the dose (nearly 3 g) would be toxic; likewise, if a child were to ingest 1 teaspoon, the dose (500 mg) would prove toxic. For NSAIDs in general, the purposeful or inadvertent consumption of excessive aggregate doses from multiple sources results in over 16,000 deaths annually in the U.S. and over 100,000 hospitalizations from NSAID-related complications [121]. Many of these poisonings result from confusion that derives from the storage of multiple medications.

Similarly, the storage of leftover drugs exacerbates the confusion caused by the well-known problem of similar-looking and similar-sounding medication names. This is another cause of accidental ingestion of incorrect medication. There are more than 15,000 formulary names in the U.S. comprising several thousand distinct drug entities. There are hundreds of instances with similar-looking names (Celebrex vs. Celexa vs. Cerebyx), similar-sounding names (Sarafem vs. Serophene), similar-looking pills (color or shape), and similar-looking packaging (numerous examples can be accessed at the U.S. Pharmacopeia's web site: http://www.usp.org/). The hazards of polypharmacy are increased by any added confusion regarding drug names or drug doses.

The hazards of polypharmacy (combined pharmacotherapy, or sometimes called "stacking") result when multiple, distinct drugs are prescribed for different therapeutic endpoints but they all impart the same side effect; a closely related problem is the prescribing of multiple medications (for a patient with undisclosed multiple specialists) that all share the same mode of action for either the therapeutic endpoint or a side-effect (e.g., anti-cholinergic syndrome or drug-induced delirium in geriatric patients). Different products containing the same API from different manufacturers are available for different intended therapies. Analogous confusion exists when a patient with undisclosed multiple physicians receives prescriptions containing the same drug entity (with different names) but marketed for different disorders. Examples include fluoxetine HCl in the form of Sarafem (for PMDD) and Prozac (for depression); and bupropion HCl for depression (Wellbutrin), smoking cessation (Zyban), and anorexia nervosa.

The main point from these examples is that reducing medication storage in the home of medications can reduce the incidence of self-medication and polypharmacy and its attendant hazards that result in part from confusion of the consumer.

1.3.2 Drug reuse and recycling

Another means that consumers sometimes employ for dealing with leftover drugs is the practice of reusing drugs by providing them to another end-user. While not always legal (e.g., for prescription substances) or medically prudent, reuse is usually accomplished by donation to charitable organizations or by sharing with family and friends. Among the available approaches for avoiding disposal or destruction of leftover drugs, is recycling. "Recycling" can be distinguished from "reuse" in that the active ingredient from the drug is reclaimed by repurifying either from the original formulation or even from excreted waste. This has been proposed, for example, in the form of "mining" drugs from excreta and other wastes, as noted by Daughton [48]. An example of such a process has been under development [122]; this is analogous to the reclamation of illicit methamphetamine from the urine of meth users. Perhaps a less confusing term for drug "re-use" would be drug "redistribution" to lessen confusion with drug "recycling," which could be reserved for the reuse of the active ingredient after it has already been used (and excreted).

The most prevalent practices of drug reuse are donation and sharing. The problems associated with charitable drug donations are discussed by Daughton [48] and can be substantial. While proper charitable drug contributions can play an essential role in humanitarian relief efforts, inappropriate charitable donations can become a significant source of drugs as environmental pollutants. For example, despite the fact that knowledgeable relief agencies for the 2005, Indonesian tsunami relief efforts attempted to avert the donation of medicines (based on their prior negative experience in Kosovo, where massive quantities of unusable medications had to be disposed at considerable cost), many companies and individuals worldwide ignored the guidelines for donations [123]. Drug redistribution is also becoming a practice among certain physicians and nursing homes who do not want medications to go unused. Redistribution is targeted to those who cannot afford medical care; for example, recent legislation [124] authorizes counties to collect unused prescriptions from nursing homes, wholesalers, and manufacturers, for redistribution to those having low-income and no medical insurance.

For individuals, relief organizations maintain that it is best to refrain from making donations of medications; monetary donations are usually more useful for relief efforts. The flood of donated medicines that arrive after disasters can create chaos of its own by necessitating that already limited resources be siphoned away from other tasks. The huge quantities of donated drugs need to be stored (often in large warehouses), cataloged, and secured; proper storage conditions for many medicines may not be available or affordable. A large percentage will eventually need to be disposed, and since proper incinerators are scarce, an unknown amount of environmental pollution can result. Significant drug diversion can also result, especially from the inability to sufficiently secure controlled substances and other drugs that can then reenter commercial distribution. Worldwide harmonized guidelines for drug donations are needed to avoid continuing problems with drug donations during relief efforts. Reference resources and guidelines for drug donations are available from WHO [125], Autier et al. [126], and from the links provided by the Pharmaceutical Journal [127].

In contrast with donations and redistribution, drug "sharing" probably does not play a significant role as an added source of environmental residues, but it does pose substantial acute risks with regard to human health and safety. Drug sharing is a practice that continues to grow as a result of frustration from not being able to make use of drugs that one person no longer wants or needs but which another person does [128]. Both consumers and physicians sometimes practice it. Drug sharing, however, is not legal (for prescription medications) and can be hazardous. The practice of self-medication is hazardous itself and is responsible for a large percentage of hospitalizations from adverse drug responses (e.g., wrong drug, wrong dosage). Drug sharing also raises the chances of introduction of counterfeit and expired drugs into the supply chain.

1.4 SUMMARY

The sources that contribute residues of human and veterinary drugs to the environment are wide in scope and tend to be diffuse in nature; some point sources for acute levels, however, are known to exist and can cause environmental damage. Although parallels exist between the origins of human drugs and those designed for animals, there are some distinct differences as well, which lead to different exposure scenarios for non-target wildlife. A vast array of approaches exists, or could be developed, for reducing these sources and thereby lessening environmental loads of PPCP residues. All sources hold the potential for control or reduction of their releases to the environment, but the associated costs and other ramifications for implementing some of these approaches could be great. Perhaps the major overlooked benefit of PPCP pollution reduction is the potential it holds for collateral improvements in the administration of health care, reducing healthcare costs, in improving therapeutic outcomes, and in lessening the consumer-acceptance problems associated with risk perception and recycled wastewater.

NOTICE

The U.S. Environmental Protection Agency (EPA), through its Office of Research and Development (ORD), funded and performed the research described. This manuscript has been subjected to the EPA's peer and administrative review and has been approved for publication.

REFERENCES

[Note regarding URLs: References in this paper rely extensively on Internet URLs (Universal Resource Locators), which can cease to function for any number of reasons. To locate information on web pages no longer accessible, an archive service such as the "Internet Archive Wayback Machine' (www.archive.org) can be useful. Other information regarding non-functional URLs can be located at: www.epa.gov/nerlesd1/chemistry/ppcp/advice.htm ("Advice on Non-Functional URLs").]

1 EPA "Everyday Choices: Opportunities for Environmental Stewardship," Technical Report prepared by the U.S. EPA Environmental Stewardship Staff Committee for the EPA Innovation Action Council, November 2005, 100 pp.; available: http://www.epa.gov/innovation/pdf/techrpt.pdf

2 P. Harremoës, D. Gee, M. MacGarvin, A. Stirling, J. Keys, B. Wynne and S.G. Vaz (Eds.), *Late Lessons from Early Warnings: The* Precautionary Principle 1896–2000. Environmental issue report no. 22. Copenhagen, Office for Official Publications of the European Communities, 2001; available: http://reports.eea.eu.int/environmental_issue_report_2001_22/en

3 J.P. Sumpter and A.C. Johnson, *Environ. Sci. Technol.*, 39 (2005) 4321–4332.

4 FDA "Approved Drug Products with Therapeutic Equivalence Evaluations: Orange Book," Center for Drug Evaluation and Research, U.S. Food and Drug Administration, 2005a (web page); available: http://www.fda.gov/cder/orange/default.htm

5 FDA "Approved Drug Products with Therapeutic Equivalence Evaluations," 25th Edition, Cumulative Supplement 3, Office of Pharmaceutical Science, Office of Generic Drugs, Center for Drug Evaluation and Research, U.S. Food and Drug Administration, March 2005b; available: http://www.fda.gov/cder/orange/supplement/cspreface.htm

6 Rote Liste Service (Ed.), "Rote Liste 2005: Drug directory for Germany including EU-approved drug products and some medical devices," ISBN 3-87193-306-6, 2138 pp., published by ECV—Editio Cantor Verlag.

7 A. Tropsha, *Curr. Opinion Drug Disc. Develop.*, 3 (2000) 310–313.

8 C.G. Daughton, *Environ. Health Perspect.*, 111 (2003) 757–774.

9 HHS "Health, United States, 2004: 28th Report on the Health Status of the Nation," Department of Health and Human Services, September 2004, DHHS Publication number 2004-1232; available: http://www.cdc.gov/nchs/data/hus/hus04.pdf

10 Medicare "Formulary Finder for Prescription Drug Plans,," Centers for Medicare & Medicaid Services, U.S. Department of Health and Human Services, 2006 (web page); available: http://www.medicare.gov/MPDPF/Shared/Include/Formulary/FormularyFinder.asp

11 EBRx "Preferred Drug List," 1 January 2006, Columbus, OH; available: http://www.ebrx.com/Docs/EBRxPreferredFormulary.pdf

12 BNF "British National Formulary 50," September 2005, BMJ Publishing Group Ltd and Royal Pharmaceutical Society of Great Britain; available (login required): http://www.bnf.org/bnf/bnf/current/noframes/index.htm

13 UM "Veterinary Formulary," University of Minnesota, 2006; available: http://www.ahc.umn.edu/rar/umnuser/formulary.html
14 NDCHealth "NDC Pharmaceutical Audit Suite (PHAST)," NDCHealth, Atlanta, GA, 2005a; available: http://www.ndchealth.com/pharmaceutical/ndc_phast.htm (see: http://www.rxlist.com/top200.htm)
15 C.G. Daughton and T.A. Ternes, *Environ. Health Perspect.*, 107(6) (1999) 907–938.
16 K. Fent, A.A. Weston and D. Caminada, *Aquatic Toxicol.*, 76(2) (2006) 122–159.
17 NDCHealth "NDC Pharmaceutical Audit Suite (PHAST)," NDCHealth, Atlanta, GA, 2005b; available: http://www.ndchealth.com/pharmaceutical/ndc_phast.htm (see: http://www.rxlist.com/top200_sales_2004.htm)
18 Med Ad News "Top 500 Prescription Drugs," PharmaLive.com Special Report, Med Ad News, May 2005; available: http://www.pharmalive.com/special_reports/sample.cfm?reportID=191
19 VFA "The Pharmaceutical Market," Association of Research-based Pharmaceutical Companies (VFA), 2005 (web page); available: http://www.vfa.de/en/statistics/pharmaceuticalmarket
20 IMS Health "Looking to China and Cancer As Cost Containment Slows Growth," IMS Health Incorporated, 2005 (web page); available: http:// www. imshealth. com/web/content/ 0,3148,64576068_63872702_70260998_73052844,00.html
21 IMS Health "IMS Health Reports 5.4 Percent Dollar Growth in 2005 U.S. Prescription Sales," IMS Health Incorporated, press release, 22 February 2006; available: http://www.imshealth.com/ims/portal/front/articleC/0,2777,6599_3665_77180090,00.html
22 FDA "CDER Approval Times for Priority and Standard NMEs and New BLAs Calendar Years 1993 – 2004," Center for Drug Evaluation and Research, U.S. Food and Drug Administration, 2005c (web page); available: http://www.fda.gov/CDER/rdmt/NMEapps93–04.htm
23 C.G. Daughton, *Renewable Resources J*, 23(4) (2005) 6–23.
24 C.G. Daughton, Illicit drugs in municipal sewage: proposed new non–intrusive tool to heighten public awareness of societal use of illicit/abused drugs and their potential for ecological consequences. In C.G. Daughton and T.L. Jones-Lepp (Eds.), *Pharmaceuticals and Personal Care Products in the Environment: Scientific and Regulatory Issues*, Symposium Series 791, American Chemical Society: Washington, DC, 2001, pp. 348–364; available: http://epa.gov/nerlesd1/chemistry/pharma/book-conclude.htm
25 E. Zuccato, C. Chiabrando, S. Castiglioni, D. Calamari, R. Bagnati, S. Schiarea, R. Fanelli, Environ. Health: Global Access Sci. Source, 4(14) 2005, 7 pp.
26 B. Turque, *Sewage Tested for Signs of Cocaine: Fairfax Participating In Federal Program To Assess Drug Use*, Washington Post, 27 March 2006,

page B01; available: http://www.washingtonpost.com/wp-dyn/content/article/2006/03/26/AR2006032600880.html
27 E. Zuccato, D. Calamari, M. Natangelo and R. Fanelli, *Lancet*, 355 (2000) 1789–1790.
28 C.G. Daughton, *Lancet*, 360 (2002) 1035–1036.
29 J.T. Sherer, *Am. J. Health Syst. Pharm.*, 63 (2006) 174–178.
30 D.L. Sedlak, J.L. Gray and K.E. Pinkston, *Environ. Sci. Technol.*, 34(23) (2000) 509A–515A.
31 J.A. Pedersen, M.A. Yeager and I.H. Suffet, *J. Agric. Food Chem.*, 51(5) (2003) 1360–1372.
32 P.M.J. Fisher and R. Borland, *J. Cleaner Prod.*, 11 (2003) 315–320.
33 Daughton/EPA, *References Relevant to PPCPs in the Environment*, US Environmental Protection Agency, Las Vegas, NV, 2005 (web page); available: http://www.epa.gov/nerlesd1/chemistry/ppcp/reference.htm
34 A.B. Boxall, L.A. Fogg, P.A. Blackwell, P. Kay, E.J. Pemberton and A. Croxford, *Environ. Contam. Toxicol.*, 180 (2004) 1–91.
35 C.G. Daughton, T.L. Jones-Lepp (Eds.), *Pharmaceuticals and Personal Care Products in the Environment: Scientific and Regulatory Issues, Symposium Series 791*; American Chemical Society: Washington, DC, 2001, 416 pp.; available: http://epa.gov/nerlesd1/chemistry/pharma/book.htm
36 M.S. Díaz-Cruz, M.J. López de Alda and D. Barceló, *TrAC Trends Anal. Chem.*, 22(6) (2003) 340–351.
37 B. Halling-Sørensen, S.N. Nielsen, P.F. Lanzky, F. Ingerslev, H.C.H. Lutzhoft and S.E. Jorgensen, *Chemosphere*, 36(2) (1998) 357–393.
38 T. Heberer, *Toxicol. Lett.*, 131(1–2) (2002) 5–17.
39 S.E. Jorgensen and B. Halling-Sørensen, *Chemosphere*, 40(7) (2000) 691–699.
40 D.W. Kolpin, E.T. Furlong, M.T. Meyer, E.M. Thurman, S.D. Zaugg, L.B. Barber and H.T. Buxton, *Environ. Sci. Technol.*, 36(6) (2002) 1202–1211.
41 K. Kümmerer, *Chemosphere*, 45(6–7) (2001) 957–969.
42 K. Kümmerer (Ed.), *Pharmaceuticals in the Environment: Sources, Fate, Effects and Risks*, 2nd ed., Springer, Germany, 2004.
43 M. Petrović, M.D. Hernando, M.S. Díaz-Cruz and D. Barceló, *J. Chromatogr. A*, 1067(1–2) (2005) 1–14.
44 T.A. Ternes, *TrAC Trends Anal. Chem.*, 20(8) (2001) 419–434.
45 J.L. Oaks, M. Gilbert, M.Z. Virani, R.T. Watson, C.U. Meteyer, B.A. Rideout, H.L. Shivaprasad, S. Ahmed, M.J.I. Chaudhry, M. Arshad, S. Mahmood, A. Ali and A.A. Khan, *Nature*, 427 (2004) 630–633.
46 C.G. Daughton, Ground water monitor, *Remed. J.*, 24(2) (2004) 127–138.
47 C.G. Daughton, *Environ. Impact Assess. Rev.*, 24(7–8) (2004) 711–732.
48 C.G. Daughton, *Environ. Health Perspect.*, 111 (2003) 775–785.
49 EPA "Profile of the Pharmaceutical Manufacturing Industry," Office of Compliance Sector Notebook Project, EPA/310-R-97-005, Office of

Compliance, Office of Enforcement and Compliance Assurance, U.S. Environmental Protection Agency, September 1997; available: http://www.epa.gov/compliance/resources/publications/assistance/sectors/notebooks/pharma.pdf

50 EPA "Permit Guidance Document: Pharmaceutical Manufacturing Point Source Category," 40 CFR Part 439, Office of Water, U.S. Environmental Protection Agency, Washington, D.C., January 2006a, EPA 821-F-05-006; available: http://www.epa.gov/waterscience/guide/pharm/guidance/pharmaceuticals-guidance.pdf

51 EPA "Development Document for Final Effluent Limitations Guidelines and Standards for the Pharmaceutical Manufacturing Point Source Category," Office of Water, U.S. Environmental Protection Agency, EPA-921-R-98-005, September 1998; available: http://epa.gov/ostwater/guide/pharm/techdev.html

52 M. Poliakoff, J.M. Fitzpatrick, T.R. Farren and P.T. Anastas, *Science*, 297(5582) (2002) 807–810.

53 IPEC, web home page, International Pharmaceutical Excipients Council, Trade Association of Manufacturers and Users of Pharmaceutical Excipients, 2006; available: http://www.ipec.org/

54 E. Eriksson, K. Auffarth, A-M. Eilersen, M. Henze and A. Ledin, *Water SA*, 29(2) (2003) 135–146.

55 EWG "Skin Deep: A Safety Assessment of Ingredients in Personal Care Products," Environmental Working Group, Washington, DC, June 2004; available: http://www.ewg.org/reports/skindeep/

56 J. Drews, *Science*, 287(5460) (2000) 1960–1964.

57 C.G. Daughton, In: K. Kümmerer (Ed.), *PPCPs in the Environment: Future Research—Beginning with the End Always in Mind," In Pharmaceuticals in the Environment*, 2nd ed., Springer, 2004c, Chapter 33, pp. 463–495; available: http://epa.gov/nerlesd1/chemistry/ppcp/images/researchneeds-chapter.pdf

58 L. Lissemore, C. Hao, P. Yang, P.K. Sibley, S. Mabury and K.R. Solomon, *Chemosphere*, 64(5) (2006) 717–729.

59 D.A. Alvarez, J.D. Petty, J.N. Huckins, T. Jones-Lepp, D.T. Getting and S.E. Manahan, *Environ. Toxicol. Chem.*, 23(7) (2004) 1640–1648.

60 M.H.M.M. Montforts "The Trigger Values in the Environmental Risk Assessment for (Veterinary) Medicines in the European Union: a Critical Appraisal," RIVM report 601500002/2005, Expert Centre for Substances of the RIVM, National Institute for Public Health and the Environment (RIVM), The Netherlands, 45 pp., 2005a; available: http://www.rivm.nl/bibliotheek/rapporten/601500002.pdf

61 M.H.M.M. Montforts, Doctoral Thesis: *Validation of the EU Environmental Risk Assessment for Veterinary Medicines*, Leiden University, 20 April 2005b; available: http://openaccess.leidenuniv.nl/dspace/bitstream/1887/648/1/thesis+montforts.pdf

62 EPA "General Information on Concentrated Animal Feeding Operations," U.S. Environmental Protection Agency, 20 October 2004a (web page); available: http://cfpub.epa.gov/npdes/afo/info.cfm
63 K. Rahn, Expired drugs headache for pharmacies, *The Korea Times*, 1 May 2006.
64 Z. Takáts, J.M. Wiseman, B. Gologan and R.G. Cooks, *Science*, 306(5695) (2004) 471–473.
65 M. Auriol, Y. Filali-Meknassi, R.D. Tyagi, C.D. Adams and R.Y. Surampalli, *Process Biochem.*, 41(3) (2006) 525–539.
66 J.M. Surian, *Am. Pharmaceut. Rev.*, 9(1) (2006) 35–39.
67 D. Bendz, N.A. Paxéus, T.R. Ginn and F.J. Loge, *J. Hazard. Mater.*, 122(3) (2005) 195–204.
68 H. Breton, M. Cociglio, F. Bressolle, H. Peyriere, J.P. Blayac and D. Hillaire-Buys, *J. Chromatogr. B*, 828(1–2) (2005) 80–90.
69 X.-S. Miao and C.D. Metcalfe, *Anal. Chem.*, 75(15) (2003) 3731–3738.
70 J.P. Bound and N. Voulvoulis, *Environ. Health Perspect.*, 113(12) (2005) 1705–1711.
71 AAPCC "Toxic Exposure Surveillance System," American Association of Poison Control Centers, 2005 (web page); available: TESS: http://www.aapcc.org/poison1.htm
72 ASPCA "Animal Poison Control Center," American Society for the Prevention of Cruelty to Animals 2005 (web page); available: http://www.apcc.aspca.org
73 ONDCP "Prescription Drug Abuse: A Dangerous Trend Among Teens," White House Office of National Drug Control Policy, National Youth Anti-Drug Media Campaign, 2005 (web page); available: http://www.theantidrug.com/drug_info/prescription_news.asp
74 P. Rozin and C.J. Nemeroff, The laws of sympathetic magic: A psychological analysis of similarity and contagion. In: J. Stigler, G. Herdt and R.A. Shweder (Eds.), *Cultural Psychology: Essays on Comparative Human Development*, Cambridge, England, 1990, pp. 205–232.
75 S.J. Khan and L.E. Gerrard, Stakeholder communications for successful water reuse operations, *Desalination*, 187(1–3) (2006) 191–202.
76 P. Berckmans, V. Dawans, G. Schmets, D. Vandenbergh and P. Autier, *New Eng. J. Med.*, 337(25) (1997) 1842–1845.
77 J.P. Bound, K. Kitsou and N. Voulvoulis, *Environ. Toxicol. Pharmacol.*, 21(3) (2006) 301–307.
78 D.L. Sedlak and L. Fono, *Use of metoprolol enantiomers to discriminate between combined sewer overflows and wastewater effluent discharges*, First Integral Virtual Symposium sponsored by the American Chemical Society [ACS] Division of Environmental Chemistry and the U.S. EPA, ACS 228th National Meeting in Philadelphia, PA, 25–26 August 2004 (http://www.epa.gov/esd/chemistry/ppcp/acs-25aug2004.htm; http://www.envirofacs.org/PPCP2.pdf)

79 L.J. Fono and D.L. Sedlak, *Environ. Sci. Technol.*, 39(23) (2005) 9244–9252.

80 C.A. Kinney, E.T. Furlong, S.L. Werner and J.D. Cahill, *Environ. Toxicol. Chem.*, 25(2) (2006) 317–326.

81 A.B.A. Boxall, P. Johnson, E.J. Smith, C.J. Sinclair, E. Stutt and L.S. Levy, *J. Agric. Food Chem.*, 54(6) (2006) 2288–2297.

82 P.K. Jjemba, *Agric. Ecosys. Environ.*, 93(1–3) (2002) 267–278.

83 M.E. Balmer, H.-R. Buser, M.D. Muller and T. Poiger, *Environ. Sci. Technol.*, 39(4) (2005) 953–962.

84 H.-R. Buser, M.E. Balmer, P. Schmid and M. Kohler, *Environ. Sci. Technol.*, 40(5) (2006) 1427–1431.

85 W. McDonough and M. Braungart, *Cradle to Cradle: Remaking the Way We Make Things*, North Point Press, April 2002; available: http://www.mbdc.com/book/index.htm

86 CARB "Consumer products program," California Air Resources Board, Sacramento, CA, 2005 (web page); available: http://www.arb.ca.gov/consprod/consprod.htm

87 Oregon DHS "Lead Found in Traditional Home Remedies," Lead Poisoning Prevention Program, Oregon Department of Human Services (DHS), Office of Environmental and Occupational Epidemiology (EOE), Oregon, 2005 (web page); available: http://www.dhs.state.or.us/publichealth/lead/homerx.cfm

88 R.B. Saper, S.N. Kales, J. Paquin, M.J. Burns, D.M. Eisenberg, R.B. Davis and R.S. Phillips, *J. Am. Med. Assoc.*, 292 (2004) 2868–2873.

89 DOHMH "NYC Health Dept. Warns Against Use of 'Skin-lightening' Creams Containing Mercury or Similar Products Which Do Not List Ingredients," New York City Department of Health and Mental Hygiene, New York, press release 27 January 2005; available: http://www.nyc.gov/html/doh/html/pr/pr008-05.shtml

90 M. Harada, S. Nakachi, K. Tasaka, S. Sakashita, K. Muta, K. Yanagida, R. Doi, T. Kizaki and H. Ohno, *Sci. Total Environ.*, 269(1–3) (2001) 183–187.

91 C.A. Smith, Infect. Control Resour.: Prevent Strat. IC Practitioners Professional Nurses, 3(4) (2006) 2–5; available: http://www.infectioncontrolresource.org/Past_Issues/IC12.pdf

92 FDA "Approved Animal Drug Products: Green Book," Drug Information Laboratory, Center for Veterinary Medicine, U.S. Food and Drug Administration, 2005d; available: http://www.fda.gov/cvm/Green_Book/greenbook.html

93 FDA "Environmental Assessments, Findings of No Significant Impact, and Environmental Impact Statements," Center for Veterinary Medicine, U.S. Food and Drug Administration, 2006 (web page); available: http://www.fda.gov/cvm/ea.htm

94 FDA "Environmental Warning Added to Animal Euthanasia Products," Center for Veterinary Medicine, U.S. Food and Drug Administration,

22 July 2003; available: http://www.fda.gov/cvm/CVM_Updates/wildup_com.htm

95 FAO "The State of World Fisheries and Aquaculture: Part 2. Selected Issues Facing Fishers and Aquaculture (Antibiotic Residues in Aquaculture Products)," Food and Agriculture Organization of the United Nations, FAO Fisheries Department, 2002; available: http://www.fao.org/documents/show_cdr.asp?url_file=//docrep/005/y7300e/y7300e00.htm

96 EPA "Aquatic Animal Production Industry Effluent Guidelines," U.S. Environmental Protection Agency, Washington, DC, 2006b (web page); available: http://epa.gov/guide/aquaculture/#rule

97 G. Hamscher, H.T. Pawelzick, S. Sczensky, H. Nau and J. Hartung, *Environ. Health Perspect.*, 111(13) (2003) 1590–1594.

98 A.C. Capleton, C. Courage, P. Rumsby, P. Holmes, E. Stutt, A.B.A. Boxall and L.S. Levy, *Toxicol. Lett.*, 163(3) (2006) 213–223.

99 B. Hileman, *Chem. Eng. News*, 81(45) (2003) 36–43.

100 R.J. Savarie, J.S. Shivik, G.C. White, J.C. Hurley and L. Clark, *J. Wildlife Manage.*, 65 (2001) 356–365.

101 USDA "Caribbean Tree Frogs in Hawaii," Animal and Plant Health Inspection Service, United States Department of Agriculture, January 2003 (web page); available: http://www.aphis.usda.gov/lpa/pubs/pa_wscoquitfrogs.html

102 R.G. Hollingsworth, J.W. Armstrong and E. Campbell, *Nature*, 417 (2002) 915–916.

103 C.G. Daughton, Environmental stewardship of pharmaceuticals: the green pharmacy, In the *Proceedings of the 3rd International Conference on Pharmaceuticals and Endocrine Disrupting Chemicals in Water*, National Ground Water Association, Minneapolis, MN, 19–21 March 2003c; available: http://www.epa.gov/nerlesd1/chemistry/ppcp/images/ngwa2003.pdf

104 NERC "Report on In-store Pilot CVS Mill Creek, South Portland, Maine, Saturday, February 5, 2005, 9–5," Northeast Recycling Council, Inc., Brattleboro, VT, 2005; available: http://www.nerc.org/adobe/CVSPilotReportFINAL.pdf

105 C.A. Kenreigh and L.T. Wagner, Medication adherence: a literature review, *Medscape Pharmacists*, 6(2) (2005); available: http://www.medscape.com/viewarticle/514164_print

106 C. Ukens, Staying the course: New gee-whiz technologies are improving patients' compliance with their drug regimens, *Drug Topics*, 12 September 2005; available: http://www.drugtopics.com/drugtopics/article/articleDetail.jsp?id=179333

107 L. Osterberg and T. Blaschke, *New Eng. J. Med.*, 353(5) (2005) 487–497; available: http://content.nejm.org/cgi/content/full/353/5/487?ijkey=3daX237PepUF6&keytype=ref&siteid=nejm

108 DEA "Controlled Substances Act," U.S. Drug Enforcement Adminstration, 2005a (web page); available: http://www.usdoj.gov/dea/agency/csa.htm

109 DEA "Definition and Registration of Reverse Distributors," 21 CFR Parts 1300, 1301, 1304, 1305, and 1307, Federal Register, 2 May 2005b, 70(83) 22591–22595; available: http://edocket.access.gpo.gov/2005/05-8692.htm

110 DEA "Can an individual return their controlled substance prescription medication to a pharmacy?" General Questions & Answers (web page), Office of Diversion Control, Drug Enforcement Administration, 2005c; available: http://www.deadiversion.usdoj.gov/faq/general.htm#pre_med

111 DEA "Can a long term care facility (LTCF) return a resident's unused controlled substance medication to a pharmacy?" General Questions & Answers (web page), Office of Diversion Control, Drug Enforcement Administration, 2005d; available: http://www.deadiversion.usdoj.gov/faq/general.htm#ltcf

112 CESAR "Current Substance Abuse Legislation," Center for Substance Abuse Research, CESAR, University of Maryland, 2005 (web page); available: http://www.cesar.umd.edu/cesar/library/statutes.asp

113 A. Wennmalm and B. Gunnarsson, *Drug Inform. J.*, 39(3) (2005) 291–297.

114 A. Wennmalm and S. Martini, *Environment and pharmaceuticals*, Department of the Environment (Sweden) and the Stockholm County Council Pharmaceutical Unit, 2005 (web page); available: http://www.janusinfo.se/imcms/servlet/GetDoc?meta_id=7236

115 EMEA "Guideline on the Environmental Risk Assessment of Medicinal Products for Human Use," Committee for Medicinal Products for Human Use (CHMP), Pre-Authorisation Evaluation of Medicines for Human Use, London, UK, 20 January 2005, CHMP/SWP/4447/00 draft (page 14, section 6: 6. Precautionary and Safety Measures to Be Taken for Administration, Disposal and Labelling); available: http://www.emea.eu.int/pdfs/human/swp/444700en.pdf

116 State of Maine "An Act To Encourage the Proper Disposal of Expired Pharmaceuticals," Unused Pharmaceutical Disposal Program, State of Maine Legislature LD 1826 (S.P. 671), 121st Maine Legislature, referred to Maine's Committee on Health and Human Services on 8 January 2004; available: http://janus.state.me.us/legis/LawMakerWeb/summary.asp?ID=280012218

117 State of Maine "Final Report of the Maine Drug Return Implementation Group," State of Maine 122nd Legislature First Regular Session, 8 March 2005; available: http://mainegov-images.informe.org/legis/opla/drugrpt.pdf

118 ASCP "Statement on the Return and Reuse of Medications in Long-term Care Facilities," American Society of Consultant Pharmacists, Board of

Directors, 11 July 2003; available: http://www.ascp.com/public/pr/policy/return.shtml

119 USP "Resolution 9: Promoting Safe Medication Use and Disposal," U.S. Pharmacopeia, 2005–2010 Resolutions, adopted at the USP Convention 2005, 13 March 2005; available: http://www.usp.org/aboutUSP/resolutions.html

120 DEA "Disposal of Controlled Substances, Section 1307.21 Procedure for disposing of controlled substances," 21 CFR part 1307.21, Federal Register, 24 March 1997; available: http://www.deadiversion.usdoj.gov/21cfr/cfr/1307/1307_21.htm

121 G. Singh, *Am. J. Therapeut.*, 7 (2000) 115–121.

122 Pharmaceuticals.org "Pharmaceuticals from Human System to the Human System," Utica, MI, Pharmaceuticals.org, Inc., 2005 (web page); available: http://www.toilets.com/pharmaceuticals.htm

123 Pharmaceutical Journal, Tonnes of inappropriate medicines arrive in SE Asia, *Pharm. J.*, 274(7335) (2005a) 139; available: http://www.pharmj.com/Editorial/20050205/news/p139asia.html

124 California "SB 798, Simitian. Prescription Drugs: Collection and Distribution Program," Senate Bill No. 798, State of California, 14 September 2005; available: http://www.leginfo.ca.gov/pub/bill/sen/sb_0751-0800/sb_798_bill_20050913_enrolled.pdf

125 WHO "General Guidelines for Drug Donations," Department of Essential Drugs and Other Medicines, World Health Organization, Geneva, Switzerland, WHO/EDM/PAR/99.4, 2nd ed., 1999; available: http://www.who.int/hac/techguidance/tools/5.6%20guidelines%20for%20drug%20donations.pdf

126 P. Autier, R. Govindaraj, R. Gray, R. Lakshminarayanan, H.G. Nassery and G. Schmets, Drug donations in post-emergency situations, Health, Nutrition and Population (HNP) Discussion Paper (Ed. Alexander S. Preker), 2002, The International Bank for Reconstruction and Development, The World Bank, Washington, DC, ISBN 1-932126-18-X; available: http://www-wds.worldbank.org/servlet/WDS_IBank_Servlet?pcont=details&eid=000160016_20040513164737

127 Pharmaceutical Journal, Charitable requests & volunteers: drug donations, *Pharm. J.*, online, 2005b (web page); available: http://www.pharmj.com/makingadifference/index.html

128 S. Strom, Old pills finding new medicine cabinets, *New York Times*, 18 May 2005; available: http://www.nytimes.com/2005/05/18/health/18reuse.html

CHAPTER 2: ANALYSIS OF PHARMACEUTICALS AS ENVIRONMENTAL CONTAMINANTS

Chapter 2.1

Analysis of antibiotics in aqueous samples

M. Silvia Díaz-Cruz and Damià Barceló

2.1.1 INTRODUCTION

Among a wide diversity of pharmaceutical compounds, antimicrobials are of particular interest; their environmental occurrence and fate has raised scientific and public concern due to the potential spread and maintenance of bacterial resistance [1] through continuous exposure, which can result in untreatable microbial infectious diseases. These effects have been reported to occur in different water bodies, such as waste effluents of pharmaceutical plants and hospitals [2–4], and more recently sulfonamide- and trimethoprim-resistant bacteria have already been detected in rivers [5].

Antimicrobial agents, also known as antibacterial or anti-infectives comprise synthetic and natural compounds. The term antibiotic years ago only designed natural substances produced by bacteria or fungi, but at present it is used to design both synthetic (or semi-synthetic), such as SAs and quinolones, as well as natural compounds such as PENs and TCs. In the following, the term antimicrobial refers to antibacterial antibiotics, since other antimicrobial agents such as antifungal or anti-parasitics are out of the scope of the present work.

Antimicrobials are extensively used in both human and veterinary medicine against microbial infections; in addition a certain fraction of antimicrobials used in livestock and poultry production is destined to increase the rate of growth by improving the feed efficiency. In Europe, two-thirds of all pharmaceutical antibotics are used in human therapy and one-third for veterinary purposes [6]. In human medicine, antibiotics pose the third biggest group among all pharmaceuticals making up more than 6% of all prescriptions [7]. In veterinary medicine, more than 70% of all consumed pharmaceuticals are antimicrobial agents [8].

Comprehensive Analytical Chemistry 50
Petrović and Barceló (Eds)
Volume 50 ISSN: 0166-526X DOI: 10.1016/S0166-526X(07)50002-4

The large quantities of antimicrobials consumed have led to their occurrence in the environment [9]. They mainly enter the environmental water bodies via wastewater effluents. Other important sources are identified in the direct application in fish farming, manure run-off, run-off from the sludge originated in sewage treatment plants and manure and slurry from confined animals either stored in waste lagoons or immediately applied to agricultural fields as fertilizers to increase crop yields, hospital and manufacturing process waste effluents and spills from pharmaceutical production plants.

The frequent detection in wastewater treatment plants (WWTPs) effluents of a large variety of residual antimicrobials reflects their incomplete removal during the treatment. Elimination rates differ considerably depending on the compound, on the environmental conditions, as well as on the process conditions, being adsorption and degradation the main processes taking place during the wastewater treatment for the removal of organic microcontaminants. But, by simply following the disappearance of a substance one cannot conclude that it was removed; it may exist in another state or form. However, identifying metabolic products is difficult both because of the great number of metabolites or degradation products potentially generated by one parent compound and because of the high cost or the lack of standards.

It is known that after consumption, antimicrobials are metabolized in the organism to different extents and are therefore excreted only partly unchanged. The inactive N4-acetylsulfamethoxazol, metabolite of sulfamethoxazole, is known to be excreted in 50% of the administered dose [10]. In lower rates, 20% and 13%, the two-step metabolites of clarithromycin, 14-OH–(R)–clarithromycin and 14-OH–(R)–N–demethyl-clarithromycin, respectively, are excreted after undergoing hydroxylation and N–demethylation.

In general, metabolism is a two-step process. In the first step, reactive functional groups are introduced into the molecule through mainly oxidation, reduction or hydrolysis reactions. In the second step, the parent drug or its first-step metabolite is covalently bound to polar molecules present in the body, such as sugars, sulfates and acids. As a consequence, metabolites are more polar than the parent compound, being more easily excreted by the organism. In some cases, the metabolites of certain drugs may be the predominant forms in the environment as would be the case of dehydro-erythromycin, and be more reactive than the parent drug. It is known that under certain environmental conditions or wastewater treatment procedures excreted metabolites may be transformed back to the parent compound. At this respect, an early study by Berger et al. [11] revealed the back transformation of the glucuronide of amphenicol and N4-acetylsulfametazine to the active parent compounds during the storage of liquid manure and suggested a possible similar behavior for other N4-acetylated SAs metabolites.

2.1.1.1 Antibiotic classification

There are different classification schemes for antimicrobials based on bacterial spectrum (narrow, broad), route of administration (oral, injectable) or activity (bacteriostatic, bactericide), but from the analytical point of view, classification by chemical structure is the most useful. Under such approach several classes result, i.e., aminoglycosides, beta-lactams, glycopeptides, macrolides, oxazolidinones, polymixins, quinolones, streptogramins, sulfonamides, tetracyclines, chloramphenicols and a group of single compounds.

2.1.1.2 Physical-chemical properties

Often information on the physical and chemical properties such as octanol/water partition coefficient (Kow), distribution coefficient (Kd), dissociation constants (pKa), vapor pressure or Henry's Law constant (K_H) of a substance helps to determine whether the compound is most likely to concentrate in the aquatic, terrestrial, or atmospheric environmental compartment. Compounds with high log Kow may show affinity to sludge or soil, high Kd values indicates the tendency for compounds to be adsorbed onto soil materials through the phenomena of adsorption, distribution or solid/liquid partition. Both factors contribute to the reduction of their concentrations in the aqueous phase. The pH of the medium and pKa of the compounds determine their ionized/non-ionized chemical form.

Antimicrobials largely vary in their molecular structure, molar mass and other physical–chemical properties, even those belonging to the same class, as is shown in Table 2.1.1. The different functional groups of the molecule are associated to their antimicrobial activity. In terms of their persistence, main antimicrobial properties are photostability, binding and adsorption to solid components, biodegradation and water solubility. Biodegradation of antimicrobials in water, soil and manure has only been addressed in few works [12,13]. Biodegradation depends upon the temperature; low values reduce the degradation rate. This fact can cause alarm in cold climate countries because of the manure

TABLE 2.1.1

Physical–chemical properties of the main antimicrobial classes. Data from literature cited

Antibiotic class	Molecular weight (g/mol)	Water solubility (mg/L)	log Kow	pKa	K_H (PaL/mol)
Tetracyclines	400–600	200–55E3	−1.5–0.05	3/8/9	1.5E-23–5E-22
Sulfonamides	150–300	7–15E2	−0.1–1.7	2–3/4–11	1.3E-12–2E-8
Amynoglycosides	300–650	1E4–5E4	−8–1	6.5–8.5	8.5E-12–4E-8
Macrolides	650–1E3	0.5–15	1.5–3	7.5–9	7.5E-36–2E-26
β-lactams	300–500	20–1E4	0.9–3	2.5	2.5E-19–1E-12
Quinolones	200–400	3–2E4	−1–1.6	2.5	5E-17–4E-8
Polyethers	650–750	2E-6–3E-3	5–9	6.5	2E-18–1.5E-18
Polypeptides	500–1E3	incomplete	−1–3	—	negligible to 3E-23

applied at low-temperature conditions can persist longer as consequence of frozen soils, and then be released to the environment through snow-melt runoff.

Tetracyclines, whose name is derived from the four rings forming their chemical structure, are photodegradable amphoteric substances stable in acids but not in bases which tend to bind divalent and trivalent metal ions, silanolic groups and proteins [14,15] as result of the presence of two ketone groups in their molecule. Sulfonamides are derivatives of sulfanilamide which have amphoteric properties, however, mainly behave as weak acids, due to the N–H bond of the sulfonamidic group, and tend to form salts in strongly acid or basic media [15]. Also susceptible of photodegradation are aminoglycoside antimicrobials, which are composed by two or three aminosugars linked between them by a glycosidic bond. The high number of amino and hydroxyl moieties are the responsible of their strong polar properties. Therefore, these basic compounds are characterized by their high water and poor lipid solubility [15]. Macrolides got their name by a common lactone macrocycle, to which one or two sugars are attached. This class of antimicrobials are weak bases characterized by their high molecular weight only comparable to polypeptides and polyether antimicrobials (PEs). β-lactams are thermo labile compounds with limited stability due to the presence of a common four-membered (β-lactam) ring in their structure; they are, unstable in alcohols and isomerize in acid medium. Quinolones are lipid soluble and prone to UV-light degradation [15], resistant to acid and basic hydrolysis and exhibit poor water solubility in the pH range 6–8. Polyether antimicrobials are composed by multiple cyclic ethers with a carboxylic acid group at one end and a terminal alcohol group to the other. These compounds are rather liphofilic, and exhibit poor solubility in water, which is even reduced when mono and divalent metal ions are present due to the formation of lipid-soluble cyclic complexes.

2.1.1.3 Occurrence in aqueous samples

The occurrence of antimicrobials in the water bodies is not restricted to surface waters only, as shows Table 2.1.2. Higher concentrations were detected in wastewaters, as expected, with maximum value up to 12 mg/L of chlortetracycline in manure lagoon wastewater [16].

All these scenarios point out the risk of highly mobile and poorly eliminated in WWTPs antimicrobials to enter the drinking water supply. Evidence is reported by Stolker et al. [17], who detected sulfamethoxazole in two drinking water samples from The Netherlands out of 22, at concentrations below 25 ng/L.

The occurrence of SAs was reported in all kind of water samples; this frequent detection may be explained on the basis of their rather poor chelating ability and low sorption to soil tendency. Other reasons for greater occurrence may be due to their relatively low elimination efficiency during sewage treatment procedures [18] and to the increase in the number of confined animal feeding operations, which often lack proper waste management practices [19].

In order to better assess the occurrence of antimicrobials in the environment Gobel et al. [20] have pointed out the need to consider the metabolites of dosed drugs. In their work, the elimination of N4-acetylsulfamethoxazol and dehydro-erythromycin are evaluated among other residues, and for the acetylated form a tentative fragmentation process is presented. In an earlier study Hilton and Thomas [21] yet included N4-acetylsulfamethoxazol among the pharmaceuticals investigated in effluent and surface water. Findings indicated that while sulfamethoxazole could not be quantified in any sample (<50 ng/L), N4-acetylsulfamethoxazol was present in all samples and at quite high concentrations (<50–2200 ng/L). In a more recent study, five N4-acetylated metabolites of sulfadiazine, sulfadimethoxine, sulfamethazine, sulfamethoxazole and sulfathiazole were analyzed. Results indicated that, in general, LODs are five fold higher than those achieved for the related unmetabolized residues [22].

Regarding erythromycin, it was never detected in its original form in environmental samples, but as a degradation product with a loss of one molecule of water. This dehydration process is known to occur in acidic aqueous solution. In order to elucidate which form is mainly present in

TABLE 2.1.2

Occurrence of antimicrobial residues in water bodies

Water sample	Compounds	Concentration range	References
Surface	Trimethoprim, dehydro-erythromycin, roxytromycin, novobiocin, clarithromycin, tylosin, chloramphenicol, ionophores, chlortetracycline, oxytetracycline, tetracycline, sulfadimethoxine, sulfamethoxazole, sulfadiazine, sulfamethazine, sulfathiazole, N^4-acetylsulfamethoxazol,	7–15,000 ng/L	[17, 21–23, 26, 27, 29, 30, 32, 49, 50, 57, 78]
Groundwater	Chloramphenicol, sulfamethoxazole, sulfadimethoxine, sulfamethazine, oxytetracycline, tetracycline, lincomycin, dehydro-erythromycin	0.05–1.4 μg/L	[16, 17, 19, 23, 26, 30, 50, 79–81]
Drinking water	Sulfamethoxazole	<25 ng/L 10–6000 ng/L	[17]

STPs/WWTPs effluents	Sulfamethoxazole, trimetoprim, N^4-acetylsulfamethoxazol, sulfadiazine, sulfacetamide, sulfisoxazole, sulfamethazine, sulfapyridine, atorvastatin, roxythromycin, novobiocin, ciprofloxacin, clarithromycin, azythromycin, ofloxacin, norfloxacin, chloramphenicol, dehydro-erythromycin, lincomycin, doxycycline, tetracycline, cephalexin, spiramycin, amoxicillin, tylosin		[18, 20, 21, 23, 27, 29, 32, 45, 57, 60, 81–84]
Hospital wastewater	Gentamycin, ciprofloxacin, metronidazole, sulfamethoxazole, trimetoprim, doxycycline	0.4–125 μg/L	[25, 72]
Manure lagoon wastewater	Tetracycline, oxytetracycline, chlortetracycline, lincomycin, sulfamethazine, sulfadimethoxine, trimetoprim, dehydro-erythromycin	2.5–12,000 μg/L	[16, 50]

environmental waters Hirsh et al. [23] carried out a simple experiment consisting of extracting spiked samples with erythromycin at different pH. The findings indicated that only the degradation product could be detected, thus the dehydration process already takes place in the natural aquatic environment contrarily to earlier explanations pointing out the formation of the degradation product during the ionization process when analyzing by MS-based methods. In this work in addition to erythromycin other antimicrobials were investigated in WWTPs effluents and surface waters. The highest concentration found in the effluents was reached by dehydro-erythromycin with a mean value of 2500 ng/L (maximun of 6000 ng/L).

The study of degradation products and epimers of TCs in the environment has scarcely been addressed. This can likely be due to the low proportion relative to the parent TC in which they are formed, especially for degradation products. These products are known to be formed through hydrolysis and photolysis reactions yielding the epi-tetracyclines, anhydro-tetracyclines and iso-tetracyclines. Nevertheless, their consideration is important because of degradation products are known to be more soluble in water phases than the parent compounds, which increases their mobility potential. In a recent work, Halling-Sorensen et al. [24] evaluated the occurrence of oxytetracycline (OTC) in soil interstitial water as regards the parent and eight degradation products namely 4-epi-OTC, α and β-apo-OTC, 4-epi-N and N-desmethyl-OTC, 4-epi-N and N-didesmethyl-OTC, and 2-acetyl-2-decarboxamido-OTC. Results indicated that OTC and 4-epi-OTC were the only compounds found to be present at significant concentrations in soil interstitial water, while all other degradation products were below 2% relative to OTC.

2.1.2 SAMPLE PREPARATION

The environmental analysis of trace pollutants constitutes a difficult task because of both the complexity of the matrices and the normally very low concentrations of the target compounds. Therefore, in essentially all cases analyte enrichment is necessary to isolate the target compounds from the matrix and to achieve the LODs required. A typical analytical procedure includes, therefore, various sample preparation steps, such as filtration, extraction, purification and evaporation; and, if the final determination is performed by GC-MS, derivatization is often necessary.

2.1.2.1 Sample preservation

Previous to extraction of target analytes from water, the sample is filtered in order to subtract the suspended matter; however, when analyzing aminoglycosides, filtration should be avoided because of significant losses occur due to their extremely high sorption ability [25]. Filtered samples are then usually pH adjusted, ranging from acidic to basic pH, depending on the acid (TCs) or alkaline (macrolides) nature of the antimicrobials. Nevertheless, in some cases, acidic pH can promote the degradation of the compound, as would be the case of penicillin G; therefore, the acidification of sample searched for penicillin G is performed immediately before extraction [26]. Similarly, when extracting other PEs, the pH of the sample has to be adjusted around 7.5 just before the extraction, since they are acid and/or base labile [27].

To avoid photodegradation, which affects especially to fluoroquinolones and TCs, samples are stored in the dark and at low temperatures (ca 4°C) until extraction. Whenever possible, samples should be analyzed immediately, since storage for long periods of time can affect analyte concentrations in the sample.

Special precautions have to be taken when analyzing TCs, SAs and PEs because they tend to form complexes with metal ions. Precautions leading to a significant improvement of extraction efficiencies are the silanization, for instance with dimethyldichlorosilane [28], of all glassware getting in contact with either the water sample or the extract, and heating all glassware at 450°C for 1 h and next rinsing with a strong chelating agent, such as Na_2EDTA [27,29,30]. Another approach to avoid complexation in the presence of metal ions is the addition of sodium chloride to the water samples before extraction. This approach has successfully been applied in PE's analysis [27] in order to obtain their single sodium adduct species in the water sample before analysis which results in more sensitive, specific and reproducible determinations.

The use of glassware has also to be rejected when analyzing strong polar compounds, such as aminoglycosides in order to prevent losses by adsorption [25] and for penicillins to avoid the formation of epimers that is catalyzed by heavy metal ions [31]. In these cases, the use of PTFE as container material is also recommended.

2.1.2.2 Enrichment and purification procedures

In the enrichment and purification of antimicrobials, the traditional soxhlet extraction [32] and lyophilization [23,29,33,34] have been almost completely replaced by solid phase extraction (SPE). This extraction technique has experienced a steady growth since the introduction of the first packed cartridge in the late 1970s. The miniaturization of SPE was called solid phase micro-extraction (SPME). This technique was introduced in the early 1990s and has been found to be useful for all kind of samples (gas, liquid and solid), being considered as a universal extraction technique because of their broad applicability. These devices consist of a sorbent phase coated on the outside of a fiber or the inside of a tube. Despite that, SPE continues to be the leading technology for the enrichment and purification of organic pollutants in water, likely due to the wide spectrum of solid phase materials, commercially available.

At present, the trend in extraction procedure development is to lessen the consumption of organic solvents together with the development of faster techniques, automation, on-line coupling (integration) and enhanced recovery and reproducibility. The improvement in the extraction selectivity is not one of the main addressed goals since the required selectivity is usually procured through separation (usually LC) and detection methods (especially tandem mass spectrometry, MS).

Different approaches can be followed in the SPE procedure according to the objective set up, the determination of (i) a wide group of antimicrobials or (ii) a limited number of compounds usually belonging to the same antibiotic class. In the first approach, broad spectrum, mixed phases or tandem solid phases are used in order to be useful in the wide range of chemical properties exhibited by the compounds. Therefore, broad-spectrum alkyl-silica or polymeric-based solid phases together with cation-exchange solid phase, mixed mode or multilayer cation exchange/alkyl-based solid phase are used. However, additional purification is usually necessary due to the co-extraction of interfering components from the sample. In the second approach, a class specific extraction using molecular imprinted polymers as extraction materials (MIP) can be used [35]; however their use in environmental analysis so far has been limited to the field of pesticide analysis [36–38].

Apart from the influence of SPE cartridges on target analyte enrichment and purification goodness, sample characteristics also play an important role. In a recent study, the matrix dependent formation of artifacts during the extraction of selected antibiotics from water has been demonstrated [39].

SPME has shown increasing applicability in drug analysis [40]. For instance, the decomposition of erythromycin A in aqueous solution was examined by LC-tandem MS after SPME extraction. Among the different fiber coatings investigated polydimethylsiloxane/divinylbenzene exhibited the best performance for erythromycin A and its degradation products [41]. SPME has also been tested for TCs extraction from water [42]. In this study, the optimization of the 'on-line' SPME–LC–MS method is described including choice of the extracting fiber and the desorption method (heating or salting out the analytes).

Despite SPME eliminates the need for lengthy sample purification and is fast and economic, the poor variety of fiber coatings available compatible with LC-MS hinders the widespread development of this technique. In contrast, the wide array of fibers commercially available for the quantification of volatile and semi-volatile organic contaminants in water and the easily interfaced has expanded their use in GC.

2.1.3 QUANTITATIVE ANALYTICAL METHODS

The requirement for analytical methods is mainly driven by the low concentration levels of antibiotics expected in the environment. Therefore, highly sensitive and selective analytical methods have to be used. In such methods, MS plays a key role. Because of the rather high polarity and, in some cases, poor thermal stability of antimicrobials, GC has been scarcely applied and usually after quite extensive analyte derivatization, for instance, through silylation or pentafluoro benzyl derivatization, which bears time consumption, non-reproducible derivatization at trace levels and formation of unwanted by-products.

LC–MS is becoming more extensively used in the identification and quantification of antibiotics because of its high sensitivity and ability to provide compound confirmation as compared to conventional LC-UV detection or LC-fluorimetric detection (LC-FD). Owing to the sensitivity of fluoroquinolones to light, FD is the technique usually employed for their detection in aqueous samples [43,44], whereas tandem MS only recently has found application in their environmental determination [45].

LC-tandem MS allows separating and detecting compounds having the same molecular mass but different product ions, even if they co-elute. Therefore, although LC-MS has been used for quantification of antimicrobial residues in the environment, tandem MS detection is preferred for enhanced sensitivity and selectivity in complex matrices such as wastewaters [46,47]. LODs reached using LC-tandem MS are slightly higher than those obtained with GC-MS. The major part of the methods published in the literature for the analysis of antibiotics in water, sediment/soil and sludge are based on LC-tandem MS have been recently reviewed by Díaz-Cruz and Barceló [31,48]. According to these reviews, the versatility and less complicated sample preparation, together with increased sensitivity and selectivity in complex matrices, such as wastewater, makes LC-tandem MS the technique of choice for environmental analysis.

As regards other analytical techniques, radioimmunoassay has also been reported as a screening method for antibiotics detection [49,50]; however, its low selectivity and sensitivity only allows semi-quantitative results at the environmental relevant concentration levels. Capillary electrophoresis is another analytical technique that has been applied by environmental researchers [51].

2.1.3.1 Gas chromatography methods

The use of GC in the analysis of antibiotics is quite limited because of the physical–chemical characteristics shown by these compounds, which are rather polar, non-volatile and in some cases thermal labile. Derivatization is then always required, which makes the analysis difficult and, in general, worsen results. Typically, derivatization of hydroxyl and amino groups is performed via trimethylsilylation; however, this method is not robust and always produces a mixture of different derivatives. These compounds are easily hydrolyzed, which makes difficult the removal of the excess of derivatizing agent as well as the extraction of the target derivatives with organic solvents. Thus, it is often unavoidable to analyze the whole mixture, which is known to affect the performances of several GC analyzers, such as flame ionization detectors and mass spectrometers as a consequence of the deposition of silicium dioxide residues. Despite that, a promising GC-atomic emission detection (AED) method for the quantitative analysis of several SAs in complex matrices was developed by Chiavarino et al. [52]. Derivatization was performed via N1-methylation through the use of an excess of diazomethane, but no application in environmental analysis was reported so far.

Since derivatization is usually a difficult task when it is unavoidable, the optimization of the derivatization reactions may be a useful tool as demonstrated by Preu et al. [53], who developed a GC-MS method for the analysis of aminoglycoside antibiotics in standard solutions using experimental design for the optimization of the process based on the Mayhew and Gorbach [54], derivatization method consisting of the silylation of hydroxyl groups using trimethylsilylimidazole and heptafluorobutyrylimidazole for acylation of amino groups.

2.1.3.2 Liquid chromatography methods

LC has become an essential technique for the determination of polar and thermal labile organic contaminants. Derivatization is not required and low LODs are achieved (in the low ng/L level) in aqueous samples, including seawater, surface water, wastewater, groundwater and drinking water. LC-tandem MS is increasingly being used with the sake of enhanced sensitivity and selectivity. Certain drawbacks associated to LC-MS, such as matrix effects, are being faced, for instance, through the use of ^{13}C-labeled [30] and deuterated internal standards.

2.1.3.2.1 Analytical columns

Complete LC separation of analytes may not always be necessary, and that allows the use of short LC columns to provide minimal separations and speed up analysis. However, good separation improves detectability. For this reason, medium-size columns are typically used, i.e., 100–250 mm length and 1–4.6 mm i.d., with pore size between 2 and 5 μm. As indicated in Table 2.1.3, the LC separation of antimicrobials has been carried out, in most instances, with reversed phase alkyl-bonded silica C18 columns, nevertheless, when analyzing basic substances, such as macrolides, on alkyl-bonded silica LC columns peak tailing is known to occur due to interaction with the residual silanol groups on the silica-gel, end-capped C18 columns based on pure silica gel are then mainly used. However, aminoglycosides are not retained in alkyl-bonded silica columns; therefore, the common approach in this case is to use ion-pair chromatography.

2.1.3.2.2 Mobile phases

The effect of LC mobile phase on the ionization process, when the further determination is performed by means of MS or tandem MS, has been the goal of several works [55,56].

As mobile phases, mixtures of water–methanol and, more frequently, water–acetonitrile with gradient elution from 10–50% to 100% organic solvent have normally been used.

Volatile organic modifiers, such as acetic/acetate buffer, formic acid or acetic acid are typically added in attempt to improve both ionization efficiencies and sensitivity, and control pH (see Table 2.1.3). The presence of non-volatile compounds in the mobile phase causes clogging in the orifice plate at the interface and a build-up of deposits in the ion source, resulting in a significant drop in signal intensity.

TABLE 2.1.3

Survey of liquid chromatography-based methods used in the determination of antimicrobial residues and metabolites in water samples

Compound	Matrix	Detection	Recovery (%)	LOD	Extraction procedure	LC conditions	Reference
Single-group analysis							
Sulfonamides	STP effluents, surface water	UV, λ = 260 nm	18–101	0.2–0.6 mg/L	NR	Supelcosil ABZ+ (250 × 2.1 mm, 5 μm), A: ACN:water (3:97), 0.05% TFA (w/v), B: ACN:water (80:20), 0.05% TFA (w/v)	[57]
		Tandem MS (ESI+)-QqQ	29–100	0.2–3.7 μg/L	SPE, pH 2.5	Supersphere RP18 ec (250 × 2 mm, 4 μm), A: ACN:water (3:97), 1% formic acid, B: ACN:water (75:25), 1% formic acid	
Penicillins	Groundwater, STP effluents, surface, drinking water	MS (ESI+)	76–105	2–24 ng/mL (river water)	SPE	Alltima (250 × 4.6 mm, 5 μm), A: MeOH, 5 mM formic acid, B: water, 5 mM formic acid	[85]
Tetracyclines	Groundwater, animal wastewater	Tandem MS (ESI+)-ion trap	87–109	0.2–0.3 μg/L	SPE, pH 2.5	BetaBasic C18 (200 × 2.5 mm, 5 μm), water: 5% formic acid: ACN:MeOH (23:40:25:12)	[16]
Nitroimidazoles	Surface water, Drinking water, Sea water	MS (ESI+)	92–104	20–30 ng/mL	SPE, pH 2	C18 (150 × 2.1, 5 μm), A: water, acetate buffer (pH 4.3), B: ACN	[86]
Tetracyclines	Groundwater	Tandem MS (ESI+)-QqQ	69–98	50 ng/L	SPE	Puresil C18 (150 × 4.6 mm, 5 μm), A: water, 0.5% formic acid, ammonium acetate (pH 2.5), B: ACN	[79]
Gentamycin	Hospital wastewater	Tandem MS (ESI+)-QqQ	49–68	0.2 μg/L	SPE, pH 7–8	Chrompack omnispher C18 (ion pair LC) (50 × 3 mm, 3 μm), A: water, B: MeOH, C: 20 mM HFBA	[25]

OTC and degradation products	Soil interstitial water	Tandem MS (ESI+)-QqQ	72–94	0.1–0.5 mg/L	No extraction pH 7.9	XTerra C18 MS (100 × 21 mm, 3.5 μm), A:water/MeOH (80:20), 0.1% formic acid, B: water/MeOH (5:95), 0.1% formic acid	[87]
Macrolides	Surface water, wastewater	Tandem MS (ESI+)-ion trap	90–96 (surface), 83–87 (waste)	0.03–0.7 μg/L (surface)	SPE	Xterra MS C18 (50 × 2.1 mm, 2.5 μm), A: water, B: ACN, 1% formic acid	[49]
Oxytetracycline, sulfachloropyridazine, tylosin	Surface water	UV, $\lambda = 285$ and 260 nm, $\lambda = 355$ and 370 nm (for tylosin)	99–100, 100–105, 72–95	0.35 μg/L, 0.25 μg/L, 0.35 μg/L	SPE, pH 2.9, McIlvine-EDTA buffer	Genesis C18 (150 × 4.6 mm, 4 μm), A: tetrahydrofuran, B: ACN, C: 0.05% trifluoroacetic acid	[58]
Sulfonamides and N^4-acetylated metabolites	Surface water	Tandem MS (ESI+)-QqQ	87–93 (SAs), 91–104 (acetyl-SAs)	1–3 (SAs), 5 (acetyl-SAs)	SPE	Nucleodur C18 gravity (125 × 2 mm, 5 μm)	[22]
Polyether antimicrobials (ionophores)	Surface water	Tandem MS (ESI+)-ion trap	82–100	30–50 ng/L	SPE, NaCl	Xterra MS C18 (50 × 2.1, 2.5 μm), A: water, 1% formic acid, B: MeOH, C: ACN	[27]
Multi-group analysis							
Penicillins, tetracyclines, sulfonamides, macrolides	Surface water	Tandem MS (ESI+)-QqQ	45–137 (Lyo), 15–120 (SPE)	20–50 ng/L (LOQ)	Lyophilization, SPE, Na_2EDTA, pH 3	Several columns and solvents	[34]
Tetracyclines, penicillins, sulfonamides, macrolides	STP effluents, surface water, groundwater	Tandem MS (ESI+)-QqQ	NR	NR	Lyophilization, SPE, Na_2EDTA	NR	[23]
Penicillins, tetracyclines, macrolides, sulfonamides	Surface water, STP effluents, drinking water	Tandem MS (ESI+)-QqQ	54–108 (Lyo.), 15–120 (SPE)	20, 50 μg/L (Lyo), 2, 5 (SPE) (LOQ)	Lyophilization, SPE, Na_2EDTA, pH 5	NR	[29]
Sulfonamides, tetracyclines	Groundwater, surface water	MS (ESI+)	84–130	0.1 ug/L (LOQ)	SPE, Na_2EDTA, pH 3	Luna C8 (100 × 4.6 mm, 3 μm), A: water:MeOH (90:10), 10 mM ammonium formate, 0.3% formic acid, B: MeOH ,10 mM ammonium formate, 0.5% formic acid	[30]
Sulfonamides, macrolides, penicillins	Groundwater	Tandem MS (ESI+)-QqQ	11–119	1–6.5 ng/L	SPE, Na_2EDTA, pH 5	Nucleosil 10-3C18 (250 × 2 mm, 3 μm), different solvents	[80]
Sulfonamides, macrolides, quinolones, tetracyclines, β-lactams	Groundwater, surface water, animal wastewater	MS (ESI+)	NR	0.5 μg/L (tetracyclines, LOQ), 0.05 ug/L (LOQ)	SPE, Na_2EDTA, pH 3	NR	[50]

TABLE 2.1.3 *(continued)*

Compound	Matrix	Detection	Recovery (%)	LOD	Extraction procedure	LC conditions	Reference
Erythromycin, trimethoprim, sulfamethoxazole, N^4-acetylsulfamethoxazol	Sewage effluents, surface water	Tandem MS (ESI+)-ion trap	56–123	10–50 ng/L	SPE, pH 3	Luna C18 (250 × 2 mm, 5 μm), A: MeOH, B: water, 40 mM ammonium acetate, formic acid (pH 5.5)	[21]
Novobiocin, roxithromycin	Sewage effluents, surface waters	Tandem MS (ESI+)-QqQ	61–93	2 pg, 3 pg	Soxhlet, SPE, pH 4	YMC ODS-AQ (100 × 1 mm, 3 μm), A: ACN, B: water, 10 mM ammonium acetate	[32]
Sulfonamides, N^4-Acetylsulfamethoxazol, macrolides, trimethoprim	Sewage effluents	Tandem MS (ESI+)-QqQ	35–124	0.4–28 ng/L (LOQ)	SPE, NaCl, pH 4	YMC pro C18 (150 × 2 mm, 3 μm), A: water, 1% formic acid (pH2.1), B: MeOH, 1% formic acid	[20]
Macrolides, quinolones, quinoxaline dioxide, Sulfonamides, tetracyclines	WWTP effluent	Tandem MS (ESI+)-QqQ	73–99	1–8 ng/L	SPE, pH 3, 6	Genesis C18 (50 × 2.1 mm, 3 μm), different methods	[45]
Sulfonamides, quinolones, trimetoprim	STP effluents	MS (ESI+)	64–114	30–90 ng/L	SPE, NaCl, pH 2.5	Zorbax SB-C18 (150 2.1, 5 μm). A: water:ACN (90:10), 10 mM ammonium formate, 0.007% glacial acetic acid, B: ACN	[60]
Sulfamethoxazole, erythromycin, chloramphenicol	Surface water, drinking water, groundwater	Tandem MS (ESI+)- Q-TOF	63–96	5–10 ng/L (LOQ)	SPE, pH 3	XTerra C18 RP, A: water, ammonium acetate 2 mM, B: MeOH, ammonium acetate 2 mM	[17]
Quinolones, tetracyclines, sulfonamides, β-lactams, trimethoprim, metronidazole	Hospital wastewater	Tandem MS (ESI+)-ion trap	49–108	1–64 μg/mL	SPE, pH 3	YMC hydrosphere C18 (150 × 4.6 mm, 5 μm), A: water, 0.1% formic acid (pH 2.1), B: ACN, 0.1% formic acid	[72]

Sulfamethoxazole, roxithromycin	Sewage effluents	Tandem MS (ESI+)-QqQ	75	6.7 ng/L	SPE	NR	[18]
Tetracyclines, sulfonamides	WWTP influent and effluent	Tandem MS (ESI+)-ion trap	80–104	30–70 ng/L	SPE, Na_2EDTA-citric acid	Xterra MS C18 (50 × 2.1, 2.5 μm) A: water, 1% formic acid, B: ACN	[27]
Sulfonamides, macrolides, quinolones, tetracyclines, lincomycin, carbadox, trimethoprim	Groundwater, STP effluents	MS (ESI+)	78–130	0.01–0.1 μg/L	SPE, Na_2EDTA, pH 3	Different columns and mobile phases	[81]
Macrolides, quinolones, sulfonamides, tetracyclines, penicillins, cephalosporins, nitroimidazole, trimethoprim	WWTP influent and effluent	Tandem MS (ESI+)-ion trap	<5–101	6–160 ng/L	SPE, pH 3	YMC hydrospher C18 (150 × 4.6 mm, 5 μm), A: ACN, 0.1% formic acid, B: water, 0.1% formic acid	[84]
Quinolones, penicillins	Surface water, groundwater	Tandem MS, (ESI+)-QqQ, (ESI+)-Q-TOF	74–123	0.4–4.3 ng/L, 50 ng/L	On-line SPE, ammonium acetate	Kromasil C18 (100 × 2.1 mm, 5 μm), A: water, 1% formic acid, B: MeOH, 1% formic acid	[26]

ACN, acetonitrile; MeOH, methanol; LOQ, limit of quantification; NR, not reported; HFBA, heptafluorobutyric acid.

Regarding other undesired phenomena, the methanolysis observed for penicillins when solutions are prepared in MeOH, is not observed during LC separations using methanolic mobile phases, on the contrary an improvement of response is observed when the detection is performed in ESI+ mode.

2.1.3.2.3 Analyzers

In recent years, UV detection has steadily been replaced by the most powerful mass spectrometric detection, because of MS and especially tandem MS, as technique for chromatographic detection, provides higher sensitivity and specificity as compared to those of UV and FD detection. Hartig et al. [57], who compared the sensitivity provided by LC-diode array detection (DAD) and LC-tandem MS in MRM detection mode, concluded that when analyzing WWTPs effluent samples, DAD can not be employed since increased LOD do not allow the quantification at the required level. Nevertheless, quite few authors still prefer developing UV-based methods for reasons of lessen costs and simplify analysis. At this respect, Blackwell et al. [58] reported LODs in the range 0.25–0.35 μg/L for oxytetracycline, sulfachloropyridazine and tylosin in groundwater and surface water by LC-UV, which evidences the suitability of this method for the environmental analysis of these samples in particular.

FD detection is successful at detecting fluoroquinolones in complex water samples and at the low concentration levels expected in the environment [59], however, is not useful when a multiresidue of different antibiotic classes is to be analyzed due to the lack of fluorophores in antibiotics other than fluoroquinolones, which would lead to the introduction of an additional derivatization step.

LC-MS can be used for quantification purposes when analytes are present in simple matrices, such as tap water and bottled water, whereas LC-tandem MS is necessary to quantify with confirmation of identity of residues in complex matrices, such as wastewaters, eliminating false positive detections. Nevertheless, recently a LC-MS method was developed by Renew et al. [60] for the determination of a number of antibiotics in WWTPs effluents in the concentration range 20–90 ng/L. In the detection, two confirming ions at two different fragmentation voltages had to be used to guarantee unambiguous analyte identification.

The analyzers used most as LC detectors are the quadrupole (Q), ion trap (IT) and time of flight (TOF), either alone or combined to give tandem mass spectrometers as the triple quadrupole (QqQ) and hybrid instruments, such as the quadrupole/time of flight (QqTOF), ion trap/time of flight (IT-TOF), and the quadrupole/linear ion trap (QqLIT).

IT instruments allow many stages of MS (MS^n) achieving extremely high sensitivity [61], since record a complete mass spectrum of each pulse of ions introduced into the trapping volume, as demonstrated by Kamel et al. [62] by performing the mass spectral characterization of selected TCs by multiple stage MS in electrospray ionization mode using IT and QqQ analyzers. Compositions of product ions and mechanism of fragmentation could be determined by comparison of the spectra of deuterated and non-deuterated species.

Sensitivity achieved by TOF and QqQ instruments is similar but for increased selectivity TOF is recommended, although the significantly lower effective linear dynamic range compared to that provided by QqQ instruments considerably limits their use in quantitative determinations.

The recently appeared hybrid QqTOF instruments are of great interest to confirm proposed analyte identities due to the accurate masses provided for both precursor and product ions and the possibility of recording a full-scan product-ion spectrum [11,63], nevertheless, its main drawback is the lower sensitivity provided as compared to that attained by QqQ instruments working in MRM mode. An attempt to improve sensitivity in QqTOF detection was performed by increasing the volume of sample extracted; however, LOD reached were still higher than those reported using QqQ detection [26]. These outcomes are in agreement with the findings by Stolker et al. [17] who compared the performances of QqQ and QqTOF detections for the screening and confirmation of selected antimicrobials belonging to different classes in surface, ground and drinking water. The method allowed screening and confirmation of a large number of trace pharmaceuticals in the range 1–100 ng/L in one run. Comparing the performances of QqQ and QqTOF, authors concluded that both techniques fully satisfactory results were obtained; however, the use of QqTOF has the advantage of enhanced selectivity due to information provided by the accurate mass measurements of product ions. Method characteristics such as linear dynamic range and repeatability were found to be similar, but LODs of QqQ resulted somewhat lower.

Hybrid IT-TOF analyzers are a variation of QqTOF spectrometers obtained by substituting the initial quadrupole by an IT, therefore combines the extremely high IT sensitivity and the excellent TOF resolution [61].

The most recently developed QqLIT mass spectrometer has revealed as a powerful tool, which provides the specificity and robustness of QqQ instruments together with the full-scan tandem MS sensitivity of IT analyzers resulting in the increase of the instrumental dynamic range [64,65]. Its ability to decouple some of the ion processing steps in the obtaining of product ion mass spectra overcomes most of the limitations of conventional IT instruments, such as the low-mass cut off, thus, it may be suitable for the analysis of small molecules. The selective detection shown for multiple charged ions over single charged ions is an attractive additional advantage. Nevertheless, no application in environmental analysis of antimicrobial residues has been reported so far.

Conventional interface systems employed years ago to link LC to mass spectrometer analyzers; thermospray and particle beam, for example, do not fulfill the requirements for environmental analysis due to the poor sensitivity and robustness shown, whereas the quite recent mild ionization interfaces working at atmospheric pressure (API), electrospray ionization (ESI) and atmospheric pressure chemical ionization (APCI), satisfy the requirements. Recently, a new API interface has been developed; the atmospheric pressure photo ionization (APPI) [66,67]. APPI is a modification of the APCI interface where the corona is replaced by a gas-discharge lamp, emitting radiation in the UV region, thus enables the selective ionization of analytes in the presence of the LC mobile phase. To the authors' knowledge, this new interface has only been applied to antimicrobial analysis in the detection of chloramphenicol residues in fish meat [68].

ESI appears to be the most employed mode of ionization in antimicrobial residue determination, since it is particularly suitable for both polar and non-polar analytes and for thermal labile substances; however, it is known to be more susceptible to signal suppression than APCI.

Positive ionization is preferred when both positive and negative ionization are possible, as would be the case of β-lactams. Therefore, ESI+ is always employed with few exceptions; cephalosporins, novobiocin and chloramphenicol, which best ionize in ESI-mode [32,34].

Tandem MS detection

The initial step in tandem MS detection is the selection of the precursor ion. In residue analysis a molecular or quasi molecular ion is preferable due to its typically single charge and negligible fragmentation. As an exception, the double charged precursor ion $[M+2H]^{2+}$ of the macrolide azithromycin was taken as precursor ion on the basis of its higher abundance under certain experimental condition [20]. Despite protonated molecular ions, $[M+H]^+$, are considered as the best precursor ions, for a better sensitivity, selectivity and reproducibility, the $[M+Na]^+$ and $[M+NH_4]^+$ ions are selected when analyzing penicillins [27,34,57]. In a recent work by Pozo et al. [26] addition of formic acid was carried out in order to avoid the formation of the sodium adduct of penicillins, responsible of deficient fragmentation in the collision cell. The formation of ammonium adducts has also been reported to occur during the ionization process, when analyzing SAs using ammonium additives in the chromatographic mobile phase, which significantly worsen sensitivity. The formation of the stable single sodium adduct species of PEs is a process somewhat difficult since other alkali metal ions present can also form adducts as result of the high affinity of PEs for alkali metal ions. To overcome this competitive process, an excess of sodium cations has to be guaranteed, for example, by the addition of sodium chloride to the water samples prior extraction and formic acid addition to the mobile phase in order to prevent deprotonation of the terminal carboxyl groups and subsequent formation of adducts with multiple metal ions [27].

Similarly, the precursor ion as the sodium adduct, $[M+Na]^+$, is preferred to $[M+H]^+$ for novobiocin, $C_{31}H_{36}N_2O_{11}$, (coumarin antibiotic class) for increased sensitivity. In the course of its fragmentation process, the noviose moiety is lost yielding the sodiated complementary fragments $[C_9H_{15}NO_5+Na]^+$ and $[C_{22}H_{21}NO_6+Na]^+$ at m/z 240 and 410, respectively. Nevertheless, recently Miao et al. [32] reported more intense signals and a higher number of fragment ions operating in negative ionization mode, with $[M\text{-}H]^-$ as the precursor ion, and proposed a fragmentation pattern of $[M\text{-}H]^-$ for novobiocin. However, the formation of adducts may not always constitute a drawback, since when internal standards are used they will show the same specific adducts formation in a given system, which may even facilitate their detection.

The application of multiresidue methods allows a large amount of data to be obtained after a single sample preparation step and in a single run. Moreover, compounds belonging to the same antimicrobial class often form common fragments under specific conditions in the fragmentation process, which may be considered as class-specific fragment ions, which contributes to a best performance (see Fig. 2.1.1). This would be the case of the four components of gentamycin, which form a common fragment detected at m/z 322 as consequence of the cleavage of the purpurosamine group. Similarly, β-lactam antibiotics show class-specific fragment ions at m/z 160 and 114 as a result of the cleavage of the common β-lactam ring during fragmentation. A typical sulfonamide fragment is detected at m/z 156, which results from the cleavage of the S-N bond yielding the stable sulfanilamide moiety. Certainly, the fragmentation process yields other group-specific ions in addition to m/z 156, i.e., at m/z 108 (rearrangement leading to loss of SO) and 92 (further loss of SO_2) as well as a number of compound-specific ions as represented in Fig. 2.1.2.

Common neutral losses also aids in the detection of antibiotics belonging to the same class, for instance the losses of desosamine and cladinose sugars in macrolides shown in Figs. 2.1.3 and 2.1.4; H_2O, CO_2 and the piperazine substituent in quinolones, the sulfanilamide moiety in SAs (m/z 156) and NH_3 and H_2O in TCs. According to Halling-Sorensen et al. [24], OTC and its degradation products (4-epi, apo and demethyl derivatives) follow similar fragmentation processes, involving neutral losses of H_2O and NH_3 for the precursor ion $[M+H]^+$. Similarly, N4-acetylated sulfonamide metabolites show a number of group-specific product ions which indicates an identical fragmentation pattern as compared to those of the unmetabolized compounds, i.e., at m/z 108, 134 (92+42) and 198 (156+42) together with characteristic compound-specific product ions, such as m/z 65 for N4-acetylsulfamethoxazol and its related unmetabolized compound [20,22].

2.1.3.2.4 *Matrix effects*

The assessment of matrix effects when developing analytical methods is extremely important in order to provide accurate and reproducible quantitative data.

Matrix effects affects to all detection techniques (UV, FD, MS, etc.), however they are known to be more problematic in MS-based methods because of extraction and clean-up procedures tend to be simplified relying in the high selectivity associated to MS and especially to tandem MS. Mass spectrometric detection is prone to matrix effects when API sources are used. Co-eluting undetected matrix components may result in the reduction or enhancement of the ion intensity of the analytes. This drawback is difficult to overcome since the extent of their effects is known to strongly depend upon both the kind of sample and the chromatographic retention time of analytes. In a recent work, Matuszewski et al. [69] suggest different approaches to face matrix effect in LC-tandem MS, including the improvement of extraction and purification procedures.

On the other hand, several papers reported on the influence of mobile-phase additives, which results in severe signal reduction, and

Aminoglycosides

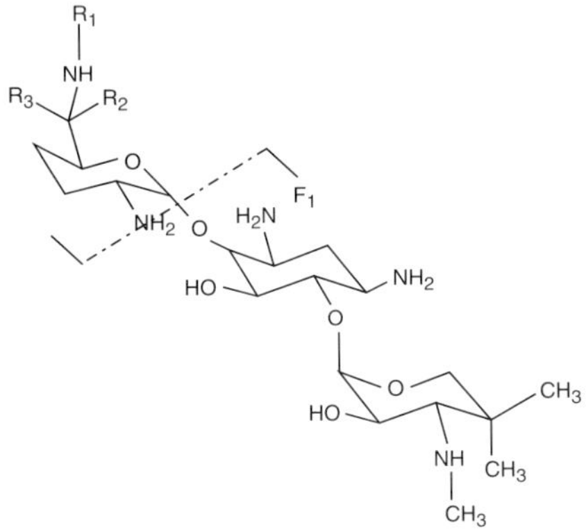

Compound	Diagnostic ions (m/z)	
Gentamycin	Precursor	Product
C_1component: $R_1=R_2=R_3=CH_3CH_2$	478 $[M+H]^+$	**322 $[F_1+2H]^+$** 158 $[M-F_1+2H]^+$
C_2 component: $R_1=R_2=H$ $R_3=CH_3CH_2$	464 $[M+H]^+$	**322 $[F_1+2H]^+$** 143 $[M-F_1+2H]^+$
C_{1a} component: $R_1=R_2=R_3=H$	449 $[M+H]^+$	**322 $[F_1+2H]^+$** 129 $[M-F_1+2H]^+$
C_{2a} component: $R_1=R_3=H$, $R_2=CH_3CH_2$	464 $[M+H]^+$	**322 $[F_1+2H]^+$** 143 $[M-F_1+2H]^+$

β-lactams

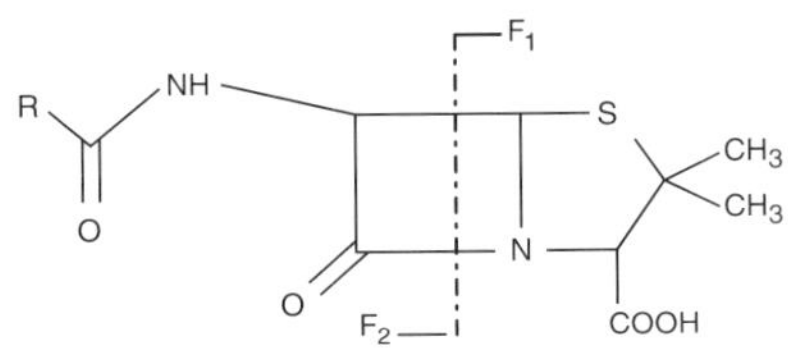

Compound	Diagnostic ions (m/z)	
	Precursor	Product
Penicillin G: R = 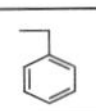	352 $[M+NH_4]^+$	**160 $[F_1+H]^+$** 176 $[F_2+H]^+$
Nafcillin: R = 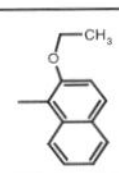	432 $[M+NH_4]^+$	171 [ethoxynaphthyl]$^+$ 199 [ethoxynaphthyl carbonyl]$^+$
Dicloxacillin: R = 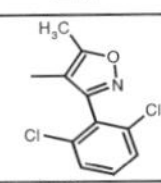	487 $[M+NH_4]^+$	**160 $[F_1+H]^+$** 311 $[F_2+H]^+$
Cloxacillin: R = 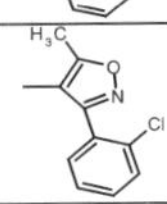	453 $[M+NH_4]^+$	**160 $[F_1+H]^+$** 277 $[F_2+H]^+$
Amoxicillin R = 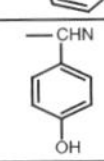	366 [M+H]+	**160 $[F_1+H]+$** 349 $[M-NH_3+H]+$
Ampicillin R = 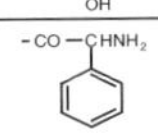	350 [M+H]+	**160 $[F_1+H]+$ 11 $[F_1-COOH]+$**

Fluoroquinolones

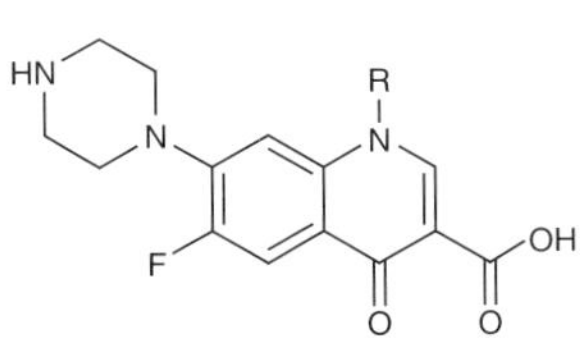

Compound	Diagnostic ions (m/z)	
	Precursor	Product
Ciprofloxacin: R= 	332 $[M+H]^+$	314 $[M-H_2O+H]^+$ 88 $[M-H_2O-CO_2+H]^+$
Norfloxacin: R= CH_2-CH_3	320 $[M+H]^+$	302 $[M-H_2O+H]^+$ 276 $[M-CO_2+H]^+$

Fig. 2.1.1. Classes, names, chemical structures and diagnostic ions of commonly used antimicrobials. Class specific product ions are in bold.

Macrolides

16-membered macrolides

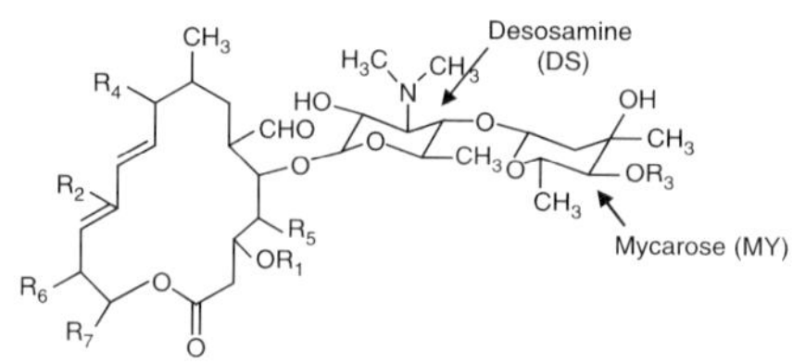

Compound	Diagnostic ions (m/z)	
	Precursor	Product
Spiramycin I: R_4= 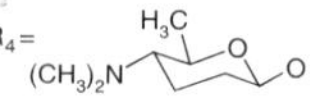 R_7= CH_3 R_1=R_3=R_5=R_6= H R_2= OCH_3	843 $[M+H]^+$	700 $[M\text{-}MY+H]^+$ 540 $[M\text{-}DS+H]^+$
Tylosin A: R_6= 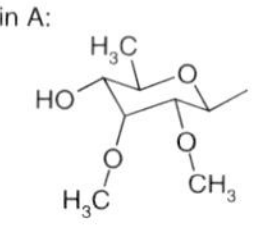 R_7= CH_2CH_3 R_1=R_3=H R_2=R_5= CH_3 R_4= =O	916 $[M+H]^+$	771 $[M\text{-}MY+H]^+$ 318 $[DS\text{-}O\text{-}MY]^+$

14-membered macrolides

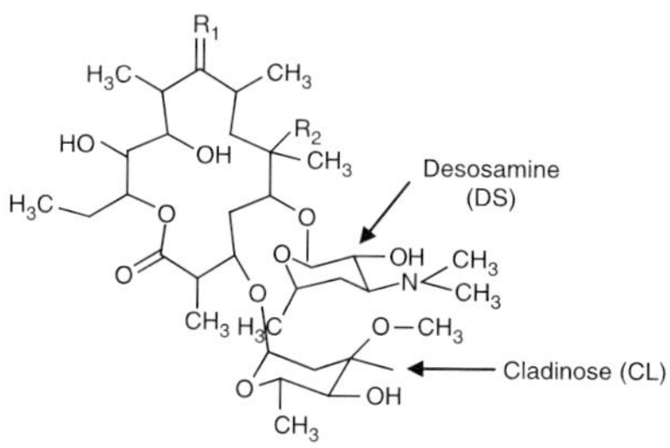

Compound	Diagnostic ions (m/z)	
	Precursor	Product
Clarithromycin: R_1=O R_2=OCH_3	750 $[M+H]^+$	116 $[CL\text{-}OCH_3+H]^+$ 592 $[M\text{-}DS+H]^+$
Erythromycin: R_1=O R_2=OH	716 $[M\text{-}H_2O+H]^+$	522 $[M\text{-}DS\text{-}2H_2O+H]^+$ 558 $[M\text{-}DS\text{-}H_2O+H]^+$
Roxithromycin: R_1=N-O-CH_2-CH_2- OCH_3 R_2=OH	838 $[M+H]^+$	158 $[DS+H]^+$ 680 $[M\text{-}DS+H]^+$

Sulfonamides

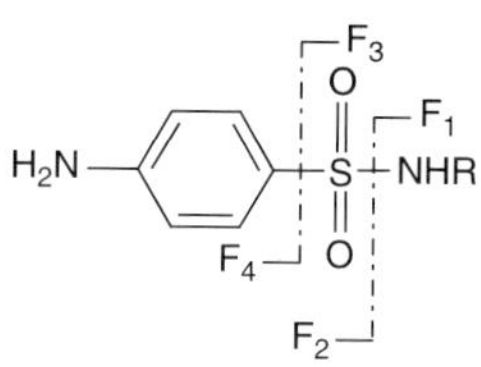

Compound	Diagnostic ions (m/z)	
	Precursor	Product
Sulfamethoxazole: R= 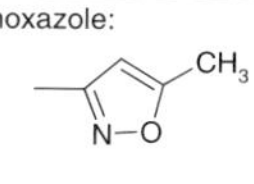	254 $[M+H]^+$	**156 $[H_2NPhSO_2]^+$** **108 $[H_2NPhO]^+$**
Sulfamethazine: R= 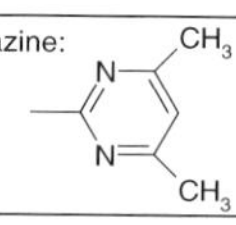	279 $[M+H]^+$	186 $[M\text{-}H_2NPh]^+$ **124 [aminodimethyl - pyridine+H]$^+$**
Sulfadiazine: R= 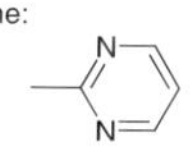	251 $[M+H]^+$	**156 $[H_2NPhSO_2]^+$** **108 $[H_2NPhO]^+$**
Sulfapyridine: R= 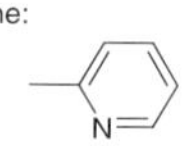	250 $[M+H]^+$	**156 $[H_2NPhSO_2]^+$** 184 $[M\text{-}H_2SO_2+H]^+$

Fig. 2.1.1. (*continued*)

Tetracyclines

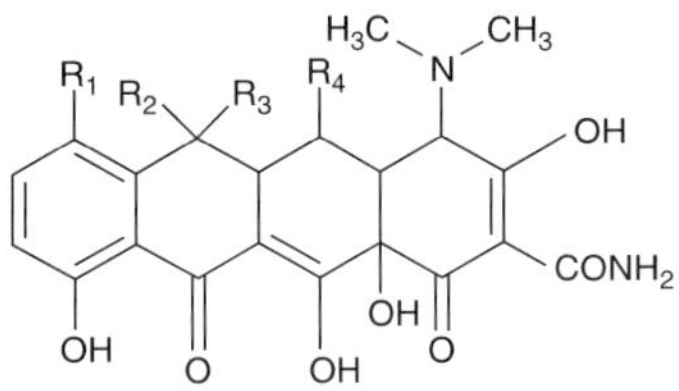

Compound	Diagnostic ions (m/z)	
	Precursor	Product
Tetracycline: R_1=H R_2=OH R_3=CH_3 R_4=H	445 $[M+H]^+$	410 $[M-H_2O-NH_3+H]^+$ 427 $[M-H_2O+H]^+$
Oxytetracycline: R_1=H R_2=OH R_3=CH_3 R_4=OH	461 $[M+H]^+$	426 $[M-H_2O-NH_3+H]^+$ 444 $[M-NH_3+H]^+$
Chlortetracycline: R_1=Cl R_2=OH R_3=CH_3 R_4=H	479 $[M+H]^+$	444 $[M-H_2O- NH_3+H]^+$ 462 $[M-NH_3+H]^+$
Doxycycline: R_1=H R_2=CH_3 R_3=CH_3 R_4=H	445 $[M+H]^+$	428 $[M-NH_3+H]$+ 410 $[M-H_2O-NH_3+H]$+

Polyether antibiotics (ionophores)

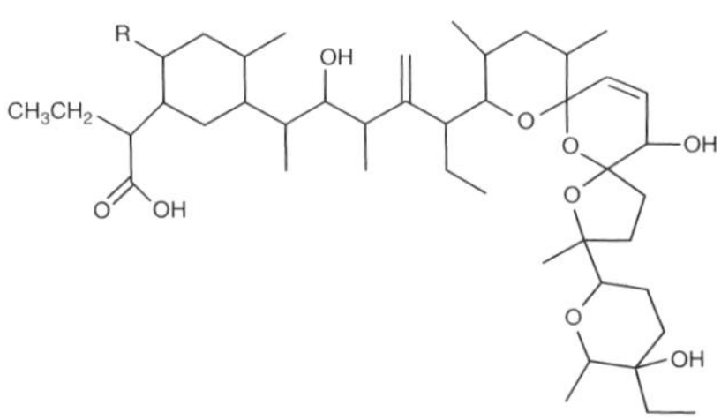

Compound	Diagnostic ions (m/z)	
	Precursor	Product
Salinomycin: R= Ø	773 $[M+Na]^+$	755 $[M+Na-H_2O]^+$ 733 $[M+Na-2H_2O]^+$
Narasin A: R= CH_3	787 $[M+Na]^+$	769 $[M+Na_2H\ O]$ 544 $[M+Na-H_2O-C_{12}H_{16}O_2]^+$

Compound	Diagnostic ions (m/z)	
	Precursor	Product
Trimethoprim 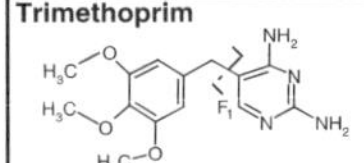	293 $[M+H]^+$	123 $[M\text{-trimetoxyphenyl}]^+$ (F1) 231 $[M-2CH_3O+H]^+$
Chloramphenicol 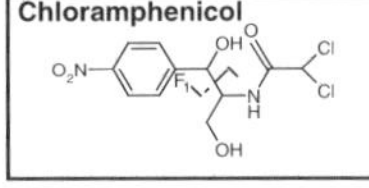	323 $[M-H]^-$	152 [nitrobenzylalcohol carbanion] (F1) 176 $[194-H_2O]^-$

Fig. 2.1.1. *(continued)*

demonstrated the importance of using suitable internal standards, structurally similar compounds that should have similar ionization properties as the analytes, or isotopically labeled standards [30,70,71], for each analyte. As a consequence, more than one internal standard is necessary in multiresidue analytical methods. However, according to

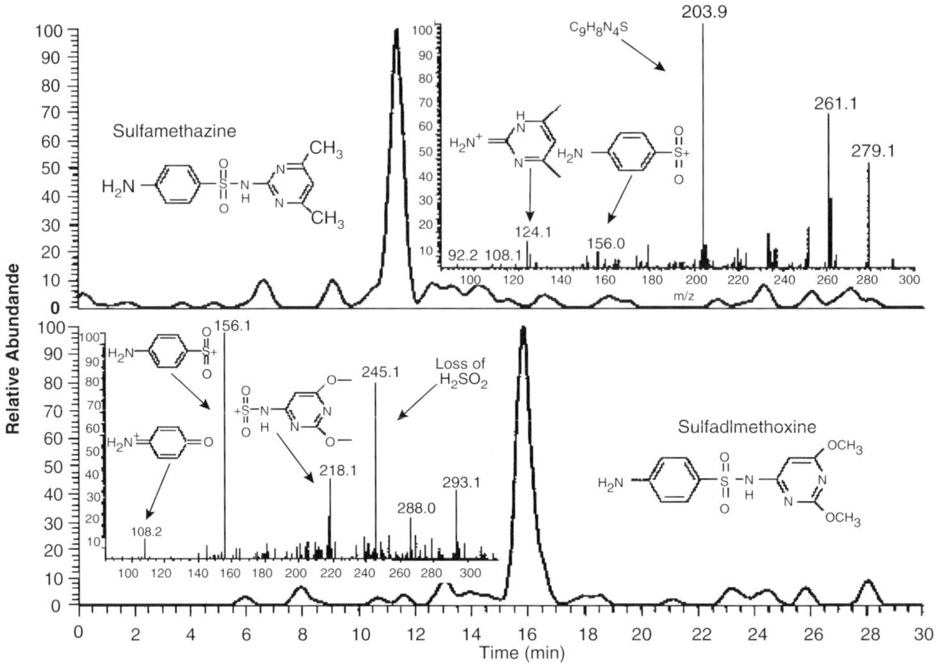

Fig. 2.1.2. Chromatogram and corresponding tandem MS spectrum of a groundwater sample from Weiser (Idaho, USA) containing sulfamethazine and sulfadimethoxine at 0.078 and 0.050 μg/L, respectively. Figure reproduced with permission from Ref. [19].

Lindberg et al. [72], the use of internal standard may compensate the variability, but the extent of this correction depends upon the analyte. Moreover, the authors pointed out the suitability of non-labeled internal standards only under no short-term matrix effects due to the different retention times at which internal standard and analytes elute. Results indicated that the use of isotopically labeled internal standards should be mandatory for the determination of ampicillin and cefadroxyl in hospital wastewaters. An alternative approach to compensate the lack of signal reproducibility and accuracy is the use of the time-consuming standard addition method [70,71].

In the particular case of antimicrobials, signal suppression is notorious when dealing with complex matrices, such as wastewaters. Several phenomena may cause signal suppression: antimicrobials may adsorb to suspended organic matter in the samples decreasing the freely dissolved antimicrobials hindering their detection, co-extracted matrix components may mask analyte peaks by increasing the baseline

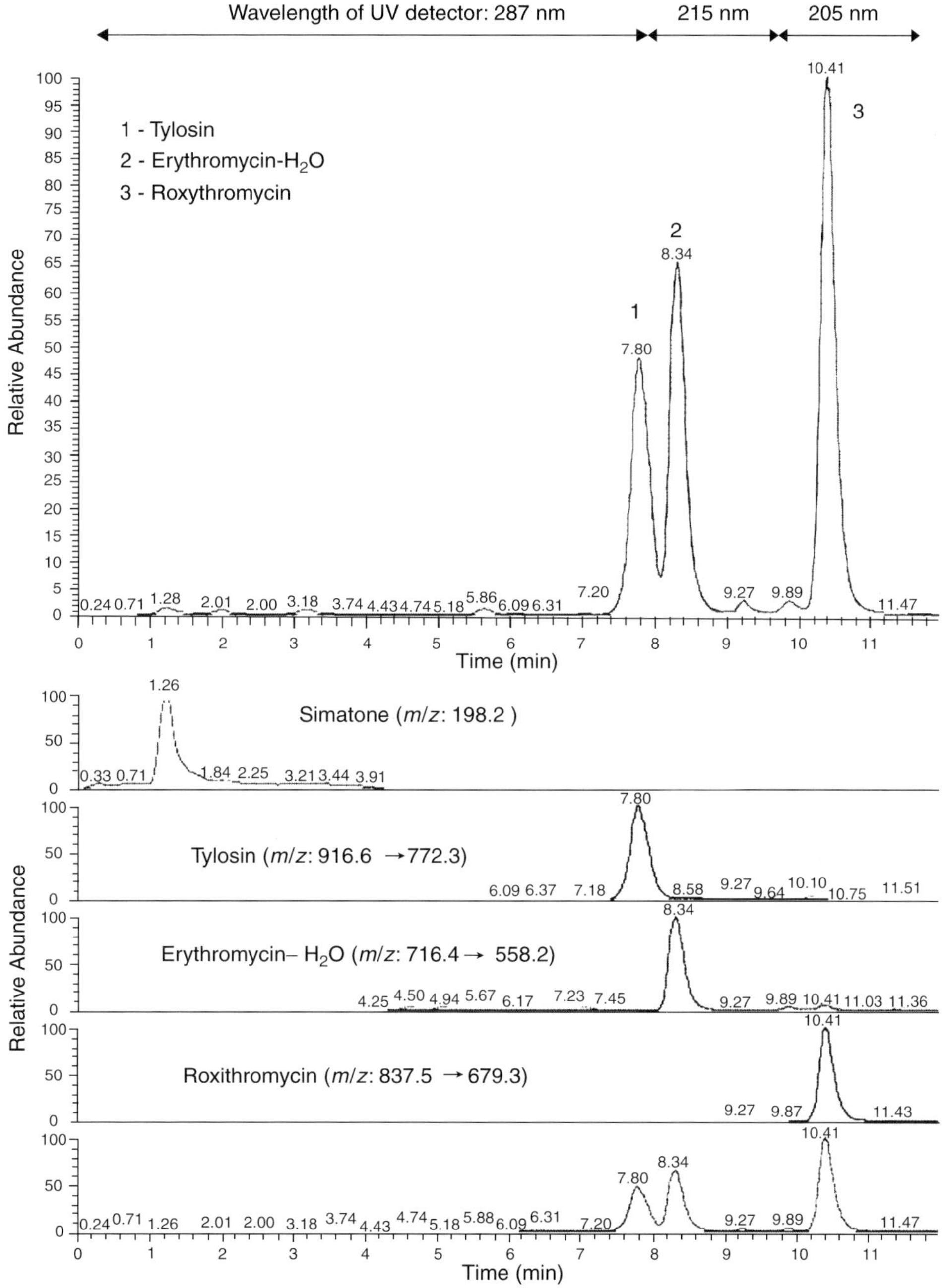

Fig. 2.1.3. Total ion current and MRM mass chromatograms and the applied UV detection programming of a Poudre River water sample spiked at 2 µg/L with tylosin, erythromycin-H_2O and roxithromycin. The MRM transitions from the precursor to the product ions are indicated in parenthesis. Figure reproduced with permission from Ref. [49].

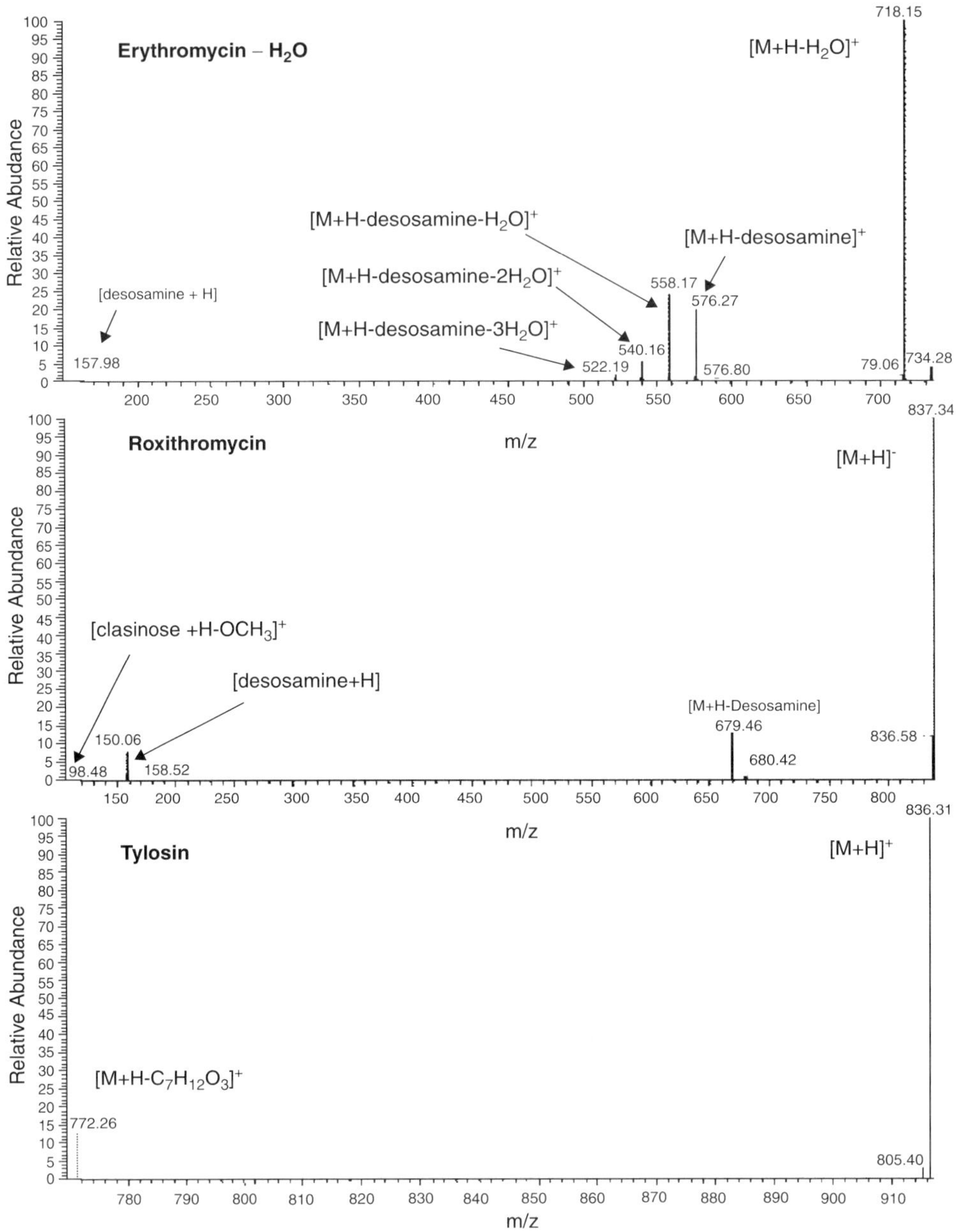

Fig. 2.1.4. Scan MS spectra under in-source CID conditions of a Poudre River water sample spiked at 2 μg/L with tylosin, erythromycin-H_2O and roxithromycin. Figure reproduced with permission from Ref. [49].

signal intensity or may reduce ionization efficiencies by taking up some of the excess charged sites on the electrosprayed droplets [73–77]. Greater matrix effects are observed when the sample contains higher amount of organic matter [30]. For example, in the LC-MS analysis of SAs and TCs, Lindsey et al. [30] reported on the loss of signal for TCs of up to 100% and the enhancement of signal up to 15% in both groundwater and surface water as compared to that in deionized water, either using ESI or APCI interfaces. In contrast, SAs did not show matrix effects in any sample. Similarly, Cha et al. [27] using LC-ESI-tandem MS observed reduction of TCs signals in WWTPs influents and effluents, pointing out the high TOC concentration as the responsible. On the contrary, no significant changes in signal intensities were observed for SAs.

The comparison of signal suppression of a set of antimicrobials revealed that compounds belonging to the same antibiotic class generally suffer of a similar signal intensity drop, as it was demonstrated for selected fluoroquinolones and SAs, and trimethoprim [60].

2.1.4 CONCLUSIONS

The requirement for quantitative data at environmental relevant concentrations (ng/L range) reinforces the need for powerful analytical techniques, which ensures low detection limits and are certain to confirm analyte identities. LC-tandem MS fulfils these criteria. Hybrid QqTOF instruments allow an ultimate identity confirmation of unknowns by the accurate mass measurement of product ion spectra. Despite that, its application in environmental analysis is still limited due to its lower sensitivity compared to QqQ analyzers. Therefore, in the analysis of antimicrobials at environmental relevant concentrations, the technique of choice so far is LC-tandem MS with QqQ instruments for both detection and confirmation purposes. For further improvement in antimicrobial environmental analysis the recently introduced hybrid QqLIT mass analyzer might be a very interesting tool.

Despite the increasing number of works devoted to the analysis of metabolites and degradation products, they are only available for some antimicrobial compounds, such as TCs. This may be due to their usually high polarity and to the lack of reference substances, both making difficult their analytical determination.

Most efforts in environmental analysis have to be focused on the minimization of matrix effects. Suppression or enhancement of the analyte signal is a complex effect, whose extent seems to be dependent on several experimental and instrumental conditions. Strategies to minimize those unwelcome effects together with improved calibration approaches are required to obtain reliable data on the occurrence and fate of antimicrobial residues in the environment. To attain these challenges, enhanced extraction and purification procedures need to be developed, and more labeled internal standards should be commercially available.

ACKNOWLEDGEMENTS

This work has been funded by the EU Framework VI Programme (Project NOMIRACLE, Co. No: 003956-2), and by the Spanish Ministry of Education and Science (Project EVITA, CTM2004-06265-CO3-01/TECNO). M.S. Díaz-Cruz acknowledges her Ramón y Cajal contract from the Spanish Ministry of Education and Science.

REFERENCES

1 J.C. Chee-Sanford, R.I. Aminov, I.J. Krapac, N. Garrigues-Jeanjean and R.I. Mackie, *Appl. Environ. Microbiol.*, 67 (2001) 1494.
2 L. Guardabassi, A. Petersen, J.E. Olsen and A. Dalsgaard, *Appl. Environ. Microbiol.*, 64 (1998) 3499.
3 M. Goñi-Urriza, M. Capdepuy, C. Arpin, N. Raymond, P. Caumette and P. Quentin, *Appl. Environ. Microbiol.*, 66 (2000) 125.
4 S.D. Costanzo, J. Murby and J. Bates, *Mar. Pollut. Bull.*, 51 (2005) 218.
5 R.J. Ash and J.L. Iverson, *NGWA Fourth Internacional Conference on Pharmaceuticals and Endocrine Disrupting Chemicals in Water*, Minneapolis, 2004.
6 FEDESA (European Federation of Animal Health). Press release Visby Sweden, 2001.
7 K. Kümmerer, *Chemosphere*, 45 (2001) 957.
8 B. Halling-Sorensen, N. Nielsen, P.F. Lansky, F. Ingerslev, L. Hansen, H.C. Lüzhoft and S.E. Jorgensen, *Chemosphere*, 36 (1998) 357.
9 M.S. Díaz-Cruz, M.J. López de Alda and D. Barceló, *Trends Anal. Chem.*, 22 (2003) 340.
10 T.B. Vree and Y.A. Hekster, *Pharmacokinetics of sulfonamides revisited*, Vol. 34, Carger Basel, New York, 1985.
11 K. Berger, B. Petersen and H. Büning-Pfaue, *Arch. Lebensmittelhyg.*, 37 (1986) 85.
12 K. Kümmerer, A. Al-Ahmad and V. Mersch-Sundermann, *Chemosphere*, 40 (2000) 701.

13 C.A. Weerasinghe and D. Towner, *Environ. Toxicol. Chem.*, 16 (1997) 1873.
14 H. Oka and H. Matsumoto, *J. Chromatogr. A*, 882 (2000) 109.
15 S. Thiele-Bruhn, *J. Plant. Nutr. Soil Sci.*, 166 (2003) 145.
16 J. Zhu, D.D. Show, D.A. Cassada, S.J. Monson and R.F. Spalding, *J. Chromatogr. A*, 928 (2001) 177.
17 A.A.M. Stolker, W. Niesing, W. Fuchs, R.J. Vreeken, W.M.A. Niessen and U.A.Th. Brinkman, *Anal. Bioanal. Chem.*, 378 (2004) 1754.
18 M. Carballa, F. Omil, J.M. Lema, M. Llompart, C. García-Jares, I. Rodríguez, M. Gómez and T.A. Ternes, *Water Res.*, 38 (2004) 2918–2926.
19 A.L. Batt, D.D. Snow and D.S. Aga, *Chemosphere*, 64 (2006) 1963.
20 A. Gobel, C.S. McArdell, M.J.S. Suter and W. Giger, *Anal. Chem.*, 76 (2004) 4756.
21 M.J. Hilton and K.V. Thomas, *J. Chromatogr. A*, 1015 (2003) 129.
22 K. Stoob, H.P. Singer, C.W. Goetz, M. Ruff and S.R. Mueller, *J. Chromatogr. A*, 1097 (2005) 138.
23 R. Hirsch, T.A. Ternes, K. Haberer and K.L. Kratz, *Sci. Tot. Environ.*, 225 (1999) 109.
24 B. Halling-Sorensen, G. Sengelov and J. Tjornelund, *Arch. Environ. Contam. Toxicol.*, 42 (2002) 263.
25 D. Löffler and T.A. Ternes, *J. Chromatogr. A*, 1000 (2003) 583.
26 O.J. Pozo, C. Guerrero, J.V. Sancho, M. Ibáñez, E. Pitarch, E. Hogendoorn and F. Hernández, *J. Chromatogr. A*, 1103 (2006) 83.
27 J.M. Cha, S. Yang and K.H. Carlson, *J. Chromatogr. A*, 1065 (2005) 187.
28 T. Reemtsma, *Trends Anal. Chem.*, 20 (2001) 500.
29 T.A. Ternes, *Trends Anal. Chem.*, 20 (2001) 419.
30 M.E. Lindsey, M. Meyer and E.M. Thurman, *Anal. Chem.*, 73 (2001) 4640.
31 M.S. Díaz-Cruz and D. Barceló, *Trends Anal. Chem.*, 24 (2005) 645.
32 X.S. Miao and C.D. Metcalfe, *J. Mass Spectrom.*, 38 (2003) 27.
33 D.W. Kolpin, E.T. Furlong, M.T. Meyer, E.M. Thurman, S.D. Zaugg, L.B. Barber and H.T. Buxton, *Environ. Sci. Technol.*, 36 (2002) 202.
34 R. Hirsch, T.A. Ternes, K. Haberer, A. Mehlich, F. Ballwanz and K.L. Kratz, *J. Chromatogr. A*, 815 (1998) 213.
35 K. Ensing, C. Berggren and R.E. Majors, *LC GC Eur.*, 15 (2002) 2.
36 I. Ferrer, F. Lanza, A. Tolokan, V. Horvath, B. Sellergren, G. Horvai and D. Barceló, *Anal. Chem.*, 72 (2000) 3934.
37 R. Koeber, C. Fleisher, F. Lanza, K.S. Boos, B. Sellergren and D. Barceló, *Anal.Chem.*, 73 (2001) 2437.
38 Q.Z. Zhu, P. DeGelmann, R. Niessner and D. Knopp, *Environ. Sci. Technol.*, 36 (2002) 5411.
39 K. Reddersen and T. Heberer, *J. Chromatogr. A*, 1011 (2003) 221.
40 T. Kumazawa, X.P. Lee, K. Sato and O. Suzuki, *Anal. Chim. Acta*, 492 (2003) 49.

41 D.A. Volmer and J.P.M. Hui, *Rapid Commun. Mass Spectrom.*, 12 (1998) 123.
42 C.M. Lock, L. Chen and D.A. Volmer, *Rapid Commun. Mass Spectrom.*, 13 (1999) 1744.
43 M.D. Prat, J. Benitoio, R. Compañó, J.A. Hernández-Artesero and M. Granados, *J. Chromatogr. A*, 1041 (2004) 27.
44 E.M. Golet, A.C. Alder and W. Giger, *Environ. Sci. Technol.*, 36 (2002) 3645.
45 X.S. Miao, F. Bishay, M. Chen and C.D. Metcalfe, *Environ. Sci. Technol.*, 38 (2004) 3533.
46 M.J. López de Alda, M.S. Díaz-Cruz, M. Petrović and D. Barceló, *J. Chromatogr. A*, 1000 (2003) 43.
47 M. Petrović, M.D. Hernando, M.S. Díaz-Cruz and D. Barceló, *J. Chromatogr. A*, 1067 (2005) 1.
48 M.S. Díaz-Cruz and D. Barceló, *Anal. Bioanal. Chem.*, 386 (2006) 973.
49 S. Yang and K.H. Carlson, *J. Chromatogr. A*, 1038 (2004) 141.
50 E.R. Campagnolo, K.R. Johnson, A. Karpati, C.S. Rubin, D.W. Kolpin, M.T. Meyer, J.E. Esteban, R.W. Currier, K. Smith, K.M. Thu and M. McGeehim, *Sci. Total Environ.*, 299 (2002) 89.
51 W. Ahrer, E. Scherwenk and W. Buchberger, *J. Chromatogr. A*, 910 (2001) 69.
52 B. Chiavarino, M.E. Crestoni, A. Di Marzio and S. Fornarini, *J. Chromatogr. B*, 706 (1998) 269.
53 M. Preu, D. Guyot and M. Petz, *J.Chromatogr. A*, 818 (1998) 95.
54 J.W. Mayhew and S.L. Gorbach, *J.Chromatogr.*, 151 (1978) 133.
55 S. Zhou and M. Hamburger, *Rappid Commun. Mass Spectrom.*, 9 (1995) 1516.
56 S. Zhou and M. Hamburger J. Chromatogr. A 755 (996) 89.
57 C. Hartig, T. Store and M. Jekel, *J. Chromatogr. A*, 854 (1999) 163.
58 P.A. Blackwell, H.C.H. Lutzhoft, H.P. Ma, B. Halling-Sorensen, A.B. Boxall and P. Kay, *J. Chromatogr. A*, 1045 (2004) 111.
59 E.M. Golet, A.C. Alder, A. Hartmann, T.A. Ternes and W. Giger, *Anal. Chem.*, 73 (2001) 3632.
60 J.E. Renew and C.H. Huang, *J. Chromatogr. A*, 1042 (2004) 113.
61 J.W. Hager, *Anal. Bioanal. Chem.*, 378 (2004) 845.
62 A.M. Kamel, H.G. Fonda, P.R. Brown and B. Munson, *J. Am. Soc. Mass Spectrom.*, 13 (2002) 543.
63 M. Ibáñez, J.V. Sancho, O.J. Pozo, W. Niessen and F. Hernández, *Rappid Commun. Mass Spectrom.*, 19 (2005) 169.
64 J.W. Hager and J.C.Y. le Blanc, *J. Chromatogr. A*, 1020 (2003) 3.
65 J.W. Hager, *Rapid Commun. Mass Spectrom.*, 16 (2002) 512.
66 A. Raffaelli and A. Saba, *Mass Spectrom. Rev.*, 22 (2003) 318.
67 K.A. Hanold, S.M. Fisher, P.H. Cormia, E. Miller and J.A. Syage, *Anal. Chem.*, 76 (2004) 842.

68 M. Takino, S. Daishima and S. Nakahara, *J. Chromatogr. A*, 1011 (2003) 67.

69 B.K. Matuszewski, M.L. Constanzer and C.M. Chavez-Eng, *Anal. Chem.*, 75 (2003) 3019.

70 M. Stüber and T. Reemtsma, *Anal. Bioanal. Chem.*, 378 (2003) 910.

71 T. Benijts, R. Dams, W. Lambert and A. De Leenheer, *J. Chromatogr. A*, 1029 (2004) 153.

72 R. Lindberg, P. Jarnheimer, B. Olsen, M. Johansson and M. Tysklind, *Chemosphere*, 57 (2004) 1479.

73 H.C.H. Lutzhoft, W.H.J. Vaes, A.P. Fredig, B. Halling-Sorensen and J.L.M. Hermens, *Environ. Sci. Technol.*, 34 (2000) 4989.

74 P. Schmitt-Kopplin, J. Burhenne, D. Freitag, M. Spiteller and A. Kettrup, *J. Chromatogr. A*, 837 (1999) 253.

75 C.G. Enke, *Anal. Chem.*, 69 (1997) 4885.

76 A.P. Bruins, *J. Chromatogr. A*, 794 (1998) 345.

77 R.J.C.A. Steen, A.C. Hogenboom, P.E.G. Leonards, R.A.L. Peerboom, W.L. Confino and U.A.Th. Brinkman, *J. Chromatogr. A*, 857 (1999) 157.

78 S. Wiegel, A. Aulinger, R. Brockmeyer, H. Haras, J. Löffler, H. Reincke, R. Schmidt, B. Stachel, W. von Tümpling and A. Wanke, *Chemosphere*, 57 (2004) 107.

79 G. Hamscher, S. Sczesny, H. Hoper and H. Nau, *Anal. Chem.*, 74 (2002) 1509.

80 F. Sacher, F.T. Lange, H.J. Brauch and I. Blankenhorn, *J. Chromatogr. A*, 938 (2001) 199.

81 K.G. Karthikeyan and M.T. Meyer, Sci. Total Environ., (2005) in press.

82 D. Bendz, N.A. Paxeus, T.R. Ginn and F.J. Loge, *J. Hazard. Mater.*, 122 (2005) 195.

83 S. Castiglioni, R. Bagnati, R. Fanelli, F. Pomati, D. Calamari and E. Zuccato, *Environ. Sci. Technol.*, 40 (2006) 357.

84 R.H. Lindberg, P. Wennberg, M.I. Johansson, M. Tysklind and B.A.V. Andersson, *Environ. Sci. Technol.*, 39 (2005) 3421.

85 F. Bruno, R. Curini, A. Di Corcia, M. Nazzari and R. Samperi, *Rapid Comun. Mass Spectrom.*, 15 (2001) 1391.

86 L.F. Capitan-Vallvey, A. Ariza, R. Checa and N. Navas, *J. Chromatogr. A*, 978 (2002) 243.

87 B. Halling-Sorensen, A. Lykkeberg, F. Ingerslev, P. Blackwell and J. Tjornelund, *Chemosphere*, 50 (2003) 1331.

Chapter 2.2

Analysis of antibiotics in solid samples

Sung-Chul Kim and Kenneth Carlson

2.2.1 INTRODUCTION

Antibiotics were originally designed for therapeutic purposes such as human bacterial infection treatment and for animal disease prevention. In addition to therapeutic uses, antibiotics are also applied for non-therapeutic reasons. The most common non-therapeutic use of antibiotics is growth promotion in animals [1,2]. Although, some European countries have banned the use of antibiotics for non-therapeutic uses in animals, large amounts of antibiotics continue to be used worldwide in the treatment of humans and animals. For instance, Jorgensen *et al.* [3] reported that the annual antibiotic consumption in Denmark for therapeutic purposes in humans is 38 tons. Meanwhile, annual consumption of veterinary antibiotics for therapeutic purposes was 3902 tons in the European Union (EU) and ranged from 7 to 625 tons in several European countries [4]. In comparison, Mellon *et al.* [5] estimated that 7%, (1360 tons) of total consumed antibiotics in the United States were used for human therapeutic purposes in 2001, whereas 5% (907 tons) and 70% (11,339 tons) of total animal-consumed antibiotics, were used for therapeutic and non-therapeutic animal purposes, respectively.

The ultimate fate of antibiotics consumed by both humans and animals is a main factor in the potential of their release into the environment. Pharmacokinetic studies have determined that administrated antibiotics are largely excreted from humans and animals in the parent compound form (up to 90%) or as metabolites. This largely depends on the type of antibiotic. Metabolites may conjugate with glucose or other polar compounds and convert back to the original parent compound in the environment with microorganism activity [6]. This information combined with

Comprehensive Analytical Chemistry 50
Petrović and Barceló (Eds)
Volume 50 ISSN: 0166-526X DOI: 10.1016/S0166-526X(07)50003-6

data reveling large worldwide antibiotic consumption supports the possibility of ultimate antibiotic contamination of the environment.

The major concern surrounding the release of antibiotics into the environment is that antibiotic-resistant bacteria can be formed in the biota. The existence of antibiotic-resistant bacteria can be harmful in human health by rendering specific antibiotics ineffective. Resistance mechanisms are often related with transposons or conjugative plasmids as mobile genetic elements and those elements can transfer the resistance genes from one bacterium to another through horizontal gene transfer [7]. Among other pharmaceuticals, resistant genes of tetracycline have been reported in lagoons and groundwater underlying two swine production facilities [8]. Other related studies include the assessment of the ecotoxicity of doxycycline in aged pig manure using a multi-species soil system and the tolerance of the soil microbial communities affected by sulfachloropyridazine [9,10].

Although there are no current regulations governing antibiotics relative to their release into the environment, a new awareness of their existence and their adverse effects in the environment has transpired. Efforts have been made not only to evaluate concentration levels of antibiotic residuals in different environmental mediums, but also to verify the transport mechanisms and ultimate fate of antibiotics in the environment. However, there is no standard method to measure approximately the 100 individual antibiotics released into the environment at relevantly low concentrations. As a result, developing analytical methods for the quantification of antibiotics in several environmental mediums is the highest priority and researchers are investigating suitable and robust analytical techniques.

For evaluating concentration levels in the aqueous phase, the solid phase extraction (SPE) method is the most common technique to clean up and concentrate target antibiotics. Detailed descriptions of the SPE mechanism and process are discussed in a later section. Conversely, developing an adequate analytical method to quantify the antibiotics in solid matrices including manure, soil, and sediment is more challenging than aqueous matrices due to the complexity of matrices and interferences. Owing to the analytical difficulty, only a few studies have been published on the quantification of antibiotics in solid matrices. To enhance the extraction efficiency in solid matrices, sorption and desorption mechanisms are generally considered. To further add to the challenges involved in developing analytical methods for the examination of solid matrices, individual antibiotics have different sorption mechanisms for binding with solid particles and also have varying sorption capacities depending on the antibiotics.

The main purpose of this chapter is to discuss a developed analytical method for quantifying antibiotics released into the environment in solid matrices including soil, sediment, and animal waste. In addition, a review of past and current trends in the development of analytical methods for quantification of antibiotics in solid matrices will be discussed. And lastly, a general review of occurrence, transport, and fate of released antibiotics in solid matrices is provided for a thorough understanding of antibiotics in the environment.

2.2.2 CLASSIFICATION OF ANTIBIOTICS

Current antibiotics in the soil are mainly derived from veterinary antibiotics. This is because excreted antibiotics from animals can sustain their potency during manure storage. The antibiotics are introduced into the soil once manure is applied to the field in the form of fertilizer. Representative veterinary antibiotics approved by the Federal Drug Administration (FDA) are listed in Table 2.2.1. Sulfonamides and ionophores are separated from the general list of antibiotics because these two groups are produced synthetically. In general, the listed antibiotics have been found in several environmental media and are used on different species for growth enhancement, feed efficiency, or disease prevention. Sulfonamides and ionophores are mainly used in poultry to prevent *coccidiostats* but are also used in beef and dairy cattle to prevent infections. Depending on the type of animal, the same compound can be used for different purposes. For example, monensin (an ionophore) is used in poultry for preventing *coccidiostats* and at the same time used as a growth promoter in beef and dairy cattle.

2.2.3 PHYSICOCHEMICAL PROPERTIES OF ANTIBIOTICS

Among hundreds of antibiotics, tetracyclines (TCs), sulfonamides (SAs), macrolides (MLs), and ionophores (IPs) are the four groups of antibiotics or antimicrobials that are the most consumed worldwide and likewise are frequently detected in the environment. Each group of medicines has different molecular structures and various functional groups that are attached to the main backbone structure and can cause diverse physicochemical properties.

TABLE 2.2.1

List of representative veterinary antibiotics approved for use in USA

	Species[a]	Growth and feed efficiency[b]	Various infections[b]
Antibiotics			
Amoxicillin	H/B&D	No	Yes
Ampicillin	H/B&D	No/Yes	Yes
Apramycin	H	No	Yes
Bacitracin	C&T/H/B&D	Yes	Yes
Bambermycin	C&T	Yes	No
Chlortetracycline	C&T/H/B&D	Yes	Yes
Erythromycin	C&T/H/B&D	No	Yes
Fluoroquinolones	C&T	No	Yes
Gentamycin	C&T/H	No	Yes
Lasalocid	B&D	Yes	No
Lincomycin	H	No	Yes
Monensin	B&D	Yes	No
Neomycin	C&T/H	No	Yes
Novobiocin	C&T	No	Yes
Oleandomycin	C&T/H	Yes	Yes/No
Oxytetracycline	C&T/H/B&D	No	Yes
Penicillin	C&T/H/B&D	Yes/Yes/No	Yes/No/Yes
Roxarsone	C&T	Yes	Yes
Spectinomycin	C&T/H	Yes/No	Yes
Streptomycin	C&T/H/B&D	No	Yes
Tetracycline	C&T/H/B&D	No	Yes
Tiamulin	H	Yes	Yes
Tylosin	C&T/H/B&D	Yes/Yes/No	Yes
Virginiamycin	C&T/H	Yes	Yes/No
Sulfonamides			
Sulfachloropyrazine	P	Used for *Coccidiostats*	
Sulfachloropyridazine	B&D	No	Yes
Sulfadimethoxine	B&D	No	Yes
Sulfamethazine	B&D	No	Yes
Sulfamethazine	P	Used for *Coccidiostats*	
Ionophores			
Lasalocid	P	Used for *Coccidiostats*	
Monensin	P		
Narasin	P		
Salinomycin	P		

[a]C&T denotes for chickens and turkeys, H for hogs, B&D for beef and dairy cattle, and P for poultry.
[b]Different usages are separated by species otherwise usage is same for all species.

2.2.3.1 Tetracyclines (TCs)

TCs are composed of four linear six-member rings. TCs have amphoteric characteristics with three acid dissociation constants (pKa) depending on polar functions, hydroxyl group and dimethylamino group (Table 2.2.2) [4,11]. In general, TCs are more stable in acidic conditions and are well known to form complexes with divalent metal ions, β-diketones, and silanolic groups. Owing to this strong binding characteristic, TCs are hard to extract from solid matrices resulting in incomplete recovery and are also difficult to separate in chromatography. In addition, TCs might undergo photo degradation and as a result produce derivatives. Aga *et al.* [12] reported four derivatives, anhydro-tetracycline, β-apo-oxytetracycline, anhydro-chlortetracycline derived from tetracycline, oxytetracyclin, and chlortetercycline in soil amended with manure.

TABLE 2.2.2

Chemical structure and physicochemical properties of representative TCs

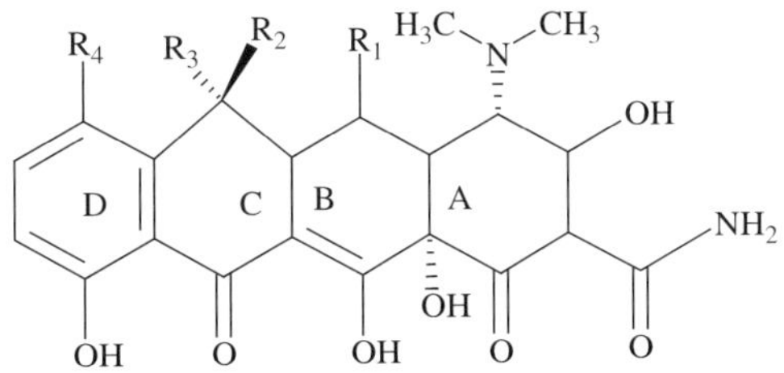

Antibiotics	Acronym	CAS number[a]	M.W	R_1	R_2	R_3	R_4
Oxytetracycline	OTC	79-57-2	460	OH	OH	CH_3	H
Chlortetracycline	CTC	64-72-2	478	H	OH	CH_3	Cl
Minocycline	MNC	10118-90-8	457	H	H	H	$(CH_3)_2N$
Demeclocycline	DMC	127-33-3	465	H	OH	H	Cl
Meclocycline	MCC	2013-58-3	477	OH	CH_2	—	Cl
Tetracycline	TC	60-54-6	444	H	OH	CH_3	H
Doxycycline	DXC	564-25-0	444	OH	H	CH_3	H

Physicochemical properties[b]

Water solubility (mg/L)	pKa	Log Kow	Koc[c] (L/kg)
230–52000	2.8–9.7	−1.3–0.05	195–93,320

[a]Reference [94].
[b]Reference [4].
[c]Values are varied depending on texture, pH, and % of organic carbon.

2.2.3.2 Sulfonamides (SAs)

SAs are synthetically produced antimicrobials to treat illnesses in humans and animals. SAs act by competing with *p*-aminobenzoic acid in the enzymatic synthesis of dihydropholic acid. This leads to a decreased availability of the reduced folates that are essential in the synthesis of nucleic acids (Table 2.2.3) [13]. SAs have two nitrogen functional groups. The amide attached to the sulfur is deprotonated at pH > 5.5–7 and the amine attached to aromatic cycle is deprotonated at pH < 2.5. Thus, most sulfonamides are positively charged under acidic conditions and negatively charged under alkaline conditions ([14]). SAs are rarely biodegradable in the environment and show fairly low sorption capacity

TABLE 2.2.3

Chemical structure and physicochemical properties of representative SAs

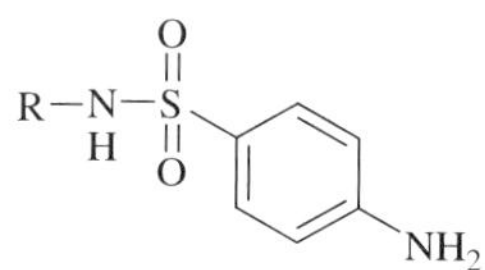

Sulfonamides (acronym, CAS number[a])	R	MW	Sulfonamides (acronym, CAS number)	R	MW
Sulfathiazole (STZ, 72-14-0)	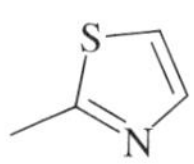	255	Sulfamethoxazole (SMX, 723-46-6)	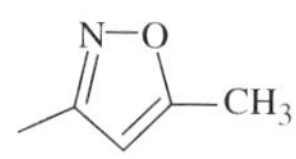	253
Sulfamerazine (SMR, 127-79-7)	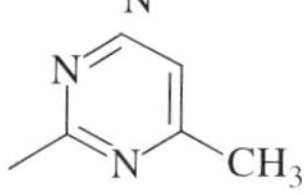	264	Sulfachloropyridazine (SCP, 80-32-0)	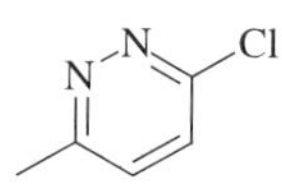	284
Sulfamethazine (SMT, 57-68-1)	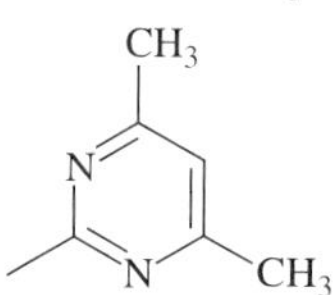	278	Sulfadimethoxine (SDM, 112-11-2)	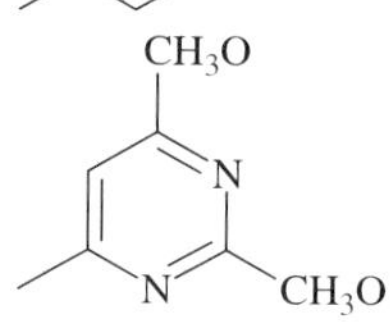	310

Physicochemical properties[b]

Water solubility (mg/L)	pKa	Log Kow	Koc[c] (L/kg)
7.5–1500	2–3, 4.5–10.6	−0.1–1.7	48–323

[a]Reference [94].
[b]Reference [4].
[c]Values are varied depending on texture, pH, and % of organic carbon.

to solid matrices compared to other antibiotics. As a result, there is an increased chance that SAs can extend into the subsurface and contaminate the groundwater.

2.2.3.3 Macrolides (MLs)

The basic structure (Table 2.2.4) of MLs is composed of a macrocyclic lactone ring substituted with hydroxyl, alkyl, and ketone groups. Neutral and amino sugars are bound to the nucleus by the substitution of hydroxyl groups [15]. MLs are mainly active against a variety of gram-positive bacteria whereas reactivity against gram-negative bacteria is minimal [16]. MLs generally have weak acidic characteristics and thus, are unstable under acidic conditions [4].

2.2.3.4 Ionophores (IPs)

IPs are naturally produced by certain strains of Streptomyces and are relatively large molecules. IPs mainly consist of a carboxylic polyether backbone that forms pseudo-macro cyclic complexes with cations [17]. This complex is formed by intermolecular hydrogen bonding between a carboxylic group at one end of the molecular and a terminal alcohol group at the other (Table 2.2.5). While IPs are beneficial to treat coccidiosis in poultry, they can be very toxic to horses when given ingested in doses over 100 mg/L [18].

2.2.4 ANTIBIOTIC EXTRACTION IN SOLID MATRICES

Extraction of antibiotics from solid matrices has been more challenging compared to aqueous media due to the complexity of solid matrices. Unlike other organic compounds such as, pesticides, no standard method has been established for the extraction of antibiotics. Consequently, there has been increasing efforts to enhance the efficiency of extraction of antibiotics not only from aqueous phases but also in solid matrices. Liquid–liquid extraction (LLE) methods were the routine analysis technique used to extract residuals of antibiotics from aqueous samples. Recently, this method has been replaced with the SPE method due to the large volume of hazardous organic solvent produced and the amount of labor involved. Similarly, the buffer solution extraction method or the pressurized liquid extraction (PLE) method have been widely used to extract antibiotics in solid matrices instead of the soxhlet extraction method. Excess amount of used organic solvents and the

TABLE 2.2.4

Chemical structure and physicochemical properties of representative MLs

Macrolides (acronym, CAS number)[a]	MW	Structure
Erythromycin (ETM, 114-07-8)	734	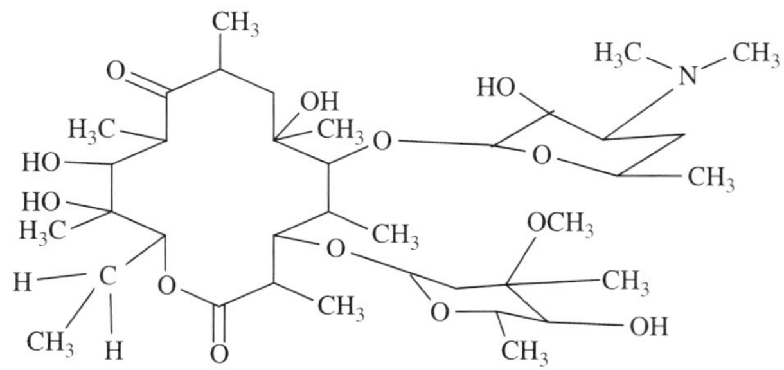
Roxithromycin (RTM, 80214-83-1)	837	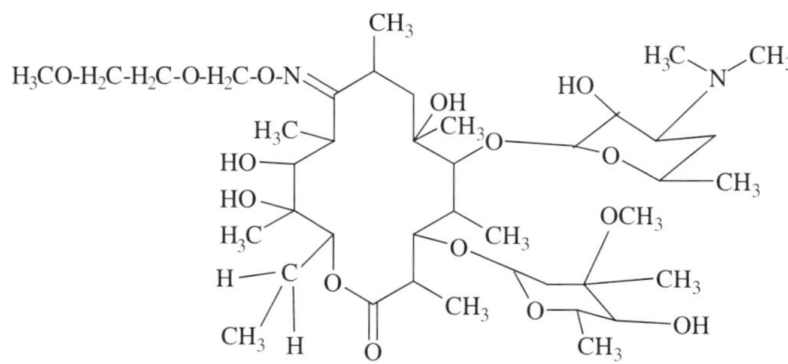

Tylosin (TYL, 1401-69-0)	916	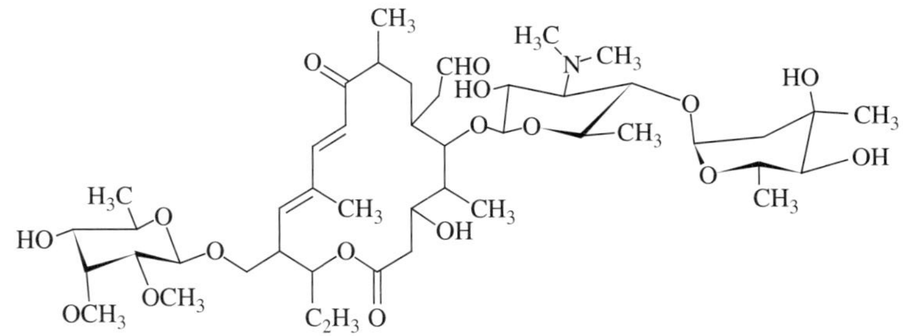

Physicochemical properties[b]

Water solubility (mg/L)	pKa	Log Kow	Koc[c] (L/kg)
0.45–15	7.7–8.9	1.6–3.1	110–7990

[a]Reference [94].
[b]Reference [4].
[c]Values are varied depending on texture, pH, and % of organic carbon.

TABLE 2.2.5

Chemical structure and physicochemical properties of representative IPs

Ionophores (acronym, CAS number)[a]	MW	Structure
Monensin (MNS, 22373-78-0)	693	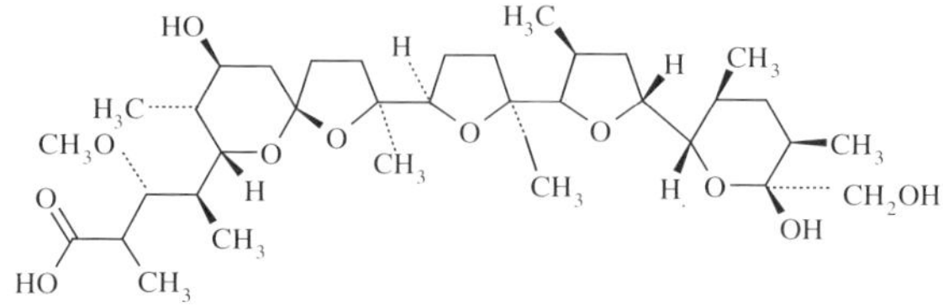
Salinomycin (SLM, 53003-10-4)	751	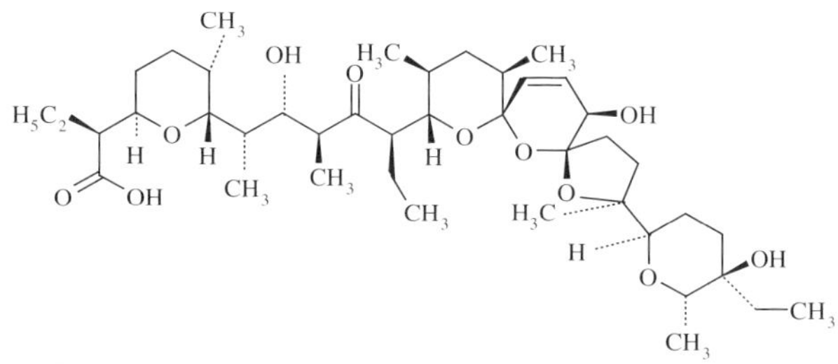
Narasin (NRS, 55134-13-9)	765	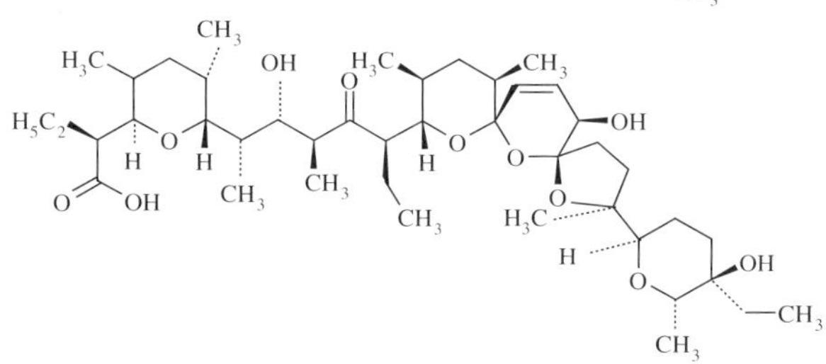

Physicochemical properties[b]

Water solubility (mg/L)	pKa	Log Kow	Koc[c] (L/kg)
2.2×10^{-6}–3.1×10^{-3}	6.4	5.4–8.5	NA

[a]Reference [94].
[b]Reference [4].
[c]Information is not available.

time consuming nature are the main reasons for this transition. A summary of recently used extraction methods for antibiotics in solid matrices is presented in Table 2.2.6 and a description of each extraction method is included.

TABLE 2.2.6

Literature review of extraction process and detection method for pharmaceuticals in LC. Reprinted from [94]

Class and compounds	Sample matrix	Pre-extration cleanup	Detection	Recovery (%)	LOQ[a] (μg/kg)	Detected level (μg/kg)	Reference
2 Fluoroquinolone	Sewage sludge	ASE[b]	FLD[c]	82–94	450	1400–2420	[57]
	Soil	SPE		75–92	180	270–400	
5 Tetracyclines	Animal feeds	ACN[d]/water (pH 3.0)	Diode array	52–96	100–400 (μg/L)	8000–57,000	[11]
4 Antibiotics	Soil	Citric buffer (pH 4.7)/EtOAc[e]	ESI/MS/MS	33–86	5	4–199	[19]
7 Antibiotics	Manure	LLE[f] (EtOAc)	ESI/MS	47–89	100	100–12,400	[14]
8 Antibiotics	Manure	LLE (EtOAc), SPE	APCI/MS/MS	75–123	1–93	11–43	[30]
18 Pharmaceuticals	Sediment	Ultrasonication	APCI/MS/MS ESI/MS/MS	56–151	0.4–20		[20,21]
2 Antibiotics	Manure	McIlvaine buffer/ methanol phosphate buffer/methanol	UV (355 nm, 282 nm)	74–80	100	2110–19,000	[96]
		Soil		81–82	5–10	6–7	
29 Antibiotics	Manure	Ultrasonication	ELISA			1000–1100	[1]
	Soil	ASE	ESI/MS/MS			15	
8 Antibiotics	Soil	PLE[g] SPE (SAX+HLB)	ESI/MS/MS	31–143	1–11	1–57	[24]

TABLE 2.2.6 (*continued*)

Class and compounds	Sample matrix	Pre-extration cleanup	Detection	Recovery (%)	LOQ[a] (μg/kg)	Detected level (μg/kg)	Reference
3 Antibiotics	Soil	Ultrasonication SPE (SAX+HLB)	UV/FLD	27–105	18–40		[49,50]
2 Antibiotics	Sediment	Mcllvaine-EDTA buffer SPE	MS/MS	88–93	0.012–0.061	0–579	[62]
Erythromycin	Soil	LLE	ED[h]				[58]
3 Antibiotics	Soil	Buffer solution/ methanol SPE (SAX+HLB)	UV (285 nm, 355 nm)	35–65	10		[81]
7 Pharmaceuticals	Sludge	Ultrasonication SPE (MCX)	APCI/MS/MS				[72]

[a]Limit of quantification.
[b]Accelerated solvent extraction.
[c]Fluorescence detection.
[d]Acetonitrile.
[e]Ethyl acetate.
[f]Liquid–liquid extraction.
[g]Pressurized liquid extraction.
[h]Electrochemical detection.

2.2.4.1 Liquid–solid extraction (LSE) method

Antibiotic residuals in the solid phase need to be extracted into the liquid phase for further cleanup or concentration process. To achieve this, the liquid–solid extraction (LSE) method that uses organic solvents or buffer solutions has been widely used. Most of the developed LSE methods have been focused on selecting proper organic solvents or buffer solutions to enhance the extraction efficiency while considering the stability of target antibiotics during the extraction process. Depending on the target antibiotics, the polarity of the organic solvent and mixture with the buffer solution are primary concerns in order to increase extraction efficiency from solid matrices. Caballero *et al.* [11] used acetonitrile and a water mixture (1:1) buffered with 0.01 M citric acid at pH 3 to extract 5 tetracyclines (TCs) in animal feed. The acidic condition of the organic solvent and buffer solution mixture was selected to increase the stability of TCs during the process. This study also evaluated the composition of the organic solvent and buffer solution and showed better recovery in the mixture of the organic solvent and buffer solution compared to the organic solvent without a buffer solution. After an adequate solution was selected, sonicating, mixing, and centrifuging steps were followed to obtain residuals of 5 TCs for further analysis.

Hamscher *et al.* [19] investigated the extraction of four antibiotics including tetracycline, chlortetracycline, oxytetracycline, and tylosin using a mixture of citric acid (pH 4.7) and ethyl acetate in soil. This study demonstrated that EDTA or a strong acid such as hydrochloric acid did not affect the chelate-binding characteristic of tetracyclines presumably due to the adsorption of target antibiotics to humic materials.

Haller *et al.* [14] also evaluated the LSE method to increase the extraction efficiency of sulfonamides (SAs) and trimethoprim in animal manure. In this study, three parameters, manure extraction time, pH, and added salts were evaluated and optimized using 6 M of NaCl at pH 9.0 with 30 s of vortexing. Longer extraction times did not improve the extraction efficiency and NaCl gave the best recovery ratio among other salts. However, optimum pH values for the extraction of SAs from purified water and animal manure showed an opposing result. In purified water as the pH increased, the recovery ratio of the studied SAs decreased due to negatively charged properties of SAs above pH 7. Meanwhile, the optimum value for animal manure extraction was pH 9.0. This study concluded that different particle sorption processes in animal manure causes the contradictory result compared to clean water.

Loffler *et al.* [20,21] studied 22 compounds including acidic pharmaceuticals, antibiotics, and parasiticides in sediment. The main extraction was conducted with different organic solvents and with the aid of an ultrasonic bath. For acidic pharmaceuticals and parasiticides, acetone/acetic acid (20/1, v/v) was initially used and ethyl acetate was used three times after. Antibiotics were extracted two times with methanol followed by acetone and ethyl acetate. The recovery efficiency of used organic solvents was assessed using a labeled standard and concluded that most of the studied compounds were extracted during the first two extraction process. Results between autoclaved and non-autoclaved sediment samples showed higher recovery in the autoclaved sediment indicating the impact of biotransformation on the recovery rates.

2.2.4.2 Accelerated solvent extraction (ASE) method

Accelerated solvent extraction (ASE) method is a newly developed analytical method for use with solid matrices. The ASE method is also referred to as the PLE and mainly enhances the extraction efficiency in solid matrices by reducing time and labor requirements. Among other benefits, the primary advantage of using the ASE method is to be able to produce much less organic solvent with the potential of automated analysis. The ASE method is mainly operated with high temperature (up to 200°C) and pressure (up to 20,000 kPa) to enhance solubility, mass transfer, and disruption of surface equilibrium [22]. The general procedure of the ASE method includes filling the solid samples in the sample cell that contains organic solvent. Typically, the organic solvent is then detained for an optimized time period to give sufficient contact between the organic solvent and the solid sample under static condition with elevated temperature and pressure. Although the ASE method is a fast and efficient extraction method for analyzing solid matrices, the method requires an expensive initial setup.

Application of the ASE method for extraction of persistent organic pollutants (POPs) including chlorobenzene, HCH isomers, DDX, PCB congeners, and PAH in soil samples has been reported [23]. This study compared the ASE method to the Soxhlet (SOX) and ultrasonic extraction (USE) methods and verified the high efficiency of the ASE method by calculating extraction yield of POPs in soil samples. This study also pointed out that temperature is the most sensitive parameter for high efficiency in the ASE method along with organic solvent selection and matrix characteristics (composition of soil components). Another study also evaluated the conventional extraction method, SOX and the ASE method to measure polychlorinated biphenyls (PCBs) in sea sediment samples possessing ranges of organic carbon content, soot carbon, sulfur, water content, and PCB concentration. The efficiency of the ASE method was highly dependent on the amount of carbon and was generally higher in lower carbon-containing sediments. The varying amount of sulfur and water in the sediment had a limited influence on the extraction efficiency of the ASE method.

Recently, the ASE method was applied in the analysis of benzalkonium chlorides (BACs) in sediment samples [22]. BACs are quaternary ammonium surfactants with detergent and antimicrobial properties and mainly originate from industrial cleansers. They are introduced in the environment through wastewater treatment plant effluents. During optimization of the ASE method for BACs, this study verified that the organic solvent alone could not improve the extraction efficiency and water is a necessary solvent component due to the fixed ionic characteristic of BACs that could form ionic bonds with humic or fulvic matter in sediment. In addition, the use of greater than 60% water content matrices and adjustment of the solvent's pH did not affect the extraction efficiency.

Furthermore, Jacobsen *et al.* [24] applied the ASE method to extract three different groups of antibiotics and metabolites in the soil and demonstrated several variables to consider for ASE extraction efficiency. This study showed that more than 10 g of soil sample caused clogging in the sampling cell and various physicochemical properties of target antibiotics needed to be considered for organic solvent selection.

2.2.5 SAMPLE CLEANUP AND CONCENTRATION

After the residuals of target antibiotics are extracted from the solid into the liquid phase, further sample cleanup processes are necessary to provide a robust analysis. A sample-concentrating step is required to measure sub micrograms per liter of antibiotics present in environment. The most widely adapted sample cleanup method following the pre-extraction step for solid matrices is SPE. The SPE method is commonly used to extract semi-volatile or non-volatile organic compounds from liquid but is often used for solid samples that have been pre-extracted with liquid solvent. The general five steps of the SPE process are presented in Fig. 2.2.1 followed by a detailed description of each step.

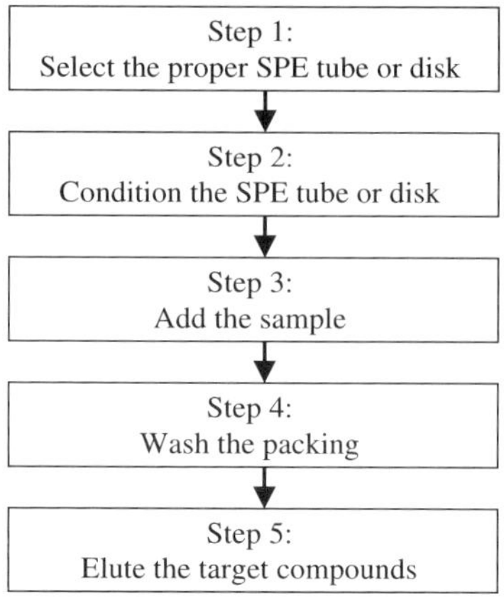

Fig. 2.2.1. Schematic diagram of SPE process.

To select the proper SPE tube or disk, the polarity of the target analytes, sample matrix, and volume of the loading solution needs to be considered. SPE phase types are categorized into four groups, reversed phase, normal phase, ion exchange, and adsorption. Among those groups, reversed phase SPE is normally used with a polar or moderately polar sample matrix. The attractive forces, known as van der Waals forces, between the carbon–hydrogen bonds in the analytes and the functional groups on the silica surface of the SPE material are the main retention mechanisms. Thus, non-polar organic solvents are used to elute adsorbed compounds from reverse phase SPE to disrupt the forces between them.

Normal phase SPE is generally applied to polar and mid-polar matrices. Interactions between polar functional groups of analytes and polar groups on the sorbent surface are involved in the retention of compounds on the SPE material. In addition, cation and anion exchange SPE is used with charged compounds in solution. In this case, electrostatic attraction of the compound's charged functional groups is the key factor in retaining analytes onto the SPE material.

When considering the tube size, the proper size should be used to prevent any breakthrough of analytes through the SPE material. In cases when the sample volume is within 1–250 mL range, a tube size of 3 mL is recommended. Larger sample sizes will require larger tube sizes. Although the pH of the solvent should be considered for maximizing retention of analytes onto the SPE material, it is not a major consideration since the SPE cartridges are disposable and are only used once.

Step 2 commonly involves pre-conditioning the SPE material with an organic solvent. The main purpose of pre-conditioning is to optimize conditions with the SPE cartridge for enhancing the retention of analytes. This is accomplished by removing any impurities that exist within the SPE cartridge. It is vital to ensure that the SPE material remains wet between the pre-conditioning and sample-loading steps. In order to ensure that the SPE cartridge remains wet between the two steps, additional amounts of the last pre-conditioning solvent are allowed to remain in the SPE cartridge.

Adding the sample into the SPE cartridge is generally conducted with the aid of a vacuum pump and typical sample loading flow rates are in the range of 2–5 mL/min. Once the samples have been loaded, the cartridges are then washed in order to remove unwanted impurities. A typical solution should contain less organic or inorganic salts than the final eluent in order to remove just the impurities and not the target compounds. The final step of SPE involves eluting the retained target compounds from the SPE material with the proper organic solvent. To increase the elution efficiency, a slow flow rate is desired and a series of two small aliquots of elution are recommended rather than one larger aliquot.

Application of SPE cleanup and concentration has been documented to extract pharmaceutical compounds in sewage treatment plant (STP), surface water, groundwater and to cleanup or purify the pre-extractants in solid matrices [15,20,21,25–33]. Most reported studies used a single SPE material. However, tandem SPE methods (strong anion exchange SAX+HLB) have been used to remove humic material with SAX and to retain antibacterial agents with the HLB cartridge in surface water and agricultural soil [24,29].

2.2.6 SAMPLE SEPARATION AND DETECTION

The present concentration of antibiotics in the environment is generally sub micrograms per liter (aqueous) or kilogram (solid) range and thus, more advanced technology has been required for sensitive and accurate detection of these low concentrations. The traditional detection and separation technique includes gas chromatography (GC) commonly combined with mass spectrometry (MS). Recently, GC/MS has been adapted to study the occurrence of pharmaceuticals and their metabolites, hormones, the metabolites of alkylphenol polyethoxylates (APEOs), N-butyl benzenesulfonamides (NBBS), and chlorinated tris-propylphosphate

(TCPPs) in wastewater-dominated river and also to evaluate natural attenuation of studied compounds in wetlands [34]. In another study, Herberer *et al.* [35,36] utilized the GC/MS technique for separation and detection of pharmaceuticals in order to verify the influence of wastewater treatment effluents on the presence of human-derived pharmaceuticals in the aquatic environment. However, there appears to be a migration from the use of GC/MS to liquid chromatography (LC) coupled with MS or even tandem mass spectrometry (MS/MS) for analysis of antibiotics. The main reason the GC/MS method is being replaced by the LC method is that most of the pharmaceutical compounds including antibiotics are not very volatile and some are highly polar compounds containing ionizable functional groups (carboxylic or amino). This scenario makes LC/MS or LC/MS/MS more suitable for separation and detection of pharmaceutical residuals or metabolites in environmental samples [37]. In addition, the GC/MS method for analyzing antibiotics typically requires an additional derivatization step that requires more labor and time, and may also introduce unwanted contamination in the sample. Also, loss of target compound can occur at the derivatization step. Therefore, more detailed description of LC/MS or LC/MS/MS is presented in this chapter.

LC and MS technology are two separate methods with different operating mechanisms. The basic principal of LC involves passing a mixture of samples in the liquid mobile phase through stationary phase column packing where the compounds will be separated by their relative affinity for partitioning between different phases. In comparison, the MS method provides the sample's mass spectrum by using different ionization techniques for the identification of particular compounds or for molecular structure confirmation. Since LC is only highly efficient in separating mixtures of compounds and MS is superior in confirming the target compounds, the two methods are commonly combined to enhance the performance of separating, positive identifying, and quantifying the complex mixtures in environmental samples.

There are several LC modes including reversed phase, normal phase, ion exchange, size exclusion, and hydrophilic interaction chromatography. These varied LC modes are used in different applications. Reversed phase is the most popular mode for HPLC due to its versatility, simplicity, and its applicability in the analysis of antibiotics. Reversed phase mode is generally adapted for a range of analytes with moderately polar to slightly hydrophobic properties, whereas this mode may not be suitable for polar or highly hydrophobic analytes. The stationary phase for the reversed mode can be composed of C_1–C_{30} hydrocarbon chains. C_1

packing is used for very non-polar compounds and C_{30} packing is for relative polar compounds. Common reverse phase stationary phase is composed with C_{18} and this carbon chain is attached to silica particles with silica–ether bonds: Si–O–CH_2–R.

Normal phase chromatography is packed with unmodified silica as stationary phase and used for small, polar compounds. Hydrophilic interaction chromatography is similar to normal phase chromatography and is suitable to analyze biomolecules that might bind irreversibly with normal phase silica packing. Ion exchange mode is useful to separate ionic compounds and size exclusion chromatography separates molecules on the basis of molecular weight. However, alternative LC modes except for reversed phase mode require high salt concentrations that can cause pump damage. As a result, few studies have reported using ion exchange or size exclusion chromatography modes. However, Ding *et al.* [38] developed an ion exchange chromatographic method using a polymeric column and acidic eluent. This application was used to measure tetracycline residuals in milk and the oxytetracycline removal rate in a biochemical WWTP process.

To enhance the sensitivity and selectivity of target compounds in LC, several variables need to be considered. Composition of the mobile phase is one of the important factors for better separation in LC and it depends on the characteristics of the compounds. The most common eluents used under gradient elution for improving the peak shape in chromatography include an acidified acetonitrile–water and methanol–water mixtures. Higher viscosity eluants can produce higher backpressure and non-volatile additives such as oxalic acid should be avoided when electrospray ionization is used. Trifluoroacetic (TFA) acid can suppress the ionization in the electrospray source and ammonium acetate or phosphate can be used as a substrate for microorganism [27]. Thus, using refrigerated acetate and phosphate or using mixtures with more than 20% of acetonitrile or methanol is necessary to prevent any possible problems. Other considerations can include injection volume, column size, and gradient of eluants.

MS is the most widely used application for detection and identification of highly polar and non-volatile compounds. A detailed and comprehensive explanation of LC/MS analysis is reviewed with a range of emerging contaminants, related pollutants, microorganisms and humic acids [39,40]. In many applications of MS, the sample solution either organic or aqueous is introduced into the front of the mass spectrometer for mass spectrum identification. During this process, excess amount of solvent will be removed as vapor in the vacuum of MS.

The most successful two techniques for selectively removing solvent from a solution without losing the target compounds are electrospray ionization (ESI) and atmospheric pressure chemical ionization (APCI). The principal of both techniques is to use high temperature and pressure to ionize target compounds for further mass analysis. The ESI system uses a coaxial ESI needle and outer capillary to introduce the sample and nebulizing gas to form highly charged microdroplets. From this droplet, ions appear in the gas phase due to emission, desorption of performed ions from the droplet surface, or soft desolvation of performed ions [41]. Meanwhile, APCI uses a different spraying method and requires another ionization region (corona discharge) for enhancing the formation of protonated molecular ions.

After target compounds are ionized through the ionization process, the resulting ions are analyzed through various mass analyzers. Through recent advances in MS technology, the capability for identifying and quantifying complex samples in less time and with better quality has been accomplished. Traditional LC/MS methods using single quadrupole MS can produce fragmented spectra using in-source collision-induced dissociation. However, this technique can prevent accurate analysis due to the fragmentation of co-eluting analytes and matrix components. To cope with this difficulty, MS/MS techniques have been adapted. LC/MS/MS methods add additional collision energy to fragment protonated or deprotonated ions formed in the several ionization sources. Although this additional fragmentation step may require more analysis time, it enhances the selectivity of the complex-matrix sample by avoiding co-elution of analytes and interferences in samples as seen in single quadrupole LC/MS analysis.

Triple quadrupole MS utilizes the multiple reaction mode (MRM), which has fixed m/z values for quadrupoles (Q) 1 and 3, while Q2 is used as the collision cell. Ion-trap MS is the innovative method that has an ability to perform multiple stages of MS/MS to isolate and fragment ions in time and to trap the product ions. While this mass analyzer method has the ability to infer the pathways easily for the identification of unknowns using MS^n, this application for pharmaceuticals in the environment has yet to be explored. Time of flight MS coupled with LC (LC/TOF/MS) is an alternative detection method for pharmaceuticals in the environment. The high-power resolving technique of the TOF/MS method removes the sample's interference signal making it easier to identify the non-target compounds in a complex environmental sample. This method was reviewed comprehensively and applied with sediment samples for identifying diphenhydramine [42,43].

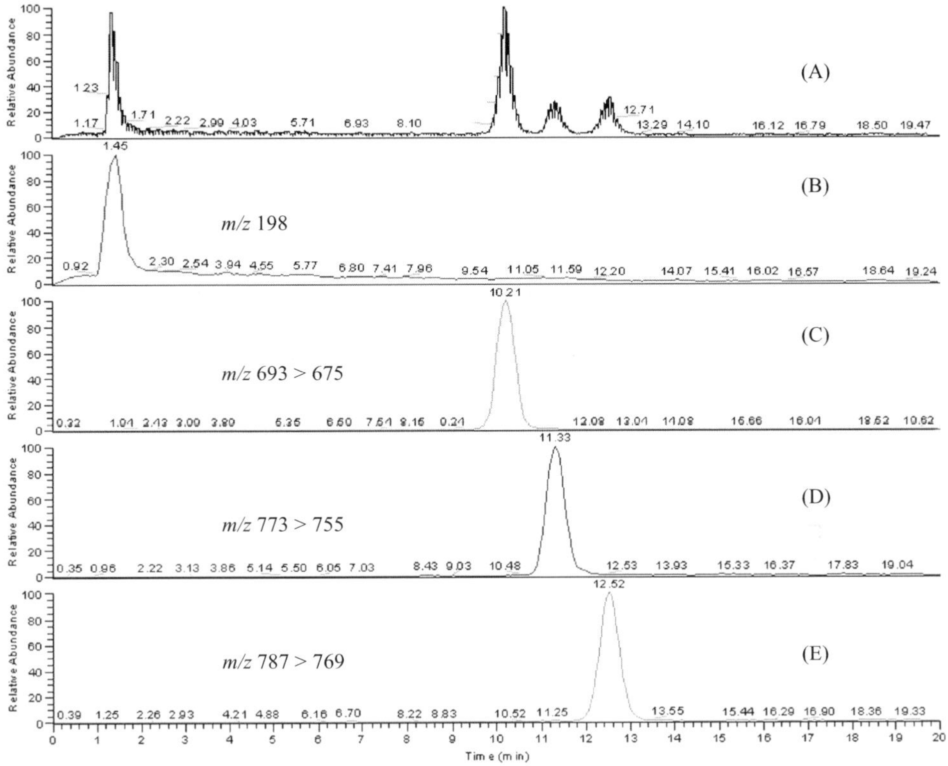

Fig. 2.2.2. Reconstructed chromatogram showing standard solution (2 μg/L) of three ionophores: (A) total ion chromatogram (TIC), (B) internal standard, (C) monensin, (D) salinomycin, and (E) narasin. Reprinted from [95].

Figures 2.2.2 and 2.2.3 show examples of chromatogram and tandem mass spectra of three ionophores, monensis, salinomycin, and narasin obtained by LC/MS/MS electrospray ionization in positive mode. Multiple target compounds are separated within 15 min (Fig. 2.2.2) and selected fragment ions showing the most abundant ions at the optimized condition are determined in the mass analyzer (Fig. 2.2.3).

2.2.7 APPLICATION OF LC/MS FOR QUANTIFYING ANTIBIOTICS IN THE ENVIRONMENT

The occurrence of antibiotics in the environment including surface water, groundwater, and various solid matrices has been detected with the use of LC combined with several ionization sources and mass analyzers. For human derived antibiotics, most wastewater treatment plants

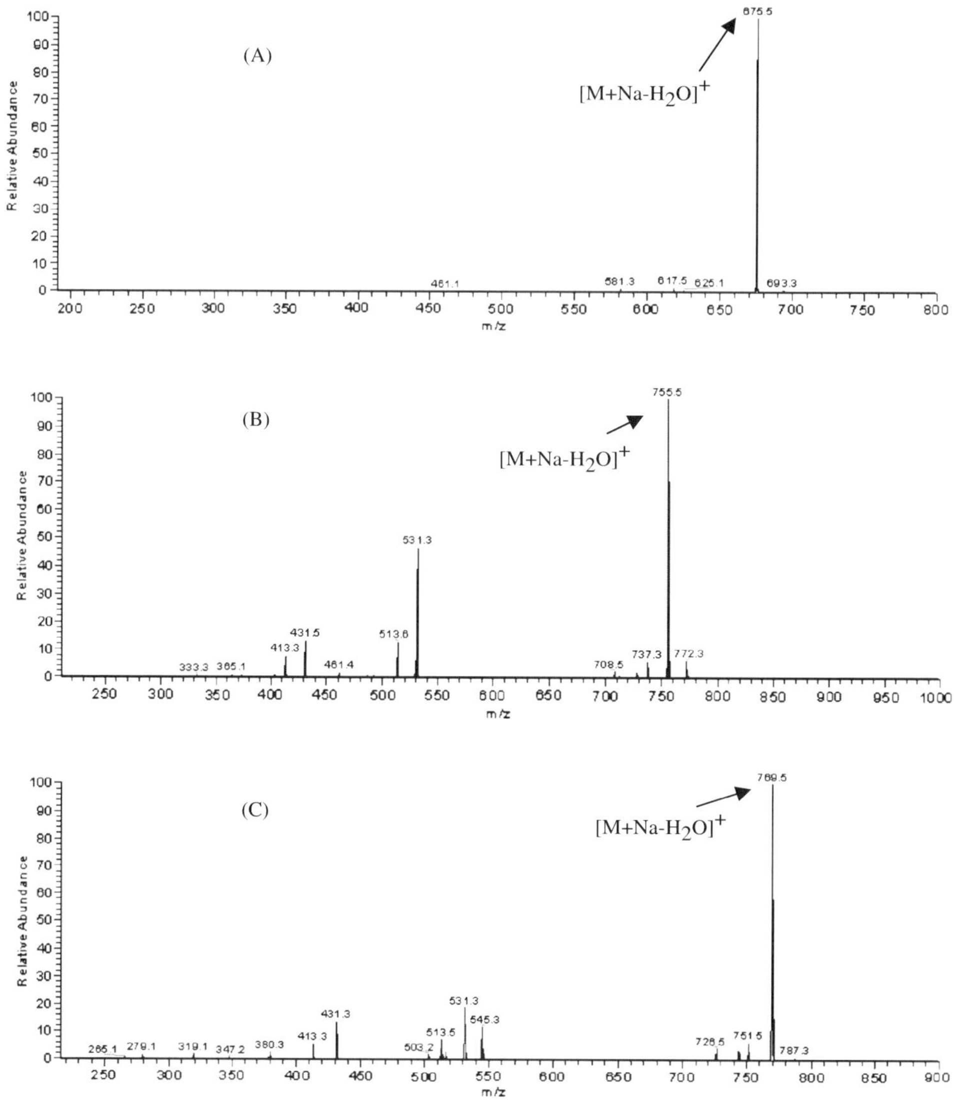

Fig. 2.2.3. Full scan tandem mass spectra of: (A) monensin, (B) salinomycin, and (C) narasin. Reprinted from [95].

can partially remove antibiotics and consequently release unaltered parent compounds or slightly modified forms of human-originated antibiotics into the watershed. As a result, residuals of human-originated antibiotics have been found in WWTP effluent and watersheds that are in close proximity to the WWTP [15,20,21,34–36,44–51]. A comprehensive and detailed review of antibiotics in the aqueous phase has been reported elsewhere [3,6,35,36,52,53]. This chapter is mainly focused on the occurrence of antibiotics in solid matrices.

As summarized in Table 2.2.6, the main solid matrices where antibiotics are present are soil, sediment, and animal waste (manure). Human-originated antibiotics discharged from WWTP effluent can be partitioned to sediment in the watershed. Additionally, biosolids containing antibiotics can be introduced as residuals in agricultural fields when biosolids are used as fertilizers. As for veterinary antibiotics, excreted antibiotics are stored as manure and parent compounds or metabolites of antibiotics, which persist during the storage period, can be introduced to the soil when manure is applied to the field. Furthermore, residual antibiotics in the fields can leach into the subsurface or be transported to the surrounding watershed as surface runoff following rain events. This will depend on antibiotic's sorption characteristics. Aquaculture is an alternative input route, which allows antibiotics to be introduced into the environment where both water sources and sediments are directly exposed to high concentrations of antibiotics. Consequently, researchers have quantified antibiotic residuals as well as verified antibiotic transport and fate mechanisms in the environment.

Recently, Xia *et al.* [54] reviewed the occurrence and fate of pharmaceuticals and personal care products (PPCPs) in biosolids. In this review, several extraction methods, ultrasonic, Soxhlet, microwave assisted extraction (MSE), and accelerated solvent extraction (ASE), were introduced for biosolid examination. ASE was deemed the most efficient extraction method in terms of required extraction time and required solvent volume. After extracting target compounds from the solid phase to liquid phase, further cleanup and concentration steps were followed due to the complexity of the biosolids' matrix and the low concentration of target compounds in the biosolids. Detected concentration levels of pharmaceuticals and PPCPs in the biosolids ranged from 1.5 μg/kg up to 1380 mg/kg depending on the compound. This study also suggested that composting biosolids prior to field application could possibly reduce the residual concentration of pharmaceuticals and PPCPs.

In the case of veterinary antibiotics in solid matrices, Campagnolo *et al.* [55] reported a broad profile of antibiotics in animal waste, surface waters, and groundwaters in close proximity to large-scale confined animal feeding operation (CAFOs) using both radioimmunoassay (RIA) and LC/ESI-MS methods. Both RIA and LC/ESI-MS methods measured the highest concentration for multiple classes of antibiotics in swine waste lagoon samples. Detected concentration levels with the RIA method ranged from below detection limit (1–10 μg/L depending on classes of antibiotics) to 540 μg/L and from below detection limit (0.05–0.5 μg/L depending on classes of antibiotics) to 1000 μg/L with LC/ESI-MS method. However, detected concentrations of the studied antibiotics in surface and groundwaters were much lower than lagoon samples. The highest concentration in groundwater was measured at 7.6 μg/L of sulfamethiazine using the RIA method. Therefore, this study generally concluded that applied animal waste as a fertilizer might contribute antibiotics to local surface or groundwaters as a non-point source. In comparing the two adapted measuring techniques in this study, RIA and LC/ESI-MS, the RIA method is only appropriate for measuring high concentrations and the LC/ESI-MS method is more suitable to detect low levels of antibiotics in the environment. A similar study focused on evaluating the occurrence of veterinary antibiotics in different environmental mediums, Christian *et al.* [1] measured 29 compounds in manure, soil, and surface water. This study applied sonication with pure water for extraction of liquid manure and the accelerated solvent extraction (ASE) method for the analysis of soil samples. For quantification of extracted samples, the ELISA method was used for liquid manure and LC/MS/MS was used for soil samples. The results of this study also indicate that higher concentrations (up to 1 mg/kg) were measured in liquid manure and much lower concentrations were detected in soil (up to 15 μg/kg) and surface waters (up to 300 μg/L).

Haller *et al.* [14] developed an extraction method to quantify six sulfonamides and their associated metabolites in animal manure. Optimized liquid–liquid extraction (LLE) method was used on KOH-buffered samples (pH 9) along with ethyl acetate. The efficiency of the developed extraction method had recovery results ranging from 51 to 89% with limit of quantification of 0.1 mg/kg. Measured concentrations of grab samples taken from cattle and pig farms were up to 12.4 mg/kg with the LC/MS method. Schlusener *et al.* [30] determined macrolides, ionophores, and tiamulin in liquid manure with the LC/MS/MS method. Sample preparation was conducted using the LLE method and APCI as the ionization technique in the positive mode. Selective reaction monitoring (SRM) was adapted for quantification of the target compounds. Detected concentrations of multiple classes of antibiotics ranged from 43 μg/kg for tiamulin and 11 μg/kg for salinomycin in manure. This research concluded that the instability of the studied antibiotics might have caused the lower concentrations in manure.

An interest in the presence of antibiotics in soil and sediment has continued to grow due to the persistence of antibiotics in solid matrices versus aqueous forms. Hamscher *et al.* [19] documented the persistence of tetracycline residuals in soil fertilized with liquid manure. Tetracycline and chlortetracycline were two of the four studied antibiotics that were detected at up to 199 μg/kg within the top 30 cm of the soil. The remaining two antibiotics, oxytetracycline and tylosin, were not found during the study period. This study emphasized the persistency of tetracyclines under conditions with repeated manure fertilization. In addition, the desorption of tetracyclines was observed during manure storage which likely was due to further degradation of organic material in the soil or variation of the pH and redox potential. In comparison, Jacobsen *et al.* [24] used pressurized liquid extraction (PLE, often called ASE) as the extraction method and LC/ESI/MS/MS technique for separation and quantification of three different classes of antibiotics, tetracyclines, sulfonamides, and macrolides in soil. In addition, the combination of SAX and ion exchange and HLB polymeric cartridges were utilized to enhance the cleanup process of the soil extractants by reducing the anionic humic material. This study applied a developed analytical method to monitor the concentration profile of target antibiotics in soil fertilized once with liquid manure. An observed general degradation over a study period of 146 days was observed for all three classes of antibiotics. Similarly, Halling-Sorensen *et al.* [56] evaluated the dissipation of multiple classes of antibiotics in soil amended with manure as a function of time using ASE combined with LC/ESI/MS/MS technique. The calculated half-life and dissipation rates of the studied antibiotics indicated no significant difference between loamy sandy soil and sand soil. Although, the results of this study revealed that antibiotics introduced into the soil could be dissipated after a certain period, antibiotic residuals continued to remain in the soil. Loffler *et al.* [20,21] used two different methods to determine residuals in river sediments, APCI/MS/MS for 10 acidic pharmaceuticals in the negative mode and ESI/MS/MS for 7 antibiotics in the positive mode. This study illustrated that compounds with varying characteristics can be determined with different ionization methods. Several alternatives to the MS method exist for the detection of pharmaceuticals. For instance, fluorescence detection (FLD) was used for fluoroquinolone determination [57], diode array spectrometry was utilized for tetracycline determination in animal feeds [11], and electrochemical detection was employed for the determination of erythromycin A degradation [58].

Aquaculture can contribute to the direct deposition of antibiotics in the sediment. Excess amount of administered antibiotics in aquaculture can be directly released into the sediment and accumulate over a long period of time. Oxytetracycline (OTC) is one of the common antibiotics used in aquaculture for the treatment of disease. Several researchers have detected OTC residuals in sediments located within areas of active aquaculture. Jacobsen *et al.* [59] observed the decomposition rate of OTC in fish farm sediments to evaluate the potential relationship between OTC concentration and hydrogen sulfide production. The analytical method utilized in this research included buffer extraction (Na2EDTA-Mcllvaine buffer solution) combined with LC/UV. Although no relationship between OTC concentration and hydrogen sulfide production was observed, this study verified that OTC is relatively persistent in anoxic conditions of sediment. Bjorkund *et al.* [60] also determined that OTC is very stable under stagnant anoxic condition and its calculated half-life was 419 days in fish farm sediments. Hektoen *et al.* [61] measured concentrations of OTC as well as six other antibacterial agents in marine sediment. Among the seven antibiotics, oxolinic acid (OA) had the strongest sorbing characteristics followed by OTC. This study also confirmed that the half-life of OTC increases as sediment is covered with additional sediment layers, hence establishing increased anoxic conditions. Recently, Lalumera *et al.* [62] investigated the occurrence and effect of antibiotics in aquaculture sediments and measured OTC and flumequine concentrations of 246.3 and 578.8 μg/kg (d.w.), respectively.

2.2.8 FATE AND TRANSPORT OF ANTIBIOTICS IN SOLID MATRICES

After antibiotics are introduced into solid matrices, the fate and degradation pathway of antibiotics can undergo several mechanisms depending on characteristics of the solid matrices, physical parameters, and physicochemical properties of the compounds. Among various mechanisms affecting fate of antibiotics in solid matrices, the compound's degree of sorption is the primary mechanism in solid matrices. Sorption is typically expressed as a sorption coefficient, K_d, which is the ratio of the compound concentration in the solid phase and the concentration in the aqueous phase at equilibrium. Aqueous phase refers to the dissolved phase and excludes suspended particles [63]. The standard method for evaluating sorption characteristics with varying parameters involves batch sorption experiments. Previous studies have reported the sorption behavior of multiple classes of antibiotics in solid matrices.

Sithole *et al.* [64] assessed the sorption of tetracycline onto peat and humic acid with varied pH and ionic strength. The results of this study indicated that sorption capacities of tetracycline decreased as the pH and ionic strength increased. The impact of pH and ionic strength indicates that hydrogen bonding is involved in the binding of tetracyclines. As pH is increased, the protonated sites on the surface of the peat and humic acid decrease and sodium ions replace the hydrogen bonding sites on the surface of the adsorbant as the ionic strength increases. Although tetracyclines are well-known to complex with metal ions, the effect of this mechanism was minimal in this study due to the acid washing of peat and humic acid for metal ion removal. Similar tetracycline sorption results were observed in research conducted by Figueroa *et al.* [65]. This study also verified that the sorption of tetracyclines were pH and ionic strength dependant in the clay portion of soil. Jones *et al.* [66] evaluated the sorption characteristics of OTC, a compound within the tetracycline group, under varied conditions including: soil texture, cation exchange capacity, and iron oxide contents. In general, higher sorption of OTC was observed as the clay portions and organic contents were increased. Surface complexation with iron also increased the sorption potential of OTC. These results were supported by a previous study that revealed tetracycline interactions with aluminum and iron hydrous oxides [67]. In contrast to the high sorption characteristics of tetracyclines, sulfonamides showed less affinity for sorption to solid matrices leading to higher mobility through solid matrices. The sorption of sulfonamides is mainly due to the presence of aromatic amino group [68] and is also dependant on pH and soil composition [69].

Furthermore, multiple classes of antibiotics were assessed together to compare sorption properties in detail [70–73]. Compound sorption characteristics were highly dependant on the compound properties, the varying chemical properties (pH, ionic strength, etc.) and the solid matrix composition. In addition, several complex processes can be involved in the sorption mechanism of pharmaceuticals in solid matrices. In addition to hydrophobicity, cation exchange, cation bridging, surface complexation, and hydrogen bonding all an important role in retaining pharmaceuticals on a solid matrix [63].

Biodegradation or biotransformation can also play an important role in the fate of antibiotics in solid matrices. Biodegradation is generally defined as molecular degradation of organic substances resulting from the complex action of organisms [74]. In order to assess the biodegradation in solid matrices, biological activity of antibiotics in solid matrices and the relationship between the solid and aqueous phase are required. Consequently, little information is available for biodegradation of antibiotics in solid matrices due to the required complex assessment and the lack of solid phase data in current literature [75–77].

Photolysis or photodegradation is one of the dissipation mechanisms of antibiotics in solid matrices. Two mechanisms, direct hydrolysis and indirect hydrolysis, can be involved in this chemical transformation. Antibiotics undergo direct photolysis when a bond within the light-absorbing molecule is cleaved and indirect photolysis generates a photosensitizer to create radicals and react with the antibiotics [78]. Eichhorn *et al.* [79] identified photooxydation products of chlortetracycline in hog lagoons and Wolters *et al.* [80] evaluated the photochemical transformation of sulfadiazine with various soil surfaces (glass and soil dust) and environmental factors (irradiation and atmospheric ozone). The results of these studies indicate that antibiotics introduced into solid matrices may react with light to undergo direct or indirect photolysis resulting in the formation of degradation products.

Field investigations along with laboratory experiments revealed that antibiotics released into the field are transported into other environmental media by leaching into the groundwater or via surface runoff. Nevertheless, few studies have documented the transport of antibiotics into the field and veterinary antibiotics are primarily discussed as non-point sources. Kay *et al.* [81] emphasized the importance of preferential flow as a transport mechanism of field-applied antibiotics by investigating three groups of antibiotics. Among them, sulfachloropyridazine (SCP) had the highest concentration in drain flow followed by OTC. Tylosin (TYL) was not detected during the two-year study period. These results concur with the known sorption capacities of the three antibiotics. Specifically,TYL was not detected due to the rapid degradation of the compound. In another study, Burkhardt *et al.* [82] evaluated sulfonamide concentrations in surface runoff from soil amended with manure. This study showed that the transport rate of sulfonamides increased in manure-amended soil compared to manure-free soil. These results indicate that the presence of manure is a critical factor controlling the transport of antibiotics in the field. Kay *et al.* [83–85] also proved that overland flow might be the potential transport route of veterinary antibiotics in the field. It has been suggested that watershed management practices should be employed in the treatment of overland flow in order to minimize the exposure of antibiotics in the watershed.

Column leaching experiments in the laboratory are a useful tool to assess the leaching behavior of antibiotics in a controlled setting. The main variables in a column leaching study are rainfall application rate, soil properties (pH, grain size, organic contents), and inhibition of microorganism activity. Similar to surface runoff, leaching behavior is also strongly correlated with the antibiotics' sorption capacity. Several researchers observed the leaching behavior of antibiotics with varied parameters and evaluated the possibility of contamination in groundwater [83–87]. In column studies, aliquots of sub-grade samples are typically analyzed to examine the mobility of antibiotics in detail. The results of the sub-grade analyses indicated that strongly sorbed antibiotics are detected well below grade. These findings imply that colloidal particles could in fact be capable of sorbing with antibiotics. Although current research has not documented colloidal-facilitated transportation of antibiotics, this transport mechanism has been evaluated in pesticide mobility studies [88–93]. Thus, antibiotics with strong sorbing characteristics have the potential to contaminate groundwater located far from the original application source.

2.2.9 CONCLUSION

Antibiotics are referred to as micro pollutants in both water and sediment. The presence of antibiotics in the environment could potentially induce the formation of antibiotic-resistant bacteria. Thus, knowing concentration levels of antibiotics in different environmental media is necessary to understand the extent of the impact on the environment. Furthermore, developing suitable and reliable analytical methods is the primary concern. In particular, analytical methods for quantifying antibiotic residuals in solid matrices has been challenging due to the complexity of matrices. This chapter has succinctly reviewed the methods adapted for analysis of antibiotics in solid matrices and described how these techniques can be used to understand occurrence as well as the fate and transport of antibiotics.

Although research concerning antibiotics in the environment has been thoroughly studied in recent years, there remains a need to develop more precise tools to improve on the risk assessment of pharmaceuticals in the environment. Thus, further research should be conducted to collect additional information on antibiotic occurrence, fate, and transport mechanisms within the environment.

REFERENCES

1 T. Christian, R.J. Schneider, H.A. Farber, D. Skutlarek, M.T. Meyer and H.E. Goldbach, Determination of antibiotics residuals in manure, soil, and surface water, *Acta Hydroc. Hydrobiol.*, 31 (2003) 36–44.
2 M.S. Diaz-Cruz, M.J.L. Alda and D. Barcelo, Environmental behavior and analysis of veterinary and human drugs in soils, sediment and sludge, *TrAC Trend. Anal. Chem.*, 22 (2003) 340–351.
3 S.E. Jorgensen and B. Halling-Sorensen, Drugs in the environment, *Chemosphere*, 40 (2000) 691–699.
4 S. Thiele-Bruhn, Pharmaceutical antibiotic compounds in soil—a review, *J. Plant Nutr. Soil Sci.*, 166 (2003) 145–167.
5 M. Mellon, C. Benbrook and K.L. Benbrook, Hogging it: estimation of antimicrobial abuse in livestock, Union of Concerned Scientists, 2001.
6 T.A. Ternes, Analytical methods for the determination of pharmaceuticals in aqueous environmental samples, *TrAC Trend. Anal. Chem.*, 20 (2001) 419–434.
7 K. Ohlsen, T. Ternes, G. Werner, U. Wallner, D. Loffler, W. Ziebuhr, W. Witte and J. Hacker, Impact of antibiotics on conjugational resistance gene transfer in *Staphylococcus* in sewage, *Environ. Microbiol.*, 5 (2003) 711–716.
8 J.C. Chee-Sanford, R.I. Aminov, I.J. Krapac, N. Garrigues-Jeanjean and R.I. Mackie, Occurrence and diversity of tetracycline resistance genes in lagoons and groundwater underlying two swine production facilities, *Appl. Environ. Microbiol.*, 67(4) (2001) 1494–1502.
9 C. Fernandez, C. Alonso, M.M. Babin, J. Pro, G. Carbonell and J.V. Tarazona, Ecotoxicological assessment of doxycycline in aged pig manure using multispecies soil systems, *Sci. Total Environ.*, 323 (2004) 63–69.
10 H. Schmitt, P.V. Beelan, J. Tolls and C.L.V. Leeuwen, Pollution-induced community tolerance of soil microbial communities caused by the antibiotic sulfachloropyridazine, *Environ. Sci. Technol.*, 38 (2004) 1148–1153.
11 R.D. Caballero and J.R.T. Lapasio, Rapid liquid chromatographic determination of tetracyclines in animal feeds using a sulfactant solution as mobile phase, *Anal. Lett.*, 35(4) (2002) 687–705.
12 D.S. Aga, S. O'connor, S. Ensley, J.O. Payero, D. Snow and D. Tarkalson, Determination of the persistance of tetracycline antibiotics and their degradates in manure-amended soil using enzyme-linked immunosorbent assay and liquid chromatography-mass spectrometry, *J. Agric. Food Chem.*, 53 (2005) 7165–7171.
13 S. Bogialli, R. Curini, A.D. Corcia, M. Nazzari and R. Samperi, A liquid chromatography-mass spectrometry assay for analyzing sulfonamide antibacterials in cattle and fish tissues, *Anal. Chem.*, 75 (2003) 1798–1804.
14 M.Y. Haller, S.R. Muller, C.S. McArdell, A.C. Alder and M.J.-F. Suter, Quantification of veterinary antibiotics (sulfonamides and trimethoprim)

in animal manure by liquid chromatography-mass spectrometry, *J. Chromatogr. A*, 952 (2002) 111–120.

15 C.S. McArdell, E. Molnar, M.J.F. Suter and W. Giger, Occurrence and fate of macrolide antibiotics in wastewater treatment plants and in the Glatt valley watershed, Switzerland, *Environ. Sci. Technol.*, 37 (2003) 5479–5486.

16 P. Edder, L. Coppex, A. Cominoli and C. Corvi, Analysis of erythromycin and oleandomycin residues in food by high-performance liquid chromatography with fluorometric detection, *Food Addit. Contam.*, 19(3) (2002) 232–240.

17 D.A. Volmer and C.M. Lock, Electrospray ionization and collison-induced dissociation of antibiotic polyether ionophores, *Rapid Commun. Mass Sp.*, 12 (1998) 157–164.

18 S.B. Turnipseed, J.E. Roybal, A.P. Pfenning, S.A. Gonzales, J.A. Hurlbut and M.R. Madson, LC/MS confirmation of ionophores in animal feeds, *J. AOAC Int.*, 84(3) (2001) 640–647.

19 G. Hamscher, S. Sczesny, H. Hoper and H. Nau, Determination of persistent tetracycline residues in soil fertilized with liquid manure by high performance liquid chromatography with electrospray ionization tandem mass spectrometry, *Anal. Chem.*, 74 (2002) 1509–1518.

20 D. Loffler and T.A. Ternes, Analytical method for the determination of the aminoglycoside gentamicin in hospital wastewater via liquid chromatography-elcetrospray-tandem mass spectrometry, *J. Chromatogr. A*, 1000 (2003) 583–588.

21 D. Loffler and T.A. Ternes, Determination of acidic pharmaceuticals, antibiotics and ivermectin in river using liquid chromatography-tandem mass spectrometry, *J. Chromatogr. A*, 1021 (2003) 133–144.

22 I. Ferrer and E.T. Furlong, Accelerated solvent extraction followed by on-line solid-phase extraction coupled to ion trap LC/MS/MS for analysis of Benzalkonium chlorides in sediment samples, *Anal. Chem.*, 74 (2002) 1275–1280.

23 A. Hubert, K.D. Wenzel, M. Manz, L. Weissflog, W. Engewald and G. Schuurmann, High extraction efficiency for POPs in real contaminated soil samples using accelerated solvent extraction, *Anal. Chem.*, 72 (2000) 1294–1300.

24 A.M. Jacobsen, B. Halling-Sorensen, F. Ingerslev and S.H. Gansen, Simultaneous extraction of tetracycline, macrolide and sulfonamide antibiotics from agricultural soils using pressurised liquid extraction, followed by solid-phase extraction and liquid chromatography-tandem mass spectrometry, *J. Chromatogr. A*, 1038 (2004) 157–170.

25 R. Hirsch, T.A. Ternes, K. Haberer, A. Mehlich, F. Ballwanz and K.L. Kratz, Determination of antibiotics in different water compartment via liquid chromatography-electrospray tandem mass spectrometry, *J. Chromatogr. A*, 815 (1998) 213–223.

26 R. Hirsch, T. Ternes, K. Hanerer and K.L. Kratz, Occurrence of antibiotics in the aquatic environment, *Sci. Total Environ.*, 225 (1999) 109–118.

27 M.E. Lindsey, M. Meyer and E.M. Thurman, Analysis of trace levels of sulfonamides and tetracycline antimicrobials in groundwater and surface water using solid-phase extraction and liquid chromatography/mass spectrometry, *Anal. Chem.*, 73 (2001) 4640–4646.

28 J. Zhu, D.D. Snow, D.A. Cassada, S.J. Monson and R.F. Spalding, Analysis of oxytetracycline, tetracycline, and chlorotetracycline in water using solid-phase extraction and liquid chromatography tandem mass spectrometry, *J. Chromatogr. A*, 928 (2001) 177–186.

29 D.W. Kolpin, E.T. Furlong, M.T. Meyer, E.M. Thurman, S.D. Zaugg, L.B. Barber and H.T. Buxton, Pharmaceuticals, hormones, and other organic wastewater contaminants in U.S. streams, 1999-2000: a National Reconnaissance, *Environ. Sci. Technol.*, 36 (2002) 1202–1211.

30 M.P. Schlusener, K. Bester and M. Spiteller, Determination of antibiotics such as macrolide, ionophores and tiamulin in liquid manure by HPLC-MS/MS, *Anal. Bioanal. Chem.*, 375 (2003) 942–947.

31 S. Wiegel, A. Aulinger, R. Brokmeyer, H. Harms, J. Loffler, H. Reincke, R. Schmidt, B. Stachel, W. Tumpling and A. Wanke, Pharmaceuticals in the river Elbe and its tributaries, *Chemosphere*, 57 (2004) 107–126.

32 S. Yang and K. Carlson, Solid-phase extraction-high-performance liquid chromatography-ion trap mass spectrometry for analysis of trace concentrations of macrolide antibiotics in natural and waste water matrices, *J. Chromatogr. A*, 1038 (2004) 141–155.

33 S. Yang, J. Cha and K.H. Carlson, Quantitative determination of trace concentrations of tetracycline and sulfonamide antibiotics in surface water using solid-phase extraction and liquid chromatography/ion trap tandem mass spectrometry, *Rapid Commun. Mass Spectrom.*, 18 (2004) 2131–2145.

34 B. Gross, J. Montgomery-Brown, A. Naumann and M. Reinhard, Occurrence and fate of pharmaceuticals and alkylphenol ethoxylate metabolites in an effluent-dominated river and wetland, *Environ. Toxicol. Chem.*, 23 (2004) 2074–2083.

35 T. Heberer, Occurrence, fate, and removal of pharmaceutical residues in the aquatic environment: a review of recent research data, *Toxicol. Lett.*, 131 (2002) 5–17.

36 T. Heberer, Tracking persistent pharmaceutical residuals from municipal sewage to drinkng water, *J. Hydrol.*, 266 (2002) 175–189.

37 S. Heinisch and J.L. Rocca, Effect of mobile phase composition, pH and buffer type on the retention of ionizable compounds in reversed-phase liquid chromatography: application to method development, *J. Chromatogr. A*, 1048 (2004) 183–193.

38 X. Ding and S. Mou, Ion chromatographic analysis of tetracyclines using polymeric column and acidic eluent, *J. Chromatogr. A*, 897 (2000) 205–214.
39 C. Zwiener and F.H. Frimmel, LC-MS analysis in the aquatic environment and in water treatment—a critical review Part1: instrumentation and general aspects of analysis and detection, *Anal. Bioanal. Chem.*, 378 (2004) 851–861.
40 C. Zwiener and F.H. Frimmel, LC-MS analysis in the aquatic environment and in water treatment—a critical review Part2: applications for emerging contaminants and related pollutants, microorganisms and humic acids, *Anal. Bioanal. Chem.*, 378 (2004) 862–874.
41 W.M.A. Niessen, Progress in liquid chromatography-mass spectrometry instrumentation and its impact on high-throughput screening, *J. Chromatogr. A*, 1000 (2003) 413–436.
42 I. Ferrer and E.M. Thurman, Liquid chromatography/time-of-fligh/mass spectrometry (LC/TOF/MS) for the analysis of emerging contaminants, *TrAC Trend. Anal. Chem.*, 22 (2003) 750–756.
43 I. Ferrer, C.E. Heine and E.M. Thurman, Combination of LC/TOF-MS and LC/Ion trap MS/MS for the identification of diphenhydramine in sediment samples, *Anal. Chem.*, 76 (2004) 1437–1444.
44 X.S. Miao and B.G. Koenig, Analysis of acidic drugs in the effluents of sewage treatment plants using liquid chromatography-electrospray ionization mass spectrometry, *J. Chromatogr. A*, 952 (2002) 139–147.
45 C.G. Barreiro and M. Lores, Simultaneous determenation of neutral and acdic pharmaceuticals in wastewater by high-performance liquid chromatography-post-column photochemically induced fluorimetry, *J. Chromatogr. A*, 993 (2003) 29–37.
46 K. Stamatelatou, C. Frouda, M.S. Fountoulakis, P. Rdillia, M. Kornaros and G. Lyberatos, Pharmaceuticals and health care products in wastewater effluents: the example of carbamazepine, *Water Sci. Technol.*, 3 (2003) 131–137.
47 B.J. Vanderford, R.A. Pearson, D.J. Rexing and S.A. Snyder, Analysis of endocrine disruptors, pharmaceuticals, and personal care products in water using liquid chromatography/tandem mass spectrometry, *Anal. Chem.*, 75 (2003) 6265–6274.
48 R. Andreozzi, V. Caprio, C. Ciniglia, M.D. Champdore, R.L. Giudice, R. Marotta and E. Zuccato, Antibiotics in the environment: occurrence in Italian STPs, fate, and preliminary assessment on the toxicity of Amoxicillin, *Environ. Sci. Technol.*, 38 (2004) 6832–6838.
49 P.A. Blackwell, H.H. Lutzhoft, H.P. Ma, B. Halling-Sorensen, A.B.A. Boxwell and P. Kay, Fast and robust simultaneous determination of three veterinary antibiotics in groundwater and surface water using a tandem solid-phase extraction with high-performance liquid chromatography-UV detection, *J. Chromatogr. A*, 1045 (2004) 111–117.

50 P.A. Blackwell, H.H. Lutzhoft, H.P. Ma, B. Halling-Sorensen, A.B.A. Boxwell and P. Kay, Ultrasonic extraction of veterinary antibiotics from soils and pig slurry with SPE clean-up and LC-UV and fluorescence detection, *Talanta*, 64 (2004) 1058–1064.
51 A. Gobel, C.S. McArdell, M.J.F. Suter and W. Giger, Trace determination of macrolide and sulfonamide antimicrobials, a human sulfonamide metabolite, and trimethoprim in wastewater using liquid chromatography coupled to electrospray tandem mass spectrometry, *Anal. Chem.*, 76 (2004) 4756–4764.
52 M. Petrović, S. Gonzalez and D. Barcelo, Analysis and removal of emerging contaminants in wastewater and drinking water, *TrAC Trend. Anal. Chem.*, 22 (2003) 685–696.
53 J.P. Bound and N. Voulvoulis, Pharmaceuticals in the aquatic environment—a comparison of risk assessment strategies, *Chemosphere*, 56 (2004) 1143–1155.
54 K. Xia, A. Bhandari, K. Das and G. Pillar, Occurrence and fate of pharmaceuticals and personal care products (PPCPs) in biosolids, *J. Environ. Qual.*, 34 (2005) 91–104.
55 E.R. Campagnolo, K.R. Johnson, A. Karpati, C.S. Rubin, D.W. Kolpin, M.T. Meyer and J.E. Esteban, Antimicrobial residuals in animal waste and water resources proximal to large-scale swine and poultry feeding operations, *Sci. Total Environ.*, 299 (2002) 89–95.
56 B. Halling-Sorensen, A.M. Jacobsen, J. Jensen, G. Sengelov, E. Vaclavik and F. Ingerslev, Dissipation and effects of chlortetracycline and tylosin in two agricultural soils: a field-scale study in southern Denmark, *Environ. Toxicol. Chem.*, 24(4) (2005) 802–810.
57 E.M. Golet, A. Strehler, A.C. Alder and W. Giger, Determiniation of fluoroquinolone antibioterial agents in sewage sludge-treated soil using accelerated solvent extraction followed by solid-phase extraction, *Anal. Chem.*, 74 (2002) 5455–5462.
58 Y.H. Kim, T.M. Heinze, S.J. Kim and C.E. Cerniglia, Adsorption and clay-catalyzed degradation of erythromycin A on homoionic clays, *J. Environ. Qual.*, 33 (2004) 257–264.
59 P. Jacobsen and L. Berflind, Persistance of oxytetracycline in sediment from fish farm, *Aquaculture*, 70 (1988) 365–370.
60 H. Bjorklund, J. Bondest and G. Bylund, Residues of oxytetracycline in wild fish and sediments from fish farms, *Aquaculture*, 86 (1990) 359–367.
61 H. Hektoen, J.A. Berge and V. Hormazabal, Persistence of antibacterial agents in marine sediments, *Aquaculture*, 133 (1995) 175–184.
62 G.M. Lalumera, D. Calamari, P. Galli, S. Castiglioni, G. Crosa and R. Fanelli, Preliminary investigation on the environmental occurrence and effects of antibiotics used in aquaculture in Italy, *Chemosphere*, 54 (2004) 661–668.

63 J. Tolls, Sorption of veterinary pharmaceuticals in soils: a review, *Environ. Sci. Technol.*, 35 (2001) 3397–3406.
64 B.B. Sithole and R.D. Guy, Models for tetracycline in aquatic environment, *Water Air Soil Pollut.*, 32 (1987) 315–321.
65 R.A. Figueroa, A. Leonard and A.A. Mackay, Modeling tetracycline antibiotics sorption to clays, *Environ. Sci. Technol.*, 38 (2004) 476–483.
66 A.D. Jones, G.L. Bruland, S.G. Agrawal and D. Vasudevan, Factors influencing the sorption of oxytetracycline to soils, *Environ. Toxicol. Chem.*, 24(4) (2005) 761–770.
67 C. Gu and K.G. Karthikeyan, Interaction of tetracycline with aluminum and iron hydrous oxides, *Envir. Sci. Technol.*, 39(8) (2005) 2660–2667.
68 S. Thiele, Adsorption of the antibiotic pharmaceutical compound sulfapyridine by a long-term differently fertilized loess chernozem, *J. Plant Nutr. Soil Sci.*, 163 (2000) 589–594.
69 A.B.A. Boxall, P. Blackwell, R. Cavallo, P. Kay and J. Tolls, The sorption and transport of sulfonamide antibiotic in soil system, *Toxicol. Lett.*, 131 (2002) 19–28.
70 M. Rabolle, Sorption and mobility of metronidazole, olaquindox, oxytetracyclune and tylosin in soil, *Chemosphere*, 40 (2000) 715–722.
71 M.L. Loke, J. Tjornelund and B. Halling-Sorensen, Determination of the distribution coefficient (log Kd) of oxytetracycline, tylosin A, olaquindox and metronidazole in manure, *Chemosphere*, 48 (2002) 351–361.
72 T.A. Ternes, N. Herrmann, M. Bonerz, T. Knacker, H. Siegrist and A. Joss, A rapid method to measure the solid-water distribution coefficient (K_d) for pharmaceuticals and musk fragnance in sewage sludge, *Water Res.*, 38 (2004) 4075–4084.
73 A. Gobel, A. Thomsen, C.S. McArdell, A. Joss and W. Giger, Occurrence and sorption behavior of sulfonamides, macrolides, and trimethoprim in activated sludge treatment, *Environ. Sci. Technol.*, 39(11) (2005) 3981–3989.
74 J. Zagorc-Koncan and M. Dular, Evaluation of biodegradation kinetics on organic wastewater in a laboratory river model, *Water Sci. Technol.*, 30(10) (1994) 229–235.
75 M. Winkler, J.R. Lawrence and T.R. Neu, Selective degradation of ibuprofen and clofibric acid in two model river biofilm systems, *Water Res.*, 35(1) (2001) 3197–3205.
76 C. Zwiener and F.H. Frimmel, Short-term tests with a pilot sewage plant and biofilm reactors for the biological degradation of the pharmaceutical compounds clofibric acid, ibuprofen, and diclofenac, *Sci. Total Environ.*, 309 (2003) 201–211.
77 R. Alexy, T. Kumpel and K. Kummerer, Assessment of degradation of 18 antibiotics in the closed bottle test, *Chemosphere*, 57 (2004) 505–512.

78 C.J. Gurr and M. Reinhard, Harnessing natural attenuation of pharmaceuticals and hormones in rivers, *Envir. Sci. Technol.*, (2006) 2872–2876.
79 P. Eichhorn and D.S. Aga, Identification of a photooxygeneration product of chlortetracycline in hog lagoons using LC/ESI-ion trap-MS and LC/ESI-time-of-flight MS, *Anal. Chem.*, 76 (2004) 6002–6011.
80 A. Wolters and M. Steffens, Photodegradation of antibiotics on soil surfaces: laboratory studies on sulfadiazine in an ozone-controlled environment, *Environ. Sci. Technol.*, 39 (2005) 6071–6078.
81 P. Kay, P.A. Blackwell and A.B.A. Boxall, Fate of veterinary antibiotics in a macroporous tile drained clay soils, *Environ. Toxicol. Chem.*, 23 (2004) 1136–1144.
82 M. Burkhardt, C. Stamm, C. Waul, H. Singer and S. Muller, Surface runoff and transport of sulfonamide antibiotics and tracers on manured grassland, *J. Environ. Qual.*, 34 (2005) 1363–1371.
83 P. Kay, P.A. Blackwell and A.B.A. Boxall, Column studies to investigate the fate of veterinary antibiotics in clay soils following slurry application to agricultural land, *Chemosphere*, 60 (2005) 497–507.
84 P. Kay, P.A. Blackwell and A.B.A. Boxall, A lysimeter experiment to investigate the leaching of veterinary antibiotics through a clay soil and comparison with field data, *Environ. Pollut.*, 134 (2005) 333–341.
85 P. Kay, P.A. Blackwell and A.B.A. Boxall, Transport of veterinary antibiotics in overland flow following the application of slurry to arable land, *Chemosphere*, 59 (2005) 951–959.
86 J. Oppel, G. Broll, D. Loffler, M. Meller, J. Rombke and T. Ternes, Leaching behaviour of pharmaceuticals in soil-testing-systems: a part of an environmental risk assessment for groundwater protection, *Sci. Total Environt.*, 328 (2004) 265–273.
87 P. Drillia, K. Stamatelatou and G. Lyberatos, Fate and mobility of pharmaceuticals in solid matrices, *Chemosphere*, 60 (2005) 1034–1044.
88 V. Chaplain and P. Mills, Simulation of the migration of colloidal particles in porous media: application to the transport of contaminants, *Sci.Total Environ.*, 123/124 (1992) 451–457.
89 H.D. Jonge, O.H. Jacobsen, L.W. Jonge and P. Moldrup, Colloid-faciliated transport of pesticide in undisturbed soil columns, *Phys. Chem. Earth*, 23 (1998) 187–191.
90 F. Worrall, A. Parker, J.E. Rae and A.C. Johnson, A study of suspended and colloidal matter in the leachate from lysimeters and its role in pesticide transport, *J. Environ. Qual.*, 28 (1999) 595–604.
91 L.A. Sprague, J.S. Herman, G.M. Hornberger and A.L. Mills, Atrazine adsorption and colloid-facilitated transport through the unsaturated zone, *J. Environ. Qual.*, 29 (2000) 1632–1641.
92 K.G. Villholth, N.J. Jarvis, O.H. Jacobsen and H.D. Jonge, Field investigations and modeling of particle-facilitated pesticide transport in macroporous soil, *J. Environ. Qual.*, 29 (2000) 1298–1309.

93 D. Grolimund and M. Borkovec, Colloid-facilitated transport of strongly sorbing contaminants in natural porous media: mathematical modeling and laboratory column experiments, *Envir. Sci. Technol.*, 39 (2005) 6378–6386.
94 S.T. Glassmeyer, E.T. Furlong, D.W. Kolpin, J.D. Cahill, S.D. Zaugg, S.L. Werner, M.T. Meyer and D.D. Kryak, Transport of chemicals and microbial compounds from known wastewater discharges: potential for use as indicators of human fecal contamination, *Environ. Sci. Technol.*, 39 (2005) 5157–5169.
95 S.C. Kim and K. Carlson, LC-MS^2 for quantifying trace amounts of pharmaceutical compounds in soil and sediment matrices, *TrAC Trend. Anal. Chem.*, 24 (2005) 635–644.
96 M.D. Liguoro, V. Cibin, F. Capolongo, B. Halling-sorensen and C. Montesissa, Use of oxytetracycline and tylosin in intensive calf farming: evaluation of transfer to manure and soil, *Chemosphere*, 52 (2003) 203–212.

Chapter 2.3

Analysis of neutral and acidic pharmaceuticals by liquid chromatography mass spectrometry

Xiu-Sheng Miao and Chris D. Metcalfe

2.3.1 INTRODUCTION

Pharmaceuticals are used in large quantities throughout the world as prescription and non-prescription drugs. In developed countries, annual consumption rates of prescribed pharmaceuticals range from a few kilograms to hundreds of metric tons. The consumption of drugs purchased without a prescription is at least an order of magnitude greater than the amounts of prescription drugs consumed. For instance, patterns of use of non-prescription drugs in the United States [1] indicate that analgesics sold over-the-counter (e.g., ASA, ibuprofen and acetaminophen) are the most highly used drugs, followed by decongestants and antihistamines.

The first data on pharmaceutical residues in the environment were mainly focused on clofibric acid, the active metabolite of three lipid regulators. Garrison et al. [2] and Hignite and Azarnoff [3] detected clofibric acid at low μg/L concentrations in treated wastewater from North America. The first studies of the occurrence of pharmaceuticals in the environment in Europe were reported by Watts et al. [4], Waggott [5] and Richardson and Bowron [6] from investigations conducted in the United Kingdom. These studies revealed that several acidic drugs that are purchased both with and without prescription were present in the aquatic environment at concentrations up to approximately 1 μg/L. To a certain extent, the focus on acidic drugs in these early studies was due to the applicability of well known analytical techniques that involved derivatization of the analytes, followed by analysis using gas chromatography mass spectrometry (GC-MS). GC-MS analysis of derivatized compounds continued to be the analytical technique used for acidic

Comprehensive Analytical Chemistry 50
Petrović and Barceló (Eds)
Volume 50 ISSN: 0166-526X DOI: 10.1016/S0166-526X(07)50004-8

drugs until relatively recently [7,8]. However, instruments for liquid chromatography and mass spectrometry (LC-MS and LC-MS/MS) are now widely available in analytical laboratories and techniques with high sensitivity and reproducibility have been developed for the analysis of acidic drugs [9,10]. LC-MS and LC-MS/MS techniques have also been applied to the analysis of pharmaceuticals that are neutral at the pH of natural waters, but are still highly polar compounds [10–17].

Table 2.3.1 provides a summary of the classes of acidic and neutral drugs that have been detected using LC-MS and LC-MS/MS analytical techniques in water and wastewater in both North America and Europe. Acidic drugs that are used as anti-inflammatories and/or analgesics and which are either prescribed or available over the counter have been detected frequently in municipal wastewater and in surface water [10,12–17]. A variety of lipid-regulating agents from the ‘‘fibrate’’ class have also been identified in wastewater and surface water by investigators in Europe and North America [10,12–16], and lipid regulators from the ‘‘statin’’ class, which are highly prescribed in North America, were detected in wastewater and surface water in Canada [13,17]. The neutral anti-epileptic drug, carbamazepine has been widely detected in wastewater, surface water and groundwater in both Europe and North America [10,11,13,18]. In Canada, several metabolites of carbamazepine were detected in wastewater and a hydroxy-metabolite of carbamazepine (i.e., 10,11-dihydro-10,11-dihydroxycarbamazepine) was detected in surface water [19]. Neutral drugs used in psychopharmacotherapy, including diazepam and fluoxetine have also been detected in wastewater and surface water [10,12–14]. Fluoxetine is a weak base, but is neutral at the pH of natural waters. Trimethoprim, which is a neutral antibiotic that is commonly prescribed in combination with antimicrobials from the sulfonamide class has been frequently detected in both Europe and North America [10–15]. The anti-asthmatic drug, pentoxyfylline and some cytostatic drugs used in chemotherapy (e.g., cyclophosphamide) have occasionally been detected in wastewater and surface water, but typically at low concentrations [10,11,13]. The stimulant, caffeine and its metabolite dimethylxanthine, and a metabolite of nicotine, cotinine have been widely detected in wastewater and in surface water [10–14]. Finally, drugs from the phenazone class that are prescribed as analgesics, anti-inflammatories and antipyretics have been detected in wastewater and surface water in Europe [20].

The appearance of commercially available liquid chromatography mass spectrometry (LC-MS, LC-MS/MS) instruments with sufficient sensitivity to detect compounds in aqueous samples at ng/L

TABLE 2.3.1

Acidic and basic/neutral pharmaceuticals that have been detected in wastewater, surface water and groundwater in Europe and North America

Pharmaceutical class	Therapeutic applications	Examples
Acidic drugs		
Analgesic/anti-inflammatory drugs	*Non-prescription*: treatment of colds, allergies, pain *Prescription*: treatment of chronic pain, arthritis, migraines, etc.	ASA (aspirin), ibuprofen, acetaminophen Indomethacin, naproxen, diclofenac
Lipid-regulating drugs	Reduce blood cholesterol	*Fibrate drugs*: clofibric acid, gemfibrozil, bezafibrate *Statin drugs*: atorvastatin
Neutral/base drugs		
Anti-epileptic drugs	Anti-convulsant	Carbamazepine
Psychiatric drugs	Psychopharmacotherapy, antidepressant	Fluoxetine, diazepam
Cytostatic drugs	Cancer chemotherapy	Ifosfamide, cyclophosphamide
Antibiotic	Treatment of bacterial diseases; generally in association with a sulfonamide antibiotic	Trimethoprim
Bronchodilator	Treatment of asthma and respiratory diseases	Pentoxyfylline
Analgesics, antipyretics and anti-inflammatories	Treatment of chronic pain, fevers and arthritis	Propyphenazone, phenazone, phenylbutazone and metabolites
Human use compounds	Stimulants, etc.	Caffeine and metabolite (methylxanthine), cotinine (metabolite of nicotine)

concentrations can be attributed to several advances in instrument design, including the development of ionization sources that can operate at atmospheric pressure in the interface between the LC and MS. These include electrospray ionization (ESI) and atmospheric pressure chemical ionization (APCI) ion sources. These instruments have the capacity to simultaneously monitor selected product ions in either positive or negative ion mode [9]. However, ESI is susceptible to modifications to ionization efficiency as a result of interference from co-extractives in samples prepared from various matrices (i.e., "matrix effects"). The more complex the sample matrix, the greater the potential for these effects. For analysis of pharmaceuticals in environmental samples, suppression of the analyte signal has been commonly reported when samples prepared from complex matrices are introduced into the ESI source. However, enhancement of ionization may also occur when certain pharmaceuticals are introduced into this source within a complex matrix. Many investigations have focused on determining the mechanisms responsible for matrix effects with ESI [21–25]. Figure 2.3.1 illustrates the signal produced by three pharmaceuticals that have been spiked into an extract prepared from surface water in comparison to the signal for the same compounds spiked into solvent. These data indicate that enhancement, suppression and no effect on ionization can be observed for individual compounds analyzed in the same chromatographic run. LC-MS with APCI has been used to analyze caffeine in surface waters [26], and LC-MS/MS with APCI has been used to analyze neutral drugs in environmental matrices [10,20]. Zhao and Metcalfe [27] investigated the performance of APCI ionization

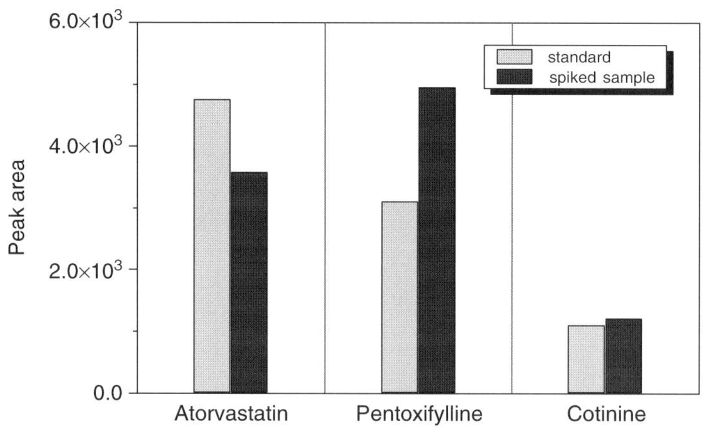

Fig. 2.3.1. The area of peaks generated by LC-ESI-MS/MS analysis (positive ion mode) of three drug analytes spiked into solvent (i.e., standard) and spiked into an extract prepared from surface water (i.e., spiked sample). The figure illustrates signal suppression for atorvastatin, signal enhancement for pentoxyfylline and no effect on the signal for cotinine.

sources for the LC-MS/MS analysis of neutral drugs and observed signal enhancement in complex matrices, rather than signal suppression.

When using LC-MS/MS instrumentation, the protonated or deprotonated precursor ions fragment into different product ions, with the pattern of fragmentation depending on the conditions in the collision cell. In order to achieve the high sensitivity required to detect ng/L concentrations of neutral and acidic drugs in environmental samples, one to three ions are typically selected through multiple reaction monitoring (MRM) for quantitation and confirmation. Typically, the protonated or deprotonated molecular ion is chosen as one of the transition ions. Combined with monitoring of the chromatographic retention times of the analytes, MRM using two or three transition ions provides acceptable specificity, while still maintaining analytical sensitivity. Figure 2.3.2 illustrates chromatograms generated by LC-ESI-MS/MS-MRM analysis of several neutral drugs in an analytical standard and in surface water samples in which pairs of transition ions were monitored in positive ion mode for each analyte.

For analysis of pharmaceuticals, there are several choices of calibration method, which vary in their ability to compensate for matrix

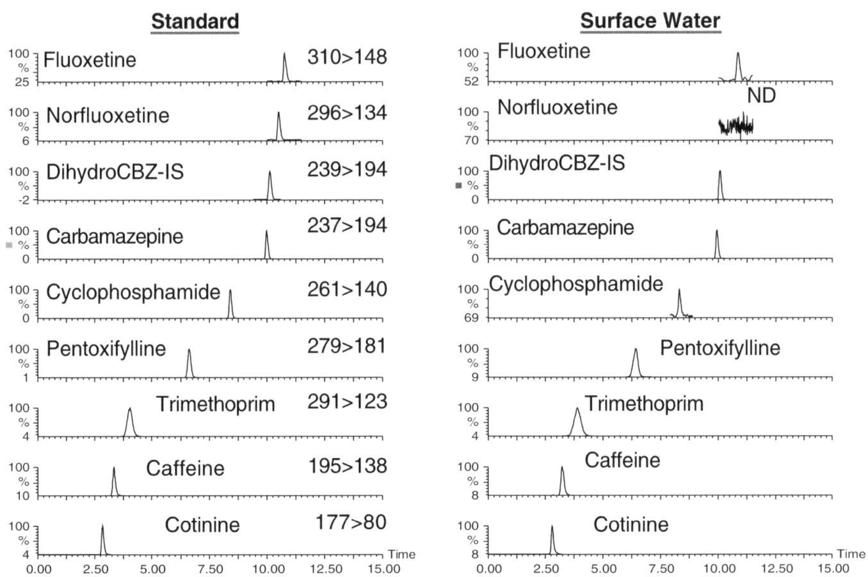

Fig. 2.3.2. LC-ESI-MS/MS-MRM chromatograms of neutral drugs in an analytical standard and surface water sample, showing the pairs of transition ions monitored in positive ion mode. Dihydrocarbamazepine (DihydroCBZ) was added as an internal standard (IS).

effects when analyzing environmental samples. Calibration using internal standards typically requires the use of isotopically labeled surrogates, which, until recently have not been available for most acidic and neutral pharmaceuticals detected in environmental samples. Calibration with an external standard can be influenced by the matrix of the solution in which the standard is dissolved. In environmental analysis, it is almost impossible to match the matrix of the external standard to the matrix in a sample, especially for the analysis of complex matrices, such as wastewater, biosolids (i.e., sewage sludge) or soil. Stüber and Reemtsma [21] recommended standard additions as a calibration method for samples containing highly complex matrices if isotope-labeled surrogates are not available. This calibration technique has been used for the analysis of neutral and acidic pharmaceuticals in water and wastewater [13,16,17,19]. However, calibration by standard additions is time-consuming and laborious because of the large numbers of spiked subsamples that must be processed and analyzed. The recent appearance of several stable isotope surrogates from commercial sources promises to improve the analytical precision for pharmaceuticals in environmental matrices. All data reported for pharmaceuticals in environmental samples that were generated using calibration techniques that did not involve the use of stable isotope surrogates or standard additions should be regarded with caution because of the potential for over- or under-estimation of concentrations because of matrix effects.

The influence of the sample matrix on the LC-MS/MS signal intensity is especially problematic when analyzing very complex environmental matrices, such as soils, sediments, biosolids and biota samples. Despite these analytical challenges and the problems associated with extracting and separating target analytes from the matrix during sample preparation, methods are now being published for the analysis of pharmaceuticals in biosolids [28], soils and sediments [29] and in biota [30–32].

2.3.2 ANALYTICAL TECHNIQUES

2.3.2.1 Sample collection and storage

Samples of water, wastewater, biosolids, sediments and soils are usually collected in glass containers, which are cleaned with organic solvents prior to use. Pharmaceuticals are relatively unstable compared to persistent organic pollutants (e.g., PCBs). Therefore, samples are typically transported to the laboratory from the collection site under cool conditions and in the dark. Samples of particulate material (e.g., biosolids, soil, sediment) have typically been stored frozen in conventional or in ultra-low (−80°C) temperature freezers. However, little work has been done to evaluate the stability of neutral and acidic drugs in particulate samples under these conditions. Typically, aqueous samples to be analyzed for neutral and acidic drugs have been stored for less than 48 h before processing. However, there are few data on the stability of pharmaceuticals in aqueous samples. Lee et al. [33] found that acidic drugs in domestic wastewater were relatively stable for periods of up to a week when the samples were stored in the dark at 4°C. Vanderford et al. [10] observed that recoveries of trimethoprim, acetaminophen and fluoxetine spiked into surface water declined rapidly over one week of storage at 4°C, but recoveries of caffeine, ibuprofen and diazepam stayed relatively constant over this time period. These authors reported that preservation of water samples with formaldehyde (1% v/v) reduced the recoveries of some drugs (e.g., acetaminophen), but adjustment of pH to 2 with sulfuric acid prevented degradation of all analytes over 14 days.

The pH will influence the speciation of acidic or basic drugs in aqueous solutions. Weak acids will be most soluble in the ionic form in solutions with a pH at least two units above the pKa (>99% ionized). However, at very high pH, acidic drugs will exist as salts. In general, the pH of natural waters or wastewater is appropriate for sample storage and no pH adjustment is necessary.

2.3.2.2 Sample preparation

2.3.2.2.1 Aqueous samples

Aqueous samples should be filtered (e.g., 0.45 μm glass fibre) or centrifuged to remove suspended materials prior to extraction of neutral or acidic drugs. The removal of suspended particulates will reduce the clogging of solid phase extraction (SPE) cartridges, and will improve analyte recoveries if other extraction methods (e.g., liquid/liquid partitioning) are used. While it has been assumed that hydrophilic acidic and neutral pharmaceuticals will not be retained to any appreciable extent on the filters, there have been few published reports on whether this is a valid assumption. Table 2.3.2 provides previously unpublished data on the amounts of selected acidic and neutral drugs adsorbed on suspended particulates centrifuged from samples of surface runoff relative to the amount dissolved in the aqueous phase and extracted

TABLE 2.3.2

Relative proportions (%) of neutral and acidic drug analytes retained on suspended particulate material (removed by centrifugation) and in the aqueous phase of samples (2 L) of runoff from an agricultural field after application of biosolids from a municipal wastewater treatment plant

Compound	Aqueous phase (%)	Suspended particulates (%)
Ibuprofen	100	ND
Carbamazepine	99.2	0.8
Cotinine	99.6	0.4
Caffeine	86.0	14.0

using SPE cartridges. These data indicate that a small proportion of acidic and neutral drugs are adsorbed onto the particulate material, indicating that the dissolved phase is the most important matrix for the distribution of these pharmaceuticals in water samples.

Solid-phase extraction (SPE)

Owing to the low concentrations of pharmaceuticals in the aquatic environment, enrichment of the analytes is necessary prior to the detection. In most instances, extraction and enrichment of neutral and acidic drugs has been performed by SPE techniques. SPE is an attractive method because it is relatively easy and rapid, requires minimal amounts of solvent, and can be tailored to a broad range of compounds. The SPE absorbent materials that have been used for acidic and neutral drugs include octadecylsilica, polymeric or hydrophilic–lipophilic balanced (HLB) stationary phases. Octadecyl (C18)-bonded silica SPE products have been widely employed [7,8,34]. However, the HLB Oasis® (60 or 200 mg) product is now the most widely used SPE cartridge because this absorbent can extract both acidic and neutral drugs with high efficiencies [10,13–17]. Immunoaffinity SPE techniques, which have been used to concentrate pharmaceuticals in biomedical applications [35], may also be useful solid phases for concentrating pharmaceuticals in environmental samples.

Prior to extraction, SPE cartridges should be conditioned by successive elution with less to more polar solvents, such as *n*-hexane, acetone, methanol and high-purity water (acidified in the case of extraction of acidic drugs). For the SPE extraction of acidic drugs, the pH of the solution should be reduced to below the pKa of the acidic analytes to maximize adsorption in the SPE cartridge. Typically, 3.5 M sulfuric acid or hydrochloric acid is used to adjust to pH 2–4, depending on the pKa values of the analytes. If surrogate standards are used, they should be spiked into the aqueous samples after pH adjustment (if necessary), and prior to extraction.

The volumes of aqueous samples extracted by SPE typically range from 50–1000 mL, depending on the concentrations of the target analytes, the size of the SPE cartridge and the complexity of the sample matrix. Typically, the smallest volume possible for detecting the analytes should be used in order to minimize matrix effects that could impact LC-MS or LC-MS/MS analysis. The samples are passed through the cartridges at flow rates of 3–20 mL/min, after which the cartridges are washed with rinsings from the sample containers and/or pure water. Excess water is removed from the cartridges by vacuum and then a stream of nitrogen. Several researchers have stored dried SPE cartridges in a freezer and in some cases, shipped the frozen cartridges to another laboratory prior to elution. While it is assumed that neutral and acidic drugs are relatively stable under these conditions, this assumption has not been adequately tested. The cartridges are eluted with a polar solvent; usually with three separate aliquots of methanol. The sample volume is then reduced with a gentle nitrogen stream or by vacuum evaporation. The samples are typically reconstituted in a solvent that is compatible with the LC-MS/MS mobile phase to volumes between 0.1 and 1 mL. Internal standards can be spiked into the final sample prior to analysis [36]. Addition of surrogates prior to SPE extraction could provide information on the recoveries of the analytes from the aqueous phase. However, suppression or enhancement of the LC-MS/MS signal intensity as a result of matrix effects may generate data that indicates "apparent" recoveries of the analytes that are low or high, respectively.

Liquid-phase microextraction (LPME)

LPME is a recently developed technique [37] that is carried out by using a membrane as an interface between the sample (donor) and an organic solvent (acceptor), which avoids mixing of the two phases and other problems encountered with traditional liquid–liquid extraction. The main advantages of LPME are very low organic solvent consumption and low cost. Quintana et al. [37] tested the suitability of LPME as a single step enrichment/cleanup technique, which could allow the extraction of acidic drugs from wastewater samples, possibly eliminating the matrix effects normally encountered by LC-ESI-MS/MS. They compared LPME results to SPE using Oasis HLB cartridges. The LPME demonstrated good selectivity, with negligible matrix effects when extracts of wastewater were analyzed for acidic drugs by LC-ESI-MS/MS. Moreover, it provides acceptable limits of detection with low sample volumes. However, the major drawback of this technique was the relatively poor precision of LPME. This was likely due to the small volume of extracts and the manual preparation of the extraction devices.

2.3.2.2.2 Particulate samples

Most studies on acidic and neutral pharmaceuticals in the environment have focussed on their distribution in aqueous matrices, such as municipal wastewater, surface water, groundwater and drinking water. However, even though these compounds typically have low Kow values or are present as anions at the pH of natural waters, there is potential for neutral and acidic drugs to adsorb to particulates. Several recent studies have investigated the levels of acidic and neutral pharmaceuticals in biosolids [28,38–40], and in soils and sediments [29].

Soils, sediments and biosolids are complex matrices and co-extracted material present in these samples can severely reduce the efficiency of extraction for pharmaceuticals. In complex environmental matrices, where the target analytes are present at very low concentrations along with higher concentrations of potentially interfering compounds, it is essential to develop effective methods for extraction and purification. Owing to the thermolabile properties of many pharmaceuticals, traditional extraction methods, such as Soxhlet extraction are not appropriate and other techniques such as pressurized liquid extraction (PLE) and ultrasonic solvent extraction are more suitable. Note that it should not be assumed that extraction methods developed for one particulate matrix will be suitable for another matrix. For instance, different ultrasonic solvent extraction methods were required to extract acidic pharmaceuticals from biosolids [38] and from sediments [41].

Miao et al. [39] used PLE to extract carbamazepine and its metabolites, as well as caffeine from biosolids using acetone/water (30:70) as the extraction medium. The wet biosolids material was mixed with a diatomaceous earth material, Hydromatrix® to absorb water from the matrix prior to extraction. Göbel et al. [40] extracted freeze-dried biosolids by PLE using methanol–water (1:1) as the extraction solvent and obtained good recoveries of trimethoprim, as well as sulfonamide and macrolide antibiotics. Kinney et al. [42] extracted wet soil by PLE using acetonitrile/water (70:30) as the extraction medium, and detected a range of pharmaceuticals in the extracts, including acidic and neutral compounds. Ternes et al. [38] used ultrasonic solvent extraction to extract both acidic and neutral drugs from activated and digested biosolids. Freeze-dried biosolids were extracted with methanol and then acetone. Using this technique, the relative recoveries of acidic drugs exceeded 70%, except for bezafibrate. River sediment was also extracted using ultrasonication to investigate the environmental distribution of acidic drugs [41]. Acetone/acetic acid (20:1) and then ethyl acetate were the solvents used for these extractions, and recoveries of acidic drugs from sediments ranged from 80% to 110%.

In order to remove co-extractives, the extracts must be further processed by using techniques such as SPE, gel permeation chromatography (GPC) or preparative HPLC. SPE has been the preferred sample purification technique for clean up of extracts containing neutral and acidic pharmaceuticals because it is fast, requires a low volume of organic solvent, presents low contamination risk and can be adapted for use on-line. Typically, the extracts from particulate matrices are purified using the same SPE method as used to extract the target pharmaceuticals from aqueous matrices. If solvents are used in the extraction process, as is usual for neutral and acidic drugs, these solvents must be evaporated off and replaced with an aqueous matrix prior to passing the extract through the SPE.

2.3.2.2.3 Biota samples

Up until relatively recently, it was assumed that pharmaceuticals released into the environment would show little potential for bioaccumulation. However, recent studies have shown that some weakly acidic, weakly basic and neutral pharmaceuticals can accumulate in the biofluids and tissues of fish. The acidic lipid-regulator, gemfibrozil, was shown to accumulate in the blood plasma of goldfish to levels approximately 100 times greater than the concentrations to which these fish were exposed in the laboratory [32]. Samples of blood plasma (100 μL) were acidified and extracted using HLB Oasis cartridges using SPE procedures which were slightly modified from methods used to extract acidic drugs from water samples, which involved washing the cartridge with HPLC-grade water and with 5% methanol in water to remove accumulated salts prior to elution with ethanol, and the extracts were analyzed by LC-MS/MS [32].

Schwaiger et al. [43] developed a method for analyzing diclofenac accumulated in the tissues of rainbow trout exposed to this compound in the laboratory, and observed residues of >2.8 μg/g wet weight in the livers of fish exposed to diclofenac at concentrations of 1 μg/L. Brooks et al. [30] developed methods for the analysis of various anti-depressents (i.e., fluoxetine and sertraline and their metabolites) in fish. Tissue homogenates were extracted into acetonitrile, and after centrifugation, the solvent was evaporated and replaced with phosphate buffer. The extract was then purified on an SPE column. The purified extract was derivatized using pentafluoroproprionic anhydride and the samples were analyzed by GC-MS-SIM. Recoveries of analytes spiked into fish tissues ranged between 49% and 107%. The concentrations of the depressants in the tissues of fish collected from an effluent-dominated stream ranged up to approximately 10 ng/g. In a more recent study, Ramirez et al [31] developed a method for the analysis of 24 pharmaceuticals in fish tissues, including several acidic, neutral and weakly basic compounds. Several extraction solvents were tested in this study, and buffer/organic mixtures over a wide range of pH were found to be efficient at recovering the majority of the analytes. Mean recoveries of acidic drugs (i.e., clofibric acid, gemfibrozil and ibuprofen) were independent of the pH of the buffer. The extracts were centrifuged, reconstituted in water and filtered, then analyzed by LC-ESI-MS/MS [31]. Although matrix effects were a challenge for these analyses, the study showed that several pharmaceuticals were accumulated to ng/g levels in fish collected from an effluent-dominated stream, including several anti-depressents (fluoxetine, norfluoxetine, sertraline), erythromycin, diltiazem, carbamazepine, dimethylxanthine and caffeine. None of the acidic drugs were accumulated in fish tissues to detectable concentrations.

2.3.2.3 LC-MS and LC-MS/MS analysis

As described in another chapter in this monograph, acidic pharmaceuticals have been analyzed by GC-MS after derivatization using several types of procedures. The specificity of the derivatization process for certain classes of compounds can be utilized to identify closely related analytes, such as drug metabolites [20]. In addition, GC-MS is less susceptible to the matrix effects that are an analytical challenge for LC-MS or LC-MS/MS analysis of pharmaceuticals in environmental samples. However, these methods have some drawbacks, including the laborious nature of the procedures, the potential for sample contamination during derivatization, and reductions to the efficiency of derivatization in complex sample matrices. Some thermolabile analytes are also susceptible to degradation during the derivatization process [7,8]. Therefore, LC-MS and LC-MS/MS techniques are now widely used to analyze both acidic and neutral pharmaceuticals in environmental samples.

2.3.2.3.1 Liquid chromatography separation

Liquid chromatography coupled to mass spectrometry (LC-MS or LC-MS/MS) is a powerful technique for the analysis of drugs in complex matrices, with high sensitivity and selectivity. However, efficient HPLC separation is crucial for successful analysis. By far the most popular stationary phases for LC-MS or LC-MS/MS analyses of neutral and acidic drugs are reverse phase columns, including C4, C8, C12 and C18 columns. Retention on the column is based on van der Waals interaction with hydrophobic components of the stationary phase. Thus, since the C8 phase has approximately 40–50% of the carbon loading of a C18 phase, its hydrophobic resolving power is less than a C18 stationary phase. Most analyses of acidic and neutral drugs have been conducted using C18 columns. Most LC-MS applications have focused on using an ESI interface, which usually handles mobile phase flow rates from 0.1 to 0.5 mL/min. Because of these low flow rates, narrow-bore LC columns (i.e., 2–3 mm i.d.) are more popular compared to the conventional 4.6 mm i.d. analytical column. Narrow-bore columns also provide high separation efficiency and sensitivity, and are less subject to producing matrix effects. Acidic and neutral drugs are usually analyzed using columns with a length of more than 15 cm. Co-extractives can cause matrix effects if they are not chromatographically separated from the target analytes. In general, retention factors greater than four are necessary to diminish matrix effects, such as ion suppression or enhancement [44].

Acetonitrile and methanol have relatively low viscosities and high vapor pressures, and so they are often used in mobile phases for the chromatographic separation of neutral and acidic drugs. Although methanol can cause higher column backpressure, it can increase retention for polar compounds such as pharmaceuticals, and backpressure is typically not a problem for narrow-bore column running at low flow rates. For the analysis of neutral drugs, methanol has been combined with acetonitrile to achieve good analyte resolution with reasonable analyte retention [19].

Figure 2.3.3 illustrates the relationship between retention factor (k) and mobile phase pH for monoprotic acidic and basic, and neutral analytes. Acidic compounds are completely dissociated at high pH values in mobile phases, and their ionization is accompanied by low

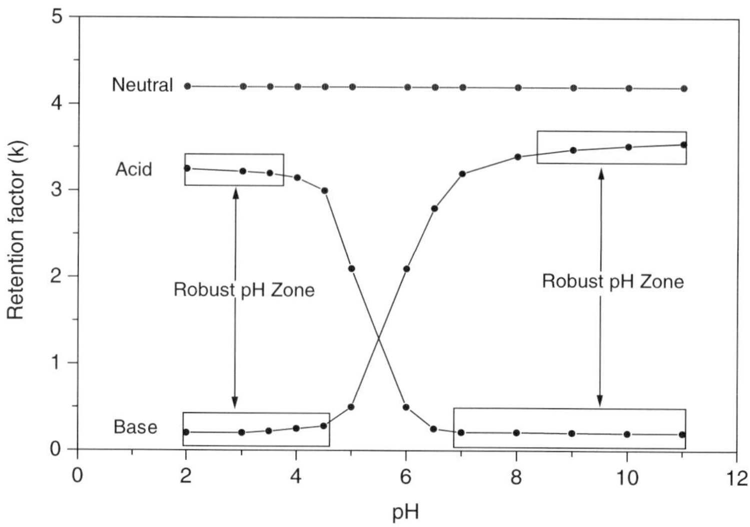

Fig. 2.3.3. Theoretical dependence of retention factor (k) on the pH in an aqueous HPLC mobile phase. The "robust pH zones" represent the optimum pH range for retention of acidic and basic analytes.

retention on reverse-phase sorbents. Conversely, decreased pH leads to reduced dissociation of weak acids and consequently, stronger retention on the stationary phase. The dissociation constant of the acidic analyte plays a key role in this process, because the most significant change in ionization and hence retention in aqueous mobile phases takes place at a pH value close to the pKa value of the analyte. The situation is the opposite for weak bases, which are neutral at high pH values and fully protonated in low pH mobile phases. In the case of neutral compounds, retention is not affected by the pH change.

Clearly, the pH of the mobile phase plays an important role in the chromatographic separation of acidic pharmaceuticals. In practice, the pH of the mobile phase should be at least 2 pH units away from the pKa of the analyte, within the "robust pH zone" (Fig. 2.3.3), to ensure good peak shape and reproducibility in retention time. For example, if the predicted pKa values of three acidic pharmaceutical analytes are 3.7, 4.2 and 4.8, the initial pH of the mobile phase should be at least 6.5 to ensure that the compounds are present in their neutral forms during chromatographic separation. In addition, pH changes can be programed over the chromatographic run in order to optimize the separation of analytes. Recent advances in LC column technology have made it possible to use pH as a tool for the separation of weakly acidic compounds.

By operating the HPLC at variable pH, there can be a 10–30-fold difference in retention that can be exploited for method development.

The main difference between an LC analysis with ESI-MS and non-MS detectors is that the mobile phase must be compatible with mass spectrometry. The mobile phase should contain only volatile components and have low concentrations of buffer and ion-pairing agents in order to maximize the MS signal. In order to obtain reproducible retention times for acidic pharmaceuticals, aqueous buffers are often used as a component in the mobile phase, but the electrolytes in the buffer usually lower the signal intensity due to suppression effects in the MS interface. In general, electrolytes used in buffers are limited to volatile compounds such as ammonium formate or ammonium acetate [45]. Optimum buffering capacity occurs at a pH equal to the pKa of the buffer. In general, most buffers provide adequate buffering capacity for controlling mobile phase pH only within ± 1 unit of their pKa. Ammonium acetate is the preferred electrolyte for analysis of acidic drugs because its ion-suppressing effect is low, and it is suitable as a buffer at pH 5.5, which is an optimum pH for the separation of most acidic pharmaceuticals. Unstable retention times have been observed at concentrations of ammonium acetate below 2 mM, but significant ion suppression is observed when ammonium acetate is present at concentrations higher than 20 mM [46]. Methylammonium acetate at 2 mM was applied as a mobile phase buffer for the analysis of blood-lipid regulators from the statin class [17], which had the added benefit of generating methylammonium adducts of the analytes that could be monitored at high sensitivity. In order to increase the sensitivity for acidic drugs in negative-ion mode, an ion-pairing agent tri-*n*-butylamine (10 mM) was used together with 0.5% acetic acid in the mobile phase [47].

2.3.2.3.2 Mass spectrometry

Owing to its high selectivity and sensitivity, liquid chromatography with tandem mass spectrometry (LC-MS/MS) is preferable to LC-MS for the measurement of acidic and neutral drugs in environmental samples. However, LC-ESI-MS has been used successfully to analyze for both acidic and neutral drugs in surface water and groundwater [12,14] and in sediments [42]. Under typical operating conditions, LC-ESI-MS generates only the protonated or deprotonated parent ions of the drug analytes. In order to increase the specificity of the analyses by LC-ESI-MS, it is possible to increase the exit voltage of the capillary interface in the electrospray ion source to produce fragment ions that often replicate the mass spectra generated by collision-induced dissociation in LC-MS/MS instruments. Cahill et al. [14] utilized this technique for the LC-ESI-MS analysis of acidic and neutral drugs in water, and referred to the programed capillary exit voltage as the "fragmentor voltage".

Atmospheric pressure ionization is the most commonly used ionization method that is coupled with liquid chromatography, and it includes all ionization techniques where the ions are formed at atmospheric pressure, such as ESI, atmospheric pressure chemical ionization (APCI), atmospheric pressure photoionization (APPI) and pneumatic-assisted sonic spray ionization (SSI). ESI and APCI are the most common ion sources for the analysis of ionic and polar substances. However, ESI is a more sensitive ionization technique than APCI for the analysis of acidic drugs [48]. Both ESI and APCI are suitable ionization techniques for neutral drugs [9,10].

Once an appropriate ionization source is chosen, the optimal parameters for LC-MS/MS analysis are determined. Owing to the carboxylic acid group, most acidic drugs provide a strong signal for the deprotonated ion ($[M-H]^-$) and are monitored in negative ion mode. However, for some acidic drugs (e.g., statin drugs) greater sensitivity is achieved by monitoring in positive-ion mode for adducts of the precursor ion [17]. Neutral drugs typically generate a strong signal for the protonated ion ($[M+H]^+$). Highest signal intensities for this precursor ion can be achieved by refining the parameters in the ionization source, such as the desolvation temperature and source voltage. Following the selection of the monitoring mode for precursor ions in the first quadrupole mass analyzer, collision-induced dissociation (CID) is optimized by adjusting the parameters in the collision cell at a fixed collision energy and gas pressure for argon or nitrogen. Product ion mass spectra are obtained at a series of collision energies to determine the optimal parameters for each analyte, based on the relative intensities of the product ions. In order to achieve the highest sensitivity, with adequate selectivity, the mass spectrometer is typically operated in multiple reaction monitoring (MRM) mode with unit resolution on the first quadrupole analyzer and the second analyzer, which can be a quadrupole or ion trap device.

Typically, ion transitions selected for MRM include the precursor ion and one or two product ions. For the analysis of many acidic and neutral drugs, there is often only one transition available in the product ion spectrum at sufficient signal intensities. However, structurally informative fragment ions can be observed in the product ion spectra of some

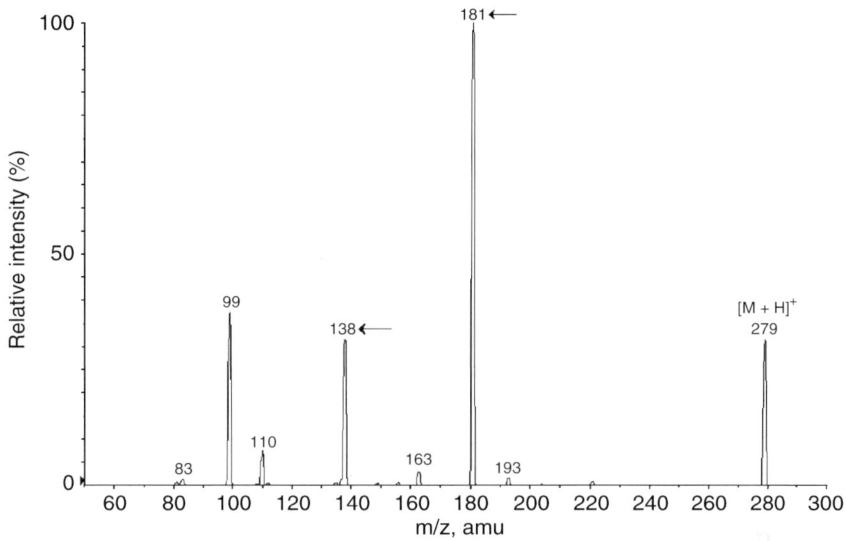

Fig. 2.3.4. Full-scan mass spectrum for pentoxyfylline analyzed by LC-APCI-MS/MS in positive ion mode showing the precursor ion ($[M+H]^+$) and the two product ions (arrows) selected for monitoring by MRM.

acidic and neutral drugs, and these can be used as a second product ion for confirmation. Figure 2.3.4 illustrates the full-scan spectrum for the neutral drug, pentoxyfylline, and indicates the precursor ion and the two product ions that were selected for MRM. Petrović et al. [9] reviewed the transitions that have been used to analyze pharmaceuticals in environmental samples, including several neutral and acidic drugs. Unlike GC-MS analytical techniques, ion ratios are not typically used for analyte confirmations in LC-MS/MS applications because of the relatively high variability observed in ion ratios (i.e., 10–20%) under even the most carefully regulated operating conditions.

As discussed previously, the pH of the mobile phase is critical for optimizing the separation of analytes on the stationary phase. However, the pH of the mobile phase also has a strong impact on the sensitivity of MS/MS detection, since the ionization efficiency for an analyte depends on its charge state. A pH value below the pKa value increases the degree of protonation of basic analytes and should therefore increase the sensitivity in positive ion mode. Conversely, for acidic analytes monitored in negative ion mode, a pH value above the pKa value should increase deprotonation and hence the sensitivity of analysis. Thus, selection of the pH of the mobile phase should take into account the operational needs of both the chromatography and the mass spectrometry. So, although acidic conditions may provide better separation of many acidic drugs, monitoring in negative-ion mode may be optimized in a basic environment and therefore, a neutral pH of the mobile phase must be selected in order to establish a sensitive method.

2.3.2.3.3 Time of flight mass spectrometry

Time of flight mass spectrometry (TOF-MS) is becoming an important analytical tool for the identification of pharmaceuticals and their degradation products in environmental samples. This instrumentation provides full-scan mass spectrometry for all compounds resolved throughout the chromatogram and high resolution mass measurements that can be used to identify analytes to an extent not possible using triple quadrupole or quadrupole/ion-trap instruments. TOF-MS has the additional advantage of a much higher mass range than triple quadrupole instruments. Stolker et al. [48] compared triple quadrupole (LC-MS/MS) to LC with quadruple and time of flight (LC-Qq-TOF-MS) mass analysers for analysis of pharmaceuticals in water samples and concluded that both analytical instruments gave comparable results, with the relative advantages of greater selectivity for the LC-Qq-TOF-MS instrument and lower limits of detection for the LC-MS/MS instrument. Marchese et al. [49] also showed that LC-MS/MS operated in MRM mode for the analysis of analgesics produced lower limits of quantitation by a factor of 3–5 than a LC-Qq-TOF-MS instrument. Stolker et al. [48] recommended using LC-Qq-TOF-MS for confirmation purposes with environmental samples in which pharmaceuticals had been detected using LC-MS/MS.

2.3.2.4 Matrix effects

One of the drawbacks of LC-MS and LC-MS/MS is the susceptibility of API interfaces to interference from co-extracted matrix components, which may lead to significant differences in the response of an analyte in a sample as compared to a pure standard solution. Since the nature and the amount of these co-eluting matrix components can be variable between samples, matrix effects in a series of samples can be highly variable and difficult to predict. The matrix effects result in suppression or enhancement of the signal of the target analyte during the ionization process. This can seriously compromise the accuracy of quantitative data and may increase or decrease detection limits when real samples are analyzed.

It has been reported that APCI is less sensitive to matrix effects than ESI [10,20]. However, our recent studies have shown that APCI is susceptible to signal enhancement for the analysis of neutral pharmaceuticals [27], and while this can improve sensitivity for environmental samples, there are still challenges for analyte quantitation. In any event, APCI cannot be used as an ionization source for all highly polar and ionic analytes. Therefore, ESI remains the interface of choice for many applications in the analysis of drugs in environmental samples despite its higher sensitivity to matrix effects. Figure 2.3.5 illustrates signal suppression for neutral pharmaceuticals (i.e., carbamazepine and metaboilites) analyzed by LC-ESI-MS/MS in samples of varying complexity. Suppression was greatest for extracts prepared from raw wastewater and declined successively in extracts prepared from treated wastewater to surface water to HPLC-grade water [19].

Because matrix effects are mainly caused by co-extracted components, sample preparation is the first step in eliminating interferences. Properly designed SPE methods may selectively enrich the analytes while eliminating much of the co-extracted material. Another approach may be the use of restricted access materials (RAM) for sample enrichment,

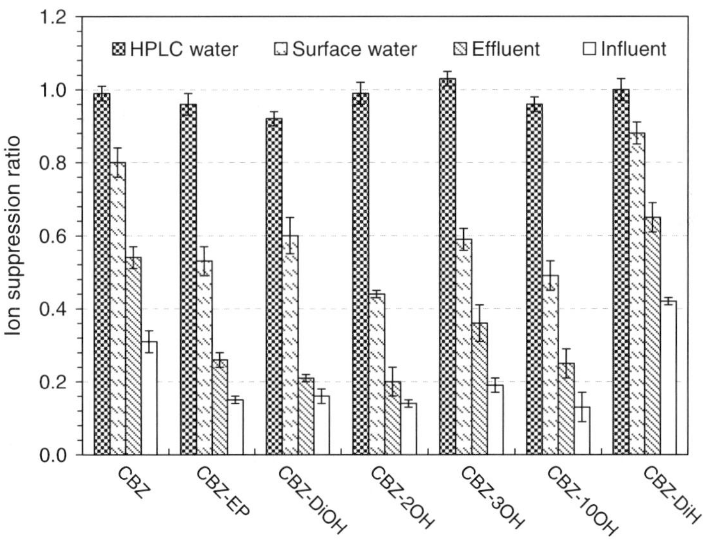

Fig. 2.3.5. Suppression of signal for carbamazepine (CBZ) and its metabolites spiked into various aqueous samples: HPLC-grade water, surface water, raw wastewater (influent) and treated wastewater (effluent) and analyzed by LC-ESI-MS/MS. Figure reproduced with permission of the American Chemical Society from the publication by Miao and Metcalfe [19].

as these sorbents exclude high molecular weight materials with nominal mass above 15 kDa. RAM was found to be useful for reducing matrix effects related to humic substances in extracts from groundwater or sediment [50]. A study using ultrafiltration for size separation of dissolved organics showed that sample clean up based on size exclusion does not seem promising to reduce matrix effects from wastewater samples, indicating that the matrix effects in LC-MS/MS analysis of acidic drugs from wastewater is primarily due to low molecular weight compounds <1 kDa [51].

High chromatographic resolution generally reduces matrix effects for LC-ESI-MS/MS applications. There is a clear tendency for decreasing signal suppression with increasing retention time, and this has been observed and interpreted as being indicative of non-specific matrix effects of moderately polar matrix components, the concentrations of which decrease with increasing retention time [52]. Owing to a gradual decrease in the matrix effect with increasing retention time it is not possible to use one surrogate for accurate quantitation of pharmaceuticals in environmental samples using LC-MS/MS. Optimally, there should be one stable isotope surrogate for each drug analyte that elutes from the chromatographic column at the same time as the native compound and therefore, is equally affected by the sample matrix.

Another useful approach for reducing matrix effects is to reduce the flow rate that is delivered to the ESI interface by using a post-column T-connection. This strategy has not gained much attention, although it was shown some years ago that decreasing the flow directed into the ESI to as low as 0.1 μL/min resulted in a substantial reduction in signal suppression [53]. However, for these applications, an ESI interface that can operate at very low flows (i.e., nano-ESI) is required. Reducing the flow into the ESI interface does not shift the ratio between analyte and the co-eluting matrix, but it significantly reduces the amount of material in the source that requires ionization at a given time. Moreover, the droplet size decreases with reduced flow and the droplet surface area increases substantially. Thus, the target analyte and sample matrix components may not compete with each other during desolvation and ionization in the ESI source.

Flows down to 20–50 μL/min can be used with a conventional ESI interface but lower flow rates result in an unstable spray, peak broadening and shifts in retention time. However, with decreasing flow, there is an increase in sensitivity for many analytes; even by one order of magnitude for some compounds. For many compounds, matrix effects can be nearly eliminated so that conventional external calibration is suitable for reliable quantitation [54].

2.3.2.5 Quantification

Three common options can be applied to obtain accurate quantitative results for samples with complex matrices: (i) using structurally similar compounds as internal standards; (ii) using isotopically labeled internal standard compounds; and (iii) using standard additions to the sample matrix. An appropriate structurally similar internal standard may compensate, over a limited retention time window, for the changes in signal that lead to inaccurate results. However, as all sample constituents are subject to chromatographic separation, the matrix effects are strongly dependent on the chromatographic retention time. In general, due to the gradual decrease in the matrix effect with retention time, it is not possible to reliably compensate for the matrix effects by using a single internal standards. The ideal standards are isotopically (i.e., deuterated, ^{13}C) labeled surrogates that have the same chromatographic retention times as the analytes [10]. Note that surrogates should be labeled at a sufficient number of sites to achieve adequate mass resolution from the native compound (i.e., >2 amu). As an alternative approach to quantitation, standard additions of each analyte into the sample may be used to compensate for matrix effects. However, this approach is time-consuming and laborious, and does not improve analytical sensitivity.

2.3.3 CONCLUSIONS

The development of methods for the analysis of neutral and acidic pharmaceuticals in environmental samples has advanced greatly over the past 10 years. However, further advancements are required to address analytical challenges, such as the matrix effects observed with LC-MS/MS analytical techniques. At present, most of the analytical effort on pharmaceuticals released into the environment has been directed at detecting the parent compounds, but it is obvious that more work is required to detect the metabolites of drugs that are excreted by both humans and animals [55], and transformation products that may be formed during the treatment of water and wastewater [56]. A particular priority is the need to develop analytical techniques to detect pharmaceuticals released as conjugated adducts to biological molecules. For many drugs, the majority of the excreted material is present as conjugated metabolites [55]. These conjugated drugs may be converted back to the free and biologically active form through microbial activity in wastewater treatment plants [57]. TOF-MS instruments may be particularly useful for detecting these high molecular weight compounds.

REFERENCES

1 D.W. Kaufman, J.P. Kelly, L. Rosenberg, T.E. Andersonand and A.A. Mitchell, *J. Am. Med. Assoc.*, 287 (2002) 337.
2 A.W. Garrison, J.D. Pope and F.R. Allen, *Identification and Analysis of Organic Pollutants in Water*, Ann Arbor Science Publication, Ann Arbor, MI, 1976 517 pp.
3 C. Hignite and D.L. Azarnoff, *Life Sci.*, 20 (1977) 337.
4 C.D. Watts, M. Craythorne, M. Fielding, C.P. Steel, In: G. Angeletti and A. Bjorseth, (Eds.), *Analysis of Organic Micropollutants in Water*, D.D. Reidel Publication, Dordrech, FDR, 1983, p. 120
5 A. Waggott, in: W.J. Cooper (Ed.), *Chemistry in Water Reuse*, Ann Arbor Publication, 1981, p. 55.
6 M.L. Richardson and J.M. Bowron, *J. Pharm. Pharmacol.*, 37 (1985) 1.
7 T.A. Ternes, *Water Res.*, 32 (1998) 3245.
8 C.D. Metcalfe, B.G. Koenig, D.T. Bennie, M. Servos, T.A. Ternes and R. Hirsch, *Environ. Toxicol. Chem.*, 22 (2003) 2872.
9 M. Petrović, M.D. Hernando, M.S. Díaz-Cruz and D. Barceló, *J. Chromatogr. A*, 1067 (2005) 1.
10 B.J. Vanderford, R.A. Pearson, D.J. Rexing and S.A. Snyder, *Anal. Chem.*, 75 (2003) 6265.
11 T. Ternes, M. Bonerz and T. Schmidt, *J. Chromatogr. A*, 938 (2001) 175.
12 D.W. Kolpin, E.T. Furlong, M.T. Meyer, E.M. Thurman, S.D. Zaugg, L.B. Barber and H.T. Buxton, *Environ. Sci. Technol.*, 36 (2002) 1202.
13 C.D. Metcalfe, X.-S. Miao, B.G. Koenig and J. Struger, *Environ. Chem. Toxicol.*, 22 (2003) 2881.
14 J.D. Cahill, E.T. Furlong, M.R. Burkhardt, D. Kolpin and L.G. Anderson, *J. Chromatogr. A*, 1041 (2004) 171.
15 D. Ashton, M. Hilton and K.V. Thomas, *Sci. Total Environ.*, 333 (2004) 167.
16 X.-S. Miao, B.G. Koenig and C.D. Metcalfe, *J. Chromatogr. A*, 952 (2002) 139.
17 X.-S. Miao and C.D. Metcalfe, *J. Chromatogr. A*, 998 (2003) 133.
18 S. Öllers, H.P. Singer, P. Fässler and S.R. Müller, *J. Chromatogr. A*, 911 (2001) 295.
19 X.-S. Miao and C.D. Metcalfe, *Anal. Chem.*, 75 (2003) 3731.
20 S. Zuehlke, U. Duennbier and T. Heberer, *Anal. Chem.*, 76 (2004) 6548.

21 M. Stüber and T. Reemtsma, *Anal. Bioanal. Chem.*, 378 (2004) 910.
22 B.K. Choi, D.M. Hercules and A.I. Gusev, *J. Chromatogr. A*, 907 (2001) 337.
23 J. Schumacher, D. Zimmer, F. Tesche and V. Pickard, *Rapid Commun. Mass Spectrom.*, 17 (2003) 1950.
24 B.J. Vanderford, R.A. Pearson, D.J. Rexing and S.A. Snyder, *Anal. Chem.*, 75 (2003) 6265.
25 T. Benijts, R. Dams, W. Lambert and A. De Leenheer, *J. Chromatogr. A*, 1029 (2004) 153.
26 P.R. Gardinali and X. Zhao, *Environ. Int.*, 28 (2002) 521.
27 X. Zhao and C.D. Metcalfe, *J. Chromatogr. A*, submitted.
28 K. Xia, A. Bhandari, K. Das and G. Pillar, *J. Environ. Qual.*, 34 (2005) 91.
29 S.-C. Kim and K. Carlson, *Trend. Anal. Chem.*, 24 (2005) 635.
30 B.W. Brooks, C.K. Chambliss, J.K. Stanely, A. Ramirez, K.E. Banks, R.D. Johnson and R.J. Lewis, *Environ. Toxicol. Chem.*, 24 (2005) 464.
31 A.J. Ramirez, M.A. Mottaleb, B.W. Brooks and C.K. Chambliss, *Environ. Sci. Technol.*, in press.
32 C. Mimeault, A.J. Woodhouse, X.-S. Miao, C.D. Metcalfe, T.W. Moon and V.L. Trudeau, *Aquatic Toxicol.* 73 (2005) 44.
33 H.B. Lee, K. Sarafin, T.E. Peart and M.L. Svoboda, *Water Qual. Res. J. Canada*, 38 (2003) 667.
34 J. Hilton and K.V. Thomas, *J. Chromatogr. A*, 1015 (2003) 129.
35 N. Delaunay-Bertoncini and M.-C. Hennion, *J. Pharmac. Biomed. Analysis*, 34 (2004) 717.
36 J.B. Quintana and T. Reemtsma, *Rapid Commun. Mass Spectrom.*, 18 (2004) 765.
37 I. Rodriguez, J. Carpinteiro, J.B. Quintana, A.M. Carro, R.A. Lorenzo and R. Cela, *J. Chromatogr. A*, 1024 (2004) 1.
38 T.A. Ternes, M. Bonerz, N. Herrmann, D. Löffler, E. Keller, B. Bagó Lacida and A.C. Alder, *J. Chromatogr. A*, 1067 (2005) 213.
39 X.-S. Miao, J.-J. Yang and C.D. Metcalfe, *Environ. Sci. Technol.*, 39 (2005) 7489.
40 A. Göbel, A. Thomsen, C.S. McArdell, A.C. Alder, W. Giger, N. Theiβ, D. Löffler and T. Ternes, *J. Chromatogr.*, 1085 (2005) 179.
41 D. Löffler and T.A. Ternes, *J. Chromatogr. A*, 1021 (2003) 133.
42 C.A. Kinney, E.T. Furlong, S.L. Werner and J.D. Cahill, *Environ. Toxicol. Chem.*, 25 (2006) 317.
43 J. Schwaiger, H. Ferling, U. Mallow, H. Wintermayr and R.D. Negele, *Aquatic Toxicol*, 68 (2004) 141.
44 I. Fu, E. Woolf and B. Matuszewski, *J. Pharm. Biomed. Anal.*, 18 (1998) 347.
45 W. Ahrer and W. Buchberger, *J. Chromatogr. A*, 854 (1999) 275.
46 W. Ahrer, E. Scherwenk and W. Buchberger, *J. Chromatogr. A*, 919 (2001) 69.

47 C.L. Andrews, C.P. Yu, E. Yang and P. Vouros, *J. Chromatogr. A*, 1053 (2004) 151.
48 A.A.M. Stolker, W. Niesing, E.A. Hogendoorn, J.F.M. Versteegh, R. Fuchs and U.A.Th. Brinkman, *Anal. Bioanal. Chem.*, 378 (2004) 955.
49 S. Marchese, A. Gentilli, D. Perret, G. D'Ascenzo and F. Pastori, *Rapid Commun. Mass Spectrom.*, 17 (2003) 879.
50 E. Caro, R.M. Marcé, F. Borrull, P.A.G. Cormack and D.C. Sherrington, *Trends Anal. Chem.*, 25 (2006) 143.
51 A. Kloepfer, J.B. Quintana and T. Reemtsma, *J. Chromatogr. A*, 1067 (2005) 153.
52 B.K. Choi, D.M. Hercules and A.I. Gusev, *Fresenius J. Anal. Chem.*, 369 (2001) 370.
53 E.T. Gangli, M. Anna, N. Spooner and P. Vouros, *Anal. Chem.*, 73 (2001) 5635.
54 A. Kloepfer, J.B. Quintana and T. Reemtsma, *J. Chromatogr. A*, 1067 (2005) 153.
55 N.H.G. Holford, In: B.G. Katzung (Ed.), *Basic and Clinical Pharmacology, Eighth edition*, McGraw-Hill Inc., New York, 2001, pp. 35–50.
56 S.A. Snyder, P. Westerhoff, Y. Yoon and D.L. Sedlak, *Environ. Engin. Sci.*, 20 (2003) 449.
57 T.A. Ternes, P. Kreckel and J. Mueller, *Sci. Total Environ.*, 225 (1999) 91.

Chapter 2.4

Multi-residue analysis of pharmaceuticals using LC-tandem MS and LC-hybrid MS

Mira Petrović, Meritxell Gros and Damià Barceló

2.4.1 INTRODUCTION

A large number of analytical methodologies have been developed for the determination of pharmaceutical residues in both surface and wastewaters. LC-MS/MS has gained popularity due to its versatility, specificity and selectivity and is applied as a method of choice for the analysis of pharmaceuticals in complex environmental and wastewater samples [1]. The vast majority of the LC-MS methods developed in the past focused on specific therapeutic classes [2–7], being antibiotics the most studied ones [8–13]. However, the general trend observed in the recent years, governed by the need for increased capabilities in environmental analysis, is in the area of development and application of generic methods that permit simultaneous analysis of multi-class compounds. The multi-residue methods are found to be a more efficient alternative to the previously developed individual methods, and they are becoming the preferred and required tools against single group analysis, as they provide wider knowledge about occurrence of contaminants in the aquatic media necessary for further study of their removal, partition and ultimate fate in the environment. Especially when results for multiple parameters are required. This trend is also recognized in the analysis of pharmaceuticals as environmental contaminants and recent analytical methodologies are aimed on the simultaneous determination of acidic, neutral and basic compounds belonging to different therapeutical classes.

It has been widely recognized that liquid chromatography-tandem mass spectrometry (LC-MS/MS) offers very good sensitivity and selectivity in trace analysis of food and environmental contaminants and also

Comprehensive Analytical Chemistry 50
Petrović and Barceló (Eds)
Volume 50 ISSN: 0166-526X DOI: 10.1016/S0166-526X(07)50005-X

is being routinely used for proteomic analysis and in the pharmaceutical industry. LC-MS/MS techniques such as triple quadrupoles (QqQ) and ion traps (IT) are in common use. More recent approaches in LC-MS/MS are linear ion traps (LIT), new generation triple quadrupoles and hybrid instruments, such as quadrupole-time of flight (QqTOF) and quadrupole-linear ion trap (QqLIT) that are gaining widespread acceptance in several application areas. These instruments offer advantages such as high scanning speeds, accurate mass measurement (QqTOF) and increased sensitivity (LIT, new generation triple quadrupoles). At the same time, recently introduced improvements in the LC side, like the use of ultra performance liquid chromatography (UPLC) or rapid resolution liquid chromatography (RRLC) makes this technology more attractive and powerful when combined with tandem MS.

A survey of the most representative recent multi-residue LC-MS/MS methods developed for the determination of regularly used pharmaceuticals in aqueous environmental matrices is given in Table 2.4.1

2.4.2 SIMULTANEOUS EXTRACTION OF MULTI-CLASS PHARMACEUTICALS FROM AQUEOUS SAMPLES

Typical problems encountered in the case of multi-residue methods are related with simultaneous extraction and pre-treatment of groups of analytes with mutually widely different polarities. The most recent analytical methodologies available are focused on the simultaneous extraction of all target compounds in one single extraction step using solid phase extraction (SPE). Another widely employed option consists in the combination of two SPE materials operating either in series or in parallel classifying target analytes in two or more groups, according to their physico-chemical properties.

Generally, Oasis HLB cartridges are the preferred ones, working at neutral pH. Owing to their chemical composition (the combination of the lipophilic divinylbenzene and the hydrophilic N-vinylpyrrolidone polymers), they are able to extract acidic, neutral and basic compounds, at a wide range of pHs, including neutral pH. As Oasis HLB, StrataX can retain a wide spectrum of analytes through both hydrophilic and lipophilic interactions.

C_{18} is another SPE sorbent widely used. When using this material, depending on the nature of the compounds included, sample pH adjustment prior to extraction is generally required. For instance, for the analysis of groups including acidic pharmaceuticals, such as

TABLE 2.4.1

Summary of the most representative multiresidue LC-MS/MS methods for the quantitative determination of pharmaceuticals in aqueous samples

Compounds	Matrix	SPE material	LC separation		MS system	Limit of quantification (ng/L)	Reference
			Column	Mobile phase			
30 compounds Antibiotics, anti-inflammatories, anticancer, bronchodilator; cardiovascular, gastrointestinal drugs; diuretics, estrogens and lipid regulators	Urban wastewaters	(A) Lichrolut EN and (B) Oasis MCX	C_8	ESI(+) Aq. formic acid/AcN and ESI(−) Aq. TEA/ACN	QqQ (ESI)	0.1–5.2	[18]
16 compounds Antibiotics, β-blockers, psychiatric drugs and anti-inflammatories	Hospital effluent wastewaters	Oasis HLB	C_{18}	ESI(+) Aq. formic acid/ACN and ESI(−) water/AcN	QqQ (ESI)	4–47 (LOD)	[19]
28 compounds Anti-inflammatories, lipid regulators, anti-ulcer agents, anti-histaminics, antibiotics and β-blockers	Urban and industrials wastewaters	Oasis HLB	Acquity UPLC™ BEH C_{18}	ESI(+) AcN/MeOH (2:1)/NH_4Ac/Hac and ESI(−) MeOH/H_2O	QqTOF (ESI)	15–500 (LOD)	[23]
28 compounds Anti-inflammatories, lipid regulators, anti-ulcer agents, anti-histaminics, antibiotics and β-blockers	River and wastewaters	Oasis HLB	C_{18}	ESI(+) AcN/MeOH (2:1)/NH_4Ac/Hac and ESI(−) MeOH/H_2O	QqQ (ESI)	River water 0.5–47 (LOD) and wastewater 1–60 (LOD)	[24]
16 compounds Illicit drugs and their metabolites	Wastewater	Oasis MCX	C_{18}	4 different gradients	QqQ (ESI)	0.63–8.7 (influent) and 0.48–3.2 (effluent)	[28]

TABLE 2.4.1 (*continued*)

Compounds	Matrix	SPE material	LC separation		MS system	Limit of quantification (ng/L)	Reference
			Column	Mobile phase			
20 compounds Psychoactive drugs and their metabolites	Surface and wastewater	Oasis HLB	Phenyl	ESI(+) ACN/Aq NH4formiate/ formic acid	QqQ (ESI)	10–100 (influent), 5–50 (effluent) and 1–10 (river water)	[27]
27 compounds Antibiotics, carbamazepine, lipid regulators and anti-inflammatories	Surface water	Oasis HLB	C_{18}	ESI(+) Aq. NH_4Ac + MeOH and HFBA/ACN and ESI(−) Aq.10 mM NH_4Ac/ACN	Qq-LIT	0.3–60	[42]
23 compounds (A) β-agonist and antagonist, anti-ulcer agent, antibiotics, lipid regulator and psychiatric drugs,(B) Antibiotics, anti-inflammatories and others and (C) Amoxycillin and omeprazole	River water	(A) Oasis MCX, (B) Lichrolut EN and (C) Bakerbond C_{18}	C_8	ESI(+) Aq. formic acid/ACN and ESI(−) Aq. TEA/ACN	*QqQ (ESI)*	0.3–10	[49]
12 compounds Anti-inflammatories, lipid regulators and triclosan	Surface and wastewater	Oasis HLB	Phenyl-hexyl	ESI(−) MeOH/ H_2O/solvent A with TrBA and acetic acid	QqQ (ESI)	0.3–5.6	[50]
13 compounds Antibiotics, lipid regulator, analgesic, anti-inflammatories, β-blocker, anti-cancer and anti-depressant	River water	StrataX	C_{18}	ESI(+) and ESI(−) water/ MeOH/40 mM NH_4Ac at pH 5.5 with formic acid	QqQ (ESI)	10–50 (LOD)	[51]

13 compounds Analgesic/anti-inflamamtory, β-blocker, lipid regulators, antibiotics and anti-epileptic	Surface water	Oasis MCX	C_{18}	ESI(+) and ESI( ) MeOH/2 mM NH_4Ac	QqQ (ESI) and QqTOF (ESI)	5–25	[52]
28 compounds Neutral and acidic pharmaceuticals and EDC and PCP	Surface and wastewater	Oasis HLB	C_{12}	ESI(+) and ESI(−) Aq. formic acid/ MeOH	QqQ (ESI/APCI)	1.0 (LOD)	[53]
60 compounds Analgesic, β-blocker, broncholytics, secretolytics, antineoplastics and lipid regulators	Ground water	PPL-bond-elut	C_{18}	ESI(+) and ESI(−) ACN/ MeOH/NH_4Ac	QqQ (ESI)	7.9–44 (LOD)	[54]
11 compounds OSPAR priority pharmaceuticals	Surface and wastewater	StrataX	C_{18}	ESI(+) Aq. 20 mM NH_4Ac and 0.1%formic acid/MeOH and ESI(−) Aq. 20 mM NH_4Ac/ MeOH	IT (ESI)	1–20 (LOD)	[55]

LOD, limit of detection.

anti-inflammatories and lipid regulators, samples are acidified, because at neutral pH target analytes exist in their ionized form, in which they are poorly retained by lipophilic sorbents. Otherwise, for neutral and basic compounds, samples are adjusted to neutral or basic pHs.

Less common cartridges used are Lichrolut ENV+, Oasis MCX and StrataX. The fist one is generally recommended for the extraction of polar organic compounds at low pH values, but it can also retain neutral drugs at pH 7, such as carbamazepine (antiepileptic) and macrolides, through hydrophobic interactions. Oasis MCX has been used to extract acidic, basic and neutral compounds, at low pHs. Therefore, basic compounds are retained due to the cation exchange properties and the acidic and neutral ones for the reversed-phase characteristics. However, to elute efficiently target analytes, a mixture of methanol-ammonia is generally used, whereas in the other cases, pure methanol is sufficient.

On the other hand, for the methods that include two or more SPE steps, classifying target analytes in different groups, the use of a single SPE material is preferred. However, sample pre-treatments (pH adjustments) are specific, depending on the nature of target compounds included in each group. The advantage of using these procedures is that better conditions are achieved for each one of the therapeutic groups, but they are time-consuming.

In some multi-residue methods available in the literature, target compounds, belonging to different therapeutic groups, are extracted either using one or various SPE protocols and afterwards, instrumental analysis is carried out by two different techniques. For instance, the most numerous are the ones where acidic pharmaceuticals are extracted in a specific SPE extraction, separated from other polar analytes, and after derivatization, they are analyzed by GC-MS, whereas the rest are determined by LC-tandem MS [14–17]. However, in the current review no more emphasis will be given to this type of methodologies as we focus on simultaneous determination of a wide spectrum of multiple-class pharmaceuticals using only LC-MS/MS as instrumental technique.

One of the advantages of developing multi-residue analytical methods is that they provide a comprehensive approach to the study of the presence of different therapeutic groups in the environment. However, simultaneous extraction of compounds from different groups with quite different physico-chemical characteristics requires a compromise in the selection of experimental conditions, which in some cases are not the best conditions for all the analytes studied. For instance, Castiglioni et al. [18] reported recoveries from 36% to 131% for 30 pharmaceuticals belonging to various therapeutic categories in urban wastewater, while Gómez et al. [19] obtained recoveries from 45% to 112% for 16 pharmaceuticals in hospital effluent wastewaters. The method of Andreozzi et al. [20] included 26 pharmaceuticals, among which some macrolide antibiotics, clofibrate and gemfibrozil yielded quite low recoveries (35%, 36% and 46%, respectively), whereas for the rest of compounds were recovered with efficiency over 75%. Similarly, Gros et al. [21] developed a method for simultaneous extraction and analysis of 29 pharmaceuticals obtaining recoveries from 60% to 102% and from 50% to 116% for surface and wastewaters, respectively, however several compounds (ranitidine, sotalol, famotidine, mevastatin and clofibric acid) yielded lower recoveries from wastewater samples (35–50%).

2.4.3 CHROMATOGRAPHIC SEPARATION

Although, one of the advantages of MS analysis is that complete LC separation of target analytes is not necessary for selective detection, it is always advisable to have good chromatographic separation in order to reduce matrix effects, which typically results in the suppression or, less frequently, the enhancement of analyte signals. As it is indicated in Table 2.4.1, for multi-residue analytical methods, reversed phase conventional high-performance liquid chromatography (HPLC) is widely used, being C_{18} column the preferred one. The size parameters of columns are typically as follows: (i) length in the range 10–25 cm, (ii) i.d. 2.1–4.6 mm and (iii) particle sizes 3–5 μm. Gradient elution represents the most common strategy in separation of multi-class pharmaceuticals producing an analysis time of 30 to 60 min. As mobile phases, acetonitrile, methanol or mixtures of both solvents are normally used, obtaining in the latter case shorter retention times and better resolution of the analytes. In order to obtain an efficient retention of the analytes in the column and to improve the sensitivity of MS detection, mobile-phase modifiers, buffers and acids are recommended and widely used. The more common ones include ammonium acetate, ammonium formiate, tri-*n*-butylamine (TrBA), formic acid and acetic acid. Typical concentrations of the salts range from 2 to 20 mM (see Table 2.4.1), since it has been observed that higher concentrations could lead to a reduction of the signal intensities [22].

Shortening the analysis times is important for attaining the high sample throughput often required in monitoring studies. The simplest way to shorten a chromatographic run is to use short columns and increased flow velocity. The second way is to decrease the particle size of stationary phase allowing high-speed analysis with high efficiency. The third way is to increase the temperature that enhances diffusivity allowing working at higher flow-rates.

These approaches were applied in two newly developed instruments. One is Acquity UPLC (ultra performance liquid chromatography) system produced by Waters Corporation (Manchester, UK) and another one is 1200 Series RRLC (rapid resolution LC) from Agilent Technologies. Both systems use rather short columns (50–100 mm, 4.6 mm i.d.) packed with sub-2 μm porous particles, allowing very short chromatographic runs. However, the negative effect of using small particle size is high back-pressure generation (reducing the particle size by a factor of 3 results in an increase in the backpressure by a factor of 27). The UPLC system is specially designed to withstand high system pressures (up to 15,000 psi (1035 bar)), while RRLC system, beside high pressure (600 bar), uses high temperatures that allow the use of small particles due to reduced mobile phase viscosity.

For the moment, only one publication describes the multi-residue analysis of pharmaceuticals in environmental samples using the Acquity UPLC system. Petrović et al. [23] developed an UPLC-QqTOF-MS method for screening and confirmation of 29 pharmaceutical compounds belonging to different therapeutical classes: analgesics and anti-inflammatories, lipid regulating agents cholesterol lowering statin agents, psychiatric drugs, anti-ulcer agents, histamine H2 receptor antagonist, antibiotics and beta-blockers. UPLC, using columns packed with 1.7–m particles and enabled elution of target analytes in much narrower, more concentrated bands, resulting in better chromatographic resolution and increased peak height. The typical peak width was 5–10 s at base, permitting very good separation of all compounds in 10 min, which represented an approximate 3–fold reduction in the analysis time in comparison to conventional HPLC as shown in Fig. 2.4.1.

2.4.4 MASS SPECTROMETRIC ANALYSIS USING TANDEM MS

2.4.4.1 Triple quadrupole (QqQ)

Triple quadrupole (QqQ) mass analyzers have become the most widely used analytical tool in the analysis of pharmaceuticals as environmental contaminants. Their application has allowed the determination of a

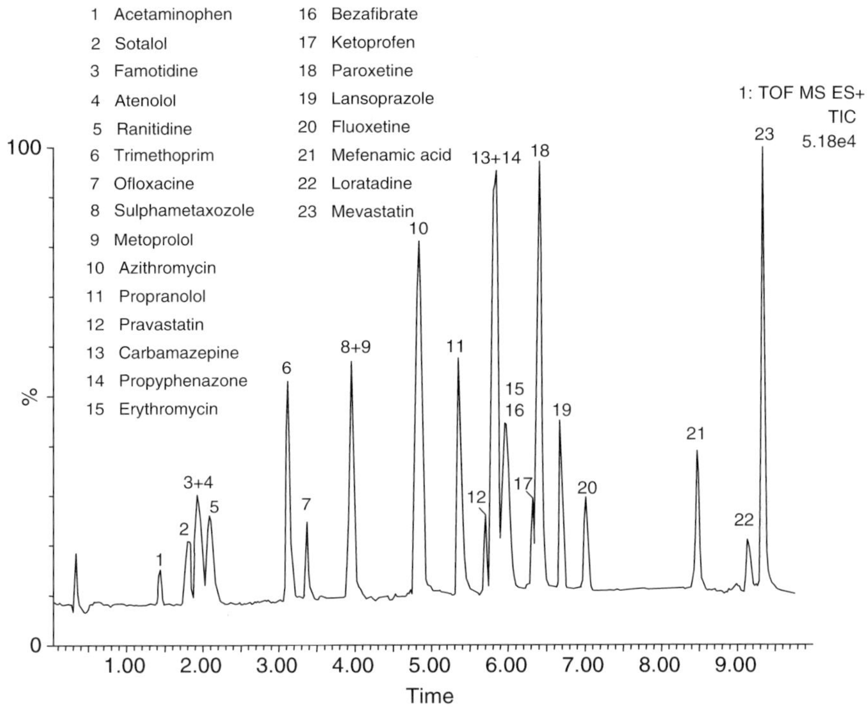

Fig. 2.4.1. UPLC-TOF total ion chromatogram showing the separation of 23 pharmaceutical compounds analyzed in PI mode (100 ng/mL standard solution). Modified from [23]

great number of compounds, especially polar ones that were previously difficult or even impossible to analyze. A number of multi-residue methods have been developed covering a wide range of compounds belonging to different therapeutical groups [24]. The selection of compounds to be monitored is mainly based on their consumption in the country, predicted environmental loads or using procedures for identification and prioritization on the bases of compound persistence, liability to bioaccumulation and toxicity and actual occurrence. Typically multi residue methods include most common painkillers and anti-inflammatory drugs, lipid regulating agents, β-blockers and selected antibiotics.

LC-MS/MS (QqQ) has been mostly applied to the determination of target analytes, using the selected reaction monitoring mode (SRM) and reaching typically ng/L detection limits. For instance, a method reported by Gros et al. [21] included 29 pharmaceuticals and yielded method LODs from 1 to 30 ng/L and from 3 to 160 ng/L for surface and wastewaters, respectively. The multi-residue method of Castiglioni et al. [18] enables the detection of 30 pharmaceuticals, yielding LOD ranging from 0.15 to 5.2 ng/L in wastewaters, whereas Sacher et al. [22] reported LODs of 2–13 ng/L for the analysis of analgesics, anti-inflammatories, lipid regulators, β-blockers and antibiotics in groundwater. Using isotope dilution LC-MS/MS Vanderford and Snyder [25] obtained method reporting limits for 15 multi-class pharmaceuticals and four metabolites between 0.25 and 1.0 ng/L based on 500 mL of sample extracted.

As it can be observed in Table 2.4.2, *anti-inflammatories and analgesics* are generally detected in NI mode, with the exception of phenazone that is detected in PI mode and acetaminophen that can be detected by both modes. For most of the anti-inflammatory/analgesic drugs analyzed under NI conditions, the main product ion corresponds to the loss of CO_2. Main product ion of acetaminophen is attributed to $[M–CH_2CO+H]^+$ in PI and to $[M–H–COCH_3]^-$ in NI. In case of mefenamic acid, the main product ion is also attributed to the loss of CO_2, whereas the second one is associated to the ion $[M–H–CO_2–CH_3]^-$. For propyphenazone, first transition corresponds to $[M+H]^+ \rightarrow [M–C_3H_7+H]^+$ ($m/z = 189$) whereas for phenylbutazone, the main product ion recorded at $m/z = 160$ is $[M–(C_6H_5–N)–(C_4H_9)]^+$.

Carbamazepine, fluoxetine, paroxetine and diazepam are the most studied *psychiatric drugs*, being carbamazepine one of the most frequently detected pharmaceuticals in the aquatic environment. Its major product ion corresponds to the loss of carbamoyl group (HNCO). Using an ESI-MS/MS in PI mode Miao and Metcalfe [26] studied degradation of carbamazepine and identified five main metabolites (10,11–dihydro-10,11–epoxycarbamazepine; 10,11–dihydro-10,11–dihydroxycarbamazepine, 2–hydroxycarbamazepine, 3–hydroxy-carbamazepine and 10,11–dihydro-10–hydro-carbamazepine). The only major ion product of carbamezapine, 2–OH-carbamezapine and 3–OH-carbamezapine corresponded to loss of the structurally characteristic carbamoyl group (HNCO, 43 Da). For other carbamezapine metabolites rather complex product ion mass spectra were observed showing different ions corresponding to losses of H_2O, NH_3 or HNCO. In the same study 10,11–dihydro-10,11–dihydroxycarbamezapine was found in Canadian sewage treatment plants (STPs) in concentrations higher than those of the parent drug. These findings were confirmed by Hummel et al. [27] who developed a multi-residue method for 19 psychoactive compounds and detected carbamazepine and its main

TABLE 2.4.2

Base peaks (m/z) of precursor and product ions used for LC-MS/MS (QqQ) analysis of pharmaceuticals in environmental samples

Compound	Precursor ion (m/z)	Product 1 (m/z)	Product 2 (m/z)
Anti-inflammatory/analgesics/antiphlogistic			
Ibuprofen	205 $[M-H]^-$	161 $[M-H-CO_2]^-$	—
2-Hydroxy ibuprofen	221 $[M-H]^-$	177 $[M-H-CO_2]^-$	133
Ketoprofen	253 $[M-H]^-$	209 $[M-H-CO_2]^-$	197
Naproxen	229 $[M-H]^-$	185 $[M-H-CO_2]^-$	170 $[M-H-C_2H_3O_2]^-$
Indomethacin	356 $[M-H]^-$	312 $[M-H-CO_2]^-$	297 $[M-H-C_2H_3O_2]^-$
Diclofenac	294 $[M-H]^-$	250 $[M-H-CO_2]^-$	214 $[M\text{-}H\text{-}ClCO_2]^-$
	296 $[M+H]^+$	278 $[M+H-H_2O]^+$	—
Fenoprofen	241 $[M-H]^-$	197 $[M-H-CO_2]^-$	93 $[M-H-C_9H_8O_2]^-$
Acetominophen	150 $[M\text{-}H]^-$	106.9 $[M\text{-}H\text{-}COCH3]^-$	—
Codeine	300 $[M+H]^+$	215 $[M+H\text{-}CH_2CHNHCH_3\text{–}CO^+$	199 $[M+H\text{-}CH_2CHNHCH_3\text{–}C_2H_4O]$
Mefenamic acid	240 $[M\text{-}H]^-$	196 $[M\text{-}H\text{-}CO_2]^-$	180 $[M\text{-}H\text{-}CO_2\text{–}CH_3]^-$
Propylphenazone	231 $[M+H]^+$	189 $[M-C_3H_7+H]^+$	201 $[M\text{-}2CH_3+H]^+$
Phenylbutazone	309$[M+H]^+$	160 $[M\text{–}(C_6H_5\text{–}N)\text{–}(C_4hH_9)]^+$	181 $[M\text{-}N\text{-}CO\text{-}NH_2+H]^{+\,=}$
Lipid regulating agents			
Fenofibrate	361 $[M+H]^+$	233	139
Bezafibrate	362 $[M+H]^+$	276	316
	360 $[M\text{-}H]^-$	274 $[M\text{-}H\text{-}C_4H_6O_2]^-$	154 $[M\text{-}H\text{-}C_{12}H_{14}O_3]^-$
Clofibric acid	213 $[M\text{-}H]^-$	127 $[C_6H_4ClO]^-$	85 $[C_4H_5O_2]^-$
	213/215 $[M\text{-}H]^-$	127/129 $[C_6H_4ClO]^-$	85 $[C_4H_5O_2]^-$
Gemfibrozil	249 $[M\text{-}H]^-$	121 $[M\text{-}H\text{-}C_7H_{12}O_2]^-$	—
Simvastatin	450 $[M+CH_3NH_3]^+$	267	199
Atorvastatin	559 $[M+H]^+$	440	—
Lovastatin	436 $[M+CH_3NH_3]^+$	285 $[436\text{-}C6H17NO2]^+$	199
Pravastatin	456 $[M+CH_3NH_3]^+$	269 $[456\text{-}C6H17NO2]^+$	—
Mevastatin	422 $[M+CH_3NH_3]^+$	185	—

TABLE 2.4.2 (*continued*)

Compound	Precursor ion (*m/z*)	Product 1 (*m/z*)	Product 2 (*m/z*)
β-blockers			
Bisoprolol	326 $[M+H]^+$	116 [(N-isopropyl-N-2-hydroxypropylamine)+H]$^+$	74
	326 $[M+H]^+$	166 [(N-isopropyl-N-2-hydroxypropylamine)+H]$^+$	56
Metoprolol	268 $[M+H]^+$	166 [(N-isopropyl-N-2-hydroxypropylamine)+H]$^+$	98 [(N-isopropyl-N-propenamine)+H]$^+$
Propanolol	260 $[M+H]^+$	116 [(N-isopropyl-N-2-hydroxypropylamine)+H]$^+$	183 $[M–H_20–C_3H_7NH]^+$
Atenolol	267 $[M+H]^+$	190 [M–H_2O–NH_3-isopropyl$^+$+2 H]$^+$	145 [190-CO–NH3]$^+$
Sotalol	273 $[M+H]^+$	255 $[M–H_2O+H]^+$	213
Pindolol	250 $[M+H]^+$	56	72
Betaxolol	308 $[M+H]^+$	166 [(N-isopropyl-N-2-hydroxypropylamine)+H]$^+$	98 [(N-isopropyl-N-propenamine)+H]$^+$
Nadolol	310 $[M+H]^+$	254 [M-tert-butyl$^+$+2 H]$^+$	201
Timolol	317 $[M+H]^+$	261 [M-tert-butyl$^+$+2 H]$^+$	244 [M-tert-butylamine$^+$+H]$^+$
Carazolol	299 $[M+H]^+$	116 [(N-isopropyl-N-2-hydroxypropylamine)+H]$^+$	222
Psychiatric drugs			
Carbamazepine	237 $[M+H]^+$	194 $[M+H_2–CONH_2]^+$	192 $[M–CONH_2]^+$ 179
Fluoxetine	310 $[M+H]^+$	44 $[M–F_3C_7H_4OC_8H_8]^+$	148 $[M–F_3C_7H_4O]^+$
Paroxetine	330 $[M+H]^+$	192 $[M–C_7H_5O_3]^+$	123 $[M–C_{12}H_{14}NOF]^+$
Diazepam	285 $[M+H]^+$	257 $[M–CO+H]^+$	193

Note: For antibiotics see Chapter 2.1.

metabolite 10,11–dihydro-10,11–dihydroxycarbamezapine in STP discharges, river water and drinking water.

For *lipid regulating agents*, the most studied substances correspond to the "fibrate" class. This group of compounds is typically analyzed in NI mode, however, some compounds, such as bezafibrate can also be detected in PI mode. Among "fibrates", bezafibrate and gemfibrozil are the ones more frequently included in multi-residue protocols. For the first one, the main product ion ($m/z = 274$) corresponds to the loss of $C_4H_6O_2$ and the second ion ($m/z = 154$) is attributed to the loss of $C_{12}H_{14}O_3$. Clofibric acid, which is a degradation product of clofibrate, is a ubiquitous compound in environmental waters. Its main fragment ion ($m/z = 127$) is attributed to $[C_6H_4ClO]^-$. "Statins" are another type of lipid regulators, particularly used to control cholesterol in blood, however, not commonly included in multi-residue methodologies. Of all "statins", simvastatin, pravastatin and mevastatin, respectively are the most frequently analyzed, while atorvastatin, which is number one prescribed drug worldwide (LIPITOR®) [28] is rarely analyzed [24]. The main reason for this is that pure standard of this compound was not available commercially until recently.

β-blockers are determined under PI conditions. The transition $[M+H]^+ \rightarrow [(N\text{-isopropyl-N-2–hydroxypropylamine})+H]^+$ ($m/z = 116$) is the predominant for propranolol. On the other hand, for metoprolol, $m/z = 133$ and 159 corresponds to $[C_6H_{15}NO_2]^+$ and $[C_8H_{17}NO_2]^+$. Atenolol produced fragment ions $m/z = 190$ and 145 corresponding to $[M–H_2O–NH_3–\text{isopropyl}+2H]^+$ and $[190\text{-}CO–NH3]^+$, respectively. Finally, for sotalol $m/z = 255$ and 213 are obtained, attributed to $[M–H_2O+H]^+$ and $[M–C_3H_9N+H]^+$.

Other frequently analyzed compounds include the *β*-agonist salbutamol, anti-ulcer agent ranitidine and omeprazole, which are all determined by PI mode. Main product ions obtained for salbutamol ($m/z = 166$ and 148) are $[M+H–(CH_3)_2C = CH_2–H_2O]^+$ and $[166–H_2O]^+$, respectively. Main transitions for ranitidine are $[M+H]^+ \rightarrow [M–C_8H_{12}NO]^+$ ($m/z = 176$) and $[M+H]^+ \rightarrow [M–C_8H_{12}NO–NO_2]^+$ ($m/z = 130$). Omeprazole was determined by monitoring MRM transition is $[M+H]^+ \rightarrow [M–H_3CO–(C_7H_4N_2)–SO–CH_2]^+$ ($m/z = 136$), and $[M+H]^+ \rightarrow [M–H_3CO–C_7H_4N_2]^+$ ($m/z = 198$).

Recently, LC-MS/MS method was also developed for the determination of illicit drugs such as cocaine, amphetamines, morphine, cannabinoids, methadone and some of their metabolites [29].

2.4.5 MASS SPECTROMETRIC ANALYSIS USING HYBRID MS

2.4.5.1 Quadrupole time-of-flight (QqTOF)

An approach for increasing the selectivity and avoiding false positive findings is the use of a hybrid instrument quadrupole time-of-flight-mass spectrometry (QqTOF-MS). In the last years its acceptance for environmental analysis has been significantly improved and the number of screening, quantitative and confirmatory methods reported in the literature is steadily increasing.

The application of hybrid QqTOF-MS technique to environmental analysis allows an unequivocal confirmation of contaminants detected. The elimination of false positives and avoiding interpretation ambiguities is due to its unique characteristic of generating full scan product ion spectra with exact masses. The main field of application is the identification of unknowns and elucidation of structures proposed for transformation products, where the amount of information obtained allows secure identification of the identity of compounds. The first attempt to identify unknown microcontaminants in surface water was published in 2001 by Bobeldijk et al. [30] using data dependent MS to MS/MS switching. By applying the developed data processing procedure, the structures of the three unknown compounds were elucidated proving that carbamazepine was one of them. Recently, Eichhorn et al. [31] reported on the structural elucidation of the metabolites of the antimicrobial trimethoprim, which were produced by nitrifying activated sludge bacteria. With absolute mass errors of <5 mDa, QqTOF allowed confirmation and structural elucidation of two metabolites. The study proved that nitrifying sludge bacteria were capable of facilitating an oxidation of trimethoprim, a pharmaceutical which is not amenable to biological degradation in a conventional activated sludge treatment.

The technique is also successfully applied for the screening of target pharmaceuticals. Usually, the technique is used as a complementary tool to confirm positive findings obtained by a QqQ screening method, but several papers also reported on sound quantitative data obtained using QqTOF.

Regarding its quantitative performances, it is clear that QqTOF, maybe is not the first, but certainly is a viable choice for quantitative analyses, especially when the application requires a high degree of certainty or is aimed on multiple tasks (target analysis combined with qualitative investigation of unknowns). The increased specificity provided by the high resolution QqTOF may provide S/N (signal-to-noise ratio) benefit in some analytical applications and enable rather sensitive quantitative determination as reported by Stolker et al. [32] (MDL of 1–100 ng/L for selected analgesics, antibiotics, lipid regulators, β-blockers and anti-epileptics after SPE preconcentration of 100 mL of surface, drinking and ground water, respectively.) and Marchese et al. [33] (LOQ of 3 ng/L in the analysis of analgesics in drinking water, by preconcentrating 1 L sample.

Recently, Pozo et al. [34] evaluated the potential of a QqTOF instrument to confirm positive findings in the analysis of 16 antibiotics belonging to the groups of quinolones and penicillins, in surface and groundwater samples. The authors concluded that the applicability of QqTOF for antibiotic analysis in the environmental field is rather limited, due to the extremely low concentrations normally present in surface waters (ng/L level), however they pointed out that the technique could be efficiently applied to other organic micropollutants that are frequently present at higher concentrations levels. This statement was proven in the work of Petrović et al. [23] who developed a multi-residue method using UPLC and QqTOF for screening and confirmation of pharmaceutical compounds in wastewaters, where typical concentrations are in the high ng/L to low μg/L range. A number of pharmaceuticals belonging to different therapeutical classes, such as analgesics and anti-inflammatories, lipid regulating agents cholesterol lowering statin agents, psychiatric drugs, anti-ulcer agents, histamine H2 receptor antagonist, antibiotics and beta-blockers were positively identified and their concentrations in wastewaters determined. An example of the analysis of selected pharmaceuticals in an urban wastewater by UPLC-QqTOF-MS is shown in Fig. 2.4.2. Traces shown on the left panel correspond to nwXICs of the $[M+H]^+$ extracted with a mass window of 20 mDa for carbamazepine (*m/z* 237.103), phenylphenazone (*m/z* 231.150), erythromycin (*m/z* 734.468), azithromycin (*m/z* 749.516), trimethoprim (*m/z* 291.146) and acetaminophen (*m/z* 152.071). In a separate experiment using the Q-TOF mode those ions were used as precursor ions to obtain accurate mass product spectra (shown on the right panel). The mass errors obtained for molecular ions were between 0.7 and 4.4 ppm (root mean square (RMS) value 2.02) and 0.2–1.2 mDa (RMS = 0.72), while the accurate masses of 49 product ions deviated from the theoretical masses by 0.2–1.3 mDa (RMS = 0.67) and 0.7–6.4 ppm (RMS = 3.53), respectively. Although, the sensitivity obtained by QqTOF permitted detection of target analytes in wastewater matrix, the reported MDLs (10–500 ng/L for STP influent) were approximately one order of magnitude higher than those reported for

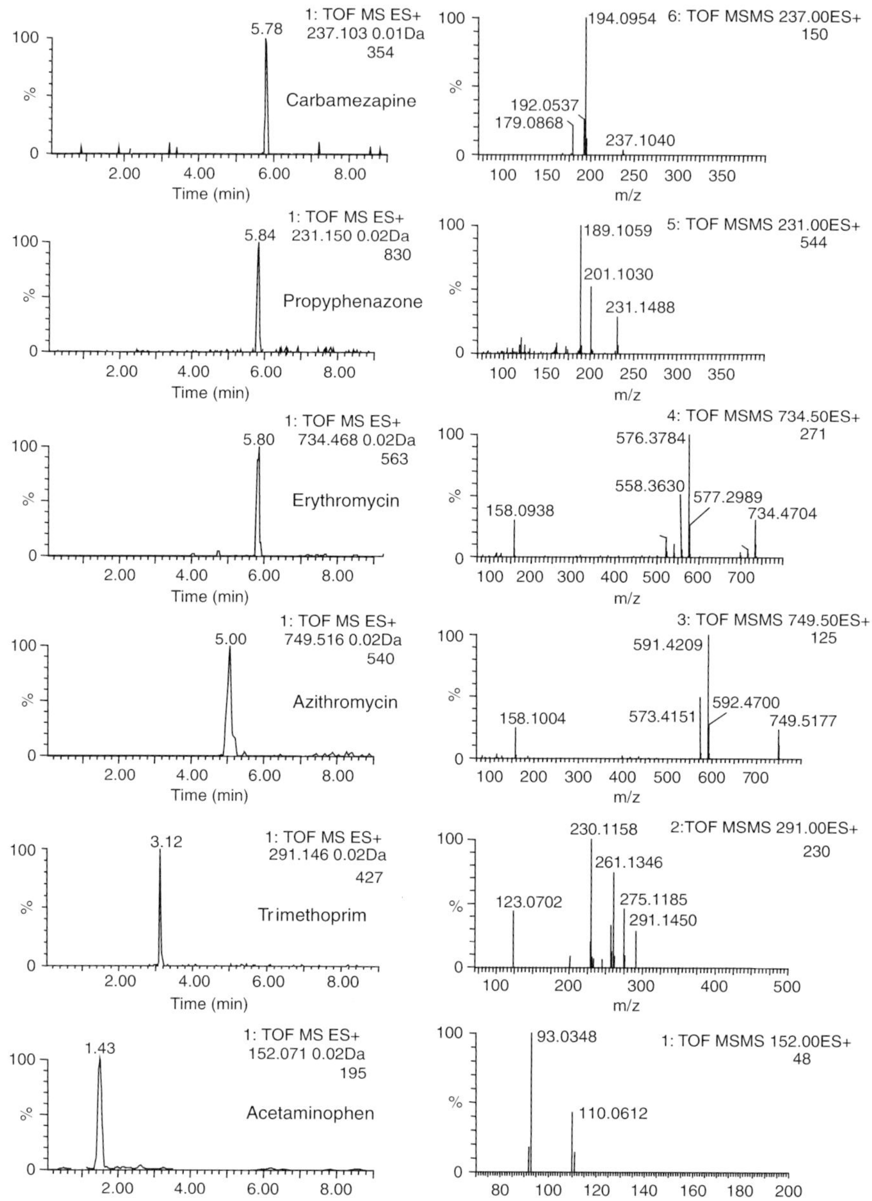

Fig. 2.4.2. Confirmation of several pharmaceuticals in an urban wastewater. Left panel: Narrow window extracted ion chromatograms (nwXICs) of [M+H] obtained in the TOF mode for *m/z* 152. 071 (acetaminophen), *m/z* 291.146 (trimethoprim), *m/z* 749.516 (azithromycin), *m/z* 734.468 (erythromcyn), *m/z* 231.150 (propyphenazone) and *m/z* 237.103 (carbamazepine). Right panel: Product ion spectra obtained in the Q-TOF mode. Reprinted with permission from Ref. [23] © 2006 Elsevier.

QqQ operating in a SRM mode [21], which hampered confirmation of some pharmaceuticals present at low ng/L level.

However, a lower linear dynamic range with respect to QqQ still remains one of the main obstacles for wider acceptance of QqTOF methods for quantitative purposes. Typically reported linear range spanned over two orders of magnitude, which is significantly lower than the dynamic range observed on QqQ instruments (typically >4 orders of magnitude). Therefore, quantitative analysis, in some cases, requires an additional adjustment of sample preparation protocol and re-analysis of samples after appropriate dilution or concentration.

2.4.5.2 Quadrupole—linear ion trap (QqLIT)

The unique feature of QqLIT is that the same mass analyzer Q3 can be run in two different modes, retaining the classical triple quadrupole scan functions such as MRM, product ion, neutral loss and precursor ion while providing access to sensitive ion trap experiments [35] (Fig. 2.4.3). This allows very powerful scan combinations when performing information-dependent data acquisition. In the case of small molecules, qualitative and quantitative work can be performed concomitantly on the same instrument. The very fast duty cycle of QqLIT provides a superior sensitivity over that of traditional QqQ and ion trap and allows to record product ion scan spectra for confirmation purposes without compromising of signal-to-noise (S/N) ratio; also the resolution and accuracy are higher and these peculiarities improves the ion selection capability for complex mixtures, i.e., improves the instrumental selectivity. The product ion scanning sensitivity is similar to that of a conventional three-dimensional IT and a QqTOF which permits obtaining useful product ion spectra over the entire quantitative dynamic range [36]. This hybrid mass spectrometer is actually considered a powerful tool for a rapid identification and confirmation of metabolites in different matrices, especially in the field of drug development [37].

Although, environmental applications are still scarce a few recent papers reported on the application of a hybrid Qq-LIT for trace level determination of emerging contaminants, such as perfluorinated chemicals, herbicides and pharmaceuticals [28,38–40]. Seitz et al. [39] developed a method for direct analysis (no sample pre-concentration) of diclofenac, carbamezapine and iodinated X-ray contrast media (among other contaminants) reaching LODs of 10 ng/L (100 ng/L for iodinated X-ray contrast media). However, the method was based on monitoring specific transitions (SRM mode) and other Q3 scan options of a QqLIT

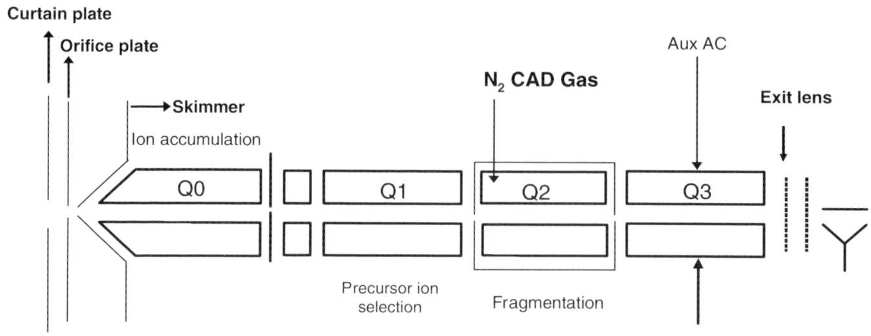

Mode of operation	Q1	Q2	Q3
Q1 Scan	Resolving (Scan)	RF-only	RF-only
Q3 Scan	RF-only	RF-only	Resolving (Scan)
Product Ion Scan (PI)	Resolving (Fixed)	Fragment	Resolving (Scan)
Precursor Ion Scan (PC)	Resolving (Scan)	Fragment	Resolving (Fixed)
Neutral Loss Scan (NL)	Resolving (Scan)	Fragment	Resolving (Scan Offset)
Selected Reaction Monitoring mode (SRM)	Resolving (Fixed)	Fragment	Resolving (Fixed)
Enhanced Q3 Single MS (EMS)	RF-only	No fragment	Trap/Scan
Enhanced Product Ion (EPI)	Resolving (Fixed)	Fragment	Trap/Scan
MS^3	Resolving (Fixed)	Fragment	Isolation/fragment trap/scan
Time Delayed fragmentation (TDF)	Resolving (Fixed)	Trap/No fragment	Fragment/trap/Scan
Enhanced Resolution Q3 Single MS (ER)	RF-only	No fragment	Trap/Scan
Enhanced Multiply Charged (EMC)	RF-only	No fragment	Trap/Scan

Fig. 2.4.3. Scheme of the QqLIT instrument (Q-TRAP, Applied Biosystems/Sciex) and description of the various triple quadrupole and trap operation modes.

instrument were not explored. Similarly, Nikolai et al. [41] used QqLIT operating in QqQ mode for stereoisomer quantification of β-blockers in wastewater. On the other hand, Gros et al. [40] developed an analytical methodology for trace analysis of eight β-blockers in both surface and wastewaters combining different functions of Q3. Quantitative analysis was performed using a 4000QTRAP tandem mass spectrometer in the SRM mode, reaching method detection limits of 0.1–5 ng/L for river water and 0.2–9 ng/L for wastewater (after preconcentration using a molecularly imprinted polymers (MIPs)). Using the information dependent acquisition (IDA) function in the software a great amount of data for unequivocal identification and confirmation of target compounds was generated at high sensitivity. The IDA uses an SRM thresholds value to trigger MS/MS analysis providing a full scan mass spectrum of the specific analyte. An example of IDA experiment for the determination of atenolol in an influent wastewater sample is shown in Fig. 2.4.4.

Similarly, Hao et al. [42] combined the SRM experiments and the IT function (IDA) for the identification and confirmation of target pharmaceutical compounds in urban and agricultural runoff samples.

2.4.6 PITFALLS IN LC-MS ANALYSIS OF PHARMACEUTICALS IN COMPLEX ENVIRONMENTAL SAMPLES

2.4.6.1 False positive results

Increasing concern about confirmation of positive data favored the development of different criteria to assure data quality and to avoid the

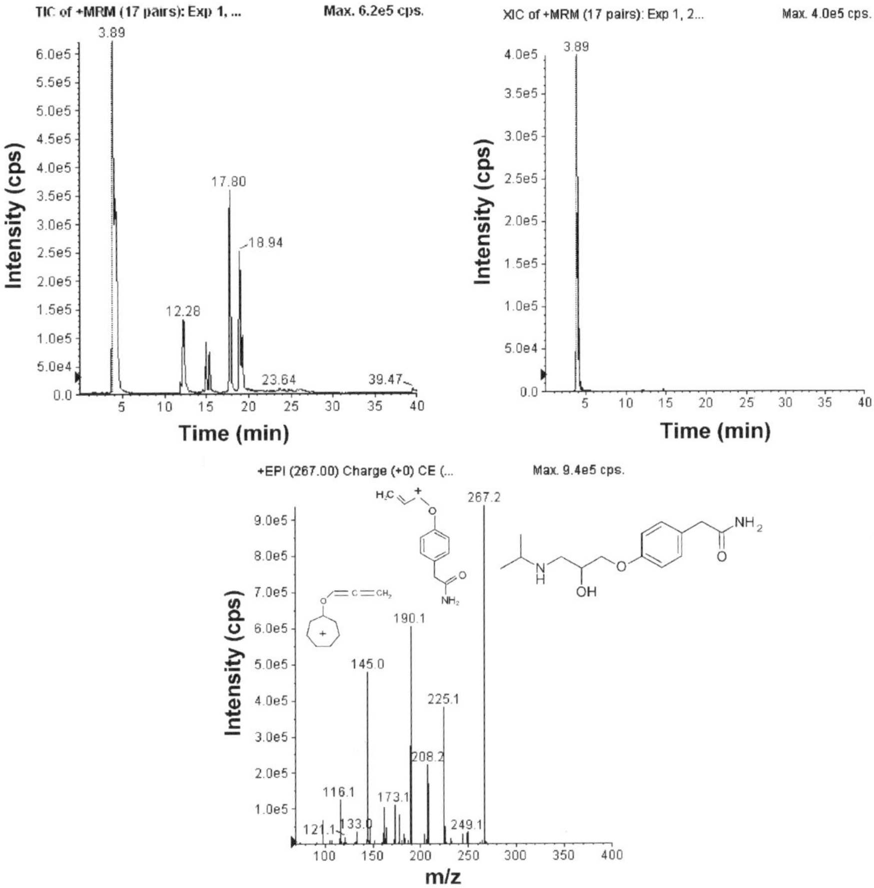

Fig. 2.4.4. Information dependent acquisition (IDA) experiment for the determination of atenolol in an influent wastewater sample. Modified from [40].

reporting of false positives. One of the relatively new concepts in environmental analysis is confirmation of contaminants based on the use of identification points (IP) proposed by a European Commission Guidelines (EU Commission Decision 2002/657/EC) for identification and quantification of organic residues and contaminants. Originally defined for the determination of organic contaminants in food samples, it has been expanded to other matrices, including environmental samples. Decision describes a set of minimum performance characteristics which have to be fulfiled by method to be used for two groups of contaminants: group A (banned compounds) stilbenes, stilbene derivatives and their salts and esters; antithyroid agents; steroids; resorcylic acid lactones including zeranol, beta-agonists and group B (compounds with established residue level) antibacterial substances, including sulfonomides, quinolones; other veterinary drugs; anthelmintics; anticoccidials, including nitroimidazoles; carbamates and pyrethroids; sedatives; non-steroidal anti-inflammatory drugs (NSAIDs); other pharmacologically active substances; other substances and environmental contaminants, such as organochlorine compounds including PCBs, organophosphorus compounds, chemical elements, mycotoxins and dyes. The decision proposes a system of IPs, where at least three IPs are required (four in the case of banned compounds) to confirm a positive finding. In addition, the deviation of the relative intensity of the recorded ions must not exceed a certain percentage of the reference standard, and the retention time must not deviate more than 2.5%. This means that in order to confirm the presence of a compound in environmental samples when using LC-MS/MS (QqQ), the application of stringent confirmation and identification criteria (two SRM transitions) is essential to ensure the correct identification of target analytes in environmental samples. However, in some cases, due to poor fragmentation of the target compound or low intensity of the fragments, only one transition is available, which is not sufficient for positive confirmation of the compounds at the level of interest. Another potential problem in the QqQ analysis is "ion shopping". If several abundant ions are available, it is necessary to choose two of them for SRM experiments, and the information available by other fragments is lost. Therefore, the quality of the transitions in tandem MS-based methods has to be carefully considered, as non-selective transitions (e.g., loss of water, carbon dioxide and hydrochloric acid) are more prone to be hampered and might lead to the reporting of false positives in the samples [43].

When using a TOF instrument a single ion would give only 2 IP, which is not sufficient to confirm the identity of environmental contaminants (see Table 2.4.3). Obtaining 3 to 4 IPs, needed for positive confirmation of target contaminants, is feasible only for compounds showing an easy in-source fragmentation or compounds having a characteristic isotopic pattern.

On the other hand, a QqTOF instrument can easily earn the required number of IP due to its unique feature of generating full scan product ion spectra acquired with high mass accuracy. One precursor and one product ion, which are not sufficient to confirm the identity of the contaminant when using a QqQ instrument, would give 4.5 IP, provided that the exact mass is obtained. If the product ion spectrum contains several fragment ions a number of IPs higher than 10 can easily be achieved.

2.4.6.2 Matrix effect

One of the main problems encountered in quantitative LC-MS/MS analysis and main source of pitfalls is the existence of matrix effects in general, and the ion suppression phenomenon in particular. The ionization suppression or enhancement may severely influence sensitivity, linearity, accuracy and precision of quantitative LC-MS/MS analysis.

TABLE 2.4.3

The number of identification points (IP) earned for the range of LC-MS techniques

Technique	Number of IP earned per ion	Example of ions	IP earned
LC-MS (Q)	1	1 ion (SIM)	1
LC-MS^2 (QqQ)	1 for precursor ion, 1.5 for product ion	1 precursor, 1 product (SRM)	2.5
		1 precursor, 2 products (2 SRM)	4
		2 precursors, each with 1 product (2 SRM)	5
LC-MS^n (IT)	1 for precursor ion, 1.5 for product ion	1 precursor, one MS^2 product and two MS^3 products	5.5
LC-TOF-MS	2	1 ion	2
LC-QqTOF-MS	2 for precursor ion	1 precursor, 1 product (MS/MS)	4.5
	2.5 for product ion	1 precursor, 2 products (MS/MS)	7.5

Ion suppression results from the presence of compounds that can change the efficiency of droplet formation or droplet evaporation in the spray chamber, which, in turn, affects the amount of charged ion in the gas phase that ultimately reaches the detector. Natural organic matter, salts, ion-pairing agents, non-target contaminants and even isotope-labeled internal standards have been shown to be responsible for ion suppression. Therefore, any study dealing with analysis of complex samples should include matrix effect study and if relevant ion suppression (or signal enhancement) occurs, additional procedures should be applied for correction and/or minimization of inaccurate quantification. Recently, several strategies are adopted as standard practices [44–47]. The most often applied approach consists of the use of suitable calibration, such as external calibration using matrix-matched samples, standard addition or internal standard calibration using structurally similar unlabeled pharmaceuticals or isotopically labeled standards. Other approaches include decrease of the flow that is delivered to the ESI interface, as well as the dilution of sample extracts.

Gros et al. [21] evaluated three different approaches (matrix-matched external calibration, internal standard calibration and extract dilution) for the analysis of pharmaceuticals in wastewater influents and effluents. The results showed that internal standard calibration is an efficient approach to assess the loss of signal intensity and therefore, can be used for quantitation purposes, with the advantage that it is less time-consuming than matrix-matched calibration or standard addition. However, it highly depends on the number and nature of internal standards used (structurally similar compounds or isotopically labeled). Since all sample components are subjected to chromatographic separation, the matrix effect also depends upon the chromatographic retention and more than one internal standards are needed. Generally, the most recommended and the most versatile approach is isotope dilution (use of an isotopically labeled standard for each target compound). Such approach was applied by Vanderford and Snyder [25] for the trace analysis of 15 pharmaceuticals and four metabolites, in various waters allowing compensation for matrix suppression and recovery loss. However, such approach is expensive and in many cases suffers from the lack of isotopically labeled compounds.

As an alternative, Gomez et al. [19] applied extract dilution as a simple approach to the matrix effect in complex samples. Fig. 2.4.5 shows extracted SRM chromatograms for fluoxetine and diclofenac from spiked hospital effluent wastewater compared to those obtained for spiked solvent, and the corresponding dilutions. Generally, the

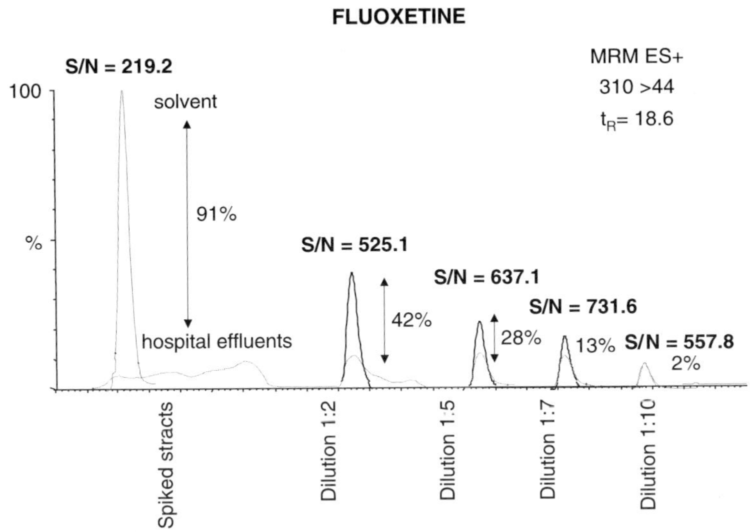

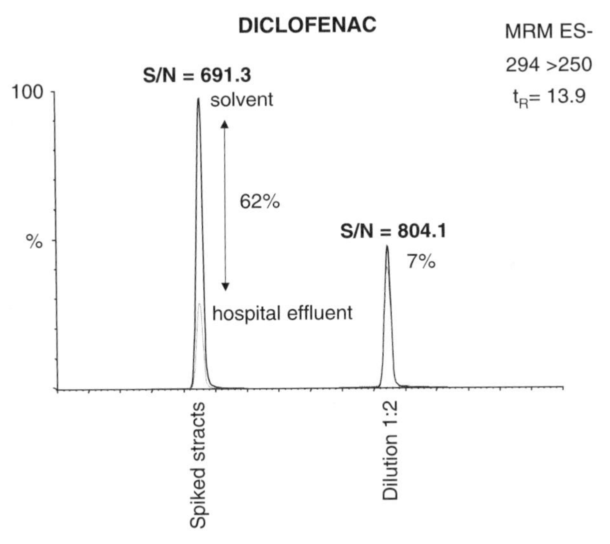

Fig. 2.4.5. Extracted MRM chromatograms for fluoxetine and diclofenac from spiked hospital effluent wastewater compared to those obtained for spiked solvent, and the corresponding dilutions. Signal-to-noise ratio from hospital extracts, and signal suppression (%) are also included. Reprinted with permission from Ref. [19] © 2006 Elsevier.

dilution 1:2 was sufficient to avoid the decrease in the analyte signal in all compounds analyzed in Ni mode, and for the compounds analyzed in PI mode, except for the compounds with severe ion suppression; erythromycin, atenolol, paroxetine and fluoxetine. For these compounds a dilution 1:5 and 1:7 were shown to be sufficient to minimize the signal suppression increasing the signal intensity of the analytes, with dilution 1:7 the signal to noise ratio is maximum. With the dilution 1:10, ion suppression was completely eliminated for these compounds, but the decrease of sensitivity was also observed.

Kloepfer et al. [46] studied whether operational modification in the LC-ESI-MS coupling are suitable to reduce matrix effect and concluded that reducing the flow directed into the ESI by a post column T-piece is helpful to increase the instrumental sensitivity and to reduce matrix effects.

2.4.7 OUTLOOK

It is estimated that of approximately 3000 compounds, approved as constituents in medicinal products, less than 5% have ever been analyzed in environmental samples [1,48]. Most of the literature reviewed on the analytical methods used to determine pharmaceuticals as environmental contaminants have focused either on one class of compounds, e.g., β-blockers, anti-inflammatory drugs, antibiotics or, in the case of multi-residue methods, includes at the maximum 30–40 target analytes. Therefore, a number of potentially relevant pharmaceuticals is waiting to be brought under scrutiny. In addition, most efforts have focused on the detection of parent compounds, while metabolites and transformation products are rarely included into monitoring programes. Main reasons for this are the lack of analytical methods, unavailability of standards and the lack of data regarding their ecotoxic effects that will help in the priotarization of the analytical efforts. In this sense, further development of generic analytical protocols that will permit simultaneous determination of parent compounds and their transformation products is required. The potential of tandem MS and hybrid MS techniques such as QqLIT and QqTOF could enable accurate mass measurement and structural elucidation that might be used to identify relevant transformation products formed by biotic and abiotic processes in the environment.

ACKNOWLEDGEMENTS

This work was financially supported by the European Union EMCO project (INCO-CT-2004-509188) and by the Spanish Ministerio de Ciencia y Tecnología (project EVITA CTM2004-06265-CO3). M. Gros acknowledges the grant from the Spanish MCyT (CTM2004-06265-CO3-01).

REFERENCES

1 M. Petrović, M.D. Hernando, M.S. Díaz-Cruz and D. Barcelo, *J. Chromatogr. A*, 1067 (2005) 1–14.
2 M.D. Hernando, M. Petrović, A.R. Fernandez-Alba and D. Barcelo, *J. Chromatogr. A*, 1046 (2004) 133–140.
3 T. Heberer, U. Dunnbier, C. Reilich and H.J. Stan, *Fresenius Environ. Bull.*, 6 (1997) 438–443.
4 J.B. Quintana and T. Reemtsma, *Rapid Commun. Mass. Spectrom.*, 18 (2004) 765–774.
5 T.A. Ternes, M. Bonerz and T. Schmidt, *J. Chromatogr. A*, 938 (2001) 175–185.
6 X.S. Miao, B.G. Koenig and C.D. Metcalfe, *J. Chromatogr. A*, 952 (2002) 139–147.
7 S. Marchese, D. Perret, A. Gentili, R. Cunni and F. Pastori, *Chromatographia*, 58 (2003) 263–269.
8 X.S. Miao, F. Bishay, M. Chen and C.D. Metcalfe, *Environ. Sci. Technol.*, 38 (2004) 3533–3541.
9 E.M. Golet, A.C. Alder, A. Hartmann, T.A. Ternes and W. Giger, *Anal. Chem.*, 73 (2001) 3632–3638.
10 S. Reverte, F. Borrull, E. Pocurull and R.M. Marce, *J. Chromatogr. A*, 1010 (2003) 225–232.
11 R. Hirsch, T.A. Ternes, K. Haberer, A. Mehlich, F. Ballwanz and K.L. Kratz, *J. Chromatogr. A*, 815 (1998) 213–223.
12 R. Lindberg, P. Wennberg, M. Johansson, M. Tysklind and B. Andersson, *Environ. Sci. Technol.*, 39 (2005) 3421–3429.
13 K.D. Brown, J. Kulis, B. Thomson, T.H. Chapman and D.B. Mawhinney, *Sci. Total Environ.*, 366 (2006) 772–783.
14 C.D. Metcalfe, B.G. Koenig, D.T. Bennie, M. Servos, T.A. Ternes and R. Hirsch, *Environ. Toxicol. Chem.*, 22 (2003) 2872–2880.
15 D. Kolpin, E.T. Furlong, M. Meyer, E.M. Thurman, S. Zaugg, L. Barber and H. Buxton, *Environ. Sci. Technol.*, 36 (2002) 1202–1211.
16 S. Weigel, U. Berger, E. Jensen, R. Kallenborn, H. Thorensen and H. Hühnerfuss, *Chemosphere*, 56 (2004) 583–592.
17 D. Bendz, N.A. Paxéus, T.R. Ginn and F.J. Loge, *J. Hazard. Mater.*, 122 (2005) 195–204.
18 S. Castiglioni, R. Bagnati, D. Calamari, R. Fanelli and E. Zuccato, *J. Chromatogr. A*, 1092 (2005) 206–215.
19 M.J. Gomez, M. Petrović, A.R. Fernandez-Alba and D. Barcelo, *J. Chromatogr. A*, 1114 (2006) 224–233.

20 R. Andreozzi, M. Rafaelle and P. Nicklas, *Chemosphere*, 50 (2003) 1319–1330.
21 M. Gros, M. Petrović and D. Barcelo, *Talanta*, 70 (2006) 678–690.
22 F. Sacher, F.T. Lang, H.J. Brauch and I. Blankenhorn, *J. Chromatogr. A*, 938 (2001) 199–210.
23 M. Petrović, M. Gros and D. Barcelo, *J. Chromatogr. A*, 1124 (2006) 68–81.
24 M. Gros, M. Petrović and D. Barcelo, *Anal. Bioanal. Chem.*, 386 (2006) 941–952.
25 B.J. Vanderford and S.A. Snyder, *Environ. Sci. Technol.*, 40 (2006) 7312–7320.
26 X.S. Miao and C.D. Metcalfe, *Anal. Chem.*, 75 (2003) 3731–3738.
27 D. Hummel, D. Löffler, G. Fink and T.A. Ternes, *Environ. Sci. Technol.*, 40 (2006) 7321–7328.
28 S. Class, *Chem. Eng. News*, 83 (2005) 15–32.
29 S. Castiglioni, E. Zuccato, E. Crisci, C. Chiabrando. R. Fanelli and R. Bagnati, *Anal. Chem.*, 78 (2006) 8421–8429.
30 I. Bobeldijk, J.P.C. Vissers, G. Kearney, H. Major and J.A. van Leerdam, *J. Chromatogr. A*, 929 (2001) 63–74.
31 P. Eichhorn, P.L. Ferguson, S. Perez and D.S. Aga, *Anal. Chem.*, 77 (2005) 4176–4184.
32 A.A.M. Stolker, W. Niesing, E.A. Hogendoorn, J.F.M. Versteegh, R. Fuchs and U.A.Th. Brinkman, *Anal. Bioanal. Chem.*, 378 (2004) 955–963.
33 S. Marchese, A. Gentilli, D. Perret, G. D'Ascenzo and F. Pastori, *Rapid Commun. Mass Spectrom.*, 17 (2003) 879–886.
34 O.J. Pozo, C. Guerrero, J.V. Sancho, M. Ibañez, E. Pitarch, E. Hogendoorn and F. Hernandez, *J. Chromatogr. A*, 1103 (2006) 83–93.
35 G. Hopfgartner, E. Varesio, V. Tschäppät, E.B. Grivet and L.A. Leuthold, *J. Mass Spectrom.*, 39 (2004) 845–855.
36 J.W. Hager and J.C.Y. Le Blanc, *Rapid Commun. Mass Spectrom.*, 17 (2003) 1056–1064.
37 R. King and C. Fernandez-Metzler, *Curr. Drug Metab.*, 7 (2006) 541.
38 M. Alaee, C. Spencer, T. Sakuma, G. Ivosev, R. Ellis, A. Schreiber, Y. Leblanc, C. Marvin, P. Crozier, E. Reiner, C. Lucaciu, N. Riddell and G. Arsenault, Proceedings of the 2nd International Workshop on Liquid Chromatography-Tandem Mass Spectrometry Screening Trace Level Quantitation Environmental Food Samples, Barcelona, Spain, October 2006, p. 23.
39 W. Seitz, W. Schulz and W.H. Weber, *Rapid. Commun. Mass Spectrom.*, 20 (2006) 2281–2285.
40 M. Gros, T. Pizzolato, M. Petrović, M.J. Lopez de Alda and D. Barcelo, *Anal. Chem.*, submitted.
41 L.N. Nikolai, E.L. McClure, S.L. MacLeod and C.S. Wong, *J. Chromatogr. A*, 1131 (2006) 103–109.

42 C. Hao, L. Lissemore, B. Nguyen, S. Kleywegt, P. Yang and K. Solomon, *Anal. Bioanal. Chem.*, 384 (2006) 505–513.
43 F. Hernandez, O.J. Pozo, J.V. Sancho, F.J. Lopez, J.M. Marin and M. Ibañez, *TrAC—Trend. Anal. Chem.*, 24 (2005) 596–612.
44 L. Alder, S. Lüderitz, K. Lindtner and H.J. Stan, *J. Chromatogr. A*, 1058 (2004) 67–79.
45 T. Benijts, R. Dams, W. Lambert and A. De Leenheer, *J. Chromatogr. A*, 1029 (2004) 153–159.
46 A. Kloepfer, J.B. Quintana and T. Reemtsma, *J. Chromatogr. A*, 1067 (2005) 153–160.
47 J.C. Van De Steene, K.A. Mortier and W.E. Lambert, *J. Chromatrogr. A*, 123 (2006) 71–81.
48 D. Löffler and T.A. Ternes, *J. Chromatogr. A*, 1021 (2003) 133–144.
49 D. Calamari, E. Zucatto, S. Castiglioni, R. Bagnati and R. Fanelli, *Environ. Sci. Technol.*, 37 (2003) 1241–1248.
50 J.B. Quintana and T. Reemtsma, *Rapid Commun. Mass Spectrom.*, 18 (2004) 765–774.
51 M.J. Hilton and K.V. Thomas, *J. Chromatogr. A*, 1015 (2003) 129–141.
52 A.A.M. Stolker, E. Dijkman, W. Niesing and E.A. Hogendoorn, In: E.M. Thurman and I. Ferrer (Eds.), *LC/MS, MS/MS and TOF/MS Analysis of Emerging Contaminants, ACS Symposium Series 850*, American Chemical Society, Washington, 2003, pp 32–49.
53 B.J. Vanderford, R.A. Pearson, D.J. Rexing and S.A. Snyder, *Anal. Chem.*, 75 (2003) 6265–6274.
54 F. Sacher, F.T. Lang, H.J. Brauch and I. Blankenhorn, *J. Chromatogr. A*, 938 (2001) 199–210.
55 P.H. Roberts and P. Bersuder, *J. Chromatogr. A*, 1134 (2006) 143–150.

Chapter 2.5

Analysis of acidic drugs by gas chromatography

José Benito Quintana, Josefina Carpinteiro and Isaac Rodríguez

2.5.1 INTRODUCTION

The acidic pharmaceuticals compound classes of major environmental relevance are summarised in Table 2.5.1. All of these compounds are carboxylic acids, having thus a pKa value of ca. 3.5 and log *P* (octanol-water partition coefficient) between 2 and 5 (see *Appendix* for details on physicochemical data). These log *P* values correspond to the neutral form, while in the aqueous media acidic drugs are presented as anionic species, due to their pKa values. This means that, in environmental samples their lipophilic character is much lower (log *P* between −1 and 1 at pH 8 [1]) and they are only associated to the particulate phase in a limited extension. As a consequence they show a high mobility and, in some cases, an excellent stability in the water phase. An example of this behaviour has been reported for clofibric acid, the metabolite and active principle of several blood lipid-regulating pharmaceuticals. This

TABLE 2.5.1

Compound classes of the most environmental relevant acidic pharmaceuticals

Compound class	Examples
Salicylates and metabolites[a]	Acetylsalicylic acid and salicylic acid
Profens and metabolites[a]	Ibuprofen and carboxy-ibuprofen
Arylacetic acids[a]	Indomethacine
Anthranilates[a]	Diclofenac
Fibrates and metabolites[b]	Bezafibrate and clofibric acid

[a]Applied as non-steroidal anti-inflammatory drugs (NSAIDs).
[b]Used as blood lipid regulators.

Comprehensive Analytical Chemistry 50
Petrović and Barceló (Eds)
Volume 50 ISSN: 0166-526X DOI: 10.1016/S0166-526X(07)50006-1

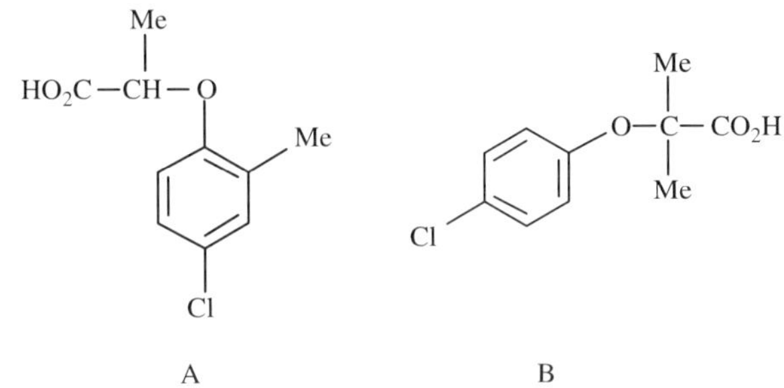

Fig. 2.5.1. Structures of mecoprop (A) and clofibric acid (B).

compound, which is not significantly removed during conventional wastewater treatments [2], has been found in many rivers from central Europe achieving even the North Sea waters [3]. Although, its potential long-term effects are not yet known they should not be underestimated due to the similarity between the structure of clofibric acid and those from some phenoxyacid herbicides such as Mecoprop (Fig. 2.5.1). Hydroxy- and carboxy-ibuprofen, the main metabolites of the parent pharmaceutical, have also been detected in sea water [4].

Other compounds that are often considered as neutral pharmaceuticals, but having in fact slightly acidic groups, such as phenolic moieties (e.g. acetaminophen) or amide groups (e.g. carbamazepine), will also be considered into this chapter, regarding the benefits of derivatisation to their GC determination.

Actually, the need for a derivatisation step prior to their GC analysis has fostered the development of several liquid chromatography-tandem mass spectrometry (LC-MS/MS) methods for the determination of acidic pharmaceuticals in the last few years [5–7]. They allow the determination of these compounds in water samples with good sensitivity and selectivity without the need for a derivatisation step and, in some cases, even avoiding the preliminary extraction/preconcentration step [5].

However, LC-MS/MS suffers from strong matrix effects that difficult quantitative analysis, requiring often the standard addition method for reliable results [5,7–9]. Furthermore, gas chromatography-mass spectrometry (GC-MS) still has the advantages of being generally available in most analytical laboratories, having higher separation efficiencies than LC and providing lower detection limits than LC combined with single MS detection, without such quantification problems.

This chapter focuses on the gas chromatographic (GC) methods (almost all of them based on GC-MS or GC-MS/MS) that have been reported in the literature for the determination of acidic pharmaceuticals

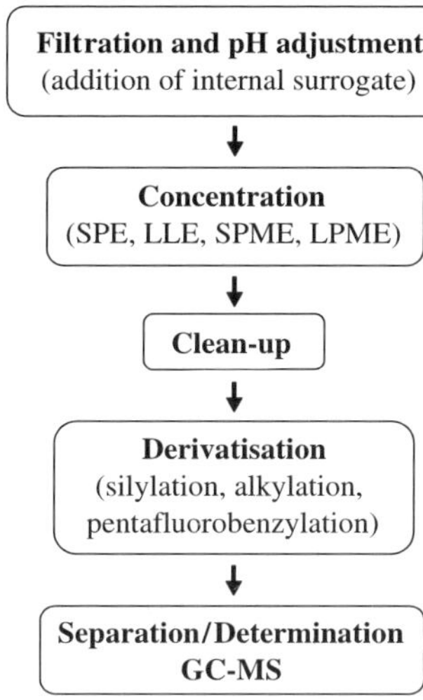

Fig. 2.5.2. Basic steps of GC methods for the determination of acidic pharmaceuticals.

in the aqueous environment. Emphasis will be put on sample preparation, including extraction/preconcentration of analytes from water samples and different derivatisation approaches. The number of methods for the determination of this class of drugs in solid matrices (sludge, sediment, etc.) is limited to very few publications and they will be also discussed.

A scheme of the different stages of the GC-based methods for the determination of acidic drugs in water samples is shown in Fig. 2.5.2. Basically, they comprise three main steps: extraction of the analytes, derivatisation and GC determination, normally using mass spectrometry detection. Other stages such as sample acidification or clean-up of the extract are not mandatory and they can be avoided depending on the type and volume of sample to be concentrated, the required enrichment factor and the selectivity of the determination technique. The three main stages (extraction/preconcentration, derivatisation and separation/determination) and some others are discussed within this chapter in individual sections.

2.5.2 PRELIMINARY TREATMENT

Sampling is an important (and the first) step in the analytical process, however, several books (e.g. [10,11]) deal with it and thus, it will not be discussed here. After sampling, several other steps have to be taken into account before extraction of the analytes, from sample preservation to pH modification. These previous steps may significantly affect the quantitative and qualitative results if not considered properly.

Stability of acidic pharmaceuticals during storage of wastewater samples is one of the aspects that have received less attention in the bibliography. The most frequent recommendation is to keep the samples in the dark at 4°C for a maximum of 24 h before being processed [12,13]. However, the only systematic study dealing with the stability of the analytes was performed by Lee et al. [14] using spiked treated wastewater samples stored at 4°C. Most of the compounds showed a good stability during at least seven days. Exceptions to this behaviour were acetaminophen and salicylic acid. The first specie was recovered in an extension around 80% from spiked distiled water samples, however, it could not be detected in fresh spiked sewage water showing an extremely low stability. In the case of salicylic acid only 12% of the spiked amount could be recovered after storing the sample at 4°C for 24 h. Furthermore, the concentration of fenofibrate decreased in a 25%, after 7 days of storage, probably due to its hydrolysis. In addition to these results, the easy biodegradability of acetylsalicylic in wastewater [15], might suggest also a low stability for this compound during the storage of the samples, however, available data have not been found.

Pinkston et al. [16] have proved that several acidic pharmaceutical compounds, containing phenolic or aromatic ether groups in their structures, react rapidly with free chlorine. Particularly, at neutral pH values, acetaminophen and indometacine show half-lives of a few minutes. As a consequence, their determination in drinking water may require the addition of reductant agents, such as ascorbic acid [17] or sodium thiosulphate [18], to the samples in order to consume the available chlorine, ensuring the integrity of the analytes during the storage step.

Another important aspect, when dealing with acidic analytes, is the pH adjustment often carried out before sample preparation. It has been reported that hydrolysis of fenofibrate to fenofibric acid and of clofibrate and etofibrate to clofibric acid is further accelerated by acidification [19]. Also, the formation of an artifact (1-(2,6-dichlorophenyl)indolin-2-one), from diclofenac, after acidification of wastewater samples [20] has been suggested. However, the extension of this last transformation is sample dependent and other processes such as natural occurring photochemical reactions could contribute also to the conversion of the parent drug into the above compound. Anyway, precaution must be taken into consideration for the determination of some of compounds and acidification avoided as far as possible. Furthermore, this may increase the hydrophobic character of the analytes, resulting in losses during filtration, which are neither found during filtration of non-acidified samples through glass fibre nor through cellulose filters [21], as compounds are in their ionic form at natural and wastewater pH values. Concentration of non-acidified samples also contributes to reduce the presence of humic and fulvic acids in the final organic extract avoiding the need of a further clean-up step and reducing the complexity of the resulting chromatograms, as well as the risk of column contamination [22]. In addition, for water samples showing a certain buffering capacity, e.g. seawater, the adjustment of the pH to values of 2–3 requires the addition of large amounts of strong acids [3].

On the other hand, conversely to all above considerations, performing the enrichment step at neutral pH requires the employment of high-efficient extraction methods to compensate for the higher water solubility of the target compounds. As a consequence, independently of the extraction technique, most authors recommend the acidification of the samples after filtration and previously to the concentration step.

2.5.3 EXTRACTION/PRECONCENTRATION

2.5.3.1 Water samples

Extraction of acidic pharmaceuticals from water samples has been carried out mainly by SPE due to its popularity and easy automation. Indeed, this is the preferred technique in screening and environmental studies dealing with a high number of samples. However, current trends in analytical chemistry are miniaturisation and reduction in the amount of organic solvents and samples to be consumed. So, several promising solid-phase microextraction (SPME) and liquid-phase microextraction alternative techniques have been developed, which can compete in performance with SPE, with a much lower consumption of sample and organic solvents. On the other hand, the use of liquid–liquid extraction (LLE) has been practically abandoned due to its obvious problems. In the following paragraphs, the use of these techniques for the extraction of acidic drugs from water samples is commented in detail. The extension dedicated to each section is proportional to the number of applications described in the bibliography, as well as, the novelty of the considered approach.

2.5.3.1.1 Liquid–liquid extraction (LLE)

Kanda et al. [23] have proposed an off-line LLE method for the determination of ibuprofen, clofibric acid and acetylsalicylic acid in wastewater samples. Samples (1 L volume) were adjusted at pH 2 and extracted twice using 100 mL of dichloromethane. After phase separation the organic extract was reduced to 0.1 mL. Recoveries from 90 to 100% were achieved for ibuprofen and clofibric acid, whereas, in the case of acetylsalicylic acid an unacceptable value (below 40%) was obtained. Continuous LLE has also been applied to the extraction of several acidic drugs from large volume (40 L) sewage water samples [17]. The process was carried out using dichloromethane as the organic phase and a water:organic solvent ratio of 10:1. Phase separation and evaporation of the organic extract was performed automatically. In spite of achieving an enrichment factor of 40000 folds, detection limits of the method for acidic drugs remained between 10 and 25 ng L^{-1}, which are in the same order of magnitude, or even higher, than those obtained by most other methods based on SPE or SPME. Another drawback of this approach was the high variability of the extraction procedure with relative standard deviations from 20 to 30%.

2.5.3.1.2 Solid-phase extraction (SPE)

As mentioned before, SPE is the most commonly employed technique for the extraction of acidic drugs. The analysis of these compounds in real samples requires the concentration of typical volumes from 0.2 to 2 L, depending on the water type and the demanded detection limits. In the case of sewage water, the maximum volume is normally limited to 1 L in order to avoid the clogging of the SPE sorbent. Main steps involved in the SPE of acidic pharmaceuticals are depicted in Fig. 2.5.3. Previously to the enrichment step, reversed-phase sorbents have to be conditioned using one or several organic solvents and ultrapure water adjusted at the same pH as the sample. Before elution, the sorbent must be dried pumping air through the cartridge or better using a gentle stream of dry nitrogen. Polar solvents such as methanol, acetone, acetonitrile or ethyl acetate have been employed in the elution step. The most appropriate solvent mainly depends on the polarities of the target compounds. In addition, its compatibility with the further derivatisation step should be also considered, otherwise the extract has to be dryness evaporated before being derivatised.

Addition of a surrogate to water samples is a common practice to compensate for the losses of the analytes during extraction and/or derivatisation steps. The use of phenoxy acid species has been proposed by some authors [14,24–26]; however, it should be kept in mind that some of these herbicides, particularly Mecoprop, have been often detected in river and wastewater [27–30], therefore, addition of these

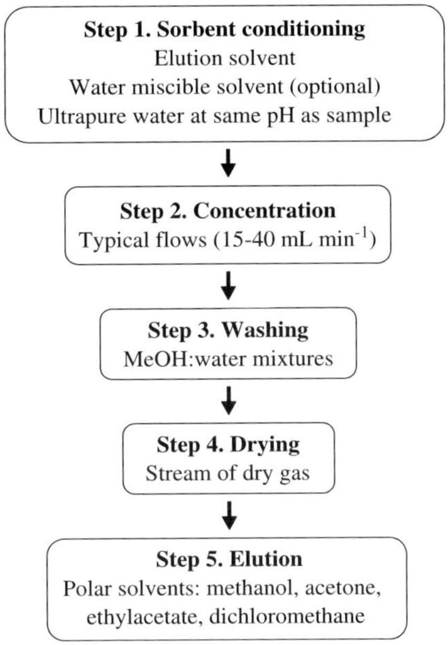

Fig. 2.5.3. Scheme of solid-phase extraction methods for the concentration of acidic drugs from water samples.

compounds to water samples, or to their extracts, might lead to underestimate the concentrations of the target species. Obviously, the problem can be overcome when the isotope labelled herbicides, instead of the native compounds, are employed [28,29,31,32]. Another compound proposed as internal surrogate is meclofenamic acid [12,21]. Although, it is also employed as a prescription drug in veterinary the parent compound is completely metabolised and therefore, it has not been detected in the aquatic environment [22].

Efficiency of SPE-based methods is normally controlled by the sorbent material. Table 2.5.2 summarised recoveries obtained using different reversed-phase materials for 1 L water samples, previously adjusted at pH 2–3. The classical octadecylsilane (C-18) cartridges, often packed manually in the laboratory and containing a typical amount of 0.5–1 g of sorbent, were the first proposed for the concentration of acidic drugs [24,25,33–37]. They have the advantages of being inexpensive in comparison with polymeric sorbents and able to provide good recoveries (over 80%) for most of the analytes, with the exception of the most polar compounds (e.g. acetylsalicylic and salicylic acid) that are recovered in an extension about 50% [15,36,38].

TABLE 2.5.2

Recoveries from acidified (pH 2–3) 1 L volume water samples using different reversed-phase sorbents

Sorbent	C-18	C-18	Lichrolut-EN	C-18 +Lichrolut-EN	Strata-X	OASIS-HLB
Amount (mg)	1000	500	200	250+100	60	200
Water type	Tap water	Ground water	Ground water	Ground water	River	Treated wastewater
Elution solvent	Acetone	MeOH	MeOH and acetone	MeOH	MeOH	MeOH
Elution volume (ml)	4	3	5	3	3	5
Reference	[24]	[15]	[72]	[15]	[41]	[14]
Salicylic acid	–	55	76	56	–	103
Clofibric acid	77	82	–	71	–	95
Gemfibrozil	49	89	70	49	–	98
Bezafibrate	93	92	–	70	–	–
Ibuprofen	67	81	91	82	92	92
Naproxen	68	91	83	54	102	93
Diclofenac	70	89	85	50	104	86

"–"corresponds to non-available data.

Polystyrene-divinylbenzene (PS-DVB) sorbents present a higher surface area than silica-based materials and, in addition to the reversed-phase mechanism, they have the capability of establishing π–π interactions with the aromatic ring contained in the structure of the acidic drugs [39,40]. As a consequence, 200 mg of these materials are enough to ensure recoveries similar to those achieved using 500 mg of C-18 (Table 2.5.2). Anyhow, differences among particle size, surface area, presence of different functional groups and introduction of additional monomers in some PS-DVB-based polymers, makes difficult a straightforward comparison of their retention efficiencies for acidic compounds.

Ternes et al. have also proposed the combination of C-18 and polymeric materials (like Lichrolut EN) for the extraction of acidic compounds from water samples [15,19,38]. The problem with this approach is that cartridges need to be packed manually in the laboratory, they did not improve the recoveries achieved using commercial available PS-DVB cartridges (Table 2.5.2) and, moreover, a high variability, especially between different batches of sorbents from the same supplier [15,38], was observed for these mixed sorbents.

Nowadays, one of the preferred sorbents for the reversed-phase extraction of organic compounds from water samples is the Oasis HLB®, a copolymer of DVB and *N*-vinylpyrrolidone, with hydrophilic and lipophilic characteristics as well as excellent wetting properties improving the transference of the analytes from the sample to the sorbent. In addition to data reported in Table 2.5.2 for 200 mg cartridges, Rodríguez et al. [21] have obtained breakthrough volumes higher than 2 L for three profens and two anthranilates drugs considering just 60 mg of this polymer. The small amount of sorbent allowed the quantitative elution of the analytes using only 2 mL of ethyl acetate. Similar recoveries have been obtained for different acidic drugs in different works using the same amount of this sorbent [5,14,28]. Other last generation sorbents, such as Strata-X, have achieved similar extraction efficiencies also using 60 mg cartridges; however, only data for ibuprofen, naproxen and diclofenac have been reported (Table 2.5.2) [41].

In addition to PS-DVB cartridges [29,42,43], PS-DVB disks [44] or broad (45 mm) packed cartridges [3] have been employed for the extraction of large volume samples (up to 20 L) in the multiresidue determination of several pollutants (including pharmaceuticals) in surface and sea water. Disks offer the advantage that samples can be enriched at a high-flow rate (100–500 mL min^{-1}) but also need higher amounts of solvents for their elution ($>$50 mL) when compared to cartridges. Therefore, they are best suited for the concentration of large sample volumes, when very low detection limits are required and when this step is to be carried out in-field (e.g. sea monitoring campaigns).

To obtain cleaner extracts, some authors have evaluated the retention of acidic drugs on different reversed-phase materials without adjusting the samples at low pHs. Lin et al. [45] have compared the retention capabilities of different sorbents using 500 mL volume samples adjusted at pH 5. The results of this comparison (including the neutral pharmaceutical carbamazepine) are shown in Table 2.5.3. The Oasis HLB polymer provided the best recoveries using only 60 mg of sorbent. Further studies, using the same amount of sorbent, have reported recoveries higher than 80% for ibuprofen, naproxen, diclofenac and mefenamic acid from 500 mL water samples adjusted at pH 7 [22]. This finding is in good agreement with those from Weigel et al. [30], who has evaluated different polymers for the extraction of acidic drugs from 1 L tap water samples adjusted at pH 7.8 (Table 2.5.4). Considering sorbent amounts from 200 to 500 mg, Oasis HLB provided better recoveries than conventional PS-DVB sorbents, Chromabond EASY (containing weak-anion exchange groups) and several other functionalised DVB copolymeric sorbents. Only carboxy-ibuprofen, the most polar metabolite of ibuprofen, could not be retained in the OASIS polymer at pH 7.8. A

TABLE 2.5.3

Recoveries and RSD (%) obtained for selected pharmaceuticals by different sorbents from spiked deionised water (500 mL, pH 5). Adapted from Ref. [45]

Compound	% Recovery (RSD)			
	C-18 (500 mg)[a]	C-18 (500 mg)[b]	Lichrolut EN (200 mg)	Oasis HLB (60 mg)
Clofibric acid	52 (10)	70 (9)	40 (10)	95 (1)
Ibuprofen	35 (20)	41 (18)	25 (8)	77 (5)
Carbamazepine	109 (15)	107 (11)	95 (10)	93 (6)
Naproxen	46 (8)	67 (10)	46 (9)	91 (10)
Ketoprofen	64 (8)	82 (10)	28 (20)	102 (6)
Diclofenac	27 (15)	56 (9)	30 (9)	92 (2)

[a]Eluted with 8 mL of acetone-ethyl acetate (1:1).
[b]Eluted with 8 mL of acetone-ethyl acetate (2:1).

TABLE 2.5.4

Comparison of polystyrene-based sorbents for the retention of selected acid drugs from 1 L tap water buffered at pH 7.8. Adapted from Ref. [30]

Sorbent type	Comercial name	Sorbent amount (mg)	% of recovery				
			Clofibric acid	Bezafibrate	Ibuprofen	Diclofenac	Paracetamol
PS-DVB	Bakerbond SDB-1	200	61	81	72	72	60
PS-DVB	Chromabond HR-P	500	36	37	15	33	72
PS-DVB	LiChrolut EN	200	38	77	72	73	37
OH functionalised PS-DVB	Isolute Env+	200	58	57	72	52	39
PS-DVB-WAX	Chromabond EASY	500	51	28	31	1	50
MA-DVB	Abselut nexus	200	23	87	68	90	0
DVB-VP	OASIS HLB	200	83	95	98	102	14

PS, polystyrene; DVB, divinylbenzene; WAX, weak anion exchancher; MA, methacrylate; VP, *n*-vinylpirrolidone.

drawback of using up to 500 mg of high retentive polymers, and pH values over the pKa of the target compounds, is that elution volumes up to 40 ml are required [30]. Marchese et al. [18] achieved to reduce this volume to only 5 mL by amending tetrabutylammonium chloride to the elution solvent (methanol) in the determination of five acidic pharmaceuticals by LC-MS/MS. However, this modifier may interfere with the further derivatisation step, when the compounds are to be detected by GC.

Apart from specific methods focussed on the determination of acid compounds, the Oasis-HLB polymer is also useful in multiresidue screening strategies. In this case, the employment of elution solvents with different polarities can be exploited to fractionate the analytes into different compound classes [3,30]. So, cartridges (200 mg) may be sequentially eluted into different fractions: first with *n*-hexane (eluting hydrophobic compounds), then with ethyl acetate (eluting polar neutral and basic compounds) and finally with methanol (to recover the acidic analytes) [30].

Finally, the use of molecularly imprinted polymers (MIP) may develop into an interesting alternative for SPE of water samples, as promising results have been obtained for the selective extraction of naproxen from urine [46]. Although, the direct application to water is still complicated, promising results have been obtained for the selective extraction of phenols [47,48]. In these cases, the analytes are first retained in an unspecific way on the MIP and selectivity is gained in the washing step by applying a slightly polar aprotic solvent, which washes out the interferences while the analytes remain retained, by the specific H-bonding and hydrophilic interactions. These findings and the fact that a company has already been created (MIP Technologies AB, Lund, Sweden) for the exclusive production and commercialisation of MIP materials may popularise them in the future.

A further application of SPE to the determination of pharmaceuticals in the environment is its use as a clean-up technique, as discussed in Section 2.4.4.

2.5.3.1.3 Solid-phase microextraction (SPME)

Since the commercialisation of SPME in 1989 [49], the number of applications has been growing year by year. This technique allows the extraction of analytes using a microfibre coated with an appropriate sorbent, e.g. polydimethylsiloxane (PDMS), by equilibrium processes. In a second step, analytes are desorbed directly into the injector port of the GC, leading to a completely solvent-free method.

The first PDMS fibres were more suitable for the concentration of non-polar compounds. But, after commercialisation of carbowax (CW) and polyacrylate (PA)-coated fibres, SPME could be applied to polar compounds. Thus, two SPME methods were developed for the analysis of ibuprofen [50], other acidic pharmaceuticals and polar compounds [51] in water and wastewater samples. The best extraction conditions were found to be acidification of samples to pH 2 without addition of salt and employing a polar PA fibre directly exposed to water samples. The kinetic of the extraction process was relatively slow (more than 90 min before reaching the equilibrium), and thus, compromising sampling times of 30 min were employed. The main limitation of these two applications was their relatively poor detection limits (0.1–0.2 $\mu g L^{-1}$ for ibuprofen) that were about one order of magnitude higher than those achieved after SPE of moderate water volumes, ca. 500 mL. The reason was not the low efficiency of the microextraction step but the direct GC determination of the (non-derivatised) acidic drugs that led to broad, tailing peaks.

The difficulty to combine SPME and derivatisation of the target acidic drugs is that available derivatisation reagents (see corresponding section) are not stable in the aqueous media. To avoid this inconvenient, on-fibre derivatisation approaches can be used. Up to now the only application of this strategy to the analysis of acidic drugs was developed by Rodríguez et al. [52]. For that, analytes are first extracted into a PA fibre for 40 min and afterwards they are derivatised by exposing the fibre (loaded with the analytes) to the vapours of a derivatisation reagent during 20 min (Fig. 2.5.4), in this case *N*-methyl-*N*-(tert-butyldimethylsilyl)-trifluoroacetamide (MTBSTFA). The limits of quantification (LOQ) obtained by this combination for five acidic drugs were comprised between 12 and 40 $ng L^{-1}$, considering a sample intake of only 22 mL. These values are similar to those achieved by same authors after SPE of 500 mL water samples using similar GC-MS detection conditions and the same silylation reagent (Table 2.5.5). In addition, the LOQs of the SPME method could be reduced, if necessary, by increasing the sampling time, due to the slow kinetics of the partition process (Fig. 2.5.5).

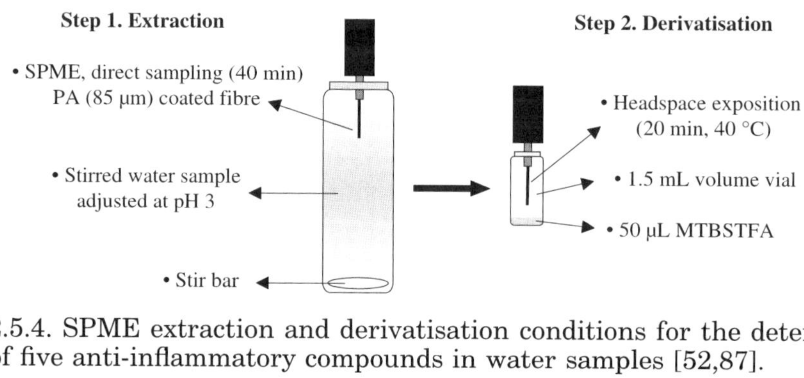

Fig. 2.5.4. SPME extraction and derivatisation conditions for the determination of five anti-inflammatory compounds in water samples [52,87].

TABLE 2.5.5

GC-MS quantification limits (S/N 10) for selected anti-inflammatory drugs, as tert-butyldimethysilyl derivatives, using SPE and SPME as concentration techniques. Adapted from Ref. [21,52]

Compound	Quantification limits (ng L^{-1})	
	SPE (500 mL samples)[a]	SPME (22 mL samples)[b]
Ibuprofen	10	18
Naproxen	10	15
Ketoprofen	25	40
Tolfenamic acid	15	12
Diclofenac	25	20

[a]Concentration using 60 mg OASIS cartridges.
[b]Direct extraction (40 min) using a PA coated SPME fibre.

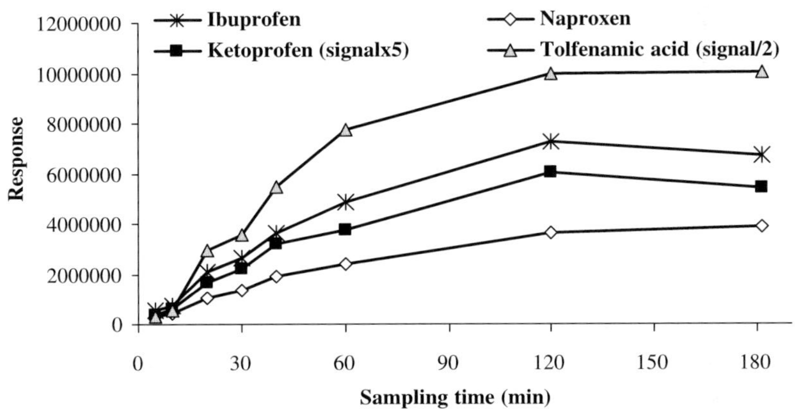

Fig. 2.5.5. SPME extraction kinetics of selected acidic drugs using a PA fibre. Direct sampling at room temperature [52,87].

However, care must be taken when SPME is employed for the quantification of water samples with a high content of dissolved organic compounds. Since it is a partition technique, thus, the presence of organic dissolved matter in the samples may change the sample-fibre distribution coefficient of the analytes and also the kinetics of the process, resulting normally in lower extraction efficiencies [53]. In the case of acidic drugs, reductions in the yield of the extraction from 20 to 30% have been reported for wastewater in comparison to model solutions prepared using ultrapure water [51,52]. Anyhow, the importance of this

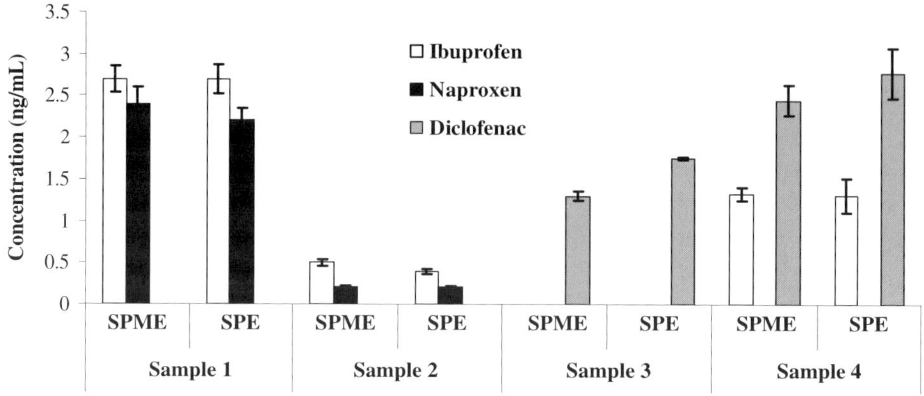

Fig. 2.5.6. Comparison of results obtained for ibuprofen, naproxen and diclofenac in non-spiked wastewater samples using SPE and SPME. Adapted from Refs. [21,52,87].

effect might change from sample to sample, making advisable to employ the standard addition technique to improve the reliability of the results for real samples. Taking in account this precaution, both SPE and SPME are equivalent techniques in terms of accuracy and precision (Fig. 2.5.6).

2.5.3.1.4 Liquid-phase microextraction (LPME)

Another equilibrium technique that is under investigation is LPME. This is based on the concentration of the analytes into a small volume (a few μL) of an acceptor solution, either by direct contact or through a porous membrane [54,55]. This last option seems to be more robust and promising. It can be carried out either in a two phases system (Fig. 2.5.7a) or in a three phases one (Fig. 2.5.7b). In the two phases system the analytes are extracted from the sample into an organic solvent impregnating the pores of the hollow membrane and also contained in its interior.

However, in the three phases system, the sample pH is adjusted adequately to the analytes acid–base character (to an acidic pH in the case of the acidic drugs discussed in this chapter) in order to obtain their neutral species. Thus, they can be extracted to the organic solvent that impregnates the membrane pores and finally transferred to the acceptor aqueous solution inside the hollow membrane. The pH of this solution is selected to trap the analytes into their ionic form (i.e. a basic pH is selected for the acidic drugs) [54,55]. Obviously, this option is best suited when the analytes will be analysed by LC or CE, rather than by GC.

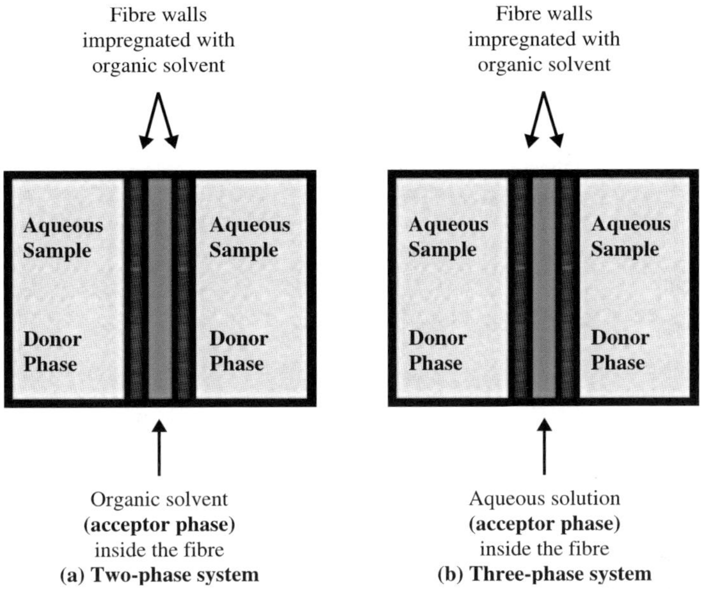

Fig. 2.5.7. Cross section of the hollow fibre inside the aqueous sample during (a) two-phases and (b) three-phases LPME. Reproduced from Ref. [55]. Copyright (2003), with permission from Elsevier.

Hollow fibre LPME has already been investigated for the determination of several acidic drugs by capillary electrophoresis (CE) [56], LC [57] and LC-MS/MS [58]. This results in a very selective extraction, which may allow to overcome matrix effects commonly observed in LC-MS/MS [58], when conducted in a three phases system.

The only application of LPME for the determination of several polar organic pollutants, including pharmaceuticals, in water samples by GC was reported by Müller et al. [59]. In that work, LPME of a 4 mL sample was carried out through a porous hollow fibre membrane to a final acceptor solution consisting of 40 μL of *n*-octanol into a semi-automated procedure (Fig. 2.5.8). The obtained LOD value for ibuprofen was $20\,\mathrm{ng\,L^{-1}}$, which can be considered acceptable, especially considering that no derivatisation was done.

Until now, the combination of LPME and derivatisation has only been explored for the GC determination of phenolic compounds [60]. This was achieved by adding a small amount of bis(trimethylsilyl)trifluoroacetamide (BSTFA) to the organic extract and then the derivatisation reaction takes place in the injection port of the chromatographic system to yield the silyl-derivatives. It is expected that a similar approach may

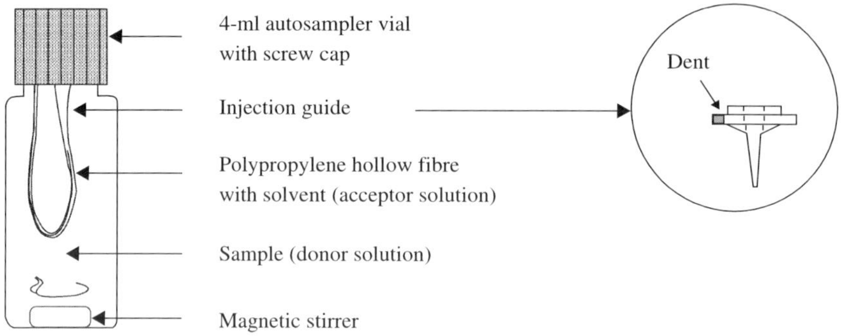

Fig. 2.5.8. Schematic set up of the hollow-fibre extraction for the semi-automated LPME of biological active compounds from water samples. Reproduced from Ref. [59]. Copyright (2003), with permission from Elsevier.

be used for acidic pharmaceuticals leading to better LODs and this technique being applicable to a wider range of compounds in the future.

Finally, a commercial LPME system based on the use of a thin (30 μm wall thickness) non-porous membrane has already been launched by the company Gerstel (Mühlheim, Germany) in cooperation with the UFZ-Environmetal Research Centre (Leipzig, Germany), were several applications have been also developed [61–63]. This is expected to aid in the popularisation of this technique, making possible its automation at the same time.

2.5.3.2 Solid samples

As it has been mentioned, the number of available methods for the determination of acidic drugs in sludge and sediment is rather limited. One explanation for this lack of information is the difficulty of such determinations, particularly in the case of sludge samples from sewage plants that contains up to 40% of carbon. Moreover, as target analytes are in their ionic form at common pHs of sewage water, their partition to particulate matter is expected to be quite low. In spite of this assumption, some authors have developed theoretical models to calculate the fraction of acidic compounds associated to solid wastes in sewage treatment plants. For example, Jones et al. [64] have estimated that about 16% of the ibuprofen amount that enters a sewage treatment plant remains associated to primary sludge. Unfortunately, the accuracy of this prediction was not verified with experimental data. Khan et al. [65] have measured concentrations of naproxen, ibuprofen and gemfibrozil in primary sludge at the $\mu g\,g^{-1}$ level (referred to dried weight) in treatment plants receiving raw influents that contain the three analytes at concentrations below 5 μg L^{-1}. On the other hand, the same study showed negligible amounts of the same compounds in digested sludge. Extraction was performed using the Soxhlet technique; however, details regarding extraction and clean-up conditions have not been reported. The interesting thing of the above work is that a reasonable agreement was found between measured concentrations and those predicted using a theoretical model [65].

Other sample preparation methods applied to the determination of acidic drugs from sludge and solid samples are based on the combination of extraction and clean-up steps. Pressurised solvent extraction using ethyl acetate containing a 0.2% (v/v) of trifluoroacetic acid has been proposed to estimate the sorption of ibuprofen, clofibric acid and carbamazepine in gravel from land-based wastewater treatments [26]. The obtained extract was evaporated nearly to dryness, reconstituted using 200 ml of water and processed as a wastewater sample. After derivatisation, determinations were carried out by GC-MS. Found concentrations ranged from 14 ng g^{-1} for clofibric acid up to 97 ng g^{-1} for carbamazepine.

Ultrasound extraction (USE) has also been employed for the extraction of several acidic drugs from sediment and sludge [66,67] using polar solvents such as methanol, acetone and ethylacetate in combination with a small percentage of an organic acid to improve the yield of the process. Again, obtained extracts were concentrated, until complete removal of the organic solvent, reconstituted with water and further processed as a sewage water sample. Then, these secondary extracts were subjected to clean-up and analysed by LC-MS/MS, resulting in acceptable relative recoveries. Finally, the application of the method to sludge showed important concentrations of diclofenac (from 200 to 450 ng g^{-1}), in both activated and digested sludge samples taken in Germany, while other acidic analytes were not detected.

The extraction of diclofenac from lake sediment has also been carried out by simply shaking-extracting the sample with methanol and submitting this extract to several clean-up steps [43]. The recovery obtained with this procedure was 90%, for samples spiked at relatively high-concentration levels (700 ng g^{-1}), and the application to lake water and sediment showed negligible adsorption.

From the above comments, it is evident that specific clean-up methods for the chromatographic analysis of acidic drugs in sediments and sludge are not available. Levels of these compounds in those solid samples are expected to be low in comparison to neutral of less polar personal care products such as musk fragrances [66] or bactericides as triclosan

[68,69], but not negligible, at least in the case of sludge samples. However, at this moment, very few experimental data have been obtained regarding the concentration of acidic drugs in bio-solid wastes. Moreover, detection limits of most reported methods (10–20 $ng\,g^{-1}$) need to be reduced in order to be able to investigate the presence of the analytes in sludge at lower concentrations. Improvements in the clean-up step and in the selectivity of the determination are key steps to achieve this aim.

2.5.4 CLEAN-UP

As the concentration of acidic pharmaceuticals in most water samples is relatively high (up to several $\mu g\,L^{-1}$ in some cases), clean-up of the extracts is considered not to be mandatory for their GC determination. However, depending on the sample nature, it may be advisable to include a clean-up step before the chromatographic separation of the target species.

This has been done by using normal phase materials, from 0.7 g up to 3 g of silica [29,44], as it is commonly carried out for the determination of hydrophobic compounds. Depending on the chosen derivatisation reaction, the clean-up step can be carried out after derivatisation [29,31,35] or previously to this step [44]. When SPE was used as the concentration technique a simple way to reduce the presence of very polar interferences in the sample extract consist on washing the reversed-phase sorbent with a few ml of water containing from 10 to 25% of methanol [14,28]. Alternatively, cleaner extracts may be obtained by a simple solvent exchange [38] or by fractionation of the SPE extracts with different solvents [30], as discussed in the section dedicated to solid-phase extraction of water samples.

Obviously, in the case of solid samples, like sediments and sludge, clean-up becomes a mandatory step in order to remove interfering compounds during GC determination, to avoid column contamination and countering matrix effects in LC-MS/MS determination. Clean-up of sediment [66] and sludge [67] has been carried out, exploiting the acidic characteristics of the analytes, by using a weak ionic exchanger for purification of the SPE of the water-diluted extracts. As stated in the previous section, analytical methods for the determination of acidic drugs in solid samples are in an early stage of development. Effective clean-up strategies for the purification of organic extracts obtained from solid matrices such as sludge using exhaustive techniques, e.g. Soxhlet or ASE, is probably the most challenging step of the whole procedure.

2.5.5 DERIVATISATION

Most acidic pharmaceuticals, with the exception on some lipid regulators (e.g. clofibrate, fenofibrate, etc.), contain carboxylic groups in their structures and therefore need to be derivatised prior to their GC analysis. Even when a few of them may be determined without derivatisation, this leads to poor detection limits and problems, especially when the GC system is not completely clean [17,51] and the column ages. For some compounds, GC determination is simple impossible without derivatisation: e.g. clofibric acid decomposes in the injector to produce 4-chlorophenol [51].

As a consequence, many different derivatisation reactions and approaches have been investigated in order to provide good chromatographic characteristics for the corresponding peaks, operational simplicity and safety for the analyst. Derivatisation is in fact the only extra step included in sample preparation strategies for the determination of acidic drugs using gas chromatography, instead of liquid chromatography or capillary electrophoresis. Table 2.5.6 shows a possible classification of the most common derivatisation reactions for acidic pharmaceuticals. In all cases, the process is carried out after extraction of the analytes from the sample. The reaction can be performed combining the sample extract with the derivatisation reagent at an appropriate temperature for a given time (off-line derivatisation mode), or it can take place in the injector of the gas chromatograph when analytes and reagent are exposed at high temperatures (on-line

TABLE 2.5.6

Summary of different derivatisation approaches for the gas chromatographic determination of acidic drugs

Derivatisation mode	Reaction	Reagent	Reference
Off-line	Methylation	Diazomethane	[13,15,23,29,31,32,92]
		Methylchloroformate	[3,4,30,73]
		BF3: MeOH	[71]
	Silylation	BSA	[34]
		BSTFA	[14,22,26,44,93]
		MSTFA	[41,81]
		MTBSTFA	[21,25,52,94]
	Pentafluorobenzylation	PFBBr	[24,25,37,77,94]
On-line (in-port)	Methylation	TMSH	[2,13,33,74,76]
		TMAH	[13]
	Butylation	TBA-HSO4	[45]

TABLE 2.5.7

Summary of off-line derivatisation conditions for acidic drugs

Reagent	Derivatisation conditions					Reference
	Vol (μL)	Solvent	Time (min)	Temperature (°C)	By-products removal	
Diazomethane (0.1 M diethylether)	75	Acetone	120	5	Evaporation	[23]
Diazomethane (3–5% diethylether)	150	Hexane	60	−20	Addition acetic acid	[15]
BF3 (14% methanol)	2000	Methanol	120	85	Evaporation and addition of K_2CO_3(1%)	[36]
Methyl chloroformate	7	Acetonitrile/ methanol/ water/ pyridine	non available	Room T	Evaporation and addition of water	[3,4]
PFBBr (2% toluene) plus 4–5 μL TEA	100–200	Toluene	60	90–110	Evaporation	[25,37,77]
PFBBr (2% cyclohexane) plus 2 μL TEA	200	Cyclohexane	120	100	Evaporation	[24]
BSA (5% TMCS)	100	None	60	120	No	[34]
BSTFA (1% TMCS)	100	Methyl tert-butylether	20	60	Addition of water	[14]
BSTFA (1% TMCS)	1000	None	20	80	No	[44]
BSTFA	200	Acetonitrile	60	70	No	[93]
BSTFA	50	Ethylacetate	20	60	No	[22]
MSTFA	50	None	45	35 min (room T), 10 min (60 °C)	Evaporation	[81]
MTBSTFA	200	Ethylacetate	60	60	No	[21]
MTBSTFA	50	Acetonitrile	60	80	No	[25]
MTBSTFA	50	"On-fibre" (SPME)	20	40	No	[52]

approach). Table 2.5.7 presents a summary of experimental conditions in the off-line derivatisation mode using different reagents.

Electron-impact mass spectra of the acidic drugs, as well as their fragmentation patterns, change depending on prepared derivatives (Fig. 2.5.9). As a consequence, using GC-MS as the determination technique, the performance of the whole method (quantification limits, resolution between peaks, elution order, etc.) is affected by the derivatisation step [25]. In addition, even when the same derivatives are obtained, quantification limits of the analytical procedure might change slightly as function of the chosen derivatisation reaction [13]. Main

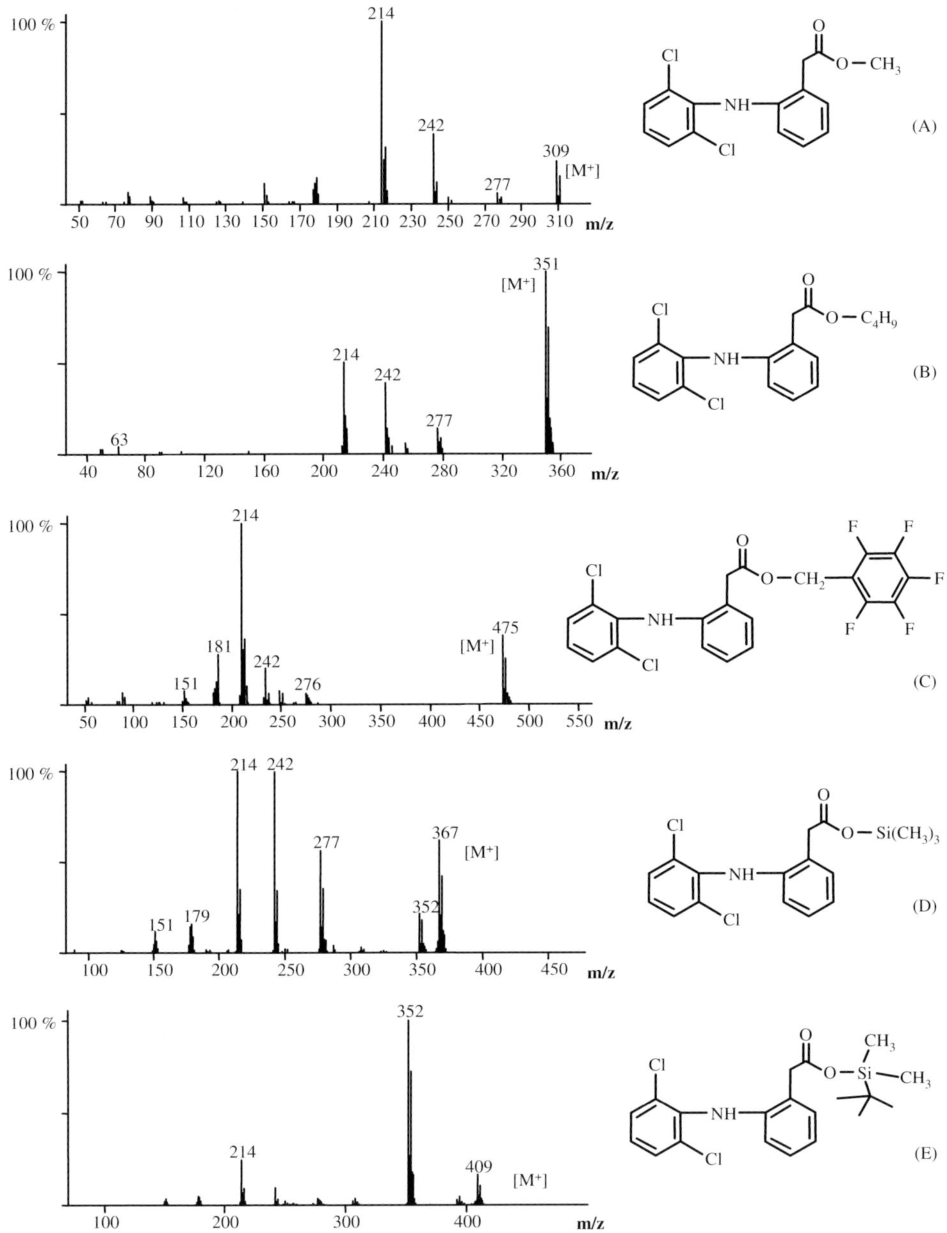

Fig. 2.5.9. Electron impact mass spectra for different diclofenac derivatives. (A) Methyl-, (B) Butyl-, (C) Pentafluorobenzyl-, (D) Trimethylsilyl- and (E) Tert-butyldimethylsilyl- derivatives [87]. Reproduced from Ref. [20,45]. Copyright (2003, 2005), with permission from Elsevier.

characteristics of derivatisation strategies and reagents presented in Table 2.5.6 are discussed in detail in the following paragraphs.

2.5.5.1 *Off-line* alkylation reactions

2.5.5.1.1 Diazomethane

Diazomethane was the first derivatisation reagent employed in the GC analysis of various acidic drugs in water samples [19] and it is still one of the most popular derivatisation reactions [15,28,29,31,34,35,38,43]. The reaction takes place in mild conditions, normally at sub-ambient temperatures, avoiding the risk of partial decomposition of the analytes during this step and leading to the quantitative formation of the corresponding methyl esters (Fig. 2.5.10a).

However, the high reactivity of diazomethane is also its major drawback. Because of its low stability, it must be generated in the laboratory, stored at –20°C and handled with care to avoid explosion risks. It is also toxic and carcinogenic [70]. In addition, the excess of diazomethane has to be removed when the reaction is completed. Normally, this implies the further dryness evaporation of the mixture [13,23]. Although diazomethane also reacts with hydroxylic groups, its applications to the derivatisation of acidic drugs containing carboxylic and hydroxylic moieties, e.g. hydroxy-ibuprofen, suggest that only the first is transformed into the corresponding methyl ester [31]. In fact, Ternes et al. [15,38] have developed a two steps method for the derivatisation of a large number of pharmaceuticals containing one or more acidic groups: diazomethane was first added to the sample extract and carboxylic groups were converted into their methylated esters, when the temperature was maintained below –15°C. Then the excess of reagent was removed and phenolic moieties were acetylated using acetic anhydride in presence of triethylamine at 80°C. Obviously, the global derivatisation procedure resulted time-consuming.

Because of all these drawbacks, other derivatisation procedures have been described in the literature.

2.5.5.1.2 Other methylation reagents

Methanol-containing boron trifluoride (MeOH:BF3) [36,71,72] and methyl chloroformate [3,4,30,73] have been applied to the methylation of different acidic drugs after their extraction from environmental samples (Fig. 2.5.10b). In the first case the process is carried out at 85°C for 2 h, whereas in the second one the reaction is performed at room temperature adding the derivatisation reagent to the sample extract

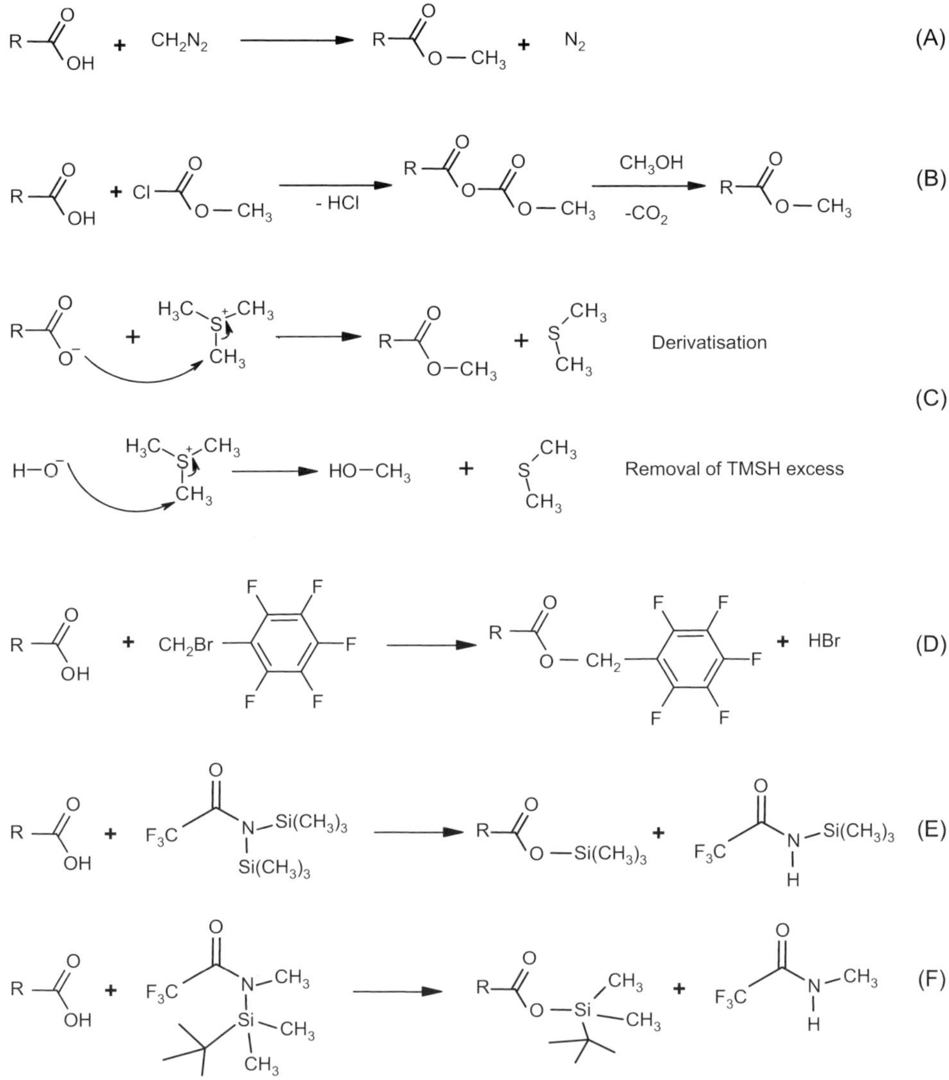

Fig. 2.5.10. Scheme of derivatisation reactions for carboxylic acids. (a) Diazomethane, (b) methanol/methyl chloroformate (c) TMSH, (d) PFBBr, (e) BSTFA and (f) MTBSTFA.

dissolved in a mixture of acetonitrile–methanol–water–pyridine. Both procedures are laborious and time-consuming requiring several evaporation and solvent exchange steps in order to remove the excess of derivatising agent, which would result into column deterioration if injected. Further drawbacks of derivatisation methods employing methanol and BF_3 are that acetylsalicylic acid and indomethacine are degraded during the process [36] and that the yield of the reaction for gemfibrozil is not quantitative [72].

2.5.5.2 On-line alkylation

A simple derivatisation approach consists on generating the derivatives of the analytes in the hot injector of the GC system, after adding an ion-pair reagent to the organic extract containing the acidic drugs. Trimethylsulphonium hydroxide (TMSH) is the most popular reagent to perform this type of derivatisation for these analytes (Fig. 2.5.10c). An aliquot of the commercial available reactive (a 0.25 M solution in methanol) is mixed with the sample extract and the mixture injected directly in the GC instrument, furnished either with a programmable temperature injector [2,74] or with a conventional split/splitless one [13,33]. The process yields the methyl derivatives of the organic acids while the excess of TMSH decomposes at the high-injection temperatures to produce methanol and dimethyl sulphide [75], which are very volatile and do not interfere with the target analytes. The only limitation of this approach is that the determination step must be carried out within the same day of addition of TMSH [2]. This method has been employed by Winkler et al. [33] and Zwiener and coworkers [2,42,76] in the analysis of diclofenac, clofibric acid, ibuprofen and its metabolites. In addition to TMSH, trimethylanilinium hydroxide (TMAH, 0.2 M in methanol) has also been proposed for the on-line methylation of several acidic pharmaceuticals [13].

Another possibility is the use of a butylation reagent instead of a methylation one. This has been done by Lin et al. [45], by employing tetrabutylammonium hydrogen sulphate as ion-pairing agent and an injection volume of 10 μL using a customised injection device. This method allowed the derivatisation of five acidic drugs, while carbamazepine (a neutral compound) was detected underivatised.

2.5.5.3 Pentafluorobenzyl-derivatives

Pentafluorobenzyl-derivatives are formed by the reaction of the acidic analytes with pentafluorobenzyl bromide (PFBBr) in the presence of triethylamine as a catalyst (Fig. 2.5.10d) at a temperature of 90–110°C for 1–2 h [20,24,25,37,77] (Table 2.5.7).

In a study of Reddersen and Heberer [25], where this derivatisation agent was compared to *N*-(tert-butyldimethylsilyl)-*N*-methyltrifluoroacetamide (MTBSTFA), it was found that PFBBr was not capable of derivatising as many compounds as MTBSTFA. However, it provided slightly better detection limits and reproducibility and could successfully be used to derivatise 19 acidic pharmaceuticals and pesticides.

A further advantage of this reaction is that the introduction of the electrophilic PFB group makes the analytes suitable for their sensitive determination by electron-capture detection (ECD) and negative chemical ionisation-mass spectrometry (NCI-MS). This procedure has already been described for the determination of endocrine disrupting compounds in drinking and surface water at the pg L^{-1} level [78]. However, only one application of this approach to acidic pharmaceuticals has been reported in the literature and it is focussed on the detection of ketoprofen in plasma and urine samples by GC-NCI-MS [79]. In that work the carboxylic group is converted to the PFB derivative but also the keto group is transformed into the O-trimethylsilyl-derivative in order to avoid the interference of fatty acids present in that kind of samples.

2.5.5.4 Silyl-derivatives

The last group of derivatisation reactions is the formation of silyl-derivatives. This is the most widely used derivatisation technique, it involves the replacement of an acidic hydrogen with and alkylsilyl group [80]. Two different silylation reactions have been applied to the determination of acidic drugs: formation of trimethylsilyl derivatives and formation of tert-butyldimethylsilyl-derivatives. Silylation agents are able to react with carboxylic and hydroxylic moieties, therefore analytes such as salicylic acid, which contains both functional groups, yield the disilylated products [14], avoiding the use of two-step approaches as those proposed by Ternes et al. [15,38] employing first diazomethane and then acetic anhydride. On the other hand, the amino group containing in the structure of species such as tolfenamic acid and diclofenac remains underivatised; however, neutral drugs with amide moieties (e.g. carbamazepine, primidone and oxazepen) also can be silylated allowing their GC determination [25]. These compounds are partially or completely decomposed when they are not derivatised previously to their injection in the GC system and, in addition, they do not react with diazomethane [32].

Trimethylsilyl-derivatives of the acidic pharmaceuticals can be obtained by reacting the (non-derivatised) analytes with a series of different reagents: bis(trimethylsilyl)acetamide (BSA) [34], bis(trimethylsilyl)triflouroacetamide (BSTFA) [14,22,44] and *N*-methyl-*N*-trimethylsilyltrifluoroacetamide (MSTFA) [81].

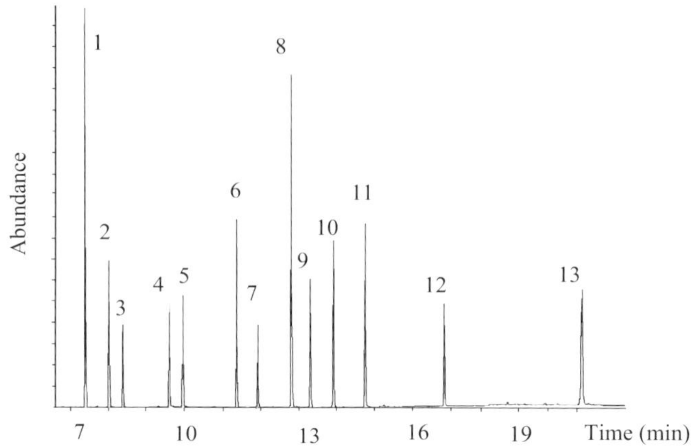

Fig. 2.5.11. GC-MS chromatogram for a group of acidic drugs as their trimethylsilyl derivatives: 1. Salicylic acid, 2. Clofibric acid, 3. Ibuprofen, 4. Acetaminophen, 5. 2,3-Dichlorophenoxyacetic acid (I.S.), 6. Gemfibrozil, 7. Fenoprofen, 8. Naproxen, 9. Triclosan, 10. Ketoprofen, 11. Diclofenac, 12. Fenofibrate and 13. Indomethacin. Reproduced with permission from ref. [14]. Copyright CAWQ, Burlington, Ontario, Canada.

BSTFA is the one most often used. Its reaction (Fig. 2.5.10e) is normally carried out at moderate temperatures (60–80°C) for 20–60 min, using in some cases TMCS as catalyser (Table 2.5.7). BSTFA is able to derivatise both the carboxylic and phenolic groups of 12 acidic drugs, at least, as shown in the chromatogram represented in Fig. 2.5.11 [14]. However, it is quite sensitive to moisture and care must be taken in order to have dry extracts. Thus, Boyd et al. [44] evaporated completely the SPE extracts to dryness before proceeding to their derivatisation as they found that anhydrous sodium sulphate failed into the complete elimination of water traces. Similar derivatisation conditions have been reported using MSTFA [81].

The second alternative (tert-butyldimethylsilyl-derivatives) is achieved by the reaction of the analytes with *N*-(tert-butyldimethylsilyl)-*N*-methyltrifluoroacetamide (MTBSTFA) (Fig. 2.5.10f). The reaction was carried out at high temperature in non-protic solvents. The experimental derivatisation conditions were optimised by Rodríguez et al. [21] for a group of six carboxylic drugs in the ethyl acetate extracts from waste water samples, a mixture of 1:4 (MTBSTFA:extract) heated at 60°C for 1 h were found as the optimal conditions. Reddersen et al. [25] have also described the use of MTBSTFA for the derivatisation of several drugs containing hydroxy, carboxy or amide groups. In this case, the reaction was performed in acetonitrile. In both applications, after finishing the derivatisation step, the reaction mixture was allowed to return to room temperature and injected directly in the GC column without removing the excess of MTBSTFA (Table 2.5.7).

The advantages of this last compound over the other silylation agents are that it is less sensitive to moisture and that the obtained derivatives exhibit a characteristic electron impact-mass spectra which provide an easy identification of the analytes [82]. As depicted for diclofenac in Fig. 2.5.9, the mass spectra of the compound prepared using MTBSTFA is completely different to that obtained for the trimethylsilyl derivative of the same compound. In the first case, the intensity of the molecular ion is relatively weak whereas an intense peak is observed at *m/z* [M-57], corresponding to the loss of the tert-butyl group. The same fragmentation pattern has been described for other profens (ibuprofen, naproxen and ketoprofen) and anthranilate (tolfenamic and meclofenamic acid) pharmaceuticals [21]. The result of this poor fragmentation, with a very intense signal in the MS spectrum, is a decrease in the minimum detectable amount of the target compounds. It is also expected that the presence of a very intense ion in the MS spectrum will lead to the achievement of very low detection limits when selected as the parent ion in MS/MS; however, this detection mode has been only employed with methyl derivatives of the analytes [29,31]. On the other hand, the main drawback of MTBSTFA is its lower reactivity in comparison to other silylation reagents [83].

In general, all silylation reagents lead to the formation of volatile derivatisation by-products. Although, in order to remove these by-products some authors have recommended the dryness concentration of the reaction mixture, normally, they do not interfere with the silylated analytes and therefore this step can be avoided.

Finally, silyl-derivatives can be used in on-fibre derivatisation reactions when SPME has been considered as the concentration technique. The only application described for acidic drugs considers MTBSTFA as the silylation reagent [52]. In addition, BSTFA and MSTFA have also been proposed for the derivatisation of different compounds previously concentrated in a microextraction fibre [84,85]. According to the bibliography, a further advantage of employing MTBSTFA in on-fibre derivatisation reactions is that the coated polymeric phase seems to be more stable when exposed to MTBSTFA vapours than to those from more reactive agents such as BSTFA [86]. In the application developed for acidic compounds fibres were used for about 70 extraction-derivatisation cycles without observing stability problems in the coated PA polymer [52,87].

2.5.6 GC SEPARATION AND DETERMINATION

The GC separation of acidic drugs, once derivatised, is simply achieved on non-polar (DB-1 or DB-5 type) open wall-coated capillary columns. Only Winkler et al. [33] reported the use of a polar, polyethylene glycol (PEG), column as a second confirmatory separation. In that work, primary separation was carried out on a non-polar column and detection was based on a flame ionisation detector (FID). Therefore, a secondary confirmatory run was needed employing the PEG column and MS detection.

Another important topic, regarding the separation of acidic drugs, is the enantiomeric separation of the chiral ''profens'' (e.g. ibuprofen). This was carried out by Buser et al. [31] by employing a homemade chiral column containing the chiral selector (2,6-O-dimethyl-3-O-n-pentyl)-β-cyclodextrin (DMPEn). After methylation of ibuprofen, that column allowed a reasonable separation of the two enantiomers (Fig. 2.5.12). This figure also points out the different behaviour of the two ibuprofen isomers (R/S): ibuprofen is sold as a racemic mixture (Fig. 2.5.12a) and the

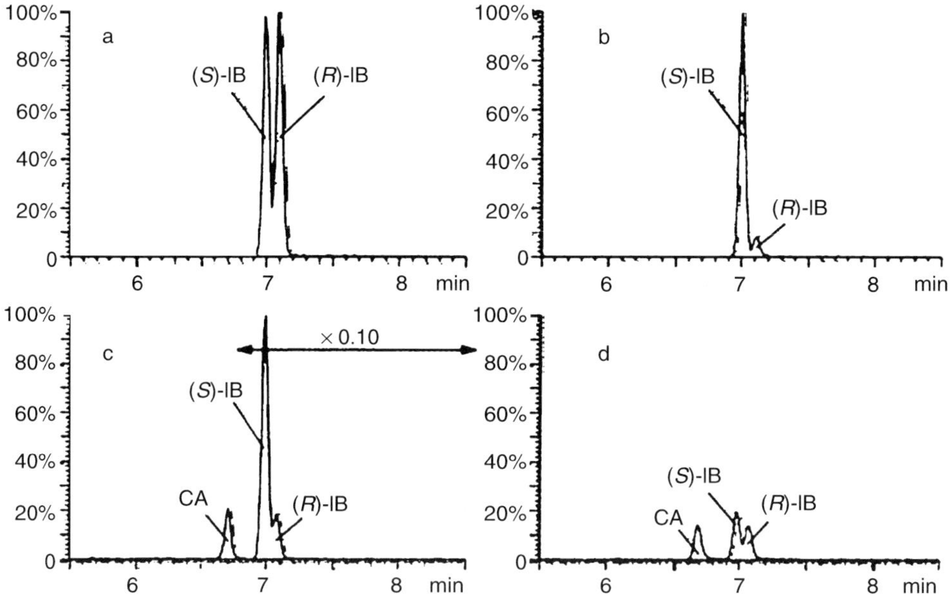

Fig. 2.5.12. GC-MS chromatograms of ibuprofen (IB) and clofibric acid (CA): (a) Racemic IB, as used as the pharmaceutical drug, (b) human urine showing enantiomeric excess of (S)-IB, (c) WWTP influent (10-fold attenuation of IB signals), and (d) WWTP effluent. Reprinted with permission from ref. [31]. Copyright (1999), American Chemical Society.

R-isomer undergoes chiral inversion to the active S-isomer, which is the most abundant in urine samples (Fig. 2.5.12b) and, therefore the main isomer reaching wastewater treatments plants (WWTPs) (Fig. 2.5.12c). But this S-isomer is also much better eliminated during WWT, whereas the R-isomer seems to be recalcitrant (Fig. 2.5.12d).

Concerning the detection step, all the published GC methods for the determination of acidic drugs in water samples (except the one cited above) rely on electron impact-mass spectrometry (EI-MS) as the detection technique, since the sensitivity and selectivity of FID is by far too low for this kind of samples. Most mass spectrometers are furnished with quadrupole [17,25,28,29,44] or ion trap analysers [2,24,37,76]. The first type of instruments is operated in the selected ion monitoring mode (SIM) to achieve the required detection limits. In the case of ion-trap systems, quantification is performed employing the selected reconstructed ion chromatograms achieving a similar sensitivity and maintaining the qualitative information contained in the mass spectra. In addition, ion trap analysers (ITD) offer the possibility of performing tandem mass spectrometric (MS/MS) analysis at an inexpensive price, compared to triple quadrupole detectors. This allows a further reduction in the noise level and thus the detection limits and can be used for the determination of these compounds at lower concentrations, like they often occur, for example, in drinking water [19] or seawater [29] and for more complex samples (e.g. sludge samples).

Other ionisation techniques, like positive (PCI) or negative chemical ionisation (NCI), have not been tested yet for the determination of acidic drugs in environmental samples, in spite of that some works have proved their usefulness for biological samples [79,88,89] and the sensitivity of NCI when combined with a previous derivatisation step using PFBBr [78,79] or MTBSTFA [90]. Moreover, it was found, that for some (non-derivatised) compounds, better sensitivity is obtained by employing an ECD or even a nitrogen-phosphorus detector (NPD) than by EI-MS, for the analysis of plasma samples [91].

2.5.7 CONCLUSIONS

Its moderate cost and capability to deal with organic extracts containing important levels of co-extracted species, without significant effects on the accuracy of determination, have become GC-MS in the workhorse, routine technique for the determination of acidic pharmaceuticals in environmental samples. The availability of safe and ready to use derivatisation reagents, such as silylation and in-port alkylation compounds, added to the previous experience acquired with the analysis of biological samples, have also contributed to the successful application of GC-MS to the analysis of acidic drugs in the aquatic environment. Once derivatised, analytes are well separated using non-polar capillary columns; however, more efficient chiral phases need to be developed for the separation of enantiomeric drugs.

SPE is clearly the preferred sample concentration technique for the determination of target compounds in water samples. Nowadays, the key issue of sample preparation is not the yield but the selectivity of the process. The use of high-efficient sorbents able to retain the analytes in samples at neutral pHs is one the ways to decrease the levels of co-extracted species.

On the other hand, microextraction techniques (SPME and LPME), based on equilibrium processes, have showed an excellent performance as sample preparation approaches in the determination of acidic drugs. Reduction in the sample intake and in the amount of generated residues are two of their main advantages. In addition SPME fibres are commercial available and the extraction can be easily automated. However, its systematic application requires the development of faster and most robust quantification strategies. Although LPME seems to be more selective than SPME, at this moment, conditioning of membranes and preparation of extraction devices is practically a work of art that requires a skilful, dedicated operator. We hope that commercialisation of membranes in different extraction formats will solve such limitations.

Determination of acidic drugs in solid samples, particularly in sludge from sewage treatment plants, is especially important to obtain a global vision of the fate and distribution of these compounds in the environment. In this field, further research should be focussed on improving the efficiency of the clean-up step.

REFERENCES

1 SciFinder Scholar Database 2004, http://www.cas.org/scifinder/scholar.
2 C. Zwiener, T. Glauner and F.H. Frimmel, *J. High Resolut. Chromatogr.*, 23 (2000) 474.
3 S. Weigel, J. Kuhlmann and H. Hühnerfuss, *Sci. Total Environ.*, 295 (2002) 131.
4 S. Weigel, U. Berger, E. Jensen, R. Kallenborn, H. Thoresen and H. Hühnerfuss, *Chemosphere*, 56 (2004) 583.

5 J.B. Quintana and T. Reemtsma, *Rapid Commun. Mass Spectrom.*, 18 (2004) 765.
6 X.S. Miao, B.G. Koenig and C.D. Metcalfe, *J. Chromatogr. A*, 952 (2002) 139.
7 B.J. Vanderford, R.A. Pearson, D.J. Rexing and S.A. Snyder, *Anal. Chem.*, 75 (2003) 6265.
8 M. Stüber and T. Reemtsma, *Anal. Bioanal. Chem.*, 378 (2004) 910.
9 T. Reemtsma, *J. Chromatogr. A*, 1000 (2003) 477.
10 J.M. Krajca, *Water Sampling*, Ellis Horwood, Chichester, 1989.
11 L.H. Keith, *Environmental Sampling and Analysis: A Practical Guide*, Lewis, Michigan, 1991.
12 P.M. Thomas and G.D. Foster, *Environ. Toxicol. Chem.*, 24 (2005) 25.
13 U. Jux, R.M. Baginski, H.G. Arnold, M. Krönke and P.N. Seng, *Int. J. Hyg. Environ. Health*, 205 (2002) 393.
14 R.B. Lee, K. Sarafin, T.E. Peart and M.L. Svoboda, *Water Qual. Res. J. Canada*, 38 (2003) 667.
15 T.A. Ternes, M. Stumpf, B. Schuppert and K. Haberer, *Vom Wasser*, 90 (1998) 295.
16 K.E. Pinkston and D.L. Sedlak, *Environ. Sci. Technol.*, 38 (2004) 4019.
17 M.A. Soliman, J.A. Pedersen and I.H. Suffet, *J. Chromatogr. A*, 1029 (2004) 223.
18 S. Marchese, A. Gentili, D. Perret, G. D'Ascenzo and F. Pastori, *Rapid Commun. Mass Spectrom.*, 17 (2003) 879.
19 M. Stumpf, T.A. Ternes, K. Haberer, P. Seel and W. Baumann, *Vom Wasser*, 86 (1996) 291.
20 K. Reddersen and T. Heberer, *J. Chromatogr. A*, 1011 (2003) 221.
21 I. Rodríguez, J.B. Quintana, J. Carpinteiro, A.M. Carro, R.A. Lorenzo and R. Cela, *J. Chromatogr. A*, 985 (2003) 265.
22 P.M. Thomas and G.D. Foster, *J. Environ. Sci. Health Part A-Toxic/ Hazard. Subst. Environ. Eng.*, 39 (2004) 1969.
23 R. Kanda, P. Griffin, H.A. James and J. Fothergill, *J. Environ. Monit.*, 5 (2003) 823.
24 F. Sacher, F.T. Lang, H.J. Brauch and I. Blankenhorn, *J. Chromatogr. A*, 938 (2001) 199.
25 K. Reddersen and T. Heberer, *J. Sep. Sci.*, 26 (2003) 1443.
26 J. Matamoros, J. García and J.M. Bayona, *Environ. Sci. Technol.*, 39 (2005) 5449.
27 I. Rodríguez, E. Rubí, R. González, J.B. Quintana and R. Cela, *Anal. Chim. Acta*, 537 (2005) 259.
28 S. Öllers, H.P. Singer, P. Fassler and S.R. Muller, *J. Chromatogr. A*, 911 (2001) 225.
29 H.R. Buser, M.D. Müller and N. Theobald, *Environ. Sci. Technol.*, 32 (1998) 188.
30 S. Weigel, R. Kallenborn and H. Hühnerfuss, *J. Chromatogr. A*, 1023 (2004) 183.

31 H.R. Buser, T. Poiger and M.D. Müller, *Environ. Sci. Technol.*, 33 (1999) 2529.
32 C. Tixier, H.P. Singer, S. Oellers and S.R. Müller, *Environ. Sci. Technol.*, 37 (2003) 1061.
33 M. Winkler, J.R. Lawrence and T.R. Neu, *Water Res.*, 35 (2001) 3197.
34 T.A. Ternes, M. Meisenheimer, D. McDowell, F. Sacher, H.J. Brauch, B.H. Gulde, G. Preuss, U. Wilme and N.Z. Seibert, *Environ. Sci. Technol.*, 36 (2002) 3855.
35 M. Stumpf, T.A. Ternes, R.D. Wilken, S.V. Rodrigues and W. Baumann, *Sci. Total Environ.*, 225 (1999) 135.
36 C.D. Metcalfe, B.G. Koenig, D.T. Bennie, M. Servos, T.A. Ternes and R. Hirsch, *Environ. Toxicol. Chem.*, 22 (2003) 2872.
37 V. Koutsouba, T. Heberer, B. Fuhrmann, K. Schmidt-Baumler, D. Tsipi and A. Hiskia, *Chemosphere*, 51 (2003) 69.
38 T.A. Ternes, *Trend. Anal. Chem.*, 20 (2001) 419.
39 I. Rodríguez, M.P. Llompart and R. Cela, *J. Chromatogr. A*, 885 (2000) 291.
40 N. Fontanals, R.M. Marcé and F. Borrull, *Trend. Anal. Chem.*, 24 (2005) 394.
41 T. Kosjek, E. Heath and A. Krbavčič, *Environ. Int.*, 31 (2005) 679.
42 C. Zwiener, S. Seeger, T. Glauner and F.H. Frimmel, *Anal. Bioanal. Chem.*, 372 (2002) 569.
43 H.R. Buser, T. Poiger and M.D. Müller, *Environ. Sci. Technol.*, 32 (1998) 3449.
44 G.R. Boyd, H. Reemtsma, D.A. Grimm and S. Mitra, *Sci. Total Environ.*, 311 (2003) 135.
45 W.C. Lin, H.C. Chen and W.H. Ding, *J. Chromatogr. A*, 1065 (2005) 279.
46 E. Caro, R.M. Marcé, P.A.G. Cormack, D.C. Sherrington and F. Borrull, *J. Chromatogr. B*, 813 (2004) 137.
47 E. Caro, R.M. Marcé, P.A.G. Cormack, D.C. Sherrington and F. Borrull, *J. Chromatogr. A*, 995 (2003) 233.
48 E. Caro, M. Masqué, R.M. Marcé, F. Borrull, P.A.G. Cormack and D.C. Sherrington, *J. Chromatogr. A*, 963 (2002) 169.
49 R.P. Belardi and J.B. Pawliszyn, *Water Pollut Res J Can.*, 24 (1989) 179.
50 N. Huppert, M. Wurtele and H.H. Hahn, *Fresenius J. Anal. Chem.*, 362 (1998) 529.
51 M. Moeder, S. Schrader, M. Winkler and P. Popp, *J. Chromatogr. A*, 873 (2000) 95.
52 I. Rodríguez, J. Carpinteiro, J.B. Quintana, A.M. Carro, R.A. Lorenzo and R. Cela, *J. Chromatogr. A*, 1024 (2004) 1.
53 J. Pawliszyn, *Solid Phase Microextraction: Theory and Practice*, Wiley-VCH, New York, 1997.
54 K.E. Rasmussen and S. Pedersen-Bjergaard, *Trend. Anal. Chem.*, 23 (2004) 1.
55 E. Psillakis and N. Kalogerakis, *Trend. Anal. Chem.*, 22 (2003) 565.

56 S. Pedersen-Bjergaard and K.E. Rasmussen, *Electrophoresis*, 21 (2000) 579.
57 X.J. Wen, C.H. Tu and H.K. Lee, *Anal. Chem.*, 76 (2004) 228.
58 J.B. Quintana, R. Rodil and T. Reemtsma, *J. Chromatogr. A*, 1061 (2004) 19.
59 S. Müller, M. Möder, S. Schrader and P. Popp, *J. Chromatogr. A*, 985 (2003) 99.
60 C. Basheer and H.K. Lee, *J. Chromatogr. A*, 1057 (2004) 163.
61 M. Schellin and P. Popp, *J. Chromatogr. A*, 1020 (2003) 153.
62 M. Schellin, B. Hauser and P. Popp, *J. Chromatogr. A*, 1040 (2004) 251.
63 B. Hauser, M. Schellin and P. Popp, *Anal. Chem.*, 76 (2004) 6029.
64 O.A.H. Jones, N. Voulvoulis and J.N. Lester, *Water Res.*, 36 (2002) 5013.
65 S.J. Khan and J.E. Ongerth, *Water Sci. Technol.*, 46 (2002) 105.
66 T.A. Ternes, M. Bonerz, N. Herrmann, D. Löffler, E. Keller, B. Bagó Lacida and A.C. Alder, *J. Chromatogr. A*, 1067 (2005) 213.
67 D. Löffler and T.A. Ternes, *J. Chromatogr. A*, 1021 (2003) 133.
68 K. Bester, *Water Res.*, 37 (2003) 3891.
69 S. Morales, P. Canosa, I. Rodríguez, E. Rubí and R. Cela, *J. Chromatogr. A*, 1082 (2005) 128.
70 J. Drozd, *Chemical Derivatization in Gas Chromatography*, Elsevier, Amsterdam, 1981.
71 C.D. Metcalfe, X.S. Miao, B.G. Koenig and J. Struger, *Environ. Toxicol. Chem.*, 22 (2003) 2881.
72 M. Farré, I. Ferrer, A. Ginebreda, M. Figueras, L. Olivella, L. Tirapu, M. Vilanova and D. Barceló, *J. Chromatogr. A*, 938 (2001) 187.
73 N. Páxeus, *Water Sci. Technol.*, 50 (2004) 253.
74 C. Zwiener and F.H. Frimmel, *Sci. Total Environ.*, 309 (2003) 201.
75 A. Zapf and H.J. Stan, *J. High Resolut. Chromatogr.*, 22 (1999) 83.
76 C. Zwiener and F.H. Frimmel, *Water Res.*, 34 (2000) 1881.
77 A. Tauxe-Wuersch, L.F. De Alencastro, D. Grandjean and J. Tarradellas, *Water Res.*, 39 (2005) 1761.
78 H.M. Kuch and K. Ballschmiter, *Environ. Sci. Technol.*, 35 (2001) 3201.
79 H.J. Leis, M. Leis and W. Windischhofer, *J. Mass Spectrom.*, 31 (1996) 486.
80 K. Blau and J.M. Halket, *Handbook of Derivatives for Chromatography*, Wiley, Chichester, 1993.
81 O.A.H. Jones, N. Voulvoulis and J.N. Lester, *Chromatographia*, 58 (2003) 471.
82 T. Heberer and H.J. Stan, *Anal. Chim. Acta*, 341 (1997) 21.
83 J.B. Quintana, J. Carpinteiro, I. Rodríguez, R.A. Lorenzo, A.M. Carro and R. Cela, *J. Chromatogr. A*, 1024 (2004) 177.
84 J. Carpinteiro, J.B. Quintana, I. Rodríguez, A.M. Carro, R.A. Lorenzo and R. Cela, *J. Chromatogr. A*, 1056 (2004) 179.

85 P. Okeyo, S.M. Rentz and N.H. Snow, *J. High Resolut. Chromatogr.*, 20 (1997) 171.
86 G. Gmeiner, P. Gärtner, C. Krassnig and H. Tausch, *J. Chromatogr. B*, 766 (2002) 209.
87 J. Carpinteiro Botana, Ph.D. Thesis, University of Santiago de Compostela, Santiago de Compostela, 2005.
88 H. Seno, H. Hattori, T. Kumazawa, A. Ishii, K. Watanabe-Suzuki and O. Suzuki, *Forensic Sci. Int.*, 100 (1999) 143.
89 R. Hsu and A.M. Au, *Bull. Environ. Contam. Toxicol.*, 66 (2001) 178.
90 W. Blum, J.W. Faigle, U. Pfaar and A. Sallmann, *J. Chromatogr. B-Biomed. Appl.*, 685 (1996) 251.
91 C. Giachetti, A. Assandri, G. Zanolo and A. Tenconi, *Chromatographia*, 39 (1994) 162.
92 R. Andreozzi, M. Raffaele and P. Nicklas, *Chemosphere*, 50 (2003) 1319.
93 S.J. Khan, T. Wintgens, P. Sherman, J. Zaricky and A.I. Schäfer, *Water Sci. Technol.*, 50 (2004) 15.
94 J.E. Drewes, T. Heberer and K. Reddersen, *Water Sci. Technol.*, 46 (2002) 73.

Chapter 2.6

Analysis of steroid estrogens in the environment

Marina Kuster, Maria J. Lopez de Alda, Sara Rodriguez-Mozaz and Damià Barceló

2.6.1 INTRODUCTION

2.6.1.1 Uses

During the last five decades, the consumption of estrogens (both natural and synthetic) for human medicine and animal farming has experienced a steady growth to the point that these substances are nowadays among the most prescribed drugs. Annual consumption of estrogens is normally in kilogram range (e.g., 119 kg of estradiol in Denmark in 1997, 29 kg of ethynyl estradiol in UK in 2000) [1]. One major medicinal application of these substances is birth control since 1960. The estrogen content in contraceptive pills is usually in the range 20–50 mg daily [2]. Other than contraception, the uses of estrogens in human medicine can be put into three main groups: management of the menopausal and post-menopausal syndrome (its widest use); physiological replacement therapy in deficiency states; and treatment of prostatic cancer and of breast cancer in post-menopausal women. The synthetic estrogens most commonly used for these purposes are 17α-ethynyl estradiol (EE) and mestranol (MES). In animal-farming estrogens are mainly applied as growth-promoters [3] and for development of single sex populations of fish in aquaculture [4–7].

2.6.1.2 Metabolism and excretion rates

After intake, estrogens absorbed by human or animal organisms, as well as the endogenous estrogens, are subject to a variety of metabolic reactions, such as hydroxylation, oxidation, methylation, or conjugation

Comprehensive Analytical Chemistry 50
Petrović and Barceló (Eds)
Volume 50 ISSN: 0166-526X DOI: 10.1016/S0166-526X(07)50007-3

TABLE 2.6.1

Daily excretion (μg) of estrogenic steroids in humans. Reprinted with permission from Ref. [8] © 2002 Elsevier

Category	E2	E1	E3	EE
Males	1.6	3.9	1.5	
Menstruating females	3.5	8	4.8	
Menopausal females	2.3	4	1	
Pregnant women	259	600	6000	
Women (contraceptive)				35

with glucuronic or sulfuric acid, although a significant amount of the original substance also leaves the organism via urine or feces.

17β-Estradiol (E2), for instance, is rapidly oxidized to estrone (E1), which can be further converted into estriol (E3). The contraceptive ingredient mestranol is converted into 17α-ethynyl estradiol by demethylation and 17α-ethynyl estradiol is mainly eliminated as conjugates [8]. Table 2.6.1 shows the daily excretions of natural and synthetic estrogens in humans.

As regards livestock, cattle excrete mostly 17α-estradiol, 17β-estradiol, estrone, and the respective sulfated and glucuronidated counterparts, whereas swine and poultry excrete mostly 17β-estradiol, estrone, estriol, and the respective sulfated and glucuronidated counterparts [9].

2.6.1.3 Sources in the environment

The main sources of environmental estrogens are thus municipal sewage discharge and livestock wastes [9,10]. Estrogens are mainly excreted as their less active sulfate, glucuronide, and sulfo-glucuronide conjugates [11]. However, in raw sewage and sewage treatment plants (STPs), as well as in the environment, these conjugates may suffer deconjugation and act as precursors of the corresponding free steroids [12–15].

In domestic wastewater, the concentrations of conjugated estrogens have been shown to be higher than those of free estrogens [16] whereas in coastal water, according to Atkinson et al. [17], one-half to two-thirds of total estrone occur as polar conjugates.

In addition to this, chlorine treatment at STPs or waterworks leads to the formation of chlorinated and brominated (when bromide ions are present) derivatives of estrogens, which, although weaker than the parent compounds, also show estrogenic activity [18].

2.6.1.4 Physico-chemical properties

The fate of these compounds, once released in the environment, is in part determined by their physico-chemical properties (see Table 2.6.2).

Natural estrogens have greater solubility in water (3.6–441 mg/L) and are less hydrophobic (log K_{ow} 2.45–4.01) than synthetic estrogens (0.97–12 mg/L, log K_{ow} 3.67–5.07) and all these are weakly acidic (pK_a 9.3–13.1) and present low volatility with vapour pressures ranging from 1.42×10^{-7} to 1.97×10^{-10} mm Hg. These properties may explain why estradiol and ethynyl estradiol, which have relatively high values of log K_{ow}, are less frequently found in environmental waters than estrone, which is the most ubiquitous compound; or why the synthetic compounds mestranol and ethynyl estradiol show greater tendency to partition to sediments than natural estrogens [20].

These properties are also important from the analytical point of view since they largely determine the solvents and sorbents more adequate for extraction or the analytical techniques best suited for their determination. Thus, for instance, although estrogens can be analyzed by gas chromatography (GC), due to their low volatility, they first need to be converted to volatile derivatives.

2.6.1.5 Environmental occurrence

The occurrence and environmental fate of estrogens have been reviewed in a few recent articles [8,9,21], all of which coincide in highlighting the very limited data available and the need for continuing working in this field to be able to appropriately assess the exposure and risks associated to these compounds.

Estrogens have been found in basically all environmental compartments. Levels reported in water have been nearly always below 100 ng/L and in most cases in the low ng/L or pg/L range [8,21]. In solid samples, the data available are very scarce but indicate the presence of these compounds in sediments at low ng/g or pg/g levels and at somewhat higher concentrations (tens of ng/g) in sewage sludge [21].

The main ways by which estrogens enter the environment are sewage treatment effuents, uncontrolled discharges, run-off of manure, and sewage sludge used in agriculture. Natural and synthetic estrogens entering wastewater treatment plants (WWTPs) are subject to a variety of treatment processes of varying efficiency and in some cases they are finally released into surface waters. In activated STPs, removal efficiencies higher than 60% have been normally reported for the most

TABLE 2.6.2

Main physico-chemical properties of the most relevant environmental estrogens [19]

Compound	CAS Number	MW	Water solubility[a,c] (mg/L)	Log P_{OW}[c]	Vapor pressure[a,b] (mm Hg)	Henry's Law Constant[a,b] (atm-m^3/mol)	pK_a
Estradiol (E2)	50-28-2	272.39	3.6 (27°C)	4.01	1.26E-008	3.64E-011	10.3
Estriol (E3)	50-27-1	288.39	441[b]	2.45	1.97E-010	1.33E-012	10.3
Estrone (E1)	53-16-7	270.37	30	3.13	1.42E-007	3.8E-010	10.3
Ethynyl estradiol (EE)	57-63-6	296.41	11.3 (27°C)	3.67	2.67E-009	7.94E-012	10.3
Diethylstilbestrol (DES)	56-53-1	268.36	12	5.07	1.41E-008	1.41E-008	9.3
Mestranol (MES)	72-33-3	310.44	0.977	4.68	9.75E-009	4.51E-009	13.1

[a]Values are at 25°C if not specified.
[b]Estimated data.
[c]Experimental data.

relevant estrogens, namely, estradiol, ethynyl estradiol, estriol, and estrone [12,22,23], of which estrone is the compound showing the lowest and more variable results [11,24,25].

In the resulting sewage sludge, which can be further used as soil amendment in agriculture, representing another source of estrogens in the aquatic environment (run-off), the most persistent compound seems to be ethynyl estradiol [13,26].

Estrogens can be eliminated from the aquatic environment by different mechanisms, the most important being sorption and biodegradation. Given the relatively low polarity of estrogens sorption to bed-sediments appears as a quite likely process [21]. According to Jurgens et al. [27], between 13% and 92% of the estrogens entering a river system would end up in the bed-sediment compartment, with the majority of sorption occurring within the first 24 h of contact, and the synthetic estrogens (mestranol, ethynyl estradiol) are shown to partition to the sediment to a greater extent than the natural estrogens [20]. Under the anaerobic, dark conditions normally present in the subsurface layers of river sediments, estrogens are expected to undergo low photodecomposition and biodegradation. Therefore, river sediments can act as sinks where estrogens may persist for long periods of time, be transported to other areas, and be eventually released back to the water column [20]. However, based on the data available, it appears that estrogens are biodegraded in the environment by many different types of organisms that require no prior adaptation [9].

The environmental half-lives of estrogens (see Table 2.6.3) have been reported to vary between hours and days depending on the particular compound and the environmental conditions (e.g. light intensity, temperature, soil and water properties, etc.), with synthetic substances being in general more persistent than natural compounds.

TABLE 2.6.3

Half-lives of some estrogens in air, water, and soil. Reproduced from Ref. [28] with permission from IUPAC

Estrogen	Air	Water	Soil/sediments
EE	10 days	20–40 days	Much longer than E2
E2	10 days	0.2–9 days	0.11 days (O), 0.37–0.66 days (R)
E1		0.2–9 days	0.4 days (O), 11–14 days (R)

(P), Photolysis (not always in air); (O), aerobic degradation; (R), anaerobic degradation.

2.6.1.6 Effects

Estrogens have often been identified as the compounds responsible for the estrogenic effects that have been observed in different wildlife species, for instance, intersex in carp, high levels of plasma vitellogenin in fish, etc. [29]. Effects in humans, although not fully conclusive, include the development of hormonal-dependent (testis, ovary) cancers, decreased fertility (associated to a decline in sperm counts to nearly half in the last decades), and increase incidence of other reproductive abnormalities, such as cryptorchidism and hypospadias in male [30–32]. However, effects have not been observed only in wildlife and humans. In plants, for instance, the irrigation of alfalfa with sewage water (containing estrogens) has been shown to increase the phytoestrogen content (coumestrol) of the plant, which in turn can affect fertility in cattle [33].

The estrogens of greatest concern, and most investigated, due to their estrogenic potency are 17β-estradiol, its metabolites estrone and estriol, and the synthetic contraceptives 17α-ethynyl estradiol and mestranol. The relative estrogenicity activity of some of these compounds is shown in Table 2.6.4.

Aqueous concentrations as low as 0.1–10 ng/L of the most potent compounds, estradiol and ethynyl estradiol, have been shown to be sufficient to exert estrogenic responses in wildlife [34–38]. However, in humans, the environmental levels normally reported for estrogens in water (including drinking water), which are in the pg/L or ng/L range,

TABLE 2.6.4

Relative estrogenic potency of various estrogens as determined by different bioassays (expressed as EEF–the molar based 17β-estradiol equivalency factor). Reproduced from Ref. [21] with permission from Springer

Compound	YES	MCF-7 assay (E-screen)	ER-CALUX
17β-estradiol	1	1	1
Estriol	3.7×10^{-1}		
Ethynyl estradiol	1.9×10^{-1}–1.2	1.25–1.9	1.2
Diethylstilbestrol	4.5×10^{-2}–1.1	2.5	
Estrone	1.9×10^{-2}–1.0×10^{-1}	1.0×10^{-2}	5.6×10^{-2}

YES, yeast-based recombinant estrogen receptor-reporter assay; ER-CALUX, estrogen receptor-mediated chemical activated luciferase gene expression assay.

the doses used for therapeutic purposes (20–50 μg/day) [2], and the estimated human dietary intake of estrogens (0.1 mg/day of E2+E1) [39] are considerably lower than the human daily production of endogenous estrogens ($0.05–0.60 \times 10^3$ μg/day estrogens) [39]. In the light of these data, it seems logical to think that the human, both environmental and dietary, exposure to these compounds is negligible and that hormonal effects cannot be expected neither from the exposure to environmental estrogens nor from the ingestion of naturally occurring dietary estrogens [39,40].

Nevertheless, neither the possible additive effect of the various routes of exposure to these compounds nor the potential synergisms that may occur because of the simultaneous presence of other substances with estrogenic activity, or the differences in sensitivity that exist during human life (fetus are believed to be more sensitive), are well-known yet. Therefore it is difficult to assess, at least in humans, whether environmental estrogens play a key role on the effects observed.

2.6.2 ANALYSIS

2.6.2.1 General remarks

In the last years, multiple articles published in the literature have reported the development and application of analytical methods for determination of esteroid estrogens in different environmental matrices. These methodologies have been reviewed in a series of papers that often cover other additional aspects, such as occurrence and fate or the analysis of other environmental contaminants. These reviews are listed, chronologically ordered from the most recent to the older ones in Table 2.6.5.

Most of the environmental programs carried out to assess the presence and impact of natural and synthetic estrogens in the environment have focused on the investigation of environmental waters, while sludge, sediments and principally soils, have received comparatively much less attention.

The analysis of estrogens in the environment constitutes a difficult task, first, because of the complexity of the matrices, and second, because of the very low concentrations at which they are present. In essentially all cases of interest, substantial analyte enrichment is necessary to isolate the target compounds from the matrix and to achieve

TABLE 2.6.5

Reviews covering the occurrence, fate, and analysis of environmental estrogens

Subject	Year	Ref.
Estrogens and progestogens in waste water, sludge, sediments and soil	2004	[41]
Analysis and distribution of estrogens and progestogens in sewage sludge, soils and sediments	2003	[42]
Analysis of steroid sex hormones, drugs, and alkylphenolic surfactants in the aquatic environment by LC–MS and LC–MS/MS	2003	[43]
Environmental behaviour and analysis of veterinary and human drugs in soils, sediments, and sludge	2003	[44]
Analysis of endocrine active substances (organochlorines, PCBs, dioxins and dioxin-like substances, polybrominated diphenyl ethers, phenolic xenoestrogens, phthalates, organotin compounds, steroidal hormones, and phytoestrogens) in food and the environment	2003	[45]
Manure-borne estrogens as potential environmental contaminants: a review	2003	[9]
Analysis of endocrine disrupting compounds in aquatic environmental samples by mass spectrometric techniques	2002	[46]
Occurrence and fate of hormone steroids in the environment	2002	[8]
Analysis of estrogens and progestogens in wastewater	2001	[47]
Analysis and environmental levels of endocrine disrupting compounds in fresh-water sediments	2001	[48]
Occurrence of hormonally active compounds in food	1999	[39]

the limits of detection (LODs) required, which translates in laborious and time-consuming procedures. A typical analytical protocol includes various sample preparation steps, such as filtration, extraction, purification, evaporation, hydrolysis, and derivatization, prior to analysis by immunoassays or by liquid or gas chromatography coupled to different detectors. All these steps and other relevant analytical aspects are discussed in the following sections. These sections are sometimes divided into two subsections to distinguish between liquid and solid environmental samples.

2.6.2.2 Standards

Chemical analysis has focused on the investigation of free estrogens, both natural (estradiol, estrone and estriol) and synthetic (basically ethynyl estradiol, mestranol and diethylstilbestrol). In contrast, conjugated estrogens and halogenated derivatives have been seldom studied, probably due to their lower estrogenic potency and, in the latter case, their still recent identification.

Pure standards of both free and conjugated estrogens have usually been obtained from Sigma. Halogenated derivatives are not commercially available yet. In case of using internal or surrogate standards for quantification, perdeuterated estrogens, which are available for the great majority of the analytes, are the compounds of choice. For the preparation of standard solutions, to be used for calibration and in recovery experiments, methanol has been the solvent most widely used.

On the other hand, estrogens are photosensible compounds; therefore, special precautions have to be taken into account when storing standards (as well as samples and sample extracts). They must be kept in brown bottles or in containers wrapped with aluminum foil in order to avoid light decomposition and the material preferred for the container is glass.

2.6.2.3 Sample collection, preservation, and handling

2.6.2.3.1 Water samples

In most instances, the analysis of environmental estrogens in water has relied on the collection of discrete samples. Discrete samples only provide punctual information, i.e., they do not allow to detect temporal variations in quality, which, in the case of rivers, may take place within minutes or hours; however, discrete sampling has the advantage of not requiring special collection systems. In contrast, the use of composite samples, representative of the contamination existing in time periods varying typically between hours and days, has only been reported in a few studies, the majority of them dealing with the investigation of wastewaters. This requires the use of, usually expensive, automatic collection devices, and another question, not clear yet, is whether during the period of collection of composite samples the compounds may degradate.

Up to now only a few works have studied the stability of estrogens in solution or during sample storage [12,49,50]. Probably because of this most authors process the samples (stored at 4°C without preservation) within 48 h of collection.

The methods most frequently used for preservation of water samples include [47]:

- storage at 4°C,
- freeze-drying at –20°C,
- acidification (with sulfuric acid)
- addition of chemical agents such as formaldehyde (1%, v/v), methanol, and mercuric chloride, and
- storage on solid supports used for extraction.

The storage of unpreserved samples at 4°C for periods varying from several days up to 1 week has been reported by various authors [51,52]; however, the stability of estrogens under these conditions is not clear yet. According to Ferguson et al. [49], who studied the potential degradation of estradiol to estrone in wastewater effluent samples spiked with estradiol deuterated, no degradation takes place in samples incubated at 4°C for periods up to 6 days. However, according with a study carried out by Baronti et al. [12], in this case with river water samples, these storage conditions lead to severe losses of the estrogens estriol, estrone, and estradiol (ethynyl estradiol was not affected), while the addition of 1% formaldehyde is useful to preserve the stability of all four compounds for periods up to 24 days. Fine et al. [50] also found that pretreatment of swine lagoon samples with formaldehyde was necessary to prevent conversion of estradiol to estrone. Notwithstanding this, the study conducted by Baronti et al. [12] concluded that the best sample storage strategy consists of passing the field sample through the extraction cartridge (Carbograph-4), washing the cartridge with methanol, and storing it at –18°C, conditions which facilitate the storage of many samples in extensive monitoring programs and under which the estrogens remain stable for periods up to 60 days.

In the case of tap water, or water samples suspected to contain residual chlorine, another precaution to be taken into account is the addition of sodium thiosulfate to the sample, immediately after collection [50].

On the other hand, the volume of sample required for analysis may be very variable. It basically depends on the sensitivity needed, the type of matrix to be analyzed and the procedure used for analysis. For instance, in procedures based on on-line solid phase extraction (SPE), where the whole sample is transferred to the analytical system, a few mL of sample may be enough to achieve the necessary sensitivity, whereas the analysis of coastal marine waters, where the concentrations of the compounds are very low, may require the collection of 50 L, as reported, for instance, by Beck et al. [53].

After collection, filtration is usually the first step in the sample preparation protocol for water matrices. The filtration step is particularly necessary when subsequent extraction of the sample is based on the use of SPE, since the suspended solids present in the sample can easily clogged the sorbent system. Removing of the organic matter is important also to avoid undesired adsorptions onto antibodies in immunochemical assays.

The filtration step can be carried out simultaneously with the sample collection and/or extraction, or as a separate step, and this fact determines, in part, the type of filter to be used, particularly with regards to the physical form (pads, filter aid powder, glass wool, etc.) and diameter, and the filter holder. As for the filter material itself, most of the studies reviewed employ glass fibre-type filters with a pore size varying between 0.22 and 1.2 μm.

Various studies have investigated the possibility that estrogens get retained in the filtering material [14,17,50,54,55]. These studies were conducted with sewage, surface and/or coastal water using glass fibre, polypropylene and/or paper filters and the conclusion of almost of all them was that estrogens remain in the aqueous phase. Fine et al. [50], however, studied the distribution of three natural estrogens (estrone, estradiol, and estriol) in lagoon samples by analyzing the aqueous phase, the undissolved phase, and the inner surface of the sample container, and found that the less-polar compounds estrone and estradiol partially adsorbed to the particulate/sediment (see Table 2.6.6).

On the other hand, according to this study [50], and that conducted by Furhacker et al. [56], sorption of estrogens onto glassware is negligible.

Despite these findings, very often analysts, after filtration of the sample, wash the filtration system with methanol (3–10 mL) to remove any analyte eventually adsorbed on the particles and add this methanolic extract to the sample [12,52,57]. Other authors, in addition to the filtration step [58], or instead of it [59], centrifuge the samples with the same aim of removing the suspended matter.

2.6.2.3.2 Solid environmental samples

Alike waters, the analysis of estrogens in solid environmental samples (sediments, sludge) has relied on discrete/bulked samples collected with

TABLE 2.6.6

Distribution of natural estrogens in aqueous phase, sediment, and on the inner surface of the sample bottle for two swine lagoon samples. Reprinted with permission from Ref. [50] © 2003 Elsevier.

	Estrogen distribution		
	Estrone	Estradiol	Estriol
SOW2 (%)			
Water sample	71	84	98
Precipitate	22	16	2
Bottle	7	0	0
SOW4A (%)			
Water sample	64	84	98
Precipitate	25	15	3
Bottle	5	0	0
SOW4B (%)			
Water sample	63	74	97
Precipitate	27	25	3
Bottle	10	0	0

the help of either cores or grab samplers. Collected samples are transported to the laboratory under cooled conditions (+4°C or –20°C) in the dark to avoid change in the composition of the analytes of interest. Thereafter, samples are usually subjected to a drying step that is commonly performed by freeze-drying. After drying, samples are ground and homogenized with a mortar and a pestle, sieved, and stored in the dark at 4°C or at –20°C until extraction. Storage of wet sediments, as performed by Peck et al. [60], requires lower temperatures (–70°C). As in the case of waters, samples must be protected from light. The amount of sample analyzed varies between 1 and 30 g [60–67].

2.6.2.4 Extraction and purification

2.6.2.4.1 Water samples

Tables 2.6.7 and 2.6.8 show the main steps and experimental conditions reported in the literature for analysis of estrogens in both aqueous and solid environmental samples by GC- and LC-based methods, respectively. These tables also include the limits of detection achieved with the various methods reviewed and the levels obtained from their application to real environmental samples.

TABLE 2.6.7

Survey of GC methods used for quantitative determination of steroid estrogens

Compounds	Matrix	Extraction	Cleanup	Derivatization	GC column	Detection method	LOD (ng/L, ng/g)	Levels	Ref. (year)
EE	Tap and river water	In-sample acetylation on-line SPE (PLRP-s)	—	Acetic anhydride	HP Ultra 2 (25 m × 0.32 mm, 0.52 μm)	GC–(EI)MS (SIM)	15	NR	[122] (1998)
E1, E2, 17α-E2, EE	Surface and drinking water, STP effluent	SPE (LiChrolut EN)	—	10% PFBCl in toluene	DB5MS (60 m × 0.32 mm, 0.25 μm)	GC–(NCI)MS	0.05–0.15	0.1–5.1 (river) 0.1–2.1 (drinking) 0.1–18 (effl.)	[121] (2001)
E1, E2 (+testoster., androstenedione, medroxyprogesterone)	Sewage and ground water	SPE (90 mm Empore C-18 disks)	—	Heptafluorobutyric anhydride (55°C, 1.5 h)	MDN-5S (30 m × 0.25 mm, 0.25 μm)	GC–MS/MS	0.1	<12 (E1), <4 (E2)	[140] (2003)
E1, E2, E3, EE	Ground water	SPE (Oasis HLB)	—	PFBBR (1 h, 60°C)+TMSI (with various complicated LLE steps with water and hexane)	DB5-XLB (60 m × 0.25 mm, 0.25 μm)	GC–(NICI)MS/MS	0.2–0.6 1[a]	4.5 (E1)	[50] (2003)
E1, E2, EE, 16α-OH-E1 (+ NP, BPA)	River water	SPE (Oasis HLB)	—	Pyridine and BSTFA with 1% TMCS (60–70°C, 30 min)	ZB5 (30 m × 0.25 m, 0.25 μm)	GC–(EI)MS (SIM)	0.3–3.4	5–10 (E1) 14–17 (E2)	[69] (2004)
E1, E2, DES (+ OP, *t*-NP, DEHA, TEST, PREG)	River water	SPME (85 μm polyacrylate fiber)	—	BSTFA (headspace on-fiber silylation, 60 min, 25°C)	HP-5 MS (30 m × 0.25 mm, 0.25 μm)	GC–(EI)MS (SIM)	1–17	In $\mu g\,L^{-1}$: 0.18 (E1), 0.10 (E_2), 0.02 (DES)	[78] (2006)
E2, EE (+other EDCs)	Surface water	SPE (C18 col. or disks or PS-DVB col.) or LLE	—	MTBSTFA (75°C, 3 h)	HP5-MS (30 m × 0.25 mm, 0.25 μm)	GC–(EI)MS	50–300	NR	[126] (2000)
E1, E2 E3, EE (+ other EDCs)	Surface and wastewater	SPE (Oasis HLB)	—	BSTFA (60°C, 15 min)	CP-Sil 8 CB (30 m × 0.25 m, 0.25 μm) BPX-5 (30 m × 0.25 m, 0.25 μm)	GC–(EI)MS (SIM), GC–MS/MS	2–10 2–20	-	[141] (2002)
E1, E2, E3, EE, DES, MES	River water	SPE (Oasis HLB)	—	MSTFA (85°C, 100 min)	BP-5 (30 m × 0.25 mm, 0.25 μm) for GC–MS	GC–(EI)MS GC–(EI)MS/MS	1–3, 2–6	30–33 (E1), 3–13 (E2), <62 (E3)	[87] (2004)
	Influent and effluent sewage water		SPE (500 mg silica)		BP-1 (30 m × 0.32 mm, 0.17 μm) for GC–MS/MS				

TABLE 2.6.7 (*Continued*)

Compounds	Matrix	Extraction	Cleanup	Derivatization	GC column	Detection method	LOD (ng/L, ng/g)	Levels	Ref. (year)
E1, E2, EE, MES, E2-val, 16α-OH-E1	River and wastewater	SPE (Lichrolut®-EN+RP-C_{18})	SPE (Silica gel)	MSTFA/TMSI/DTE (1000:2:2; v/v/w; 0.5 h, 60°C)	XTI-5 (30 m × 0.25 mm, 0.25 μm)	GC–MS/MS	0.5 (river) 1 (wastewater)	In river: <1.6 (E1), In ww:<70 (E1), <64 (E2), <42 (EE), <4 (MES), ND (E2-val), <5 (16α-OH-E1)	[22] (1999)
E1, E2, EE, 17α-E2 (SS)	River water and STP effluent	Continuous LLE	—	BSTFA/10% TMCS	HP Ultra II (25 m × 0.2 mm, 0.33 μm)	GC–(EI)MS (SIM)	58 (only E2)	ND	[142] (2000)
E2, EE, E2gluc, E2sulf	Surface water, STP effluent	SPE (C18 disk)	Hydrolysis HPLC fraction.	Heptafluorobutyric anhydride (50°C, 1.5 h)	Rtx-5 (30 m × 0.25 mm, 0.25 μm)	GC–(EI)MS-MS (only E2)	0.2–0.4	0.27–3.9 (E2)	[55] (2001)
E1, E2, 17α-E2, EE, glucuronides	Surface and waste water	SPE (SDB-XC disk)	Hydrolysis SPE (C18/ NH2 col.) HPLC fraction.	SIL A reagent	DB-5MS (30 m × 0.25 mm, 0.25 μm)	GC–(EI)MS/ MS	0.1–2.4	In ww: 0.1–47 (E1), 0.5–12 (E2), 0.1–5 (αE2), 1.4–7.5 (EE), In surface: 0.1–3.4 (E1), 0.3–5.5 (E2), 0.1–3 (αE2), 0.1–4.3 (EE)	[15] (1999)
E1, E2, EE	Reservoir and river water, STP effluent	SPE (C18 disk)	—	MTBSTFA with 1% TBDMCS	5% phenyl methylsilicone (30 m × 0.25 mm, 0.25 μm)	GC–(EI)MS-MS	1	10–55 (E1), 2–48 (E2), 1–55 (EE)	[143] (2000)
E1, E2, EE	STP effluents	SPE (C18 col.)	SPE (C18 col.) HPLC fraction. HPLC fraction. LLE	—	—	GC–(EI)MS	0.2	1.4–76 (E1) 2.7–48 (E2) 0.2–7 (EE)	[14] (1998)
E1, E2, EE	STP effluents	SPE (ENV+ col.)	LLE, LLE, GPC (BioBeads SX-3), hydrolysis, LLE	Acetic anhydride	—	GC–(EI)MS (SIM)	0.5 (E1, E2), 2 (EE)	5.8 (E1), <1.1 (E2), <4.5 (EE)	[81] (1999)

E1, E2, EE	STP effluents	SPE (C18 col.)	HPLC fraction.	—	—	GC–MS	0.5-1	15–220 (E1) 4–88 (E2) 1.7–3.4 (EE)	[82] (2000)
E1, E2, E3, EE, MES, (+LEV, NOR-acetate)	STP effluent	SPE (LiChrolut EN/ Bondesil C18)	Silicagel 60	MSTFA:TMSI:DTE (1000:4:2, v:m:m, 60 °C, 30 min)	DB5MS (60 m × 0.32 mm, 0.25 μm)	GC–(EI)MS (SIM)	1[a]	3–13 (E1), 1–13 (E2), <9 (E3), <5 (EE), <8 (MES)	[51] (2000)
EE	STP effluent	SPE (Empore C18 disk)	—	—	DB5MS (30 m × 0.25 mm, 0.25 μm)	GC–(EI)MS (SIM)	74	ND	[58] (2000)
E1, E2, E3	STP effluent	SPE (C18 col.)	—						
Pentafluoropropionic acid anhydride	HP5-MS	(30 m × 0.25 mm, 0.25 μm)	GC–(EI)MS (SIM)	5–10	6–109 (E1) 5–15 (E2) 10–250 (E3)	[144] (1998)			
E1, E2	Influent and efluent wastewater	SPE (C18)	SPE (Silica)	MSTFA/TMSI/DTE	JW DB-5 (60 m × 0.25 mm, 0.25 μm)	GC–HRMS	0.7 (E1), 0.8 (E2)	In influent: 19–78 (E1) 2.4–26 (E2) In effluent:1–96 (E1), 0.2–14.7 (E2)	[25] (2005)
EE (+BPA, t-NP)	Wastewater	SPME (85 μm polyacrylate)	—	—	HP-5 MS (30 m × 0.25 mm, 0.25 μm)	GC–(EI)MS (SIM)	0.02–0.04 μg/L	—	[77] (2003)
E1, E2, E3, EE (+ NP, NP1/2EO)	WWTP effluent, river and lake water	SPE (LiChrolut RP18+LiChrolut EN)	SPE (silica gel)	MSTFA/TMSI/DTE (1,000:2:2, v:v:w)	XTI-5 (30 m × 0.25 mm, 0.25 μm)	GC–(EI)MS (SIM)	0.1–1.0 (E1) 0.3–0.9 (E2) 0.3–1.5 (E3) 0.2–1.0 (EE)	<51 (E1), <6 (E2), <2 (EE)	[88] (2004)
E1, E2, E3, EE	Swine lagoon	SPE (Oasis HLB)	LLE	PFBBR (1 h, 60 °C)+TMSI	DB5-XLB (60 m × 0.25 mm, 0.25 μm)	GC-(NICI)MS/ MS	40[a]	28–74700 (E1), 19–3000 (E2), 175–10900 (E3), ND (EE)	[50] (2003)
E1, E2, α-E2, E3	Flushed dairy manure wastewater	SPE (graphitized carbon black)	SPE (C-18+C-18)	BSTFA in DMF (16 h, room T)	HP-5MS (30 m × 0.25 mm, 0.25 μm)	GC–(EI)MS (SIM) 2 ions	75	370–2356 (E1), 351–957 (E2), 1750–3270 (α-E2), ND (E3)	[83] (2006)
E1, E2, α-E2, E3, MES (+BPA, NP)	River sediment	Ultrasonication (acetone:DCM, 1:1)	LLE with DCM+Silica gel fraction.	PFPA (60 °C, 2 h)	HP-5 MS (30 m × 0.25 mm, 0.25 μm)	GC–(EI)MS (SIM)	0.6 (E1, α-E2), 0.8 (E2), 1.5 (E3), .5 (MES)[a]	ND	[63] (2006)
E1, E2	River sediment	Ultrasonication (MeOH)	SPE (Oasis HLB)+prep. HPLC	MSTFA:pyridine (1:1, 60 °C, 15 min)	HP5-MS (30 m × 0.25 mm, 0.25 μm)	GC–MS (magnetic sector, SIM)	Instrumental (in ng/μL): 0.05 (E2), 0.10 (E1)	In ng/kg: 24.9–52.4 (E1) 6.3–14.6 (E2)	[60] (2004)

TABLE 2.6.7 (*Continued*)

Compounds	Matrix	Extraction	Cleanup	Derivatization	GC column	Detection method	LOD (ng/L, ng/g)	Levels	Ref. (year)
E1, E2, EE	River sediment	Ultrasonication ($DCM:H_2O$, 1:1)	Prep. HPLC	MTBSTFA +TBDMCS 1%+ACN	DB5.625 (30 m × 0.32 mm, 0.25 μm)	GC–MS/MS (QqQ)	0.04–0.1	<0.04–0.388 (E1), ND (E2, EE)	[62] (2003)
E1, E2, EE, MES		Ultrasonication (MeOH+Acetone)	SPE (silica gel)+SPE (C18)+prep. HPLC	MSTFA/TMSI/DTE (1000:2:2; v/v/w)	XTI-5 (30 m × 0.25 mm, 0.25 μm)	GC–(IT)MS/MS	0.2–0.4	<2 (E1), <1.5 (E2), <0.9 (EE), ND (MES)	[61] (2002)
E1, E2, EE, MES	Sludge	Ultrasonication (MeOH+Acetone)	GPC (Biobeads SX-3)+SPE (silica gel)	MSTFA/TMSI/DTE (1000:2:2; v/v/w)	XTI-5 (30 m × 0.25 mm, 0.25 μm)	GC–(IT)MS/MS	2–4	16–37 (E1), 5–49 (E2), 2–17 (EE), ND (MES)	[61] (2002)

[a]Limit of quantification.

TABLE 2.6.8

Survey of LC methods used for quantitative determination of steroid estrogens

Compounds	Matrix	Extraction	Cleanup	LC column	Mobile phase	Detection method	LOD (ng/L, ng/g)	Levels	Ref. (year)
E1, E2, E3, EE, DES (+progestins)	Drinking, surface and waste water	SPE (LiChrolut RP-18)	-	LiChrospher 100 RP-18 (250 × 4 mm, 5 μm, Merck) LiChrospher 60 RP-Select B (250 × 4 mm, 5 μm, Merck)	ACN/water	DAD-ESI(NI)-MS	25 500 (MS) 50 (DAD)	—	[129] (2000)
E1, E2, E3, EE, DES, MES (+ progestogens)	Drinking, ground, surface, and waste water	Fully automated on-line SPE (PLRP-s or HySphere-Resin-GP)		LiChrospher 100 RP-18 (250 × 4 mm, 5 μm)	ACN/water	DAD ESI(NI)–MS	10–20 <1	—	[67,70] (2001)
E1, E2, EE	Drinking, ground, surface, and waste water	SPE (Bakerbond C18)	SPE (silica gel, WWTP infl.)	RP-C8 Hypersil MO5 (100 × 2.1 mm, 5 μm, Agilent)	ACN/(MeOH)/water	ESI(NI)–MS/MS	0.1–2[a]	0.1–188 (E1) 0.5–17.8 (E2) 0.4–8.8 (EE)	[84,85] (2004, 2005)
E1, E2, E3, EE, E1-3S, E1-3G, E3-16aG, EE-3S, EE-3G	Drinking, lake and waste water	SPE (Sep-Pak Vac tC18, 2 cartridges raw sewage)	—	Synergi 4μ Hydro-RP (75 × 2 mm, 4 μm, Agilent)	MeOH/water	ESI(NI)–MS	0.6–3.9 (free estrog.) 0.8–7.1 (conjugated estrogens)	Up to ≈70	[93] (2005)
E1, α-E2, E2, EE, DES, MES, (+3 progestins)	Well water and urine	SBSE	—	Tracer 120 ODS-A (150 × 4.0 mm, 5 μm, Teknokroma)	ACN/water	DAD	0.3–1 μg/L	<LOD	[79] (2006)
E1, E2, E3, EE, DES, E2-17G, E1-3S, E2-17-Acet.	Ground, river and treated water	Fully automated on-line SPE (PLRP-s)	—	Purospher STAR-RP-18e (125 × 2 mm, 5 μm, Merck)	ACN/water	ESI(NI)-MS/MS	0.01–0.38	0.22–0.33 (E1-3S), 0.68 (E1)	[71] (2004)
E1, E2, E3, EE (+PROG, +6 androgens)	Ground and river water	SPE (Carbograph-1)	—	Alltima C_{18} (250 × 4.6 mm, 5 μm, Alltech)	ACN/water (5 mM ammonium acetate)	APCI(PI)–MS/MS	0.5–1[a]	—	[137] (2001)
E2, EE	Aquaria water	SPE (Sep-Pak C18)	—	Prodigy ODS (150 × 2 mm, 5 μm)	Water/methanol (0.2% formic acid)	APCI(PI)–MS	0.6–1	—	[145] (2002)
E1, E2, E3, EE	Purified and river water (photodegradation studies)	Direct injection-	—	RP-C18 (50 × 2.1 mm, 5 μm, Higgins Analytic)	MeOH/water	APPI(PI)–MS/MS	0.02–0.05 μg/L	—	[132] (2005)

TABLE 2.6.8 (*Continued*)

Compounds	Matrix	Extraction	Cleanup	LC column	Mobile phase	Detection method	LOD (ng/L, ng/g)	Levels	Ref. (year)
E1, E2, E3, EE (+genistein, daidzein, NP, OP, BPA)	Coastal marine water	SPE (Oasis HLB)	Column chromatography (Silica gel)	Synergi™ Hydro-RP (150 × 2.0 mm, 4 μm, Phenomenex)	Methanol/water (both with 2.5–5 mM NH_4Ac)	ESI–MS/MS	0.02 (E1)-1 (E3)	0.10 (E1)-17 (EE), ND (E2, E3)	[53,86] (2005, 2006)
E2, EE (+ ANDR, PROG, TEST, pharmac., PPCP)	Surface water	SPE (HLB)	—	Synergi Max-RP C12 (250 × 4.6 mm, 4 μm, Phenomenex)	MeOH/water (0.1% formic acid)	APCI(PI)–MS/MS	1	<1–44	[131] (2003)
E2	River water	SM-MIPs	—	Inertsil ODS-3 (150 × 2.1 mm)	ACN/water	ESI(NI)–MS (+ECD, +UV)	1.8	—	[74] (2006)
E1, E2, E3, EE, DES	River water	SPE (C18)	—	Purospher STAR RP-18 (55 × 2 mm, 3 μm, Merck)	ACN/water	ESI(NI)–MS	3.2–10.6	<LOD	[136] (2002)
E1, E2, E3, EE	River and waste water	SPE(Carbograph-4)	—	Alltima C-18 (250 × 4.6 mm, 5 μm, Alltech)	ACN/water (post column addition of methanolic ammonia)	ESI(NI)–MS/MS	0.08–0.6[a]	1.5–132 (E1) 0.11–25 (E2) 0.33–187 (E3) 0.04–13 (EE)	[12] (2000)
E1, E2, E3, EE, DES	Surface and waste water	Fully automated in-tube SPME (Supel-Q PLOT)	—	XDB-C_8 (50 × 2.1 mm, 5 μm, Agilent)	ACN/water (0.01% ammonia)	ESI(NI)–MS/MS	2.7–11.7	35.7 (E3, effluent)	[76] (2005)
15 Free and conjugated estrogens	River and lake water and STP effluents	SPE (Autoprep EDS-1)	SPE (Florisil, free fraction)	Zorbax Extend-C_{18} (150 × 1 mm, 3.5 μm, Agilent)	ACN/water/100 mM TEA	ESI(NI)-MS/MS	0.1–3.1 (conj. estro.) 0.1–1.5 (free estro.)	0.3–2.2 (E1-3S) 0.2–1 (E2-3S) 0.2–34 (E1) 0.3–2.5 (E2)	[68] (2003)
E2, α-E2, EE, E1, DES, MES (+3 Aps, + BPA)	River and waste water	SPME (85 μm polyacrylate fiber)	-	LiChrospher 100 RP-18 (250 × 4.6 mm, 5 um, Agilent)	ACN/water (1% AcH, 0.5 g/L KCl)	UV-ECD	0.3–0.7 (UV) 0.07–0.08 μg/L (ED)	1.9–2.2 μg/L (E2, WW)	[75] (2002)
E1, E2, E3, EE, E3-3G, E2-3G, E1-3G, E3-16G, E2-17G, E3-3S, E2-3S, E1-3S	River and waste water	SPE (Carbograph-4)	-	Alltima C18 (250 × 4.6 mm, 5 μm)	ACN/water (post-column addition of ammonia)	ESI(NI)–MS/MS	0.003–15	2–100 (infl.) 3–5 (effl.)	[138] (2002)
E1, E2, EE	River and wastewater	Hemimicelle-based SPE	—	Hypersil ODS (150 × 4.6 mm, 5 μm)	ACN/water	DAD (E1) FL (E2, EE)	In μg/L: 0.1 (E1), 0.02 (E2, EE)	41 (E2), 209 (E1), <LOD (EE)	[72] (2006)
E1, E2, E3, E3-3G, E2-3G, E1-3G, E3-16G, E2-17G, E3-3S, E2-3S, E1-3S	Waste water and female urine	SPE (Carbograph 4)	—	Alltima C-18 (25 × 4.6 mm, 5 μm, Alltech)	ACN/water (both 10 mM formic acid)	ESI(NI)–MS/MS	0.3–60[a]	0.7–72 (mean conc.)	[146] (2003)

E1, E2	STP effluent	SPE (LiChrolut EN+C18)	Immunoaffinity extraction	Betasil C18 (150 × 2.1 mm, 3 μm, Keystone Scientific)	ACN/water	ESI(NI)–MS	0.07–0.18	0.77–6.4 (E2) 1.6–18 (E1)	[49] (2001)
E1, E2, E3, EE	STP influent and effluent	SPE (ENVI-CARB)	—	Alltima C-18 (250 × 4.6 mm, 5 μm, Alltech)	ACN/water	APCI(PI)–MS/MS	0.5–1[a]	13–70 (E1), 4–47 (E2), 24–87 (E3), 2–28 (EE)	[57] (2000)
E1, E2, EE	STP effluents	SPE (C18)	LLE+SPE (Florisil)	Hypersil BDS C_{18} (250 × 2.1 mm, 3 μm)	ACN/water	ESI(NI)–MS/MS	2 (E2, EE) 1 (E1)	1–5 (E1) <LOD (E2, EE)	[80] (2003)
E1, E2, E3, EE, E1-3S, E2-3S, E3-3S, E1-3G, E2-17G, E3-3G, E2-3S-17G, E2-3,17diS	Domestic waste water	SPE (Oasis HLB)	SPE (Sep-Pak Plus Florisil and NH_2)	Zorbax Extend-C18 (150 × 2.1 mm)	ACN/water (1 mM NH_4OH)	ESI(NI)-MS/MS	0.1–1.4	<LOD (EE)-1500 (E2-3,17diS)	[16] (2004)
E1, E2	Estuary sediment	Sonication (MeOH)	+ SPE (Lichrolut EN+BondElut C_{18}) + NP-LC fractionation + IA extract.	Betasil C18 (150 × 2.1 mm, 3 μm, Keystone Scientific)	ACN/water	ESI(NI)–TOF–MS	0.03 (E1) 0.04 (E2)	0.07–2.52 (E1) 0.05–0.53 (E2)	[64] (2005)
E1, E2, E3, EE, DES (+PROG, NOR, APs, APEOs, APECs and halogenated derivatives, BPA)	River sediment	PLE(acetone: MeOH 1:1)	On-line SPE (RAM, LiChrospher ADS alkyl diol silica)	LiChrospher 100 RP-18 (250 × 4.0 mm, 5 μm, Merck)	ACN/water (NI), MeOH/water (PI)	ESI(NI/PI)–MS	0.5–5 ng/g (steriods)	–	[65] (2002)
E1, E2, E3, EE, DES (+ progestins)	River sediment	Sonication (acetone: methanol 1:1)	SPE (C18)	LiChrospher 100 RP-18 (250 × 4 mm, 5 μm, Merck)	ACN/water	ESI(NI)–MS	0.04–1	1.0–11.9 (E1), 0.07–3.37, (E3), 4.16–22.8 (EE), 0.28–2.01 (DES), <LOD (E2)	[66] (2002)

[a]Limit of quantification.

Extraction of estrogens from water has usually been carried out by off-line solid-phase extraction (SPE) with either disks or, most frequently, cartridges (see Tables 2.6.7 and 2.6.8), using a variety of SPE sorbents (octadecyl (C18)-bonded silica, polymeric, graphitized carbon black (GCB) and combinations of them).

Various authors have compared different sorbent materials in terms of extraction efficiency for SPE, off-line [50,67–69] or on-line [67,70,71], of estrogens from water. In many cases, Oasis HLB has been found to be the most suitable [50,68,69].

Liu et al. after concluding that Waters Oasis HLB (0.2 g, 6 mL) was the best among nine different SPE cartridges tested for the GC–MS analysis of estrogens (and other endocrine disrupting compounds, EDCs) in water, evaluated various solvents (acetone, methylene chloride, ethyl acetate and methanol) as eluents and concluded that the best ones were methanol and ethyl acetate and that the recovery of some EDCs was enhanced by the addition of salt, but reduced by the increase in pH value and humic acid concentration [69].

Isobe et al., after examining several commercial SPE columns for trapping estrogens and conjugates from waters, found that Autoprep EDS-1 (Showa Denko) and Oasis HLB (Waters) were the best among others such as ODS, PS2, and Polyamide. To improve the efficiency of elution from the cartridge, 5 mM of TEA was added to 10 mL of MeOH as an ion pair reagent. Without ion pair reagents, conjugates were not eluted effectively from the SPE cartridge by MeOH. With this approach recoveries were higher than 80% for all compounds [68].

In another work, García-Prieto et al. [72] evaluated the use of hemimicelle-based SPE for the concentration and purification of E1, E2, and EE from sewage and river water samples followed by LC-fluorescence/diode array detection (FL/DAD) analysis. In this work, hemimicelles and admicelles of sodium dodecylsulfate (SDS) on alumina, and cetryltrimethyl ammonium bromide (CTAB) on silica, were evaluated. The approach based on the use of hemimicelles of SDS-coated alumina was found to be the most efficient and was recommended to simplify the sample preparation procedure for the analysis of estrogens in environmental waters. However, the limits of detection achieved with the method proposed (0.02–0.1 μg/L) were higher than those frequently reported in the literature for methods based on traditional SPE followed by LC–MS analysis.

Other sample treatment procedures have relied on the use of molecularly imprinted polymers (MIPs) [49,73,74]. One of the most recent works involving MIPs describes the development of a fully automated

SPE–LC–MS method based on the use of surface modified-MIPs for the determination of estradiol in river water. Surface modification of the MIP particles packed in the pretreatment column provided selective affinity and on-line concentration of low levels of estradiol while simultaneously eliminating sample matrix interferences, resulting in a significant increase in sensitivity and reproducibility. Fifty millilitres of water sample was sufficient to achieve a limit of detection of 1 ng/L and the analysis time per sample was only 50 min [74].

Some recent works have also proposed the use of solid-phase micro-extraction (SPME) in combination with either LC or GC analysis for the determination of estrogens in water. An example is the method developed by Peñalver et al., based on SPME coupled to HPLC with both ultraviolet (UV) and electrochemical detection (ED), to determine estrogenic compounds in water samples (river water and wastewater). With this method, which makes use of a modified liquid chromatograph, polyacrylate fibres (85 km) and static desorption, LODs were between 0.3 and 1.1 μg/L using UV detection and between 0.06 and 0.08 μg/L using ED. The levels of E2 found with this method in samples from a WWTP were fairly high, between 1.9 and 2.2 μg/L [75].

Much lower detection limits, between 2.7 and 11.7 ng/L, were achieved by Mitani et al. with a fully automated method based on in-tube SPME coupled in this case with LC–MS/MS in the determination of five estrogens, namely, E1, E2, E3, EE, and DES, in environmental water samples without any other pretreatment. These authors tested four different columns and found that the extraction efficiency of the porous polymer-type capillary column Supel-Q PLOT was better than that of the liquid-phase type capillary columns Omegawax 250, DB-17, and DB-1. The optimum in-tube SPME conditions were 20 draw/eject cycles of 40 µL of sample. The extracted compounds were easily desorbed from the capillary by passage of the mobile phase, and no carryover was observed [76].

Another automated SPME method coupled in this case with GC–MS has been developed by Braun et al. for the analysis of EE plus technical nonylphenol and bisphenol A in water. The extraction performance of different SPME fibre coatings (85 µm polyacrylate (PA), 100 µm polydimethylsiloxane (PDMS) and 65 µm polydimethylsiloxane/divinylbenzene (PDMS/DVB)) was examined and polyacrylate was found to be the most suitable. However, the LODs achieved were fairly high, in the lower µg/L range, and matrix influence was found to be a sensitive parameter in method development [77].

In a more recent work, SPME has been used, also in combination with GC–MS, for the analysis of various exogenous and endogenous endocrine disrupting chemicals, including E1, E2, and DES, in river water and blood serum, with the advantage that derivatization is performed by on-fibre silylation. The proposed SPME method was compared with a traditional SPE procedure based on the extraction of 50 mL of water with C18 cartridges and the results found using both methods were quite agreeable. However, the SPME procedure was comparatively less time-consuming and labor-intensive, has less sample volume requirements (3 mL) and is a solvent-free method. In addition, the LODs achieved for the target estrogens with the SPME method proposed were a bit lower (0.001–0.017 μg/L) than those obtained by SPE (0.001–0.101 μg/L). Application of the method developed to the analysis of river water samples gave considerably high values (0.02 μg/Lfor DES, 0.18 μg/L for E1, and 0.10 μg/L for E2) as compared with those frequently reported in the literature, which, in most cases, are in the low ng or pg/L range [78].

Almeida et al. [79] tested for the first time the technique of stir bar sorptive extraction (SBSE), which is a novel sample preparation technique based on the same principles as those of SPME, for the determination by HPLC–DAD of nine steroid sex hormones (E1, E2, 17α-estradiol, EE, DES, mestranol, progesterone, 19-norethisterone, and norgestrel) in water and urine matrices. These authors tested the most important experimental parameters that can affect the efficiency of extraction, including extraction time (1–6 h), agitation speed (750, 1000, and 1300 rpm), pH (2.0, 3.2, 7.2, and 10.2), ionic strength (NaCl, 5–30%), organic modifier (MeOH, 5–20%), and back extraction solvents for liquid desorption (MeOH, ACN, and equimolar mixtures of both), and found that the most relevant parameters were the equilibrium time, the ionic strength and the back extraction solvents. However, even under the conditions selected as optimal for use with stir bars coated with 126 μL of polydimethylsiloxane (extraction time, 2 h; agitation speed, 750 rpm; sample pH, 7.2; ionic strength, 20% NaCl; and back extraction solvent, acetonitrile) the recoveries achieved for the steroidal estrogens, were, except for mestranol (100.2 ± 10.4%), lower than 50% (spiking concentration 10 μg/L) and the method detection limits were higher than 0.3 μg/L [79].

After extraction, some analytical methods, especially those developed for the analysis of complex environmental samples such as wastewaters, proceed with further purification. This step has been carried out by liquid–liquid extraction (LLE) [14,80,81], HPLC fractionation

[14,15,52,55,82], gel permeation chromatography (GPC) [81], immunoaffinity (IA) extraction [49], and/or SPE using Florisil [68,80], C18 sorbents [14,15,52,83] and silica gel [22,25,51,53,84–88]. Silica gel has been the sorbent most widely used for SPE purification of extracts from water samples. Quintana et al. [87] evaluated the use of alumina, florisil, cianopropyl, and amino sorbents, as alternatives to silica in the clean-up step for analysis of estrogens in river and wastewater, but they led to higher levels of interferences.

Taking into account aspects such as labor, sensitivity, and throughput, one of the most advantageous methods proposed in the literature is that described by Rodríguez-Mozaz et al., based on on-line SPE–LC–MS/MS, for the fully automated analysis of estrogens at the low pg/L level in water, with an analysis time per sample of 1 hour and minimum (filtration) or no sample pretreatment [71].

2.6.2.4.2 Solid environmental samples

The analysis of steroid sexual hormones and related synthetic compounds in solid environmental samples (soils, sediments, sludge and manure) demands the use of complicated, time- and labor-consuming analytical procedures, and high analytical skills. This is due to both the complexity of the environmental matrices and the requirement of low-detection limits. There is scarce documentation on their analysis in solid samples and very often methods applied to these matrices are a simple adaptation (additional purification steps are incorporated) of those used for the analysis of water samples [42].

As shown in Table 2.6.5, the analytical methods described so far for the analysis of estrogens in solid environmental samples have been reviewed in some recently published papers [41,42,44,48].

Extraction of estrogens from solid samples has been performed by pressurized liquid extraction (PLE) [65,89], microwave assisted extraction (MAE) [90] and, more frequently, ultrasonication [60–64,66], using methanol [60,64], methanol/acetone [61,65,66,89], acetone/dichloromethane [63], ethyl acetate [91,92], and dichloromethane/water [62] as extracting solvents.

Peng et al. [63] compared the extraction efficiency of ultra-sonication, mechanical shaking, and Soxhlet extraction with acetone/dichloromethane (1:1, v/v) in the analysis of various steroids (plus bisphenol A and nonyphenol) in river sediments by GC and found that the recoveries attained with Soxhlet and ultrasonication were pretty similar between each other and higher than those achieved by mechanical shaking. Differences were also observed depending on the compound: E1 showed high recoveries (141% with Soxhlet and 134% with ultrasonication), while E2 and mestranol showed relatively low recoveries with both Soxhlet extraction (30% and 35%, respectively) and ultrasonication (68% and 70%, respectively). These relatively low recoveries were attributed to the possible partial transformation of E2 into E1, and mestranol into EE, during extraction, which would explain the higher extraction efficiencies achieved with ultrasonication (with shorter extraction times) as compared to Soxhlet (with longer extraction times).

For pressurized liquid extraction (PLE) of estrogens and other EDCs from sediments the following experimental conditions were used by Petrović et al. [65] and Cespedes et al. [89]: mixture of acetone–methanol (1:1, v/v) as extraction solvent, temperature of 50°C, pressure of 1500 psi, heating time of 5 min, and two cycles of static extraction (5 min each).

For MAE of EDCs, including E1, E2, EE, and 16α-hydroxyestrone, from river sediments and further analysis by GC–MS Liu et al. investigated the effects of various parameters on the extraction efficiency and the most efficient extraction (recovery $>74\%$) of the target compounds was achieved by using methanol as solvent, an extraction temperature of 110°C and 15 min of holding time.

Further purification of the extracts obtained has been carried out by liquid–liquid extraction (LLE), SPE (with restricted access materials (RAMs), reversed-phase silica and/or polymeric materials), HPLC fractionation, immunoaffinity extraction or combinations of them (see Tables 2.6.7 and 2.6.8). Methods reported differ largely in the number of clean-up steps applied, and thus, in the overall analysis time and labor. Sample preparation procedures used before analysis by GC and GC–MS/MS are in general more complicated and time-consuming than those reported for LC-based methods. However, the sensitivity achieved with the various methodologies is, in all instances, in the same range: low ng/g or subng/g in the case of sediments and slightly higher (2–4 ng/g) in the case of sludge. The most laborious method described was that developed by Ternes et al. for the analysis of sediment samples [61]. With this method that includes solvent extraction combined successively with silica gel clean-up, SPE, HPLC fractionation, derivatization, and final analysis by GC–MS/MS, limits of detection varied between 0.2 and 0.4 ng/g of sediment [61]. The simplest method described was that developed by López de Alda et al. [66], consisting of solvent extraction, further SPE clean-up by means of C18 cartridges, and analysis by LC–MS. This method was applied to the analysis of estrogens (and progestogens) in a monitoring program conducted in some Spanish rivers and the limits of detection achieved varied between 0.04 and 1 ng/g of sediment.

2.6.2.5 Evaporation

Most of the analytical methodologies described include one or various evaporation steps that are performed to either improve the sensitivity of the method and/or for solvent exchange. According to Baronti et al. [12] this step may result in losses of E1, E2, E3, and EE if in the process (evaporation under nitrogen in a bath at 40°C) the eluate is completely dried. This, however, is in contrast with the conclusions obtained by Huang et al. in experiments carried out specifically to investigate this phenomenon, according to which losses during blow-down are minimal, and with the, in general, good recoveries obtained for most authors. Nevertheless, the following general precautions can be useful to help minimize potential losses: (i) control the flow rate of the nitrogen and the temperature, (ii) protect the sample solutions from light to prevent photolytic degradation, and (iii) avoid the extract to go completely dry for extended periods of time.

2.6.2.6 Analytical determination

2.6.2.6.1 General remarks

Several reviews have covered the analysis of steroid sex hormones and related synthetic compounds, along with other classes of chemicals, in aquatic environmental samples (see Table 2.6.5).

In the past, the techniques most commonly employed for the environmental analysis of estrogens have been immunoassays and, to a greater extent, GC–MS. Immunoassays are simple and sensitive but tend to overestimate the concentrations, because of the influence of coexisting materials. GC–MS and GC–MS/MS are also highly sensitive methods, but time-consuming sample pretreatment and derivatization steps are required prior to analysis [76]. On the other hand, both immunoassays and GC-methods are limited to the analysis of unconjugated (i.e., free) estrogens, unless intermediate hydrolysis steps are performed [68,93].

LC–MS and LC–MS/MS have gained in popularity in the last years and nowadays is considered as the most promising analytical technique for the determination of steroids [43,94]. The main advantage of using LC is that the enzymatic hydrolysis, required for the immunoassay analysis of both conjugated (glucuronides, sulfates, etc.) and unconjugated estrogens [93], and the derivatization step that normally precedes subsequent GC–MS analysis, can be avoided. Its main drawback is the potential presence of matrix effects that translate in enhancement or suppression of the analyte signal and, consequently, in over- or underestimation of results [93].

2.6.2.6.2 Bioassays

Immunochemical methods of detection, and within them enzyme-linked immunosorbent assays (ELISA) and radioimmunoassays (RIA), are by far the most common bioassays used for the determination of estrogens.

Several recent works have reported their application in the analysis of estrogens in environmental matrices, such as water [17,54,95–99], sludge [95], and manure [100], although they have been more extensively used in the analysis of biological samples in clinical studies. Their main advantages are ease of use, relatively simple protocol and fairly good sensitivity.

Takigami et al. [95], for example, employed an ELISA approach to determine E2 in hydrophobic fractions of water and sludge samples collected from a night soil treatment plant at different treatment phases.

Hintemann et al. [96] developed two immunoassays for the determination of E2 and EE in various STP effluents and river waters. In tap water, recovery rates for E2 and EE2 were 94% and 110%, in surface water 115% and 151%, and in STP effluent 113% and 125%, respectively. Detection limits could be established at 0.05 ng/L for E2 and 0.01 ng/L for EE2, taking a 50-fold enrichment into account. The main advantages were relative ease of automation and the possibility of running ELISA tests *on site* due to robust instrumentation, fast measurements, and a low demand of skilled personnel.

Quantitative enzyme immunoassays have also been used for measuring estrogens (E2 and E1) in dairy manure waste solids [100]. In press cake samples spiked with E2, ELISA, and GC–MS E2 concentrations from all experiments were well correlated ($r(2) = 0.93$), although the ELISA values were higher than the GC–MS values.

Hanselman et al. 2004 [98] compared three commercially available enzyme immunoassay (EIA) kits by measuring E2 in two samples of flushed dairy manure wastewater. The concentrations measured differed according to the immunoassay used and the differences were attributed to a matrix interference associated with coextracted humic substances.

Schneider et al. [54] developed and validated a chemiluminescence ELISA for the direct detection of EE in water at sub-ppt levels. The detection limit achieved was 0.2 ± 0.1 ng/L. Cross-reactivities for E2 and E1 were lower than 0.2%. The ELISA was tested in four different matrices, namely, ultrapure water, tap water, surface water, and effluents of STPs. All measurements were validated using an LC–MS/MS method. Results were consistent in both methods below 1 ng/L. However, the immunoassay showed false-positive results with some surface and wastewater samples probably due to matrix effects.

Majima et al. [97] developed a sensitive time resolved fluoroimmunoassay based on polyclonal antibodies for detecting E2 and E3 in river water. Detection was facilitated by a europium streptavidin–BSA conjugate. Detection limits (2.3 and 4.3 pg/mL for E2 and E3, respectively) were in the same orders of magnitude as those of ELISA for E2, and 1–2 orders of magnitude better for E3.

Atkinson et al. [17] used a highly specific radioimmunoassay for analysis of E1 and its conjugates in coastal marine and sewage waters after extraction of the samples with Sep-Pak C_{18} SPE cartridges. Tested C_{19} and C_{21} compounds yielded $<0.1\%$ cross-reactivities; E1-sulfate and E1 were 100% immunoreactive. Estrone-3-glucosiduronate was 51%, whereas all forms of E2 and E3 were $<0.1\%$ immunoreactive. EE was cross-reactive at 0.1%. The method detection limit for E1 was as low as 40 pg/L. However, the determination of both conjugated and unconjugated E1 required the performance of duplicated analyses, one of them including an enzymatic hydrolysis step.

Barel-Cohen et al. [99] investigated the occurrence of E2, E1, E3, EE, and testosterone along a 100 km course of the Lower Jordan River. After SPE of the samples (1 L) with C-18 columns testosterone and estrogen (E2 and E1) were measured by radioimmunoassay and the limit of detection was 0.3 ng/L (the antibody for E2 and E1 cross-reacted 50% with E1). EE and E3 were measured with commercial ELISA kits and the limit of detection was 0.1 ng/L.

As pointed out by some of the above-mentioned authors the main drawbacks of immunoassays are that they are often inaccurate and prone to interferences, and suffer from relatively low reproducibility and cross-reactivity. In addition, RIAs require handling of radioactive materials, organic extraction, chromatography, and prolonged incubation; they are sometimes difficult to quality control in a routine laboratory and are very susceptible to artifacts caused by nonspecific binding of radioactivity; and, most importantly, there is often poor agreement among results obtained by different RIAs, sometimes even assays from the same manufacturer [101].

Bioassays are also used to measure, apart from target compounds, the estrogenic (or endocrine disrupting) activity of a sample or extract, or of chemicals. The in vitro and in vivo assays available for this purpose have been recently reviewed by Soto et al. [102] and Clode [103], respectively.

On the other hand, many bioassays show potential for development as biosensors [104]. A biosensor is a self-contained (all parts being packaged together), usually small, integrated device, which is capable of providing specific quantitative or semi-quantitative analytical information using a biological recognition element which is retained in direct spatial contact with a transduction element that converts the biological recognition event into a useable output signal [105].

Several reviews have covered the availability and application of biosensors for environmental monitoring of contaminants, including estrogens [105–107].

Some biosensors, such as those based on estrogen receptors (ER), which measure the binding ability of the chemicals towards the ER, are useful to obtain information about the estrogenic potency of a sample. The advantage of this kind of assays is that they are quite simple to perform and allow the identification of all endocrine disrupters that act through the estrogen receptor. The natural sensing element most commonly used is the human ER. Table 2.6.9 shows some of these ER-based biosensors, as reviewed by Rodríguez-Mozaz et al. [105].

TABLE 2.6.9

Estrogen receptor (ER)-biosensors for detection of estrogens and xenoestrogens

Analytes	Transduction element	Reference
Estrogens, progestogens, bisphenol A, 4-nonylphenol and tamoxifen	SPR (BIAcore)	[108]
17β-Estradiol, synthetic estrogens and xenoestrogens	SPR (BIAcore^{TM})	[109]
Estrogens and xenoestrogens	SPR (BIAcore)	[110]
Estrogens	Piezoelectric	[111]
17β-Estradiol	Cyclic voltametry	[112]

Source: Reproduced from Ref. [105] with permission from Springer.

Other biosensors respond to the presence of a specific substance (or group of substances) based on the specific recognition of a biomolecule [105]. An example of such kind of biosensors is the optical immunosensor RIANA, which was used for fast and simultaneous multi-analyte determination of estrone (plus atrazine and isoproturon) in real water samples from a drinking water treatment plant [113]. The performance of this immunosensor was evaluated against a well-accepted traditional method based on solid-phase extraction followed by liquid chromatography–mass spectrometry (LC–MS). The chromatographic method was superior in terms of linearity, sensitivity and accuracy, and the biosensor method in terms of repeatability, speed, cost, and automation.

Estrone has also been determined in water samples with a fully automated optical immunosensor, which enables rapid, simultaneous and high-sensitivity fluorescence detection of up to 32 pollutants in water, with a detection limit for E1 below 1 ng/L [114].

Very recently, Butler et al. [115] have developed a disposable amperometric immunosensor using screen-printed carbon electrodes, based on a direct competition assay with monoclonal antibodies, for rapid, sensitive and selective detection of E2 in water without sample pretreatment. The limit of detection achieved was 0.25 pg/mL. Structurally related compounds, such as 17-α-E2, E1, E3, and progesterone, showed no cross-reactivity. The suitability of the sensor was assessed through the analysis of various distilled, river, and tap water samples.

Finally, Wozei et al. [116] are currently in the process of developing a biosensor based on live cells of the estrogen-sensitive yeast strain RMY/ER-ERE for the analysis of estrogens in water.

2.6.2.6.3 Gas chromatography-(tandem)mass spectrometry

Gas chromatography coupled to mass spectrometry (GC–MS) or tandem mass spectrometry (GC–MS/MS) has been widely used for the determination of estrogens in environmental samples (see Table 2.6.7) [46]. However, owing to the poor volatility of these compounds, a derivatization step aimed to produce more volatile products is required to improve the sensitivity of the analysis.

GC separation has been performed with a variety of capillary columns (DB5-MS, XTI-5, HP Ultra II, etc.), using helium as carrier gas, and temperature programs from approximately 45–300°C. Both conventional MS and MS/MS detection has been accomplished in most instances in the electron impact (EI) mode at 70 eV. The use of negative ion chemical ionization (NICI) has been reported in fewer occasions [50,51,117–120]. However, according with Fine et al. [50], the highest sensitivity for the GC–MS methods is obtained when NICI is used to determine estrogens having pentafluorobenzyl (PFB) [117,118], pentafluorobenzoyl [119,121], and other fluorine-containing derivatives [120].

As previously mentioned, to improve the stability of the compounds and the sensitivity and precision of the GC–MS or GC–MS/MS analysis, the analytes are usually derivatized in the –OH groups of the steroid ring.

Although some authors have reported the direct derivatization of the analytes in the aqueous samples, e.g. using acetic anhydride [122], in most cases the native species are derivatized after their extraction from the matrix.

The derivatization is usually performed by silylation with reagents such as *N*,*O*-bis-(trimethylsilyl)-acetamide (BSA), *N*-methyl-*N*-trimethylsilyltrifluoroacetamide (MSTFA), *N*,*O*-bis-(trimethylsilyl)-trifluoroacetamide (BSTFA), or *N*-(*tert*-butyldimethylsilyl)-*N*-methyl-trifluoroacetamide (MTBSTFA), which lead to the formation of trimethylsilyl (TMS) and *tert*-butyldimethylsilyl (TBS) derivatives, respectively [123]. To enhance derivatization catalysts such as trimethyl-chlorosilane (TMCS) together with trimethylsilyl-imidazole (TMSI), or *tert*-butyldimethylsilyl-chlorosilane (TBCS) and *tert*-butyldimethylsilyl-imidazole (TBSI), are usually added. Other factors affecting the effectiveness of the derivatization procedure, apart from the nature of the derivatization reagent and the catalyst, are the solvent and the reaction conditions, basically the temperature and the time of reaction. Solvents commonly used in silylation procedures with TMS or TBS reagents include ethyl acetate, dichloromethane, acetonitrile, toluene, pyridine, and dimethyl formamide.

The derivatization step is very critical. Very few studies have systematically optimized the derivatization reaction conditions [87,123]. However, in the last years various authors [124,125] have alerted about the incomplete efficiency of some frequently used derivatization procedures and have suggested that some of the current methods may need re-evaluation.

Mol et al. [126], for instance, found that the derivatization of E2 and EE with MTBSTFA was not complete in their analysis.

Shareef et al. [123,124] and Labadie and Budzinski [125] reported breakdown of TMS or TBS-EE2 derivatives to the corresponding E1 derivatives with various solvent-reagent combinations. According with these authors pyridine and dimethyl formamide are the most suitable solvents, to be used in combination with TMS reagents under appropriate reaction conditions.

Peng et al. [63] tested BSTFA, BSTFA+TMCS (1%) and pentafluropropionic anhydride (PFPA) under the same condition (60°C for 2 h) to derivatize E1, E2, α-E2, E3, and mestranol (plus bisphenol A and nonylphenol). Estrogens obtained much better separation and higher sensitivity after being derivatized by PFPA than by silylation reagents (BSTFA and BSTFA+TMCS (1%)) whereas mestranol appeared only partially derivatized by all three derivatizing reagents tested.

Finally, Quintana et al. [87] investigated the reactivity of different silylation reagents versus the aromatic and aliphatic hydroxyl groups contained in the structure of the estrogenic species (E1, E2, E3, EE, DES, mestranol) by adding a fixed amount (100 μL) of the considered reagent (alone or in combination with a catalyzer) to a standard of the analytes in 100 μL of ethyl acetate. The MTBSTFA reagent was able to react only the hydroxyl groups in the position 3 of the aromatic ring of all considered analytes, forming the *tert*-butyldimethylsilyl derivatives. The BSTFA reagent was able to react with the aromatic hydroxyl groups of all compounds and the aliphatic hydroxyl groups of 17β-estradiol and estriol; however, the aliphatic hydroxyl groups of mestranol and ethynyl estradiol remained underivatized. Similar results were obtained using a mixture of BSTFA and TMCS (1%). Silylation of all hydroxyl groups contained in the considered analytes was achieved using MSTFA, without the addition of any catalyser; probably because of its smaller size, it can approach effectively to the hindered hydroxyl groups of mestranol and EE which did not react with BSTFA.

The ion masses selected for quantitation in each case vary depending on the derivatizating reaction performed. As an example, Fig. 2.6.1 illustrates the GC–MS/MS spectra of estradiol after derivatization with pentafluoropropionic acid anhydride and the purported fragmentation scheme [127].

On the other hand, GC–MS/MS has received much less application than GC–MS. However, various authors have evidenced the need for using this technique, especially in the analysis of complex environmental samples such as wastewater or sludge. Thus, according to Quintana et al. [87], who developed a method for the analysis of estrogens in river and wastewater, quantification of estrogens in STP influents, in spite of the thorough sample preparation procedure applied (see Table 2.6.7), must be accomplished by GC–MS/MS and not by single GC–MS, due to the complexity of this kind of samples.

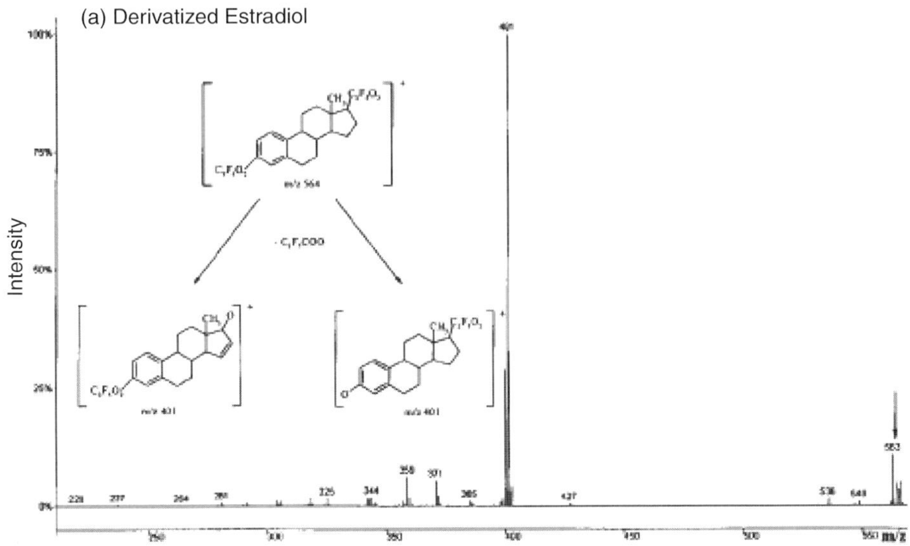

Fig. 2.6.1. GC–MS/MS spectra of estradiol after derivatization with pentafluoropropionic acid anhydride and the purported fragmentation scheme. Reprinted with permission from Ref. [127] © 2000 John Wiley & Sons Ltd.

Clear indications of the benefits of using GC–MS/MS, instead of GC–MS, were also provided by Ternes et al. [22]. These authors developed a GC–MS/MS method for the analysis of estrogens in wastewater. With this technique the baseline could be enormously reduced and hence the signal to noise ratio increased, as compared to GC–MS, and, more importantly, improved confirmation could be achieved. These authors found two compounds, EE and an unknown impurity, with exactly the same retention time and EI spectra showing the *m*/*z* values of 440 (molecular weight of silylated EE) and 425 (M_W–CH_3), however, with a different ratio. Using MS/MS-detection of the target ion *m*/*z* 425, a confirmation with regard to identification and quantification of EE could be carried out. Owing to the fact that the MS/MS-spectra of the contraceptive and the unknown impurity were different, a precise quantitation was possible using the product ions *m*/*z* 193 and *m*/*z* 231 of the precursor ion 425 *m*/*z*. For the unknown compound the excitation energy used was not appropriate to decompose the target ion *m*/*z* 425 completely and the product ions *m*/*z* 193 and *m*/*z* 231 were not formed. Using single MS detection the probability to determine excessive concentrations of EE is then relative high.

2.6.2.6.4 Liquid chromatography-(tandem)mass spectrometry

Table 2.6.8 shows a selection of the most recent LC analytical methods published in the literature for determination of estrogens in environmental samples. This table includes information on the main steps used in the sample preparation procedure, the column employed for chromatographic separation, the chromatographic mobile phase, the detection technique, the limits of detection achieved, and the environmental levels found from the application of the methods reviewed to real environmental samples.

LC separation of both conjugated and unconjugated estrogens has almost invariably been performed on octadecyl silica stationary phases. Both classical, long (250 mm) columns with 4.0–4.6 mm ID and 5 μm particle size, and short (55–150 mm) columns with smaller (2.0–2.1 mm) ID and smaller (3 μm) particle size have been used for separation. As mobile phases, mixtures of water/methanol and, more frequently, water/acetonitrile, with gradient elution from 10–50% to 100% organic solvent have normally been used.

As shown in Table 2.6.8, apart from a few works that report the use of fluorescence (FL) [128], ultraviolet (UV) or diode array detection (DAD) [70,75,79], and/or electrochemical detection (ED) [75], most methodologies are based on the use of MS and tandem MS/MS. The wavelength of choice for analysis of estrogens by UV or DAD is 200 nm [70,79,129].

In the past, one of the obstacles to the routine analytical application of LC–MS was the unavailability of rugged and reliable LC–MS interfaces. However, during the last ten years, LC–MS has gained in popularity, due to the sensitivity, ruggedness and ease of use given by the newer atmospheric pressure ionization (API) interfaces. Today, the API interfaces most widely used for the LC–MS analysis of environmental estrogens are electrospray (ESI) in the negative ionization (NI) mode and, to a lesser extent, atmospheric pressure chemical ionization (APCI) in the positive ionization (PI) mode. Various authors have evaluated APCI and ESI in both modes of ionization for the determination of environmental estrogens [76,80,129–131] and most of them have concluded that the sensitivity achieved with ESI(NI) is greater than that of APCI(PI) [76,80,129,130]. Very recently, however, Lin et al. [132] have reported the use of the recently developed atmospheric pressure photoionization interface to investigate the photodegradation rates of four estrogens (E1, EE, 17α-E2, and EE) in river water using toluene as dopant. The method allowed the detection of the four compounds through direct injection (50 μL) of spiked purified or river water at levels of 20–50 ng/L.

These API technologies have been interfaced with a variety of MS analyzers, including quadrupole, ion-trap, orthogonal-acceleration time-of-flight (oaTOF), and combinations of them. Single and triple quadrupole instruments have been the most widely used for analysis of estrogens.

LC–MS and LC–MS/MS have been mostly applied in the selected ion-monitoring (SIM) mode and in the selected reaction-monitoring mode (SRM), respectively. Nowadays, LC–MS/MS is considered as the technique of choice for the determination of environmental estrogens [43]. The very good selectivity and sensitivity offered by LC–MS/MS makes possible the determination of estrogens in environmental water and solid samples at the ng/L and ng/g level, and even at the pg/L and pg/g level, after appropriate sample preparation. Additional benefits are that analytes do not have to be fully resolved to be identified and quantitated, as is required using conventional diode array detection (DAD), and that chemical derivatization is not needed, as in GC–MS. However, in the case of very complex matrices, such as wastewater and sludge, even when using SRM detection, both false negative results, due to matrix ionization suppression effects, and false positive results, due to insufficient selectivity, can be obtained [133].

An approach for increasing the selectivity, and avoiding false positive findings is the use of oa-TOF-MS. Oa-TOF-MS instruments became commercially available some 6–7 years ago and combine the ability to perform accurate mass determination with an excellent full-scan sensitivity [134]. An even more powerful technique results from the combination of a quadrupole front-end and an oa-TOF back-end for MS-MS, in the so-called Q-TOF, where accurate mass determination at excellent sensitivity can be achieved after conventional low-energy collision induced dissociation (CID) in a hexapole collision cell. However, neither of these two advanced techniques (LC–oa–TOF–MS and LC–Q–TOF–MS) have been routinely employed yet for the qualitative or quantitative determination of steroids in environmental samples, probably due to their, at the moment, high price compared to ion-trap and triple quadrupole instruments [43]. The resolving power of TOF-MS has been demonstrated by Reddy and Brownawell for the analysis of E2 and E1 in estuary sediments [64] and by Farre et al. for the analysis of estrogens and fitoestrogens in ground, surface, and wastewaters [135] (see Fig. 2.6.2).

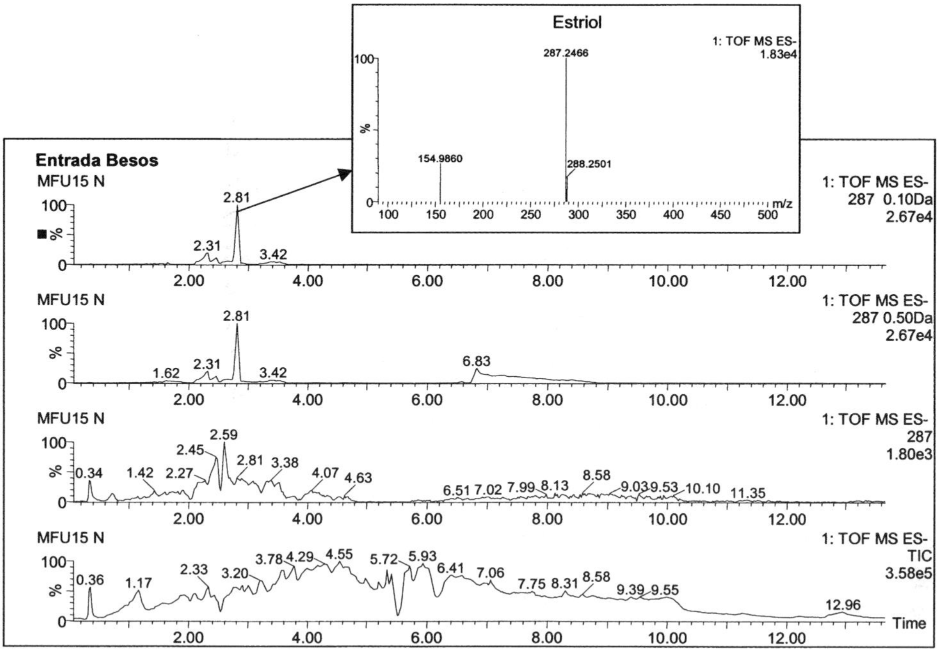

Fig. 2.6.2. Total ion current (TIC) and reconstructed ion chromatograms ($m/z = 287$) obtained at different mass windows from the LC-TOF-MS analysis of estriol in a wastewater influent. Both the selectivity and the sensitivity increase when decreasing the mass interval. Inset shows the spectrum of estriol.

Modification of the mobile phase, when performed in an attempt to improve the sensitivity of MS detection, has been accomplished with various reagents, such as acetic acid 0.1%, formic acid 0.2%, and ammonium acetate 10 mM (see Table 2.6.8).

According to Benijts et al. [136], who studied in detail the influence of different mobile phase compositions on the ionization efficiency of an ionspray interface, a mixture of water and acetonitrile, without addition of bases or buffer systems is the best choice for optimal ionization of estrogens, which is in agreement with the findings of other authors [129]. However, as it is commented in more detail below, there are great discrepancies in this respect between authors, discrepancies that are believed to originate in the different interface designs.

In a first series of experiments, Benijts et al. investigated the effect of acetonitrile and methanol as organic modifiers on the ionization of estradiol. It was observed that an increasing amount of organic modifier gradually increases the ionization efficiency of the ion source for estradiol, this effect being more apparent with acetonitrile than with methanol, which is in agreement with the observations made by other authors [80,129,137]. Nonewithstanding this, Gomes et al. [93], for example, chose methanol over acetonitrile because conjugates and free steriods co-eluted when using acetonitrile.

The use of mobile phase additives such as ammonium hydroxide, isopropylamine, and triethylamine (TEA), commonly employed to both improve LC separation and ionization efficiency in LC–(NI)MS, was also investigated by Benijts et al. [136], who found that none of these volatile bases improved the estradiol signal. On the contrary, TEA even had a negative impact on the ionization of the analyte. However, according to Isobe et al. [68] the use of TEA increases the sensitivity in the analysis of steroid estrogens and metabolites by LC–ESI(NI)–MS/MS and other authors [12,138] have reported a drastic increase in the ESI-MS response with methanolic ammonia.

Volatile buffers such as formic acid/ammonium formate or acetic acid/ammonium acetate are also usually added to the LC eluent for chromatographic purposes, in spite that numerous researchers have observed that even low concentrations of these buffers result in ionization suppression [68,84,136,137]. However, this phenomenon seems to be even more complicated since, as it is shown in Fig. 2.6.3, increasing NH_4Ac concentrations have been observed to cause an decrease in the response of EE when measured in a standard solution and a increase when measured in a spiked matrix sample [53].

The quantitation and diagnostic ions and transitions used for the LC–MS and LC–MS–MS analysis of free and conjugated estrogens, along with explanations on the purported fragmentation pathways that take place in both ESI(NI) and APCI(PI) can be found in a recent review published by López de Alda on the analysis of estrogens (and other emerging pollutants) in the aquatic environment by LC–(tandem)MS [43]. In most cases, the base peak selected for quantitation of estrogens in the SIM mode, or as precursor for collisionally induced dissociation in the SRM mode, corresponds to the deprotonated analyte molecule $[M-H]^-$ in ESI(NI) and to the $[M+H-H_2O]^+$ ion ($[M+H]^+$ for estrone) in APCI(PI). However, as pointed out by Vanderford et al. [131], this seems to be a function of the source design itself, as researchers using the same instrumentation report similar predominant ions whereas researchers using different instrumentation do not, and the same argument is used to explain the strong discrepancies observed among authors when investigating the sensitivity of different interfaces

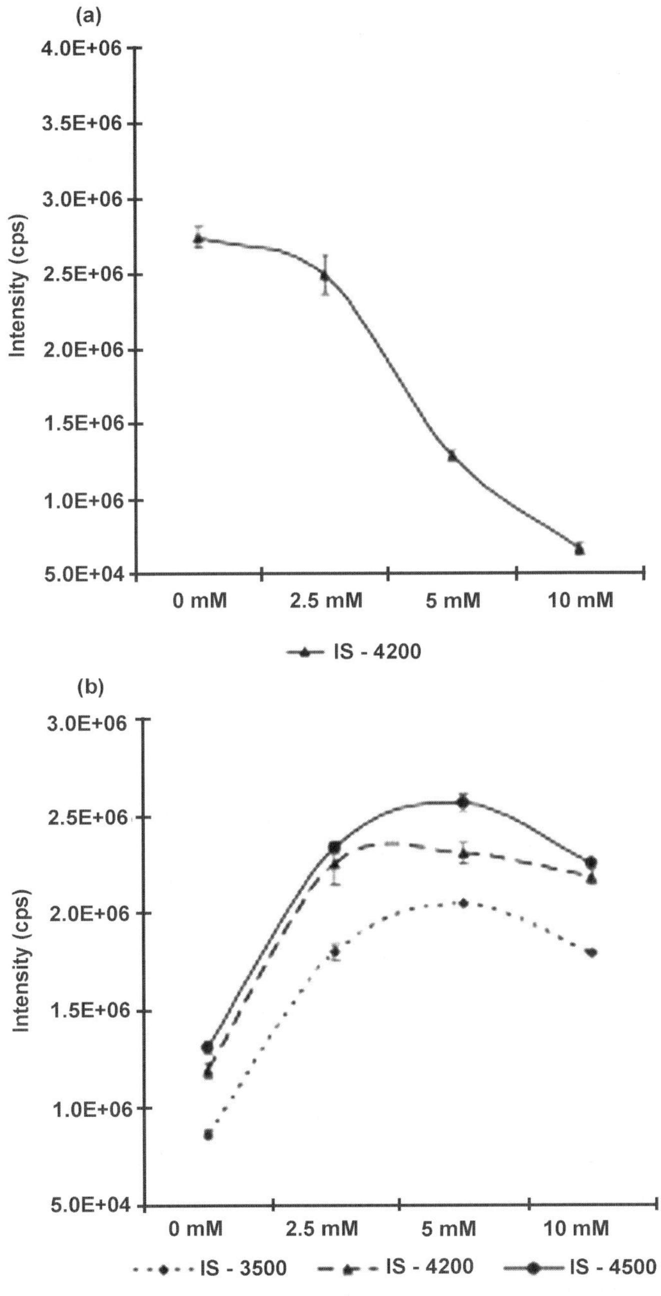

Fig. 2.6.3. Influence on signal intensities of EE2 using different modifier (NH_4Ac) concentrations for (a) a standard solution and (b) a spiked field sample. In part (b) the impact of varying ion spray voltages is shown. Reprinted with permission from Ref. [53] © 2005 Elsevier.

and ionization modes and the effect of different mobile compositions (organic modifier, buffers, etc.) on the ionization efficiency of estrogens.

The main drawback of LC–MS methods is the existence of matrix effects, which translate in either suppression or enhancement of the analyte signal and have a negative influence on the reproducibility and accuracy of the analyses. Matrix effects vary between compounds and from sample to sample and it is generally accepted that ESI is more prone to matrix effects than APCI.

Various approaches, such as selective extraction followed by efficient sample clean-up [49,65,74], matrix matched calibration, correction using matrix spike recoveries [53], the use of isotope labeled surrogate standards [64], or the method of standard addition, have been suggested to overcome matrix effects.The use of molecularly imprinted polymers (MIPs) [49], surface modified molecularly imprinted polymers (SM-MIPs), [74] and restricted access materials (RAMs) [65] has been found to efficiently remove most of the matrix-related interferences that would otherwise cause severe ionization suppression in the ESI analysis of estrogens for example in wastewaters [49], river water [74], and sediments [65], while allowing a significant increase in sensitivity. However, these techniques are not yet consolidated as robust extraction/purification techniques for routine analysis.

On the other hand, the use of matrix matched calibration solutions does not compensate for differences between samples, and correction using matrix spike recoveries and the method of standard addition are very laborious and time-consuming. Stable isotope-labeled surrogate standards are considerably expensive but they are available for basically all estrogens of environmental interest and their use appears as the most convenient means to compensate for matrix effects in routine analysis.

2.6.2.6.5 LC–(tandem)MS vs GC–(tandem)MS

Various authors have compared the techniques of GC and LC coupled to MS and MS/MS in terms of sensitivity and general performance for the determination and quantitation of steroid estrogens in environmental matrices [127,130] and according to them, the sensitivity increases in the order LC–ESI(NI)MS $<$ GC–(EI)MS $<$ GC–(EI)MS/MS $\leqslant$ LC–ESI (NI)MS/MS.

In terms of accuracy and repeatability, both techniques are in general satisfactory, although the presence of matrix effects in LC–(tandem)MS and the derivatization step usually carried out prior to GC–(tandem)MS can constitute a source of inaccuracy [130].

Advantages of GC–(tandem)MS over with LC–(tandem)MS are the lower cost of the instrumentation and the availability of extensive libraries of mass spectra for identification of unknown peaks in estrogenically active fractions. Advantages of LC–(tandem)MS over GC–(tandem)MS are shorter run times, easy on-line coupling with on-line SPE, and capability for analysis of both free and conjugated estrogens.

2.6.3 CONCLUSIONS

In the light of the occurrence and the estrogenic potency and effects observed for estrogens in the environment, it seems logical to think that these compounds will be subjected to regulation in the near future and that they will have to be monitored on routine basis. In this context, methods for their analysis will have to be reliable, accurate, sensitive, precise, robust, and as fast and cheap as possible. Owing to the fact that estrogens are the most potent environmental EDCs, already active at levels in the 0.1–10 ng/L range, their determination in real-world environmental samples requires, apart from very sensitive and selective analytical techniques, such as LC and GC coupled to MS/MS, efficient sample preparation procedures.

The technique most widely used for extraction of water samples is SPE, although other techniques such as SPME, MIPs, and RAMs also show promise. For extraction of solid samples, ultrasonication and PLE seem to be the most adequate techniques. Both are fairly fast and provide good recoveries with appropriate extracting solvents although an advantage of PLE is that it does not require subsequent filtration or centrifugation of the extracts. Further purification of the extracts is often necessary, especially in the case of complex environmental matrices, such as wastewater, sediment, or sludge. The reported purification procedures are in general laborious and time-consuming and include intermediate evaporation steps. However, they cannot be obviated at present, to prevent matrix interferences in the subsequent analysis. For analysis, GC and LC coupled to MS are the most adequate techniques because, in contrast to bioassays, they allow the simultaneous determination of multiple analytes. In addition, the recent introduction of tandem mass spectrometric detection has largely improved the performance of the chromatographic methods by reducing the detection limits and helping analyte identification. The use of tandem MS (instead of single MS) and the acquisition of two SRM transitions per compound is considered nearly indispensable to achieve reliable identification and quantification of the analytes and to avoid false positive results. To this end, the application of stringent confirmation and identification criteria [139], based on the agreement of retention times, base peak and diagnostic ions, and relative abundances, between the standards and the analytes, is essential. The next few years will no doubt see the general application of these advanced techniques integrated in completely automated, on-line systems, which will increase sample throughput, sensitivity and reproducibility, and will diminish sample manipulation, operating costs and contamination risks. As compared to GC, LC offers a series of advantages: it has shorter analyses times, is easily coupled on-line with SPE, does not require derivatization and allows the simultaneous analysis in the same run of both free and conjugated estrogens without an intermediate hydrolysis step [93]. In contrast, the main drawback of this technique is the possibility of matrix effects. At present, the use of isotope labeled surrogate standards appears as the most convenient means to overcome this problem whereas future developments in the field of selective materials, such as MIPs or immunoaffinity sorbents, may help avoid matrix effects and simplify sample preparation.

A promising alternative to traditional methods can be found in the biosensors, which are expected to allow the continuous, field monitoring of estrogens in the environment, although most of them are currently under development and have not been validated yet in the analysis of real samples.

While more field and laboratory studies are needed, the application of these (and others to come) advanced sample preparation and detection techniques, will help expand our knowledge about the presence, fate, and persistence of estrogens and their degradation products in the environment, and to estimate the degree of human and wildlife exposition, all of which will help environmentalists (i) assess potential risks, (ii) define priority pollutants, (iii) propose quality criteria, and (iv) suggest remediation actions.

REFERENCES

1 J. Beausse, *TrAC-Trend. Anal. Chem.*, 23 (2004) 753.
2 Martindale, *The Extra Pharmacopoeia*, The Pharmaceutical Press, London, 1982.
3 A.M. Andersson and N.E. Skakkebaek, *Eur. J. Endocrinol.*, 140 (1999) 477.

4 J.A. Beardmore, G.C. Mair and R.I. Lewis, *Aquaculture*, 197 (2001) 283.
5 M.A. Hurley, P. Matthiessen and A.D. Pickering, *J. Theor. Biol.*, 227 (2004) 159.
6 C.A. Strussmann, F. Takashima and K. Toda, *Aquaculture*, 139 (1996) 31.
7 F. Yamazaki, *Aquaculture*, 33 (1983) 329.
8 G.-G. Ying, R.S. Kookana and Y.-J. Ru, *Environ. Int.*, 28 (2002) 545.
9 T.A. Hanselman, D.A. Graetz and A.C. Wilkie, *Environ. Sci. Technol.*, 37 (2003) 5471.
10 A.C. Johnson, R.J. Williams and P. Matthiessen, *Sci. Total Environ.*, 362 (2006) 166.
11 A.C. Johnson and R.J. Williams, *Environ. Sci. Technol.*, 38 (2004) 3649.
12 C. Baronti, R. Curini, G.D'. Ascenzo, A. Di Corcia, A. Gentili and R. Samperi, *Environ. Sci. Technol.*, 34 (2000) 5059.
13 T.A. Ternes, P. Kreckel and J. Mueller, *Sci. Total Environ.*, 225 (1999) 91.
14 C. Desbrow, E.J. Routledge, G.C. Brighty, J.P. Sumpter and M. Waldock, *Environ. Sci. Technol.*, 32 (1998) 1549.
15 A.C. Belfroid, A. Van der Horst, A.D. Vethaak, A.J. Schafer, G.B.J. Rijs, J. Wegener and W.P. Cofino, *Sci. Total Environ.*, 225 (1999) 101.
16 K. Komori, H. Tanaka, Y. Okayasu, M. Yasojima and C. Sato, *Water Sci. Technol.*, 50 (2004) 93.
17 S. Atkinson, M.J. Atkinson and A.M. Tarrant, *Environ. Health Persp.*, 111 (2003) 531.
18 H. Nakamura, T. Shiozawa, Y. Terao, F. Shiraishi and H. Fukazawa, *J. Health Sci.*, 52 (2006) 124.
19 http://esc.syrres.com/interkow/physdemo.htm
20 K.M. Lai, K.L. Johnson, M.D. Scrimshaw and J.N. Lester, *Environ. Sci. Technol.*, 34 (2000) 3890.
21 M. Petrović, E. Eljarrat, M.J.L. de Alda and D. Barcelo, *Anal. Bioanal. Chem.*, 378 (2004) 549.
22 T.A. Ternes, M. Stumpf, J. Mueller, K. Haberer, R.-D. Wilken and M. Servos, *Sci. Total Environ.*, 225 (1999) 81.
23 M. Nasu, M. Goto, H. Kato, Y. Oshima and H. Tanaka, *Water Sci. Technol.*, 43 (2001) 101.
24 A.C. Johnson and J.P. Sumpter, *Environ. Sci. Technol.*, 35 (2001) 4697.
25 M.R. Servos, D.T. Bennie, B.K. Burnison, A. Jurkovic, R. McInnis, T. Neheli, A. Schnell, P. Seto, S.A. Smyth and T.A. Ternes, *Sci. Total Environ.*, 336 (2005) 155.
26 A.C. Layton, B.W. Gregory, J.R. Seward, T.W. Schultz and G.S. Sayler, *Environ. Sci. Technol.*, 34 (2000) 3925.
27 R.J.W.M.D. Jurgens and A.C. Johnson, Research and Development Technical Report P161, Environment Agency, Bristol (1999).

28 M. van den Berg, T. Sanderson, N. Kurihara and A. Katayama, *Pure Appl. Chem.*, 75 (2003) 1917.
29 J.P. Sumpter and A.C. Johnson, *Environ. Sci. Technol.*, 39 (2005) 4321.
30 J. Toppari, J.C. Larsen, P. Christiansen, A. Giwercman, P. Grandjean, L.J. Guillette, B. Jegou, T.K. Jensen, P. Jouannet, N. Keiding, H. Leffers, J.A. McLachlan, O. Meyer, J. Muller, E. RajpertDeMeyts, T. Scheike, R. Sharpe, J. Sumpter and N.E. Skakkebaek, *Environ. Health Persp.*, 104 (1996) 741.
31 R.M. Sharpe and N.E. Skakkebaek, *Pure Appl. Chem.*, 75 (2003) 2023.
32 N.E. Skakkebaek, N. Jorgensen, K.M. Main, E. Rajpert-De Meyts, H. Leffers, A.M. Andersson, A. Juul, E. Carlsen, G.K. Mortensen, T.K. Jensen and J. Toppari, *Int. J. Androl.*, 29 (2006) 2.
33 L.S. Shore, Y. Kapulnik, M. Gurevich, S. Wininger, H. Badamy and M. Shemesh, *Environ. Exp. Bot.*, 35 (1995) 363.
34 R. Lange, T.H. Hutchinson, C.P. Croudace, F. Siegmund, H. Schweinfurth, P. Hampe, G.H. Panter and J.P. Sumpter, *Environ. Toxicol. Chem./SETAC*, 20 (2001) 1216.
35 E.J. Routledge, D. Sheahan, C. Desbrow, G.C. Brighty, M. Waldock and J.P. Sumpter, *Environ. Sci. Technol.*, 32 (1998) 1559.
36 P.A.H.C.E. Purdom, V.J. Bye, N.C. Eno, C.R. Tyler and J.P. Sumpter, *Chem. Ecol.*, 8 (1994) 275.
37 C. Pelissero, G. Flouriot, J.L. Foucher, B. Bennetau, J. Dunogues, F. Legac and J.P. Sumpter, *J. Steroid Biochem.*, 44 (1993) 263.
38 P.D. Hansen, H. Dizer, B. Hock, A. Marx, J. Sherry, M. McMaster and C. Blaise, *TrAC-Trend. Anal. Chem.*, 17 (1998) 448.
39 S. Fritsche and H. Steinhart, *Eur. Food Res. Technol.*, 209 (1999) 153.
40 I. Shaw and S. McCully, *Int. J. Food Sci. Tech.*, 37 (2002) 471.
41 M. Kuster, M.J. López de Alda and D. Barceló. In D. Barceló (Ed.), *The Handbook of Environmental Chemistry*, Vol. 50; *Water Pollution: Emerging Organic Pollution in Waste Waters and Sludge*, Vol. 2, Springer-Verlag, Heidelberg, 2004.
42 M. Kuster, M.J. Lopez, M.J.L. de Alda and D. Barcelo, *TrAC-Trend. Anal. Chem.*, 23 (2004) 790.
43 M.J.L. de Alda, S. Diaz-Cruz, M. Petrović and D. Barcelo, *J. Chromatogr. A*, 1000 (2003) 503.
44 M.S. Diaz-Cruz, M.J.L. de Alda and D. Barcelo, *TrAC-Trend. Anal. Chem.*, 22 (2003) 340.
45 P.T. Holland, *Pure Appl. Chem.*, 75 (2003) 1843.
46 M. Petrović, E. Eljarrat, M.J.L. de Alda and D. Barcelo, *J. Chromatogr. A*, 974 (2002) 23.
47 M.J.L. de Alda and D. Barcelo, *Fresen. J. Anal. Chem.*, 371 (2001) 437.
48 M. Petrović, E. Eljarrat, M.J.L. de Alda and D. Barcelo, *TrAC-Trend. Anal. Chem.*, 20 (2001) 637.

49 P.L. Ferguson, C.R. Iden, A.E. McElroy and B.J. Brownawell, *Anal. Chem.*, 73 (2001) 3890.
50 D.D. Fine, G.P. Breidenbach, T.L. Price and S.R. Hutchins, *J. Chromatogr. A*, 1017 (2003) 167.
51 H.M. Kuch and K. Ballschmiter, *Fresen. J. Anal. Chem.*, 366 (2000) 392.
52 A.C. Johnson, A. Belfroid and A. Di Corcia, *Sci. Total Environ.*, 256 (2000) 163.
53 I.C. Beck, R. Bruhn, J. Gandrass and W. Ruck, *J. Chromatogr. A*, 1090 (2005) 98.
54 C. Schneider, H.F. Scholer and R.J. Schneider, *Anal. Chim. Acta*, 551 (2005) 92.
55 C.H. Huang and D.L. Sedlak, *Environ. Toxicol. Chem.*, 20 (2001) 133.
56 M. Furhacker, A. Breithofer and A. Jungbauer, *Chemosphere*, 39 (1999) 1903.
57 A. Lagana, A. Bacaloni, G. Fago and A. Marino, *Rapid Commun. Mass Sp.*, 14 (2000) 401.
58 R. Siegener and R.F. Chen. In: L.H. Keith, T.L. Jones-Lepp, and L.L. Needham (Eds.), *Environmental Endocrine Disruptors*. ACS Symposium Series 747, American Chemical Society, Washington D.C., 2000, p. 125.
59 L.S. Shore, M. Gurevitz and M. Shemesh, *Bull. Environ. Contam. Toxicol.*, 51 (1993) 361.
60 M. Peck, R.W. Gibson, A. Kortenkamp and E.M. Hill, *Environ. Toxicol. Chem.*, 23 (2004) 945.
61 T.A. Ternes, H. Andersen, D. Gilberg and M. Bonerz, *Anal. Chem.*, 74 (2002) 3498.
62 R.J. Williams, A.C. Johnson, J.J.L. Smith and R. Kanda, *Environ. Sci. Technol.*, 37 (2003) 1744.
63 X.Z. Peng, Z.D. Wang, C. Yang, F.R. Chen and B.X. Mai, *J. Chromatogr. A*, 1116 (2006) 51.
64 S. Reddy and B.J. Brownawell, *Environ. Toxicol. Chem.*, 24 (2005) 1041.
65 M. Petrović, S. Tavazzi and D.A. Barcelo, *J. Chromatogr. A*, 971 (2002) 37.
66 M.J.L. de Alda, A. Gil, E. Paz and D. Barcelo, *Analyst*, 127 (2002) 1299.
67 M.J.L. de Alda and D. Barcelo, *J. Chromatogr. A*, 938 (2001) 145.
68 T. Isobe, H. Shiraishi, M. Yasuda, A. Shinoda, H. Suzuki and M. Morita, *J. Chromatogr. A*, 984 (2003) 195.
69 R. Liu, J.L. Zhou and A. Wilding, *J. Chromatogr. A*, 1022 (2004) 179.
70 M.J.L. de Alda and D. Barcelo, *J. Chromatogr. A*, 911 (2001) 203.
71 S. Rodriguez-Mozaz, M.J.L. de Alda and D. Barcelo, *Anal. Chem.*, 76 (2004) 6998.
72 A. Garcia-Prieto, L. Lunar, S. Rubio and D. Perez-Bendito, *Analyst*, 131 (2006) 407.

73 C. Tozzi, L. Anfossi, G. Giraudi, C. Giovannoli, C. Baggiani and A. Vanni, *J. Chromatogr. A*, 966 (2002) 71.
74 Y. Watabe, T. Kubo, T. Nishikawa, T. Fujita, K. Kaya and K. Hosoya, *J. Chromatogr. A*, 1120 (2006) 252.
75 A. Penalver, E. Pocurull, F. Borrull and R.M. Marce, *J. Chromatogr. A*, 964 (2002) 153.
76 K. Mitani, A. Fujioka and H. Kataoka, *J. Chromatogr. A*, 1081 (2005) 218.
77 P. Braun, M. Moeder, S. Schrader, P. Popp, P. Kuschk and W. Engewald, *J. Chromatogr. A*, 988 (2003) 41.
78 L.H. Yang, T.G. Luan and C.Y. Lan, *J. Chromatogr. A*, 1104 (2006) 23.
79 C. Almeida and J.M.F. Nogueira, *J. Pharmaceut. Biomed.*, 41 (2006) 1303.
80 V. Ingrand, G. Herry, J. Beausse and M.R. de Roubin, *J. Chromatogr. A*, 1020 (2003) 99.
81 D.G.J. Larsson, M. Adolfsson-Erici, J. Parkkonen, M. Pettersson, A.H. Berg, P.-E. Olsson and L. Forlin, *Aquat. Toxicol.*, 45 (1999) 91.
82 T.P. Rodgers-Gray, S. Jobling, S. Morris, C. Kelly, S. Kirby, A. Janbakhsh, J.E. Harries, M.J. Waldock, J.P. Sumpter and C.R. Tyler, *Environ. Sci. Technol.*, 34 (2000) 1521.
83 T.A. Hanselman, D.A. Graetz, A.C. Wilkie, N.J. Szabo and C.S. Diaz, *J. Environ. Qual.*, 35 (2006) 695.
84 S. Zuehlke, U. Duennbier and T. Heberer, *J. Sep. Sci.*, 28 (2005) 52.
85 S. Zuehlke, U. Duennbier, T. Heberer and B. Fritz, *Ground Water Monit. R.*, 24 (2004) 78.
86 I.-C. Beck, R. Bruhn and J. Gandrass, *Chemosphere*, 63 (2006) 1870.
87 J.B. Quintana, J. Carpinteiro, I. Rodriguez, R.A. Lorenzo, A.M. Carro and R. Cela, *J. Chromatogr. A*, 1024 (2004) 177.
88 H.-R. Aerni, B. Kobler, B.V. Rutishauser, F.E. Wettstein, R. Fischer, W. Giger, A. Hungerbuhler, M.D. Marazuela, A. Peter and a. Schonenberger et, *Anal. Bioanal. Chem.*, 378 (2004) 688.
89 R. Cespedes, M. Petrović, D. Raldua, U. Saura, B. Pina, S. Lacorte, P. Viana and D. Barcelo, *Anal. Bioanal. Chem.*, 378 (2004) 697.
90 R. Liu, J.L. Zhou and A. Wilding, *J. Chromatogr. A*, 1038 (2004) 19.
91 G.G. Ying and R.S. Kookana, *Environ. Sci. Technol.*, 37 (2003) 1256.
92 G.G. Ying, R.S. Kookana and P. Dillon, *Water Res*, 37 (2003) 3785.
93 R.L. Gomes, J.W. Birkett, M.D. Scrimshaw and J.N. Lester, *Int. J. Environ. An. Ch.*, 85 (2005) 1.
94 K. Shimada, K. Mitamura and T. Higashi, *J. Chromatogr. A*, 935 (2001) 141.
95 H. Takigami, N. Taniguchi, T. Matsuda, M. Yamada, Y. Shimizu and S. Matsui, *Water Sci. Technol.*, 42 (2000) 45.
96 T. Hintemann, C. Schneider, H.F. Scholer and R.J. Schneider, *Water Res*, 40 (2006) 2287.

97 K. Majima, T. Fukui, J. Yuan, G. Wang and K. Matsumoto, *Anal. Sci.*, 18 (2002) 869.
98 T.A. Hanselman, D.A. Graetz and A.C. Wilkie, *J. Environ. Qual.*, 33 (2004) 1919.
99 K. Barel-Cohen, L.S. Shore, M. Shemesh, A. Wenzel, J. Mueller and N. Kronfeld-Schor, *J. Environ. Manage.*, 78 (2006) 16.
100 D.R. Raman, A.C. Layton, L.B. Moody, J.P. Easter, G.S. Sayler, R.T. Burns and M.D. Mullen, *Transactions of the Asae*, 44 (2001) 1881.
101 R.E. Nelson, S.K. Grebe, D.J. O'Kane and R.J. Singh, *Clin. Chem.*, 50 (2004) 373.
102 A.M. Soto, M.V. Maffini, C.M. Schaeberle and C. Sonnenschein, *Best Pract. Res. Cl. En.*, 20 (2006) 15.
103 S.A. Clode, *Best Pract. Res. Cl. En.*, 20 (2006) 35.
104 S. Rodriguez-Mozaz, M.J. Lopez de Alda and D. Barceló, *J. Chromatogr. A* (in press).
105 S. Rodriguez-Mozaz, M.P. Marco, M.J.L. de Alda and D. Barcelo, *Anal. Bioanal. Chem.*, 378 (2004) 588.
106 S. Rodriguez-Mozaz, M.P. Marco, M.J.L. de Alda and D. Barcelo, *Pure Appl. Chem.*, 76 (2004) 723.
107 S. Rodriguez-Mozaz, M.J. Lopez de Alda and D. Barceló, *Anal. Bioanal. Chem.*, 386 (2006) 1025.
108 M. Usami, K. Mitsunaga and Y. Ohno, *J. Steroid Biochem.*, 81 (2002) 47.
109 B. Hock, M. Seifert and K. Kramer, *Biosens. Bioelectron.*, 17 (2002) 239.
110 M. Seifert, S. Haindl and B. Hock, *Anal. Chim. Acta*, 386 (1999) 191.
111 M. Zhihong, L. Xiaohui and F. Weiling, *Anal. Commun.*, 36 (1999) 281.
112 M. Murata, M. Nakayama, H. Irie, K. Yakabe, K. Fukuma, Y. Katayama and M. Maeda, *Anal. Sci.*, 17 (2001) 387.
113 S. Rodriguez-Mozaz, M.J.L. de Alda and D. Barcelo, *Talanta*, 69 (2006) 377.
114 P. Hua, J.P. Hole, J.S. Wilkinson, G. Proll, J. Tschmelak, G. Gauglitz, M.A. Jackson, R. Nudd, H.M.T. Griffith, R.A. Abuknesha, J. Kaiser and P. Kraemmer, *Optics Express*, 13 (2005) 1124.
115 D. Butler and G.G. Guilbault, *Sensor. Actuat. B-Chem.*, 113 (2006) 692.
116 E. Wozei, S.W. Hermanowicz and H.Y.N. Holman, *Biosens. Bioelectron.*, 21 (2006) 1654.
117 S. Nakamura, T. Hwee Sian and S. Daishima, *J. Chromatogr. A*, 919 (2001) 275.
118 S. Cathum and H. Sabik, *Chromatographia*, 53 (2001) S394.
119 X.-Y. Xiao, D.V. McCalley and J. McEvoy, *J. Chromatogr. A*, 923 (2001) 195.
120 O. Lerch and P. Zinn, *J. Chromatogr. A*, 991 (2003) 77.
121 H.M. Kuch and K. Ballschmiter, *Environ. Sci. Technol.*, 35 (2001) 3201.
122 D. Jahr, *Chromatographia*, 47 (1998) 49.

123 A. Shareef, M.J. Angove and J.D. Wells, *J. Chromatogr. A*, 1108 (2006) 121.
124 A. Shareef, C.J. Parnis, M.J. Angove, J.D. Wells and B.B. Johnson, *J. Chromatogr. A*, 1026 (2004) 295.
125 P. Labadie and H. Budzinski, *Environ. Sci. Technol.*, 39 (2005) 5113.
126 H.G.J. Mol, S. Sunarto and O.M. Steijger, *J. Chromatogr. A*, 879 (2000) 97.
127 T.R. Croley, R.J. Hughes, B.G. Koenig, C.D. Metcalfe and R.E. March, *Rapid Commun. Mass Sp.*, 14 (2000) 1087.
128 S.A. Snyder, T.L. Keith, D.A. Verbrugge, E.M. Snyder, T.S. Gross, K. Kannan and J.P. Giesy, *Environ. Sci. Technol.*, 33 (1999) 2814.
129 M.J.L. de Alda and D. Barcelo, *J. Chromatogr. A*, 892 (2000) 391.
130 M.S. Diaz-Cruz, M.J.L. de Alda, R. Lopez and D. Barcelo, *J. Mass Spectrom.*, 38 (2003) 917.
131 B.J. Vanderford, R.A. Pearson, D.J. Rexing and S.A. Snyder, *Anal. Chem.*, 75 (2003) 6265.
132 A.Y.C. Lin and M. Reinhard, *Environ. Toxicol. Chem.*, 24 (2005) 1303.
133 T. Reemtsma, *TrAC-Trend. Anal. Chem.*, 20 (2001) 533.
134 W.M.A. Niessen, *J. Chromatogr. A*, 856 (1999) 179.
135 M. Farré, M. Kuster, R. Brix, F. Rubio, M.J. López de Alda and D. Barceló, *J. Chromatogr. A* (in press).
136 T. Benijts, R. Dams, W. Gunther, W. Lambert and A. De Leenheer, *Rapid Commun. Mass Sp.*, 16 (2002) 1358.
137 A. Lagana, G. Fago, A. Marino and D. Santarelli, *Anal. Lett.*, 34 (2001) 913.
138 A. Gentili, D. Perret, S. Marchese, R. Mastropasqua, R. Curini and A. Di Corcia, *Chromatographia*, 56 (2002) 25.
139 Commission decision 2002/657/EC, Official Journal of the European Communities L221/8.
140 E.P. Kolodziej, J.L. Gray and D.L. Sedlak, *Environ. Toxicol. Chem./SETAC*, 22 (2003) 2622.
141 R. Jeannot, H. Sabik, E. Sauvard, T. Dagnac and K. Dohrendorf, *J. Chromatogr. A*, 974 (2002) 143.
142 L.B. Barber, G.K. Brown and S.D. Zaugg. In: L.H. Keith, T.L. Jones-Lepp, and L.L. Needham (Eds.), *Environmental Endocrine Disruptors*, ACS Symposium Series 747, American Chemical Society, Washington D.C., 2000, p. 97.
143 C. Kelly, *J. Chromatogr. A*, 872 (2000) 309.
144 H.B. Lee and T.E. Peart, *J. AOAC Int.*, 81 (1998) 1209.
145 J. Rose, H. Holbech, C. Lindholst, U. Norum, A. Povlsen, B. Korsgaard and P. Bjerregaard, *Comp. Biochem. Phys. C*, 131 (2002) 531.
146 G. D'Ascenzo, A. Di Corcia, A. Gentili, R. Mancini, R. Mastropasqua, M. Nazzari and R. Samperi, *Sci. Total Environ.*, 302 (2003) 199.

Chapter 2.7

Analysis of iodinated X-ray contrast media

Anke Putschew and Martin Jekel

2.7.1 INTRODUCTION

Iodinated X-ray contrast media (ICM) are applied to enhance the contrast between vessels or organs which otherwise could not be examined by X-ray. Figure 2.7.1 shows the structure of selected ICM, which are all derivatives of 2,4,6-triiodo benzoic acid and some possible transformation products.

The iodine atoms are responsible for the absorption of X-ray, the properties are controlled by side-chains in position 1, 3, and 5. The compounds are designed to be very polar and persistent. Both properties result in a rapid excretion of the metabolically stable ICM via urine within 24 h after application [1]. ICM are applied, in most cases, intravenously with a dose of up to 100 g ICM for one examination [1]. The maximum iodine concentration in urine can be found already within the first hour after application varying between 20 and 70 g/L [2]. The worldwide consumption of the ICM is about 3500 t/a [3]. Up to 50% of the ICM are applied in hospitals, at least in Berlin [4] and thus, hospital wastewater is characterized by high adsorbable organic halogen (AOX) concentrations, e.g. 0.41 mg/L AOX [5]. With the hospital wastewater, the ICM are transported to municipal wastewater treatment plants (WWTP) that are not able to remove the compounds [6–9]. This behaviour is in agreement with the classification as not readily biodegradable [10–12]. Owing to the properties and the negative log P_{OW} values [13] ICM occur preferential in the water phase. In surface waters, which are influenced by WWTP effluents, several ICM can be detected with concentrations in the lower µg/L range, e.g. in the river Rhine 0.15 µg/L Iopromide [14]. In Berlin, the surface water concentrations are in general higher because of less dilution (Iopromide

Comprehensive Analytical Chemistry 50
Petrović and Barceló (Eds)
Volume 50 ISSN: 0166-526X DOI: 10.1016/S0166-526X(07)50008-5

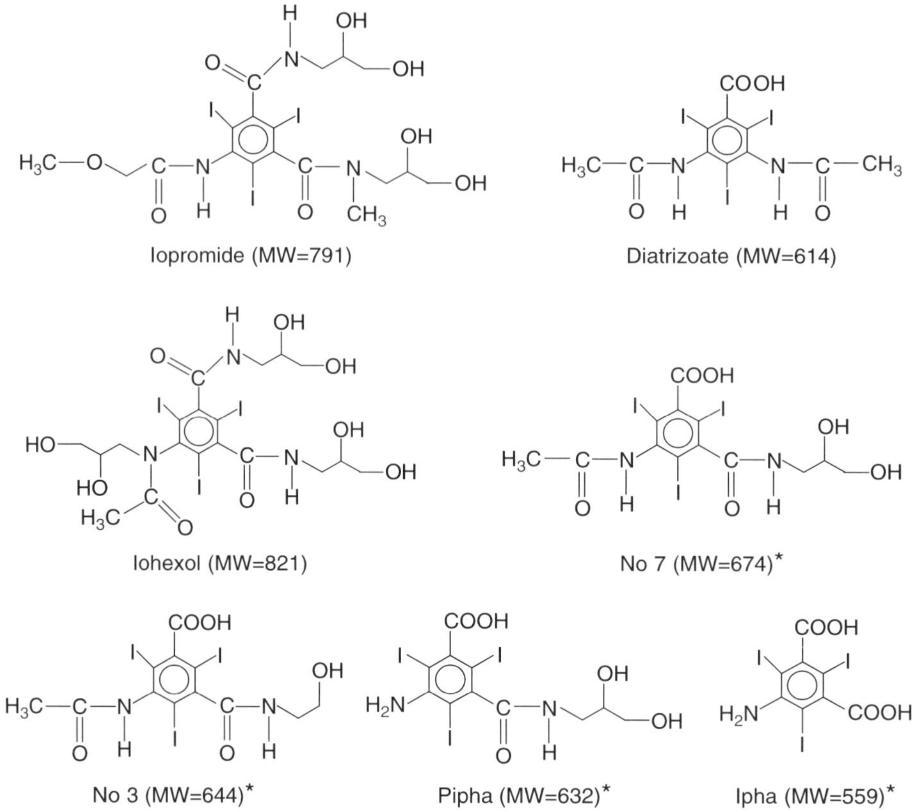

Fig. 2.7.1. Structures of selected ICM and possible transformation products (marked by an asterisk).

0.86 μg/L) [15–18] due to a short water cycle within city. Even groundwater which is influenced by surface water is contaminated with ICM [17–21]. The concentrations are in the lower/mid ng/L range. During bank filtration decreased concentrations of ICM are recognized. The decreased concentrations are a result of dilution and transformation. Under anoxic condition a (partly)deiodination is possible [17,19] but (partly)deiodinated ICM as well as other transformation products are not identified up to now, except in laboratory studies where transformation products of Iopromide could be detected [12,22].

Conventional wastewater treatment techniques are not able to remove the ICM. The potential of ozonation and advanced oxidation processes for the oxidation of the iodinated compounds in drinking [21] and wastewater were investigated [23–24], but the removal efficiencies are low. In case of ozonation the concentration of the ICM can be reduced but the amount of organic bound iodine (AOI) remains nearly constant [25–26]. A further treatment of ICM suited for the treatment of highly contaminated wastewater like hospital wastewater, and urine is the reductive dehalogenation by zero-valent iron [27]. ICM can be deiodinated completely but the biodegradability of the deiodinated compounds needs still to be investigated.

Today, the knowledge about the occurrence and behaviour of the iodinated X-ray contrast media is comprehensive. To receive that knowledge different analytical techniques were used. In the following, the analytical methods are described and information about the occurrence and behaviour of the ICM are given.

2.7.2 ADSORBABLE ORGANIC BOUND IODINE (AOI)

Iodinated X-ray contrast media can be detected via the AOI. For that reason the AOX method was modified permitting the differentiation of the AOX into AOCl, AOBr, and AOI [7]. For the AOI analysis the filtered samples (0.45 μm) are acidified (conc. HNO_3) and then sodium sulfite is added. Sulfite reduces any inorganic iodine species like polyiodide (I_3^-) or iodine (I_2), which can be adsorbed onto activated carbon, to iodide. The water samples are then adsorbed on activated carbon. Inorganic halogens are displaced from the activated carbon by nitrate ions. The loaded activated carbon is oxidized at 1000°C in an oxygen stream and the combustion gas is trapped in an adsorption solution (ultra pure water containing a trace of sodium sulfide to reduce any produced iodate (IO_3^-) and iodine to iodide). An aliquot of the absorption solution is analysed by ion chromatography and the halogenides are quantified. The limit of quantification for the AOI is 0.5 μg/L (UV detection at 226 nm).

The method permitting to distinguish between the halides was used for the quantification of iodinated organic compounds released by a small hospital (300 beds) in Berlin [28]. Over one week 24 h mixed hospital wastewater samples were analysed for iodinated organic compounds (AOI). The wastewater flow was monitored and it was possible to calculate the discharged AOI load. Figure 2.7.2 shows the AOI concentration as well as the AOI load/d. The highest AOI concentration is 1136 μg/L and the maximal discharged AOI load about 507 g/d. During the studied time period the consumed amount of ICM (Diatrizoate, Ioversol, Iopromide, Iomeprol) was documented and the equivalent

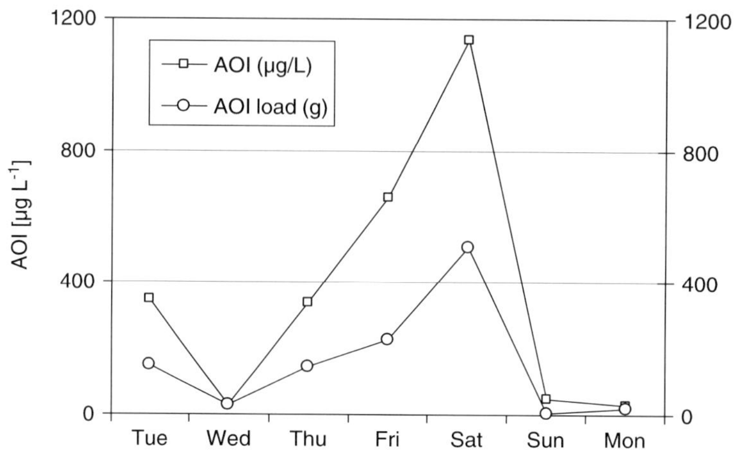

Fig. 2.7.2. AOI (µg/L) and AOI load (g/d) of a hospital wastewater. AOI was determined in 24 h mixed samples.

TABLE 2.7.1

ICM applied in a hospital over one week and calculated AOI

ICM	Amount applied (g)	AOI calculated (g)
Diatrizoate	60.37	37.5
Iopromide	1464.05	704.3
Iomeprol	244.96	116.4
Ioversol	190.80	118.4
Sum		976.6

amount of organic bound iodine was calculated (Table 2.7.1). The sum of the measured AOI load (see Fig. 2.7.2) is 1084.6 g and slightly higher than the calculated AOI, which is 976.6 g. Ninety per cent of the measured AOI is produced by ICM, which means that the parameter AOI represents the ICM.

Based on single compound analysis it is known that WWTP are not able to remove ICM [8–9]. The result was verified by the AOI, because single-compound analysis of ICM can be affected by the complex matrix of such water samples. Figure 2.7.3 shows the AOI of a WWTP in and effluent and as expected the AOI respectively the ICM are just poorly removed.

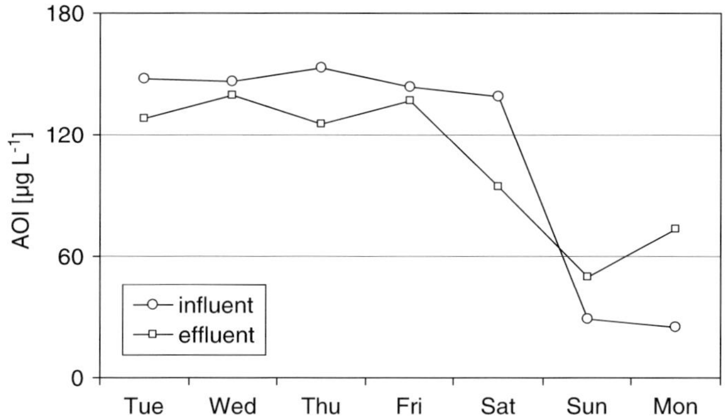

Fig. 2.7.3. AOI (µg/L) of a wastewater treatment plant in- and effluent. 24 h mixed samples.

2.7.3 LIQUID CHROMATOGRAPHY–TANDEM MASS SPECTROMETRY (LC–ESI–MS/MS) AND AOI

A selective and sensitive method for the qualitative and quantitative determination of ICM is liquid chromatography (LC) coupled with tandem mass spectrometry (MS/MS) [8–9,14,16–18]. The water samples have to be enriched. For the enrichment solid-phase extraction (SPE) is used. In case of the very polar ICM, a sequential SPE is advisable [15–16], especially for the ionic ICM or transformation products that are usually more polar than the parent compound. The sequential SPE uses for the first extraction step an EN cartridge (Merck) and for the second step EnviCarb material (Supleco). The second step was introduced because the EnviCarb material is like activated carbon and the AOI analysis showed that activated carbon is suitable for the adsorption of the very polar iodinated compounds [7]. The recovery was determined by spiking tap water with selected ICM as well as some possible transformation products (see Fig. 2.7.1). The results show (Table 2.7.2) that for the enrichment of ionic triiodinated benzoic acid derivatives the second extraction step is required. Other SPE methods with good recoveries for ICM are published [14], but if ionic compounds and/or transformation products are under investigation the sequential method should be used.

The enriched compounds are detected after LC by MS/MS. After separating by reversed-phase chromatography [14,16,18] the ICM are

TABLE 2.7.2

Recovery of triiodinated benzoic acid derivatives (1 μg/L each compound) in the EN and EnviCarb extract and the sum of both extracts

	Recovery (%) EN extract	Recovery (%) EnviCarb	Sum of both extracts (%)
Iopomide	82	19	101 ± 5
Iohexol	49	37	86 ± 5
Diatrizoate	22	77	99 ± 10
Pipha	37	49	86 ± 10
No. 7	0	66	66 ± 10
No. 6	48	26	74 ± 10
No. 3	12	72	85 ± 10

Note: $n = 3$; extraction volume 500 ml; final volume 1.0 ml.

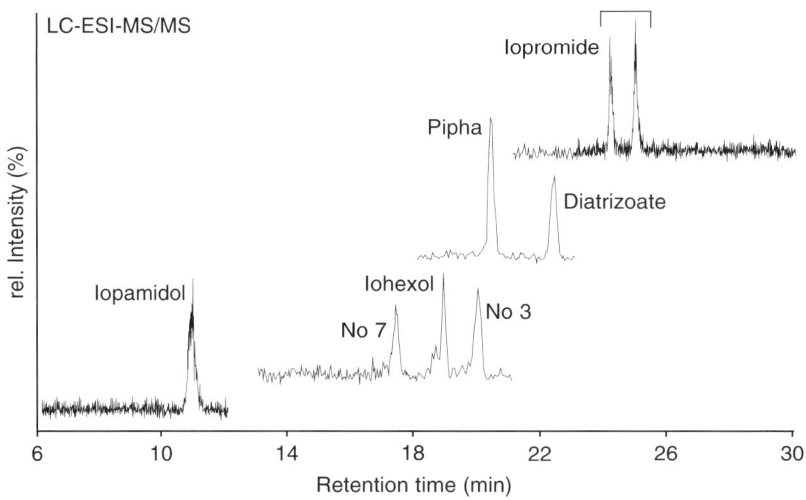

Fig. 2.7.4. Chromatogram (LC–ESI–MS/MS) of a standard mixture (each 0.01 μg/ml).

transferred into the MS where they are ionized by positive-ion (PI) electrospray ionization (ESI) and are detected after collision-induced dissociation (MS/MS). The ICM calibration curves are linear over a range, e.g. from 0.1 to 10 ng absolute of each compound. Figure 2.7.4 shows a chromatogram of a standard mixture of 7 triiodinated benzoic acid derivatives, separated on a C18-column (3μm particle size) using water and methanol both acidified with formic acid. If native water samples are analysed the quantification is influenced by (I) losses during sample clean-up and (II) by the matrix which can reduce or enhance the ionization yield. These two factors can be compensated by using an internal standard and an internal calibration or if an external calibration in used, the samples must be analysed with and without addition of the analytes, whereby the concentration added must be in the same range as the concentration in the water sample.

Using LC–ESI–MS/MS the concentration of ICM was determined in different surface and groundwaters [8,14–18,20]. In Berlin ICM were quantified in a partly closed water cycle influenced by a WWTP effluent [15–18]. The effluent of a municipal WWTP, receiving hospital wastewater, is conducted via a receiving channel into a lake. At the lake a waterworks is located producing drinking water from bank-filtered lake water. The distance between the lakefront and the monitored drinking water wells is about 100 m. The quality of the bank-filtered groundwater was examined via monitoring wells (for more details, see [17–18]). In selected samples the concentration of ICM was quantified and the AOI was determined. The annual average AOI in the lake is 9.6 μg/L respectively, 939 ng/L Iopromide and 1044 ng/L Diatrizoate. During bank filtration the AOI as well as the concentration of Iopromide and Diatrizoate decreased (Table 2.7.3). For the decreasing concentrations dilution and/or transformation could be responsible. With LC–ESI–MS it is not easy to analyse unknown trace compounds and thus not possible to prove if ICM are transformed to other compounds. But the comparison between AOI and the ICM concentration gives an idea about the behaviour. The AOI decrease is not as high as the decrease of Iopromide and Diatrizoate indicating that ICM are definitely not mineralized, and compounds which are still iodinated must be produced. Iodinated transformation products were detectable (see below)

TABLE 2.7.3

AOI and ICM concentrations of surface water and influenced groundwater

Distance from the lake bank (m)	AOI (μg/L)	Iopromide (ng/L)	Diatrizoate (ng/L)
Lake	9.6	939	1044
0	6.9	245	503
40	3.6	44	285
60	3.8	67	696
80	4.2	39	201
100	2.3	129	182

Note: Data given are mean values over nine months.

but it was not possible to identify the compounds due to the low concentrations.

2.7.4 SELECTIVE DETECTION OF ORGANIC BOUND IODINE (LC–ESI–IISF–MS), LC–ESI–MS/MS, AND AOI

As known LC–ESI–MS/MS is a powerful tool for the analysis of trace amounts of polar organic pollutants in water. A disadvantage is that the compounds to be analysed must be known and standards must be available for quantification. For the analysis of ICM PI–ESI is used, which implicates that only positive ions can be detected. The detection of iodine, respectively iodide as a product of fragmentation cannot be monitored and used to trace, e.g. unknown iodinated compounds. Negative-ion (NI)–ESI, which allows the detection of iodide is not applied to the analysis of the triiodinated benzoic acid derivatives, because the ionization yield is very low. LC inductively coupled plasma (ICP)–MS is an analytical method allowing the quantification of organic bound iodine as well as the detection of unknown iodinated compounds by monitoring *m/z* 127 [29]. It could be shown that even LC–NI–ESI–MS, with an induced in-source fragmentation (IISF) generated by a high-cone voltage and detection of *m/z* 127 is a useful tool for the detection of iodinated organic compounds via detection of iodide as a product-ion of the in-source fragmentation [30]. In contrast to LC–ICP–MS the response depends on the structure and probably on the composition of the eluent used for LC separation. A generic and independent signal can just be obtained for derivatives of a specific substance class measured, with respect to the eluent composition, under nearly the same conditions. In Fig. 2.7.5, the area of six different triiodinated benzoic acid derivatives are plotted against the iodine concentration. The calibration curves are linear and the correlation coefficients are better than 0.999. The iodine response is not equal for all compounds but the relative ration to Iopamidol varying between 0.8 and 1.2 shows that the signals correlated well with iodine amount. Figure 2.7.6 shows a chromatogram of *m/z* 127 of eight triiodinated benzoic acid derivatives. Although the retention times a slightly different as compared to those of the peaks in Fig. 2.7.4, it is obvious that the signals are produced by the ICM and that the method is very sensitive, same concentrations were injected for the analysis by LC tandem MS and LC–ESI–IISF–MS.

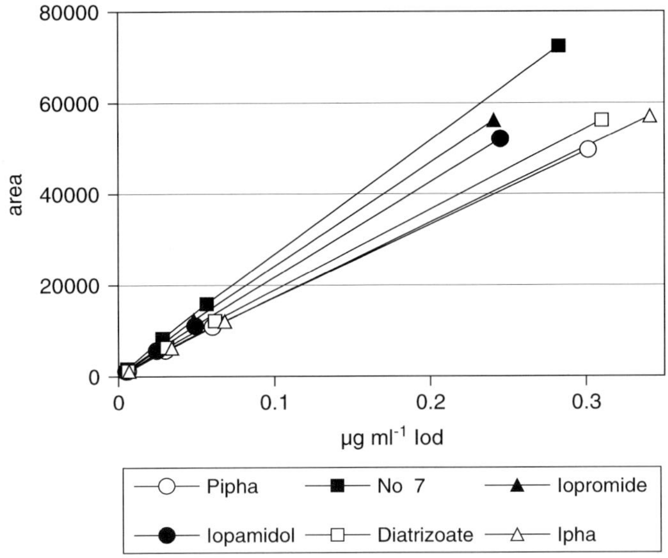

Fig. 2.7.5. Calibration curves of the iodine concentration of triiodinated benzoic acid derivatives.

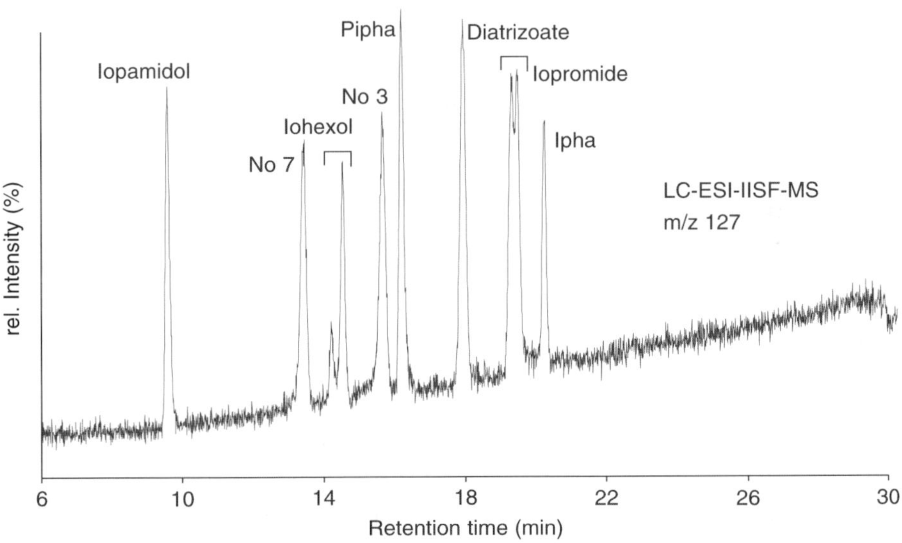

Fig. 2.7.6. Chromatogram (LC–ESI–IISF–MS; m/z 127) of a standard mixture (each 0.01 µg/ml).

TABLE 2.7.4

Contribution of ICM and OI–MS to the AOI in %

	ICM contribution to the AOI (%)	OI-MS contribution to the AOI (%)
Receiving water	80	70
Lake	12	34
Ground water	7	20

The AOI, the concentration of ICM as well as the amount of organic bound iodine, detected by LC–ESI–IISF–MS (OI–MS) were determined in selected samples of the above-described water cycle. Table 2.7.4 shows the contribution of ICM to the AOI in %. In the receiving water, 80% of the AOI is produced by the quantified ICM, following up the water cycle the percentage of the AOI produced by these compounds decreases to just 7% in the influenced groundwater. As already mentioned above, within the water cycle the ICM are transformed, most probably to still iodinated organic compounds, explaining the decreased contribution of ICM to the AOI. By LC–ESI–IIFS–MS analysis of a groundwater sample (Fig. 2.7.7) many unknown iodinated organic compounds are detectable, supporting the assumption of ICM transformation. It was possible to determine some *m/z* values, representing most probably molecular ions of the unknown compounds by classical LC–ESI–MS using the scan mode. But owing to the low concentration it was not possible to identify the compounds. The amount of organic bound iodine (OI–MS) was also quantified by LC–ESI–IISF–MS. For the quantification of the OI–MS, the triiodinated benzoic acid derivatives (see Fig. 2.7.1) were used, whereby the amount of organic bound iodine was used instead of the compound concentrations. As observed before for the contribution of the ICM to the AOI, the contribution of the OI–MS to the AOI decreases within the water cycle (Table 2.7.4). The reason for the low contribution of the detected unknown iodinated compounds to the AOI is not clear. One reason could be that the transformation products of the ICM are very polar and not extractable by SPE. A second reason could be that the quantification of the OI–MS is not accurate, because an independent signal can just be obtained for derivatives of a specific substance class measured, with respect to the eluent composition, under nearly the same conditions (see above).

The results of another study will be summarized, showing the advantages of LC–ESI–IISF–MS, even based on quantification. Common

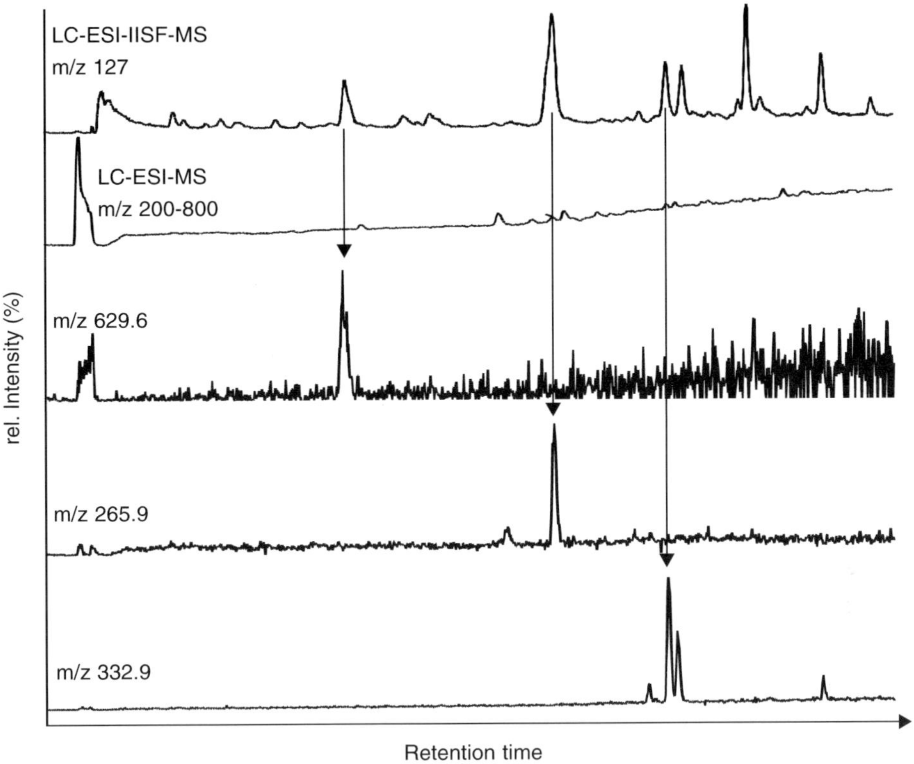

Fig. 2.7.7. Chromatogram of a groundwater sample. LC–ESI–IISF–MS: selective detection of organic bound iodine m/z 127; LC–ESI–MS: m/z 200–800 (scan) and mass chromatograms.

wastewater treatment is not efficient for the removal of ICM. The concentration of ICM can be reduced by advanced wastewater treatment techniques like ozonation [21,23–24]. But if transformation products are investigated, then it is questionable if advanced oxidation processes (AOP) are the method of choice. 10 mg/L Iopromide dissolved in tap water was ozonated in a semi-batch reactor. For the first 30 min the ozone influent concentration was 10 mg/L there after 30 mg/L. Over the reaction time of 1 h the batch reactor was sampled and the concentrations given in Table 2.7.5 were determined. The amount of Iopromide and OI–MS was quantified by LC–ESI–IISF–MS, the AOI was determined and the concentration of I-Iodate as an oxidation product was quantified by ion chromatography. 4.82 mg/L I-Iodate should be detectable if the Iopromide is completely deiodinated by ozonation. Already after 10 min the concentration of Iopromide is reduced by 90%

TABLE 2.7.5

Ozonation of 10 mg/L iopromide in tap water

Time (min)	Ozone consumption (mg/L)	Iopromide (mg/L)	AOI (mg/L)	OI–MS (mg/L)	I-Iodate (mg/L)	Sum (AOI+I-Iodate, mg/L)
0	0	11.1	4.5	7.0	0	4.5
5	9.5	5.1	4.3	4.5	0.85	5.15
10	13.5	1.1	3.5	2.8	1.45	4.95
15	14.6	0.4	2.9	2.4	1.46	4.36
30	15.9	n.d.	2.4	2.2	1.79	4.19
45	35.9	n.d.	1.5	1.3	2.66	4.16
60	36.9	n.d.	1.1	0.9	3.00	4.10

n.d., not detected.

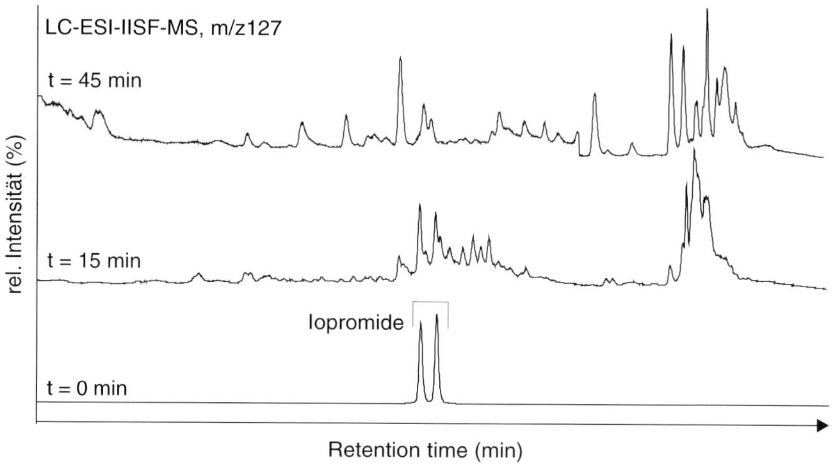

Fig. 2.7.8. LC–ESI–IISF–MS chromatograms of ozonated Iopromide.

but at the same time 70% of the initial AOI is still detectable and 30% of the maximal inorganic iodine. The iodine balance shows that Iopromide is transformed by ozonation to organic compounds that are still iodinated. The OI–MS determined by LC–ESI–IISF–MS is comparable with the measured AOI and thus nearly all iodinated transformation products are included. Figure 2.7.8 shows the chromatograms for different reaction times. By ozonation many iodinated transformations products with relatively low concentrations are produced. All concentrations were determined in the original samples (not in extracts), this seems to be the reason why the AOI and the OI–MS are comparable. In another experiment performed with just 100 µg/L Iopromide the samples were extracted. Here, the OI–MS values are in general much lower as the

AOI and thus the transformations products are not extractable by the used method.

REFERENCES

1 U. Speck and H. Hübner-Steiner, Röntgenkontrastmittel. In: Oberdisse, Hackenthal and Kuschinsky (Eds.), *Pharmakologie und Toxikologie*. Springer-Verlag Berlin, Heidelberg, New York, 2. Auflage, 1999, 597–601.
2 M. Sprehe, in: *Photochemische Oxidation iodierter Röntgenkontrastmittel im Krankenhausabwasser-Verfahrens und Reaktorentwicklung*, Dissertation, TU Clausthal, CUTEC-Schriftenreihe, 2000, Vol. 44
3 M. Sprehe and S.U. Geissen, in: *ATV-DVWK Schriftenreihe, Verfahrensauswahl zur AOX-Eliminierung im Krankenhausabwasserbereich, Herausgeber* ATV-DVWK, 2000, Vol. 18, p. 257.
4 C. Pineau and B. Heinzmann, *GWF Wasser Abwasser*, 146/9 (2005) 646.
5 T. Hündesrügge, *Krankenhauspharmazie*, 19 (1998) 245.
6 J. Oleksy-Frenzel, S. Wischnack and M. Jekel, *Vom Wasser*, 85 (1995) 59.
7 J. Oleksy-Frenzel, S. Wischnack and M. Jekel, *Fresenius J. Anal. Chem.*, 366 (2000) 89.
8 T.A. Ternes and R. Hirsch, *Environ. Sci. Technol.*, 34 (2000) 2741.
9 M. Carballa, F. Omil, J.M. Lema, M. Liompart, C. Garcia-Jares, I. Rodrigues, M. Gomez and T.A. Ternes, *Wat. Res.*, 38 (2000) 2918.
10 T. Steger-Hartmann, R. Länge, H. Schweinfurth, M. Tschampel and I. Rehmann, *Wat. Res.*, 36 (2002) 266.
11 S. Gartiser, L. Brinker, T. Erbe, K. Kümmerer and R. Willmund, *Acta Hydrochim. Hydrobiol.*, 24 (1990) 90.
12 W. Kalsch, *Science Total Envion*, 255 (1999) 143.
13 T. Steger-Hartmann, R. Länge and H. Schweinfurth, *Ecotoxicol. Environ. Saf.*, 42 (1999) 274.
14 R. Hirsch, T.A. Ternes, A. Lindart, K. Haberer and R.-D. Wilken, *Fresenius J. Anal. Chem.*, 366 (2000) 835.
15 A. Putschew and M. Jekel, *Vom Wasser*, 97 (2001) 103.
16 A. Putschew, S. Schittko and M. Jekel, *J. Chromatogr. A*, 930 (2001) 127.
17 S. Schittko, A. Putschew and M. Jekel, *Water Sci. Technol.*, 50/5 (2005) 261.
18 S. Grünheid, G. Amy and M. Jekel, *Water Res.*, 39 (2005) 3219.
19 E.J. Drewes, P. Fox and M. Jekel, *J. Environ. Sci. Health*, A36/9 (2001) 1633.
20 F. Sacher, S. Gabriel, M. Metzinger, A. Strez, W. Wenz, F.T. Lange, H.-J. Brauch and I. Blankenhorn, *Vom Wasser*, 99 (2002) 183.
21 M.M. Huber, S. Canonica, G.Y. Park and U. von Gunten, *Environ. Sci. Technol.*, 37 (2003) 1016.

22 S. Perez, P. Eichhorn, M. Dawan Celiz and D.S. Aga, *Anal. Chem.*, 78 (2006) 1866.
23 M.M. Huber, A. Göbel, A. Joss, N. Hermann, D. Löffler, C.S. Mcardell, A. Ried, H. Siegrist, T.A. Ternes and U. von Gunten, *Environ. Sci. Technol.*, 39 (2005) 4290.
24 T.A. Ternes, J. Stüber, N. Herrmann, D. McDowell, A. Ried, M. Kampmann and B. Teiser, *Water Res.*, 37 (2003) 1976.
25 J. Schumacher, Y.Z. Pi and M. Jekel, *Proceedings of 3rd International Conference on Oxidation Technologies for Water and Wastewater Treatment*, Goslar, Germany, 2003, p. 255.
26 A. Putschew and M. Jekel, *Proceedings of Jahrestagung der Wasserchemischen Gesellschaft in der GDCh*, Celle, Germany, 2006, p. 38.
27 A. Putschew, A. Solis Tellez and M. Jekel, *Proceedings of Jahrestagung der Wasserchemischen Gesellschaft in der GDCh*, Bad Mergentheim, Germany, 2005. p. 70.
28 A. Putschew and M. Jekel, Iodinated X-ray contrast media. In: T. Reemtsma and M. Jekel (Eds.), *Organic Pollutants in the Water Cycle*, Wiley-VCH, Weinheim, Germany, 2006, p. 87.
29 B.-O. Axelsson, M. Jörnten-Karlsson, P. Michelsen and F. Abou-Shakra, *Rapid Commun. Mass Spectrom.*, 15 (2001) 375.
30 A. Putschew A. and M. Jekel, *Rapid Commun. Mass Spectrom.*, 17 (2003) 2279.

Chapter 2.8

Application of bioassays/biosensors for the analysis of pharmaceuticals in environmental samples

J.-Pablo Salvador, Javier Adrian, Roger Galve, Daniel G. Pinacho, Mark Kreuzer, Francisco Sánchez-Baeza and M.-Pilar Marco

2.8.1 INTRODUCTION

Pharmaceuticals are diverse groups of chemically active compounds used in humans, animals and plants for the treatment or prevention of different diseases. They include human and veterinary drugs, such as antibiotics, hormones, analgesics, cytostatics and β-blockers. Their presence in the environment has become an important parameter of the impact of human activity in the environment due to their frequent use by the general population or by farming. Thousand of tons of pharmacology active compounds are used yearly and excreted to the waste. As an example of this use, of the 50,000 registered drugs for human use in Germany, 2700 are responsible for 90% of the total consumption and which, in turn, contain 900 different active substances [1]. Also, these drugs have been employed in farming as a usual practice for the prevention of the animal health or as growth promoters (i.e. antibiotics, feed additives and hormones). Regarding hormones, it is important to mention that in the latest decades the animal population has greatly increased, the endogenous substances excreted by the animal can have an important environmental impact. In this context, several reviews have discussed and presented real data, environmental fate and occurrence of these drugs [1–4].

Pharmaceuticals are inherently biologically active. Often they are resistant to biodegradation since they have been designed to show certain metabolic stability for their pharmacological action. Pharmaceuticals can remain unaltered in the environment or as persistently active

Comprehensive Analytical Chemistry 50
Petrović and Barceló (Eds)
Volume 50 ISSN: 0166-526X DOI: 10.1016/S0166-526X(07)50009-7

metabolites. As well as deconjugated drug, human and veterinary drugs are also excreted as glucoronide or sulphate conjugates that can be easily hydrolyzed to obtain the active parent compound in the environment. Distinct drugs have different behaviour depending on their hydrophility or hydrophobicity, moreover combined with a low biodegradability, their removal in wastewater treatment plants (WWTPs) can be intricate [3,5]. Additionally, certain drugs, such as steroids, are lipophilic, thus can be bioaccumulated in organisms or deposited in sediments in the environment [6]. However, the presence of antibiotics in the aeration tanks of sewage treatment plants (STPs) may produce resistant bacteria [1].

In light of this emerging problem, through early warnings by the scientific community, the authorities and governmental bodies have established several regulations [7,8]. Thus, Directive 2001/82/EC regulates the requirements for ecotoxicity testing of pharmaceuticals. Moreover, a joint effort is being made to establish an uniform risk assessment criteria by the International Cooperation on Harmonization of Technical Requirements for Authorization of Veterinary Medicinal Products (VICH) formed by the EU, USA and Japan (Australia and New Zealand participate as observers) [9].

Because of the bioactivity of these pharmaceuticals and the impact they can cause on the environment and the public health, it is necessary to provide analytical methodologies to control them. Environmental monitoring of pharmaceuticals requires efficient methodologies to detect trace levels of contamination in the environment by the parent drugs and the still active metabolites. In the food safety field Directives 96/23/EC and 2377/90/EC regulate the maximum residue limits (MRLs) and the requirements of the analytical methods that should be used by the veterinary and public health control laboratories to detect residues. In addition to the usual chromatographic methods coupled to mass spectrometry detectors analysis can also be carried out using bioassays, biochemical assays and biosensors. These types of analytical methodologies can offer important advantages as screening methods due to their simplicity and high-throughput capabilities. This chapter intends to provide information on existing bioanalytical methods for the determination of these emerging pollutants in environmental samples.

Antibiotics, hormones, analgesics, NSAIDs (non-steroidal antiinflammatory drugs), cytostatics agents and β-blockers (see Fig. 2.8.1) are some of the drug families that, attending to their actual use and activity, may have a more strong negative environmental impact. In this chapter, we will give some examples on the bioanalytical methods

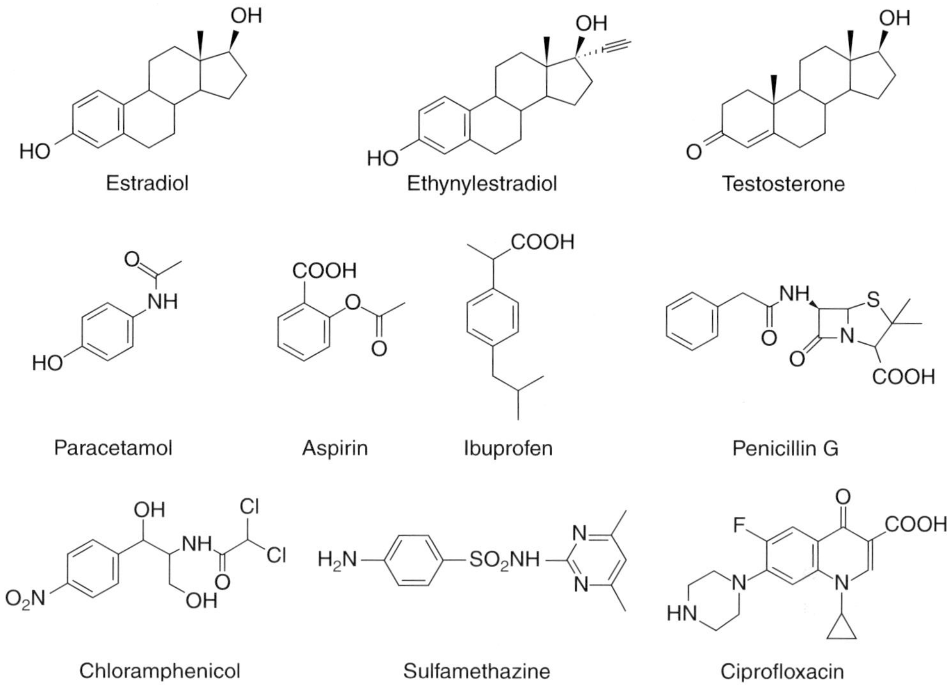

Fig. 2.8.1. Most relevant pharmaceuticals with environmental interest.

available to analyse these substances. For some of these substances there are bioassays/biochemical methods or biosensors available that have never been applied in the environmental analytical field. However, the availability of the necessary bioreagents opens up the possibility to develop new methodologies for the determination of pharmaceuticals in environmental samples. Furthermore, often these bioreagents have been applied to complex biological matrices, therefore the application to environmental water samples can be considered straightforward.

Antibiotics are chemical substances that are able to suppress or kill the growth of bacteria. They have been widely used in human and veterinary medicine as well as in aquaculture for the treatment of infectious diseases produced by bacteria. Furthermore, some of these antibiotics have also been used as growth promoters in cattle. After administration in humans or animals, these substances pass to the environment. The most important impact is related with the development of resistance mechanism. It has been reported that more than 70% of bacteria are insensitive against at least one antibiotic. This situation causes a problem for public health, because certain infections cannot be treated with traditional antibiotics. Antibiotics are classified into several families such as penicillins, fluoroquinolones, sulfonamides, macrolides, tetracyclines and chloramphenicol.

Steroid hormones are a group of biologically active compounds controlling human body functions related to the endocrine system and the immune system. Steroids are synthesized from cholesterol and have in common a cyclopentan-*o*-perhydrophenanthrene ring. Natural steroids are secreted by the adrenal cortex, testis, ovaries, and placenta in humans and animals, and include *progestagens*, *corticoids*, *androgens* and *estrogens* [10]. As a result of the continuous growth of the population and of livestock farming, the level of endogenous hormones excreted into the environment has gradually increased. However, non-ethical human and veterinary practices related to the use of the natural and synthetic sex hormones as anabolic substances and growth promoters are of great concern. The use of hormones to enhance growth and as reproductive aids for synchronization of the ovarian cycle has been regulated for animal drugs because they alter the structure or function of the animal. For these reasons, the EU has banned the use of these compounds as growth promoters in food-producing animals (Directive 2003/74/EC and related). However, hormone implants are widely used in the USA, Australia and Canada where their use is regulated but allowed. Thus, the use of progesterone, testosterone, estradiol, zeranol and trenbolone acetate for animal food production has been regulated by the US Food and Drug Administration (FDA) and by the Food and Agriculture Organization of the World Health Organization (FAO/WHO).

From the broad range of pharmaceuticals that reach the environment, drugs such as *analgesics* and NSAIDs are regularly employed, often even without prescription. However, *cytostatic agents* are of concern not because of their production volume but for their high-pharmacological potency. For instance, in Germany the total quantity of *acetylsalicylic acid* sold per year has been estimated to be greater than 500 tons, 75 tons for *diclofenac* and 180 tons for *ibuprofen* [11]. The same occurs in other EU countries where common drugs such as *paracetamol* or *aspirin* are sold in quantities comparable to high-production volume materials – close to or exceeding 1000 tons per year [12]. Ibuprofen, which is in the top ten list of pharmaceuticals used in Denmark in 1995, was used annually to the extent of 33 tons and analgesics 28 tons [2]. During the same year, *psychiatric* drugs were used to the tone of 7.4 tons [11]. *Antineoplastics* (*cytostatic agents*) differ from the other groups by the fact that they are mainly utilized in the hospital sector and by their intrinsic mutagenic action. About 13–14 kg of *cyclophosphamide* is used in hospitals per year [13]. In addition, 5969 kg are prescribed for sale at private pharmacies.

As mentioned previously, throughout this chapter we will present some of the bioassays, biochemical assays and biosensors currently used or described in the literature for the determination of pharmaceuticals. Often some literature reports apply the terms biosensor, biochemical assay or bioassay indistinctly, however, according to our point of view there are clear differences between these methods. Thus, in this chapter the following definitions have been used as criteria to identify each of these techniques.

A *bioassay* is a tool for the determination of a biological activity, or the quantification of a target analyte based on this activity, using as a recognition element bacteria, cells or tissues. This recognition event is mainly determined by physical or indirect measurement methods. For instance, in the case of the determination of antibiotics, most of the bioassays are based in the measurement of the diameter of the growth inhibition zone produced for the antibiotic in a bacteria culture. In other bioassays, antibiotics are determined by measuring the CO_2 production rate in relation of microbial growth. In the food industry, the majority of antibiotic residues are determined through bioassays and in particular microbial tests are the preferred methods for quality control.

By *biochemical assay* we understand an assay where the biorecognition element is a biomolecule such as an enzyme, an oligonucleotide or a protein, that can be either a nuclear or a membrane protein, or an antibody. Several types of biochemical assays have been described for the determination of small organic molecules. Regarding pharmaceuticals, biochemical assays exist for the detection of antibiotics, hormones, cytostatics and analgesics.

A *biosensor* is a self-contained integrated device, consisting of a biological recognition element in direct contact with a transduction element, which converts the biological recognition event into a useable output signal. Biosensors are usually classified into various basic groups according either to the method of signal transduction or to the biorecognition principle. Accordingly, biosensors can be categorized as electrochemical, optical, piezoelectric and thermometric sensors on the basis of the transducing element, and as immunochemical, enzymatic, non-enzymatic receptor, whole-cell and DNA biosensors on the basis of the biorecognition principle.

2.8.2 BIOASSAYS

A bioassay is defined as a procedure for determining the concentration and/or biological activity of a substance (i.e. vitamin, hormone, plant growth factor, antibiotic, enzyme) by measuring its effect on an organism, tissue, cell, enzyme or receptor preparation compared to a standard preparation [14]. Depending on the type of observed effect and on the biorecognition element chosen, there are a large number of bioassays. Bioassays are widely used for drug development, environmental monitoring and pollutant detection. In the pharmaceutical field, bioassays are employed in different stages of the drug discovery process, mainly to test the biological activity of libraries of potential therapeutic agents for lead identification and optimization. Each target must be screened against different libraries that can exceed one million compounds, for this reason it is suitable for high-throughput approaches are necessary. The major type of bioassays being conducted in pharmaceutical laboratories today are cell-based assays [15]. For environmental monitoring, bioassays have been used to assess toxicity of different chemicals substances on environmental living organisms. Thus, bioassays have been used to assess toxicity of an effluent or a pollutant by exposing a standard test organism to several concentrations of the suspected samples and observing the resulting effect established by the US Environmental Protection Agency [16]. The test organisms incorporated in that type of assays include representatives from four groups: microorganisms, plants, invertebrates and vertebrates. In the case of microbial bioassays the test are based on microbial transformations, growth and mortality, respiration inhibition and luminescent [17]. For the plant and algae bioassays the tests are based on growth responses of plant (length measurement of root and shoot of the plants) and the detection gaseous agents (oxygen, carbon dioxide) or fluorescence emission of photosynthetic processes. The two main freshwater toxicity tests with invertebrates, which are routinely used, are survival and reproduction tests. In the case of vertebrate, the tests are usually based on larval growth and survival [18]. Metals, pesticides, herbicides, surfactants, endocrine disrupting compounds, toxic gases and hydrocarbons [17,18] can be detected on these bioassays. Moreover, pharmaceutical drugs present in environmental and food samples are also seldom analysed with bioassays, used as a screening methodology to minimize the number of samples to be assessed by the official control routines.

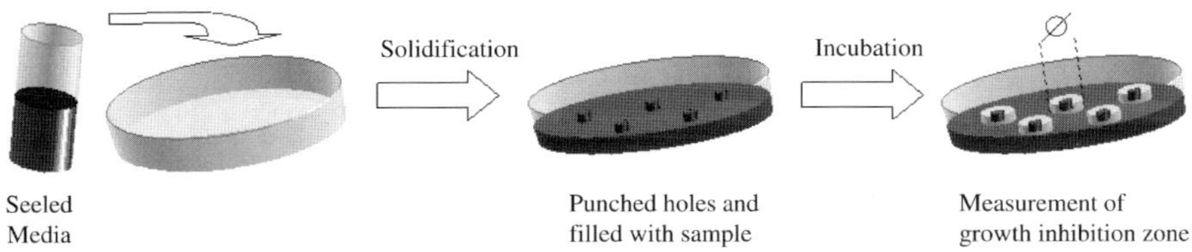

Fig. 2.8.2. General procedure for growing inhibition test.

The most commonly employed bioassays are the microbiological inhibition tests. Agar test plates with the medium seed for a relevant bacterial test strain (see Fig. 2.8.2) are prepared. Holes are punched out of the agar and filled with the antibiotic. After the appropriate incubation time the growth inhibition diameters around the holes are measured with a caliper [19–23]. Table 2.8.1 summarizes bioassays for the detection of pharmaceuticals, classified according to the biorecognition element and notices on the assay length, detectability and matrix application.

On another type of bioassays for antibiotics, measurements are made through a colour change. A solid agar medium that contains a standardized number of spores of the test microorganism is used. The potential inhibitors (or antibiotics) present in the unknown samples reduce or prevent the metabolic activity and thereby also the growth of the microorganism. During the incubation, growing test culture reduces a specific indicator or modifies the pH of the medium, leading to a colour change. Normally, after a 1-h diffusion period at 4°C, the test was floated on a water bath and incubated at 64°C for 3 h and visual interpretation is carried out. For the statistical calculations, those visual results that presented at least two similar interpretations are considered [24,25].

Microbiological tests are non-expensive, easy to perform on a large scale and do not require specialized equipment or toxic solvents. The main disadvantages are the possibility of false-positives and -negatives for the influence of the matrix, so the use of confirmatory techniques is necessary [26]. Specific bioassays, based on bacteria cells, genetically modified to produce a particular enzyme or receptor for a particular family of antimicrobials have been developed and applied to different matrices. A bacterial whole-cell assay has been used to measure chlortetracycline in pig feces using *Escherichia coli* GM10.1 that produces measurable β-galactosidase in a quantitative response to the presence of tetracyclines. The limit of detection obtained was around 0.03 mg/kg [27]. On a different study, a bioluminescent *E. coli* K-12 strain is prepared for the detection of the tetracycline antibiotics in fish samples. In

TABLE 2.8.1

Bioassays employed for the detection of pharmaceuticals classified according to the biorecognition element (type of cell). It is also considered the matrix studied and the time of the assay

	Type of cells	Analyte	Sensitivity	Matrix	Time	Reference	Miscellaneous
Bacteria							
	Recombinant E. Coli (B-Galactosidase)	Chlortetracycline	LOD: 0.03 mg/kg	Pig feces		[27]	Bacterial whole-cell bioassay. Specificity not determined
	Escherichia coli S30 (Transcription – Translation reactions)	Tetracycline–HCl	Concentrations lower than 10 ng/mL could be detected	Aqueous Solution		[30]	A cell-free bioassay compared to whole-cell bioassay. Better sensitivity for the detection of Tetracycline. The system omits cell cultivation and bacterial membranes as molecule passage inhibitors. Use of genetically modified organisms not required
	E. coli K-12 (Luciferase)	Tetracycline Oxytetracycline	LOD Tc: 20 μg/kg OxyTc: 50 μg/kg	Fish samples	2 h per assay	[29]	Bioluminescent sensor strain incorporating bacterial luciferase reporter genes. Results correlated well with those obtained by conventional HPLC ((R) 0.81)
	E. coli (Tc inducible promoter P_{tet}-lac)	Tetracycline Oxytetracycline	LOD: 0.01 μg/mL	Milk		[175]	Three Tn5 plasmids, Ptet and a regulatory gene, tetR, in operon fusions with areporter gene system (lacZYA, luxCDABE, or gfp), were constructed. The bioassay responds to Tc producing L-galactosidase, light or green fluorescent protein

Micrococcus luteus ATCC 9341	Ofloxacin	12–27 μg/mL	Commercial formulation	19 h	[19]	It is an agar diffusion bioassay. Measurement of diameters of growth inhibition zones (mm), using the cylinder-plate method. Laboratory quality control
Bacillus subtilis ATCC 6633	Erythromycin Thiocyanate	0.3–2.5 μg/mL	Medicated premix and mixtured with feed	18 h	[20]	Test-agar was no. 1 Difco laboratories (commercial). Measurement of diameters of growth inhibition zones (mm). Validation of the method
Bacillus stearothermophilus var. *calidolactis* C 953	Penicillins cephalosporins	LOD 2–20 μg/L 50,40 μg/L	Raw milk	3 days	[21]	Microbiological multiplate system Measurement of diameters of growth inhibition zones (mm)
Bacillus cereus var. *mycoides* ATCC 11778	Tetracyclines	10–30 μg/L				
Micrococcus luteus ATCC 9341	Macrolides	6–100 μg/L				
Bacillus subtilis BGA	Aminoglycosides	25–150 μg/L				
	Sulfonamides	20–150 μg/L				
E. coli ATCC 11303		5–150 μg/L				
Staphylococcus epidermidis ATCC 12228	Quinolones Novobiocin, rifamycin	25 μg/L 60 μg/L				
E. coli ATCC 11303	Tetracycline quinolone	25 μg/mL	Milk	3 h	[176]	Measurement of carbon dioxide production rate in relation to inhibition of microbial growth. Based on the use of a high-performance, non-invasive, specific carbon dioxide sensor (CO2-S)

TABLE 2.8.1 (*continued*)

Type of cells	Analyte	Sensitivity	Matrix	Time	Reference	Miscellaneous
B. stearothermophilus var. calidolactis C953	 Penicillins aminoglycosides Macrolides Sulfonamides Tetracyclines chloramphenicol	LOD 2–230 μg/kg 69–1300 μg/kg 120–6000 μg/kg 3500–6500 μg/kg 390–6200 μg/kg 22000 μg/kg	Ewes' milk	4 h	[25]	Commercially available version of brilliant blackreduction test, BRT Inhibitor Test with prediffusion AiMs (BRT AiMs), (microbiological inhibition test). Interpreted by visually assessing the colour change of an indicator (brilliant black)in the test medium
Staphylococcus aureus ATCC 6538P	Enrofloxacin	WR 3.2 –12.8 μg/mL	Commercial formulation	18 h	[23]	Measurement of diameters of growth inhibition zones (mm). Laboratory quality control
Bacillus subtilis ATCC 9372	Gatifloxacin	WR 4–16 μg/mL	Tablets and raw material	18 h	[22]	Grove Randall number 1 agar (Merck) Measurement of diameters of growth inhibition zones (mm). Quality control
B. stearothermophilus var. *calidolactis*. Eclipse 100®	Penicillins aminoglycosides Macrolides Sulfonamides Tetracyclines Fluoroquinolones	LOD 5–68 μg/kg 28–15 μg/kg 230–10,100 μg/kg 170–750 μg/kg 260–1500 μg/kg 400076,200 μg/kg	Ewe milk	3.5 h	[24]	Eclipse 100® (Commercial) interpreted by visually assessing the colour change of an acid–base indicator (bromocresol purple) in the test medium
Vibrio Fischeri	Acetaminophenol Diclofenac Ibuprofen Ketoprofen Ketoprofen	EC_{50} (μg/mL) 173 13.5 12.1 15.6 21.2	Saline solution		[177]	Bioluminescence assay

Daphnia magna	Ibuprofen Diclofenac Naproxen	EC_{50} (μg/mL) 108 68 174	ADaM (natural fresh water imitation for cultures)	48 h	[178]	Immobilisation Test
Lemna minor	Ibuprofen Diclofenac Naproxen	EC_{50} (μg/mL) 22 7.5 24.2	Steinberg-medium+2 phosphate species added	7 days	[178]	Growth Inhibition Test
Desmodesmus subspicatus	Ibuprofen Diclofenac Naproxen	EC_{50} (μg/mL) 315 72 >340	Water+algal chemicals	4 days	[178]	Algal Growth Inhibition Test
Microtox® *Daphnia magna C. Dubia*	Diclofenac	EC_{50} (μg/mL) 11.45 22.43 22.70	Buffer	30 min 48 h 48 h	[179]	Assessment endpoint Bioluminescence at 15°C Mobility inhibition
P. subcapitata	Naproxen	IC_{50} (μg/mL) 31.82		96 h	[180]	Algae test, measured endpoint by IC50
B. calyciflorus	Naproxen	EC_{50} (μg/mL) 0.56		48 h	[180]	Rotoxkit (rotifer), measured endpoint by EC50
C. dubia	Naproxen	EC_{50} (μg/mL) 0.33		7 days	[180]	Ceriodaphnids (crustaceans) measured endpoint by EC_{50}

this case, the strain contains a plasmid incorporating a bacterial luciferase under the control of the tetracycline responsive element from transposon Tn10 [28]. The optimized assay, using the appropriate luminometer, is able to detect oxytetracycline residues below the MRLs established by the UE, in 2 h, with a well correlation of results with the ones obtained by conventional HPLC [29]. However, a cell free assay for the detection of transcriptional inducers has been developed to compare it with common whole-cell assays reported. Thus, *E. coli* S30 extract containing firefly luciferase, detectable by a luminometer, as reporter gene was prepared for coupled transcription–translation reactions. This methodology omits cell cultivation and bacterial membranes as molecule passage inhibitors making possible to carry out assays in much shorter times and without the use of genetically modified organism. Results obtained for detection of tetracyclines in aqueous solution presented LOD around 10 ng/mL improving sensitivity observed in whole-cell configurations [30].

Antimicrobial compounds are not the only substances analysed by this type of determination procedure, there are also bioassays described to detect different hormones such as androgens, estrogens and thyroids. The main purpose of this sort of assays is to analyse how chemicals and environmental extracts are capable of mimic or block human hormone receptors.

The E-screen is a cell-culture based assay developed to assess the estrogenicity of environmental chemicals, using the proliferative effects of estrogens on their target cells (MCF-7) as an endpoint. Because this cell line proliferates in response to estrogens, compounds that cause proliferation are considered estrogenic. Different compounds and water samples have been tested with that methodology [31]. The quantitative assay compares the cell number achieved by MCF-7 cells, using a coulter count, in the absence of estrogens (negative control) and in the presence of 17β-estradiol (positive control) and a range of concentrations of the chemicals suspected to be estrogenic. The assay is very sensitive; proliferation can be detectable in the picomolar range of estradiol but require between 4 and 6 days to be performed. However, the assay called A-screen uses the same MCF-7 cells but transfected with the androgen receptor. These cells will respond to androgens (such as testosterone) by inhibiting proliferation. Configuration of the assay is very similar to E-screen assay described above with a similar range of detectability being also tested jointly with river water samples [31].

The Yeast Assay is another useful tool to assess potential impacts to the endocrine system capable to produce infertility, endometriosis and certain cancers such as breast, uterine and prostate. Yeast cells not only are transformed to contain human estrogen (hER) and androgen receptors but also contain expression plasmids carrying reporter genes, such as lac-Z (encoding β-galactosidase), which is used to measure receptor's activity. As example, surfactants have been tested with that methodology to know if there are estrogenic or not [32]. When the estrogen receptor is occupied, β-galactosidase is synthesized and secreted to the medium, where it causes a color change from yellow to red that can be measured by a spectrometer after 3 days incubation. All the results obtained where compared to the effects presented by a 17β-estradiol, arriving to a limit of detection near 2 ng/L. Only alkylphenol polyethoxylate surfactants were weakly estrogenic.

Another in vitro bioassay is the T-Screen that is based on thyroid hormone dependent cell proliferation of a rat pituitary tumor cell line (GH3) in serum-free medium. As the other assays mentioned above, it has been used to study interference of compounds or sediment extracts with thyroid hormone at the cellular level [33]. Detectability of the assay is around the nanomolar range. For concluding, all the hormone screen assays described have very good sensitivity but time required to perform the analysis is too high in some occasions.

The fact that β-blockers may induce insulin resistance in obese animal models through the expression of the tumor necrosis factor-α (TNF-α) by adipocytes, have lead to set up and assay to detect this type of activity. The assay uses LM cells, a cell line of TNF-α-sensitive murine fibrosarcoma and recombinant human TNF-α as a standard. The TNF-α, immunoreactivity in the culture cell is then measured with a commercial enzyme-linked immunosorbent assay (ELISA) kit (R&D Systems, Minneapolis, MN, USA) [34]. A new way of assaying β-adrenoreceptors antagonists has been proposed by measuring their ability to prevent catecholamine-induced myotoxicity [35].

2.8.3 BIOCHEMICAL ASSAYS

As mentioned above, on a biochemical assay, the biorecognition element has been isolated. The biorecognition element consists in a biomolecule such as an enzyme, a nuclear or membrane receptor or an antibody that recognizes selectively or specifically the analyte of interest. The mode of action of each biomolecule depends on different mechanism. In the case of enzymes, the mechanism involves the catalytic transformation of the pollutants. Regarding the nuclear receptors, their affinity versus particular endogenous and exogenous substances is exploited. For instance, the affinity of the estrogen receptor (ER) for estrogenic compounds such as estradiol, estrone and ethynylestradiol has been used to develop a variety of methods. One of the most important biorecognition elements are the antibodies. Because of the broad variety of specificities that can be achieved, several immunochemical assays have been developed for a great variety of substances, such as pharmaceuticals. However, the use of these biochemical assays for the detection of pharmaceuticals in the environment has not been frequently reported. Following we will describe some of the most frequently described biochemical assays available for the detection of pharmaceuticals with a great potential for environmental analysis.

2.8.3.1 Biochemical assays based on receptors

Many biochemical processes, essential for the functioning and survival of cells (and the organism), are regulated by hormones, neurotransmitters, cytokines and other "messenger" molecules. This regulation proceeds by interaction of these naturally occurring molecules with receptors that are either embedded in the cell membrane (membrane-bound) or present in the cytoplasm (soluble receptor) or the nucleus of the cell. The membrane-bound receptors can be subdivided into G-protein coupled receptors (GPCRs), ion channels and receptors with a single transmembrane segment. Nuclear or soluble receptors are represented by the group of steroid receptors (i.e. the estrogen receptor) and the non-steroidal receptors (i.e. Vitamin D receptor) that regulate biological functions by controlling gene expression. This class of receptors consists of a DNA-binding and a ligand-binding domain.

Receptor-screening methodologies can be based on either the determination of a functional response (i.e. cell proliferation), the production of second messengers (i.e. Ca^{2+}) or the interaction of a ligand with its receptor. While in the first two cases, we would consider those methods as bioassays, the third case can readily be considered as a biochemical assay. Moreover, it is sometimes still costly and difficult to obtain stable eukaryotic cell lines to perform these types of functional measurements, for which reason receptor biochemical assays can be contemplated as excellent alternatives. Binding of a ligand (agonist or antagonist) to its cognate receptor is the initial and indispensable step in the cascade of reactions that finally cause a pharmacological effect and many successful and widely used techniques are thus based on measuring ligand binding.

As with the well-known immunochemical assays (see below), receptor–ligand binding assays may be classified according to the need for separation of bound from free ligand or the detection technique. According to the first criterion, the assay types can be heterogeneous (use of a solid phase for separation) or homogeneous (no need of separation steps). Regarding detection methods, receptor assay formats usually require labelling of either the ligand or the receptor. Radio-isotopic labels such as ^{3}H, ^{125}I and ^{32}P have been used (RRA, radio receptor assay; SPA, scintillation proximity assay), however because of the disadvantages of disposal of radioactive waste, costs, health hazards, the requirement for special licenses, etc., efforts have increased to develop new technologies based on either colorimetric (ELRA, enzyme-linked receptor assay), fluorescence (i.e. FRET, fluorescence resonance energy transfer; FP, fluorescent polarization, etc.) or (chemo-/bio-) luminescence detection systems. The ideal assay should be specific, sensitive, easy to perform, reliable and reproducible, unexpensive, rapid and suitable for automation. Moreover, the possibility to quantify multiple analytes in a single assay (multiplexed assays) is becoming one of the important goals in this area. For more information on these types of assays the reader is addressed to recent reviews [36].

RRA assays have been reported for the determination of benzodiazepines [37,38], neuroleptics [39,40], opioids [41], antipsychotic [42] and antihypertensive drugs [43,44]. SPA has been developed for a range of receptors including the α_1- and α_2-adrenergic receptors (α-AR) [45,46].

The first receptor assays that made use of fluorescence was described by McCabe et al. [47] for the benzodiazepine receptor using a fluorescein-labelled ligand. The significant background signal presented was reduced in the assay developed by Takeuchi et al. [48], who made use of time-resolved fluorescence (TRF), by labelling the benzodiazepine ligand with a europium chelate. Neuroactive drugs have been determined by means of a FRET assay through their competitive binding to the labelled human M1 muscarinic receptor (hM_1-R) in the presence of a labelled antagonist [49] or also by measuring the biding to the ligand-gated ion channel $GABA_A$ receptor α_1-subunit, using the same assay format [50]. The luminescent variant of FRET, where energy transfer occurs between a luminescent donor and a fluorescent acceptor, is called bioluminescence resonance energy transfer (BRET). The enzymatic oxidation of a substrate results in the emission of energy from the donor, which means that no excitation light is needed in contrast to FRET. Moreover, the enzyme reaction does not produce a background signal and the assay is therefore more sensitive than

FRET [51]. Because of the fact that there is no requirement of a light source, the instrumentation for BRET assays is simpler and cheaper [52] which makes these assays very valuable in high-throughput screening. BRET has been mainly used in protein–protein interaction research, for example in studying the β_2-adrenergic/β-arrestin interaction [53] and the determination of insulin receptor activity [54,55], where the latter is governed by a conformational change in the β-subunits of the receptor, bringing them into close proximity. The FP technology has been applied to, i.e. the soluble estrogen receptor [56], the G-protein coupled delta-opioid receptor [57] and the ligand-gated ion channel serotonin $5HT_3$ receptor [57,58]. This receptor is involved in rapid signal transduction in the central nervous system and the peripheral nervous system. Strong interest for this receptor has been provoked by the ability of 5HT3 receptor antagonists to treat emesis caused by anticancer chemotherapy. Moreover, antagonists for this receptor show promise for the treatment of colonal dysfunction. Fluorometric microvolume assay technology (FMAT) has also been used on few cases to set up receptor assays. This technology makes use of a scanner that measures multiwell plates. It is a mix-and-measure assay where the small molecule ligand is labelled with a fluorophore and the receptor is immobilized on beads or in the bottom of special multiwell plates (either 96-, 384- or 864-well with a clear bottom and black sidewalls). The FMAT scans a $1\,mm^2$ area at the bottom of the multiwell plate where the generated images indicate the size and amount of bound fluorescence. The capillary-based scanner uses as an excitation source a Helium–Neon (He–Ne) red laser ($E_x = 633\,nm$) and makes simultaneous detection of two independent red dye emissions, i.e. Cy5 and Cy5.5, possible via two photomultiplier tubes with band-pass filters for the respective labels (multiplexing). Multiplexing minimizes reagent consumption and increases the throughput [59]. A different format of a homogeneous bead-based assay, called AlphaScreen^{TM} (Amplified Luminescence Proximity Homogeneous Assay [60]) makes use of singlet oxygen ($^1O_2$, half-life $4\,\mu s$) production on donor beads, and a chemiluminescent reaction on the acceptor beads is observed. This assay allows probing interactions over longer distances than FRET and BRET. An example of a receptor–ligand-binding assay, which made use of the AlphaScreen^{TM} methodology, was described for the estrogen (ER_α)-receptor by Rouleau et al. [61]. Flow cytometry has also been used to discriminate between agonist and antagonist binding using the solubilized β_2-adrenergic receptor fused to green fluorescent protein (β_2AR-GFP) [62,63].

Finally, a variety of ELRA assays have been established in parallelism with the well-known ELISAs (see below). The amplification provided by the enzyme allows reaching excellent detection limits for a variety of drugs and environmental contaminants with specific activity on the isolated protein receptor. In some examples, the detectability has been increased by substituting the chromogenic substrate by a luminescent one (ELBRA, enzyme-linked bioluminescent receptor assay). As a result of the increasing concern regarding the hormonal effects of a wide variety of pollutants (endocrine disrupter compounds, EDC), several research groups have invested great efforts on developing biochemical assays based on the use of the nuclear receptors responsible of such type of bioactivity. Steroid hormone receptors are the members of the nuclear receptors family, which are ligand-dependent transcriptional modulators. These kind of receptors can be produced by genetic engineered bacteria, for instance by fusion of glutathione-S-transferase (GST) with the D, E and F domains of native receptors in *E. Coli* [64]. With this technology human receptors for estrogens (hER), androgens (hAR) and progestagens (hPR) have been produced. These receptors have been used to set up ELRAs with excellent detectability limits. As an example, the androgenic receptor has been used assess binding of a variety of pesticides and industrial pollutants [65–67]. Similarly, hER has been used to detect 17β-estradiol (E2), tamoxifen, bisphenol-A and resveratrol. A detection limit of 20 ng/L has been found for E2 [68–70]. Recently, an ELRA and a yeast estrogen screen (YES) assay has been applied to determine whether automobile tires immersed in fresh water can leach chemicals, which display estrogenic activity [71]. Table 2.8.2 shows some of the most frequently used biochemical assays for the detection of hormones using nuclear receptors.

Other receptor assays have been established for the determination of antibiotics. One of the most well known is based on the use of a specific β-lactam receptor (β-Star^{TM}). By using appropriate labels, the assay can detect a wide range of penicillins and cephalosporins with excellent detectability. For instance, the limit of detection for the determination of Penicillin G and Amoxicilin is between 2–4 ppb in milk samples [72]. The assay that is commercialized as a test strip is commonly used to determine antibiotic contamination of dairy products. However, to our knowledge the application to the analysis of environmental samples has not been described.

TABLE 2.8.2

Biochemical assays for the detection of pharmaceuticals classified by the biorecognition element and the type of assay developed. It is described the matrix applied and the sensitivity achieved

	Assay	Analyte	Sensitivity	Matrix	Ref.	Miscellaneous
Antibodies						
	SPIA	Sulfadimidine	LOD	Urine	[85]	Soil particle immunoassay (SPIA) based on the use of dyed colloidal particles as labels.
			10 ng/mL 20 ng/mL	Milk		One-step strip test. Use of Pabs
	ELISA	Sulfachloropyridazine	Easy detection at the MRL levels	Porcine tissues (kidney, liver, muscle and fat)	[181]	Good cross-reactivity against several Sulfonamides in buffer
	ELISA	Sulfadimethoxine	IC_{50} 1.5 μg/mL	Liver tissue	[182]	Specificity not described
	EIA	Tetracycline	Detection level of 20 μg/kg	Honey	[183]	Enzyme immunoassay (EIA)
	ELISA	Tetracycline	Detection between 0.1–6 ng/mL	Aqueous solution	[184]	Use of polyclonal antibodies Chlortetracycline, rolitetracycline and minocycline also recognized
	EIA	Tetracyclines	LOD (μg/L) Tetracycline: 1.5 OxyTc: 15.5	Milk	[86]	Kit commercially available (RIDASCREEN EIA) Can also recognise rolitetracycline and minocycline Also used for matrices such as meat and honey

RIA	Penicillins	LOD: 2 μg/L	Water	[84]	Charm II RIA test in water samples proximal to a US farm
Electrohemical ELISA	Macrolide	0.4–4 ng/mL 1.4–13 ng/mL	Milk	[87]	MAb (mouse) competitive indirect assay Electrochemical detection (μA)
ELISA	Quinolone	LOD 2–6 μg/kg IC_{50} 11.7–30 μg/kg	Bovine Milk, Ovine Kidney	[88]	PAb competitive direct assay Generic and specific
ELISA	Chloramphenicol	12.5/50 μg/kg	Milk, milk powder, honey, shrimps, meat, fishmeal and eggs	[86]	Commercial Kit
ELISA Dip strip	Enrofloxacin	LOD 1–10 ppb ELISA LOD 50–100 ppb Dip strip	Chicken liver Cattle milk	[89]	Mab Competitive direct assay
ELISA	Ciprofloxacin	LOD 0.32 ng/mL IC_{50} 50 ng/mL	Milk, chicken, and pork	[91]	Pab Competitive indirect assay
ELISA Dip-Strip	Bacitracin	LOD 0.1 ppb, 0.28 ppm ELISA LOD 100 μg/L dip strip	Chicken plasma Chicken serum	[185]	PAb Competitive direct assay
ELISA	Flumequine	LOD 12.5 μg/kg IC_{50} 90 μg/kg^{-1}	Raw milk	[90]	IgY Competitive indirect assay
ELISA	Ciprofloxacin	LOD 0.32 ng/mL	Pork	[186]	PAb competitive indirect assay
ELISA	Sarafloxacin	IC_{50} 7.3–48.3 ppb	Liver	[187]	MAb (mouse) competitive indirect assay
ELISA	Tilmicosin	LOD 9.2 and 4.45 ng/mL IC_{50} 48 and 32 ng/mL	Buffer	[188]	MAb (mouse) competitive indirect assay
ELISA	Macrolides	LOD 0.3 ng/mL IC50 8 ng/mL	Buffer	[189]	
RIA	Erythromicyn	10 μg/L	Water	[84]	

TABLE 2.8.2 (*continued*)

	Assay	Analyte	Sensitivity	Matrix	Ref.	Miscellaneous
	RIA	Chloramphenicol	LOD 0.5–0.3 ng/g	Tissue	[190]	(1) Ridascreen EIA Competitive enzyme immunoassay
	EIA		LOD 5–20 ng/mL	Urine		(2) Charm II assay Radioimmunoassay
	ELISA	Chloramphenicol	3 ng/mL	Muscle	[191]	
	ELIFA Dipstick EIA	Chloramphenicol	0.7 17 ng/mL	Milk	[92]	
	ELISA	Chloramphenicol	2 mg/kg	Meat	[192]	Le carte test
	EIA Kit	Chloramphenicol	LOD 0.1 μg/kg	Shrimp tissue	[193]	5091CAP1p PAb Competitive direct assay
	ELISA	Testosterone	LOD 10 pg per well	Human Serum	[194]	
	ELISA	Boldenone	LOD 26 pg per well	Urine	[195]	
	IA	Stanozolol		Water	[196]	
	ELISA	Trenbolone	LOD 0.1 ppb	Meat samples	[197]	
	ELISA	Trenbolone	LOD 0.1 ng/mL	Urine	[198]	
			LOD 0.02 ng/g	Muscle tissue		
	ELISA	Nandrolone	LOD 1 ng/mL	Equine urine	[199]	
	ELISA	Estradiol	LOD 5 pg per ml	Waste water	[200]	
	ELISA	Estradiol Ethynylestradiol	LOD 0.1 ng/L	Waste water effluent	[93]	
			LOD 0.05 ng/L	Surface water		
			LOD 0.1 ng/L	Waste water effluent		
			LOD 0.05 ng/L	Surface water		
	RIA	Ethynylestradiol	LOD 5 ng/L	Water samples	[94]	
		Progesterone	LOD 4 ng/L			
	ELISA	Estriol	LOD 12 pg per well	Saliva	[201]	
	ELISA	Progesterone	LOD 3.8 pg per tube	Human serum	[202]	
	EIA	Norethindrone	LOD 10 ng/L	Water samples	[94]	
	ELISA	Cortisol	LOD 2.8 ng/mL	Human Serum	[203]	
	ELISA	Betamethasone	LOD 12.5 ng/mL	Urine	[204]	
		Dexamethasone	LOD 3.1 ng/mL			
		Flumethasone	LOD 2.5 ng/mL			
	ELISA	Dexamethasone	LOD 4 ng/mL	Urine	[205]	

Nuclear receptors						
	RRA hER	Estradiol	IC_{50} 2 nM	Buffer	[206]	Radio-Receptor assay using hER, hAR, hPR and hGR
	RRA hAR	Dihydrotestosterone (DHT)	IC_{50} 50 nM	Buffer	[206]	
	RRA hPR	Progesterone	IC_{50} 50 nM	Buffer	[206]	
	ELRA	Estradiol	LOD 0.1 μg/L	Buffer	[68]	High cross-reactivity with ethynylestradiol and no matrix interferences observed in the ELRA with lake water and sewage plant outflow
	ER-CALUX	Estradiol	IC_{50} 6 pM	Buffer	[207]	Estrogen receptor-mediated, chemical-activated luciferase reporter gene expression
	Androgen receptor assay	DHT Trembolone Testosterone	IC_{50} 2.23 nM IC_{50} 2.75 nM IC_{50} 15.9 nM	Buffer	[208]	

2.8.3.2 Antibodies

Immunochemical techniques are based on the affinity of the antibody against an antigen. The formed complex has a high-affinity constant (k_a) that can reach values around $10^{-10}\,M^{-1}$. This interaction is specific between the antigen and the corresponding antibody. The immunochemical techniques use this characteristic as a powerful tool for the detection of pollutants at low concentrations. Several immunochemical techniques have been developed for the determination of small molecules. The reader can be addressed to recent reviews to find more information on immunochemical technologies for residue analysis [73–76].

Immunoassays (IAs) are the most frequently used methodologies for the detection of pollutants [77–79] such as pesticides and other industrial residues at trace levels. They have been applied to the analysis of environmental samples (wastewaters, river water, sediments and other kinds of matrices) and also to complex biological matrices, such as urine, serum and saliva. IAs have found wide application in forensic, clinic and veterinary analysis. In immunoassays for small organic molecules such as pharmaceuticals, the reaction Antigen–Antibody (Ag–Ab) is quantified under competitive conditions. As with the receptors, most of these techniques relay on the use of labels that are responsible of the signal generated. There are several kinds of labels for the indirect determination of the analyte. In the firsts, in IAs developed, the label was always a radioisotope and the assay was called RIA (Radioimmunoassay). However, the drawbacks of using radioisotopes are not friendly prompted to the employment of other kind of labels less hazardous. Thus, fluorescent labels, such as rodhamine, fluoresceine or more recently lanthanides, are used in fluoroimmunoassays (FIA). The use of enzyme labels (EIA, enzyme immunoassay) offers the possibility to increase detectability, by amplifying the signal produced by a substrate. Enzymes like horseradish peroxidase (HRP), alkaline phosphatase (AP) and glucose oxidase (GOx) are the most used labels.

Most of the immunoassays developed for pharmaceuticals have been applied in biological samples. Certain drugs show significant toxicity that claims for therapeutic drug monitoring (TMD) of the real internal doses attending to individual idiosyncrasies. TMD significantly reduced risks associated to unappropriate doses or treatment protocols in particular patients. For this reason high-throughput screening immunochemical methods have been develop to provide efficiency to these monitoring programs. As an example several EIA or PFIA (Polarization

Fluorescent Immunoassay) do exist available to routinely monitoring antiarrhythmic drugs (see review by Campdell et al. [80]) but not so much to the analysis of environmental matrices. However, pharmaceutical residues in the environment have become a matter of concern recently. The significant immunochemical methods already available can therefore offer a great potential on environmental monitoring programs due to the important benefits of this type of methods. As an example, Table 2.8.2 summarizes some of the different biochemical assays described for the determination of pharmaceuticals.

As mentioned before, IAs can work under homogeneous or heterogeneous conditions. Enzyme-linked immunosorbent assays (ELISAs) are the most well known and frequently used heterogeneous IA formats. One of the immunoreagents is immobilized onto a solid support and, for the case of small molecules, the assay takes under competitive configuration. The most usual are the direct and the indirect formats (see Fig. 2.8.3). In the direct format, usually the immunoreagent immobilized onto the well is the antibody. Then, equilibrium is established between the antibody, the analyte and the enzymatic tracer (both in solution). After a washing step, the unbound reagents are removed; the amount of label bound to the Ab is measured, the signal being inversely proportional to the amount of analyte in the sample. In the indirect format (see Fig. 2.8.3), the coating antigen is coated on the plate, but in this case the amount of analyte present in the sample is indirectly measured by measuring the bound Ab with a second Ab that is conveniently labelled (Anti-IgG-enzyme). ELISAs have been developed for the antihypertensive agents enalapril

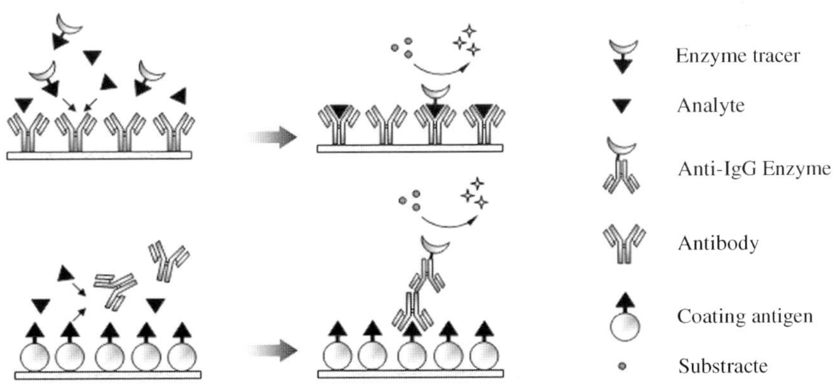

Fig. 2.8.3. Scheme of two of the ELISA formats most frequently used. Above is described the competitive direct format and below the competitive indirect format.

and amlodipine in human plasma [81,82]. Amlodipine is detected using a biotin–avidin-based ELISA in just 50 μL of plasma for TMD pharmacokinetic studies and pharmaceutical analysis. The free form of amlodipine is detected down to 0.1 ng/mL and the intra- and interassay coefficient of variation ranged from 1.6–10.2%. Regarding multiplexed methods, a sensitive and rapid multi-residue technique method for the detection of five tranquillizers and one β-blocker using a single ELISA plate has been reported to detect sedative misuse [83]. The assay uses three polyclonal antibodies raised against azaperol, propionylpromazine and carazolol conjugates and taking advantage of the cross-reactivity with related substances. Azaperol, azaperone, carazolol, acepromazine, chlorpromazine and propionylpromazine are detected in liver at 5, 15, 5, 5, 20 and 5 μg/kg, respectively. ELISAs have also been developed for the detection of antibiotics. However, the matrices that have been developed are biological samples such as milk, tissues, urine or plasma. It is remarkable the limit of detection using the Charm II RIA test for the determination of penicillins in water [84]. A LOD of 2 μg/L was achieved using the commercial test, value closed to the regulation established for the FDA. The main application for the detection of antibiotics is measuring in milk. The detection limits obtained were very low for sulfadimidine [85], tetracyclines [86], macrolide [87], quinolones [88–91] and chloramphenicol [86,92], examples were summarized in Table 2.8.2. Huang et al. [93] have been developed an ELISA for the determination of estradiol and ethynylestradiol in wastewater effluent and surface water with a LOD of 0.1 and 0.05 ng/L, respectively. Aherne et al. [94] have been analysed by immunoassay ethynylestradiol (LOD of 5 ng/L), progesterone (LOD of 4 ng/L) and norethindrone (LOD of 10 ng/L) in water samples.

Fluoroimmunoassays (FIAs) for pharmaceuticals are based on the same competitive principle described above, but with the difference that a fluorescent label despite of an enzymatic label is used. FIAs have also been developed on heterogeneous and homogenous formats, being now the last case the most frequently employed. Thus, homogeneous FIAs have been reported for the determination of different types of pesticides and industrial residues [95–97]. FIAs have been developed for the determination of β-lactam antibiotics by automated fluorescent immunoassay [98]. It has been achieved an IC_{50} of 30 ng/mL for Penicillin G. A high cross-reactivity value was observed for Penicillin V (145%) and Amoxicillin (50%). The determination of cortisol in saliva [99] has been measured by DELFIA^{TM} (dissociation-enhanced lanthanide fluoroimmunoassay) method achieving sensitivity of 0.5 nmol/L.

2.8.4 BIOSENSORS

Biosensors are integrated analytical devices, usually small in size, consisting of a biological component in intimate contact with a physical transducer that converts the biorecognition process into measurable signal (see Fig. 2.8.4).

A biosensor should be capable of providing specific quantitative or semi-quantitative analytical information using a biological recognition element that translates information from the biological domain into a chemical or physical output signal [100,101]. In recent years many efforts have been made to develop biochemical techniques, integrating specific recognition elements and electronic components to obtain small devices with the ability to carry out direct, selective and continuous measurements of one or several analytes present in the samples (see Table 2.8.3 for the main biosensor features in comparison with conventional analytical methods). The development of new techniques such as

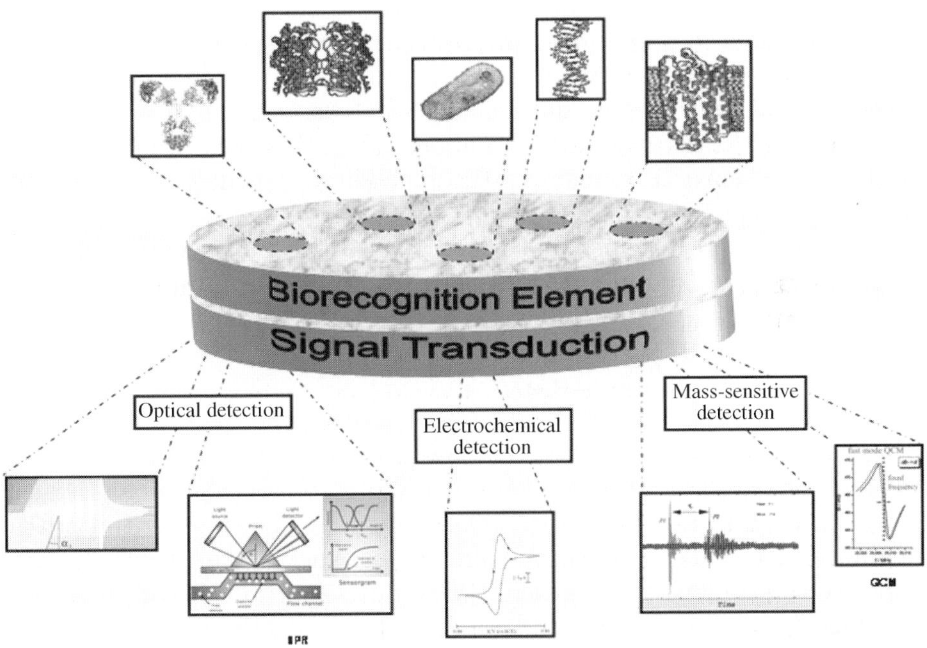

Fig. 2.8.4. Schematic view of a biosensor, classified according the biorecognition element (antibodies, enzymes, bacteria, DNA or membrane receptors) or the signal transduction method (optical, electrochemical or piezoelectric detection).

TABLE 2.8.3

Advantages and disadvantages for biosensors vs. conventional methods for the detection of pharmaceuticals in real samples

Biosensors	Conventional methods
Direct Analysis	Pretreatment procedures
Simplicity, user friendly	Trained Personnel Required
Portability	Laboratory Methods
Small volume sample	High volume sample
No organic solvents	Organic solvents consumption
Cost-effective	Expensive
Fast results in real time detection	Long analysis time
Low biological material stability	No biological stability restrictions
Single analyte determination	Multianalyte determination
Mostly prototypes	Mostly commercial

array configurations has recently allowed multianalyte determinations. To accomplish the analytical requirements of these TMD, clinical, food safety programs and environmental studies, there is a demand for new analytical devices, able to be integrated into automated devices to provide efficiently fast and reliable data. Table 2.8.4 summarizes the biosensors developed for the determination of pharmaceuticals according to the type of transducer used to transform the biochemical signal into a physical signal.

Biosensors can be classified according to the biological recognition element or to the signal transduction principle (see Table 2.8.4).

2.8.4.1 Transducing principle

Biosensors also can be classified into four different basic groups on the basis of the signal transduction principle: electrochemical, optical, piezoelectric and thermometric. The transducer is an essential part of the biosensor since it is responsible of converting the (bio)-chemical response to an electrical signal. It may have a dramatic effect on both the specificity and sensitivity of the system. Thus, selecting the appropriate transducer giving the optimum detection for a particular system is crucial. Following a brief description of the main transducing principles is given, although the reader is addressed to other reviews for more information [101–106].

TABLE 2.8.4

Biosensor classification according to the biorecognition element and the signal transduction

Recognizing biomolecule	
Antibodies (Immunosensors)	Monoclonal or polyclonal
Protein receptors	Metallotropic receptors, Ionotropic receptors
Whole cells:	Microbial sensors, Mammalian cells, Tissue
Nucleic acids:	Hybridization, Low weight compound interaction
Enzymes	Oxidases, Esterases, etc.
Signal transduction	
Electrochemical	Amperometric, Conductimetric, Impedimetric, Potentiometric
Optical	Absorption, Fluorescence or Phosphorescence, Bioluminescence or Chemiluminescence, Reflectometric Intereference Spectroscopy (RIFS) and Surface Plasmon Resonance (SPR), Evanescent wave.
Piezoelectric	Surface acoustic wave, Bulk acoustic devices, Cantilever
Thermometric	Calorimetric

2.8.4.1.1 *Electrochemical sensors*

Owing to its simplicity, electrochemical transduction constitutes a successful route to create low-cost biosensors when coupled to enzymes. However, electrochemical detection of just a biorecognition process is difficult. Catalysis leading to the formation of electroactive substances is frequently necessary. Electrochemical sensors may work under amperometric, potentiometric, conductimetric and impedimetric transducing principles.

For *amperometric* devices the current generated by oxidation or reduction of redox species at the electrode surface maintained at the appropriate electrical potential is measured. The current observed has a linear relationship with the concentration of the electroactive species. The electrode is usually constructed of platinum, gold or carbon. The potential applied to the working electrode is largely dependent on the enzyme-substrate system utilized. Adjacent to the electrode, bound by a membrane or directly immobilized, is placed one of the bioreagents involved in the recognition event. Enzymes used in such electrochemical assays are usually oxidoreductases such as horseradish peroxidase (HRP), or hydrolytic enzymes such as alkaline phosphatase (AP), that yield electroactive species as products of the enzymatic reactions. Other enzymes commonly used as sensing catalysts for environmental monitoring are tyrosinase, laccase, aldehyde deshydrogenase, etc. Sometimes the substrate or the product of the enzymatic reaction can be monitored amperometrically, without the need of a mediator. These electrodes are called *unmediated amperometric enzyme biosensors*. However, a number of factors must be taken into account when assessing the suitability of an enzyme substrate to be used on an electrochemical detection system: the electrochemistry of the substrate, the electrochemistry of the product of the enzymatic reaction, the medium in which the measurements will be performed and the electrochemistry of endogenous materials in the test sample. A problem often encountered with unmediated sensors is that other species present in the samples being analysed are also electroactive at the potential applied. For example, ascorbic acid and uric acid, present in many biological samples, are oxidized at an anodic potential of +0.35 V [107]. AP combined with *p*-aminophenyl phosphate (pAPP) as substrate has been shown to be a good alternative when measuring with such a kind of system [108–111]. Although *p*-aminophenyl phosphate has an irreversible wave in cyclic voltammetry at around 0.45 V vs Ag/AgCl, its hydrolysis product *p*-aminophenol (pAP) shows a reversible electrochemistry with a half wave potential of –0.065 V vs Ag/AgCl. Consequently, measurements on biological matrices can take place at lower potentials avoiding interference of endogenous compounds. Choosing an alternative electron transfer acceptor can also circumvent these problems. Usually the *mediator* is a species of low MW that shuttles electrons between the redox centre of the enzyme and the working electrode. These sensors are called *mediated amperometric enzyme biosensors*. A mediator should react rapidly with the enzyme, exhibit reversible heterogeneous kinetics, possess a low overpotential for regeneration and to be stable at certain range of pH, temperature, redox state and dioxygen. Some mediators frequently used are I^-, $[Fe(CN)_6]^{-4}$, *o*-phenylenediamine, diaminobenzidine, hydroquinone and 5-aminosalicylic acid [112–118]. Joseph et al. [119] developed a biosensor based on the redox properties of an immobilized human CYP3A4, a member of the P450 enzyme superfamily, to directly monitor electron transfer to the heme protein. Addition of substrates such as verapamil, midazolam, quinidine or progesterone, at low concentrations (μM), to the oxygenated solution cause a concentration-dependent increase in the reduction current in cyclic voltammetric and amperometric experiments.

In *potentiometric* biosensors changes in potential after the specific binding of the target to the immobilized partner, under zero-current conditions, are detected. For example, proteins in aqueous solution are polyelectrolytes and consequently the electrical charge of the antibody can be affected by binding the corresponding antigen due to charge redistribution. The potential difference is measured between the sensing electrode, where the specific antibody has been immobilized, and the reference electrode. The main disadvantage of this system is that variations in the potential due to the antibody–antigen interaction are too small (1–5 mV) and therefore the reliability and sensitivity of the analysis are limited to background effects. The potential of a single electrode in solution is caused by the tendency of the solution to either donate or accept electrons and may be calculated by the Nernst equation. Thus, the potential of an ion-selective electrode is a logarithmic function of ionic activity. The best-known potentiometric device is the ion-selective electrode (ISE). Coupling an ISE with a field-effect transistor (FET) result in an ISFET, which is a sensitive device for measuring ions such as H^+, Na^+ and K^+. Because of the problems encountered when directly measuring affinity reactions, attention shifted towards systems detecting the products of an enzymatic reaction. Thus, ISFETs are frequently associated to the use of urease as label. Marshall et al. [120] described a pH-sensitive holographic sensor for penicillin detection. The holograms have been used as transducer systems to monitor the pH changes associated with specific enzymatic reactions.

Conductimetric and *Impedimetric* biosensors are essentially based on the same physical principle. *Conductimetry* describes the dependence of the current generated by ions in solution while *impedance* refers to the voltage as a function of the current. In both cases, the conductimetric or impedimetric properties are influenced by a sensing layer placed between the two electrodes. Conductimetric detection is well suited to chemical measurement as many chemical reactions produce or consume ions, thus altering the overall electrical conductivity of the solution. A drawback may be the fact that the overall conductivity determined is the sum of all the conductivities. Therefore, unless, this background response can be quantified, the method lacks specificity. Impedance-based devices in the 1990s was largely developed by the work of McNeil and Martelet. Formation of complex on a conductive or semiconductive surface alters the capacitance and the resistance at the surface–electrolyte interface. Furthermore, the build-up of the sensing biomaterial film on the conductive or semiconductive support alters the capacitance and resistance properties of the solid support-electrolyte interface. Impedance spectroscopy is a very powerful tool for the analysis of interfacial properties changes of modified electrodes upon biorecognition events occurring at the modified surfaces. Impedance measurements provide detailed information on capacitance/resistance changes occurring at conductive or semiconductive surfaces. Thus, impedance spectroscopy, including non-Faradaic impedance measurements resulting in capacitance sensing, is becoming an attractive electrochemical tool to characterize biomaterial films associated with electronic elements, thus, allowing transduction of biorecognition events at the respective surfaces [121]. Yagiuda et al. [122] developed a simple immunosensor based on a conductivity method for determination of methamphetamine in urine. This sensor uses anti-MA antibody immobilized onto the surface of a pair of platinum electrodes. The working range of this biosensor was found between 1 and 10 μg/mL.

2.8.4.1.2 Optical transducers

First developments of these kinds of sensors took advantage of the flexibility and low cost of the optical fibres measuring the absorption or emission of light of one of the components of the bioreaction. Many optical transducers exploit properties such as simple light absorption, fluorescence/phosphorescence, bio/chemiluminescence, reflectance, Raman scattering and refractive index [123]. Apart from speed, sensitivity and robustness, other attractive features of optical sensors include their suitability to component miniaturization, remote sensing and their multi-analyte sensing capabilities. A variety of sophisticated optical sensing principles have been developed mainly based on the evanescent wave (EW) or on the surface plasmon resonance (SPR) phenomena.

An *evanescent wave* is produced in the external media (refractive index, n_2) of a waveguide (n_1) by the electromagnetic field associated to the light guided by total internal reflection (TIR). The electromagnetic field does not abruptly switches to zero at the interface between the two media ($n_1 > n_2$), but decays exponentially with the distance from the interface. The penetration depth of the evanescent field is defined as the distance where its strength is reduced to 1/e of its value at the interface and generally has a value around one hundred of nanometers. The penetration depth is dependent of the incidence angle at the interface and is proportional to the wavelength of the excitation light.

When molecules with an absorption spectrum including the excitation wavelength are located in the evanescent field, they absorb energy leading to attenuation (attenuated total reflection, ATN) in the reflected light of the waveguide. One of the advantages of the biosensors based in this principle is that possible interferences from the bulk media are avoided since only directly absorbed substances interfere with the electromagnetic field. However, as mentioned before, the sensitivity reached with this simple setup is often not sufficient to accomplish the necessary detection limit. For this reason most of the immunosensors reported make use of labelled molecules that are able to re-emit the absorbed evanescent photons at a longer wavelength as fluorescence. Part of this emission is coupled back to the waveguide and in this way is transmitted to the receptor. This phenomenon is known as total internal reflection fluorescence (TIRF). The RIANA immunosensor (see Fig. 2.8.5) based on the EW principle allows measuring small organic molecules, such as estrone or certain pesticides, in water samples. Particularly, estrone can be detected with a LOD of 0.20 ng/L and a LOQ 1.40 ng/L with the aid of fluorescent labels [124]. Recently, a new prototype called AWACSS [125,126] has been developed based on the previous experience with the RIANA prototype for the determination of antibiotics, hormones, endocrine-disrupting chemicals and pesticides in real in water and sediment samples (see Fig. 2.8.6). Theoretically, the prototype would allow detection of 32 different analytes in a single analysis. The emitted light is collected for detection with 32 polymer

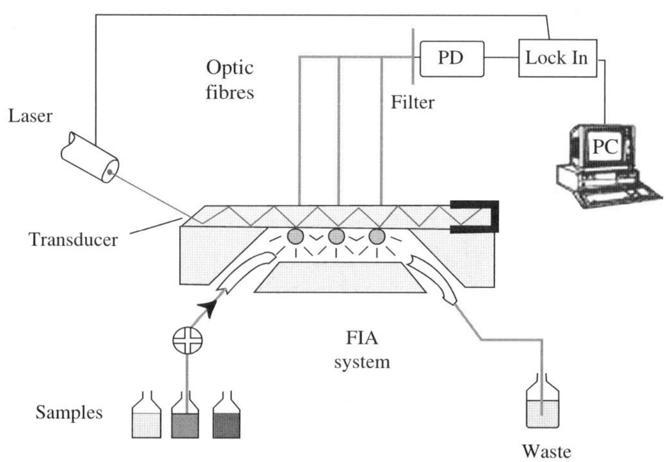

Fig. 2.8.5. Scheme of the River Analyser immunosensor (RIANA). Reprinted from Ref. [124]. Copyright (2004), with permission from Elsevier.

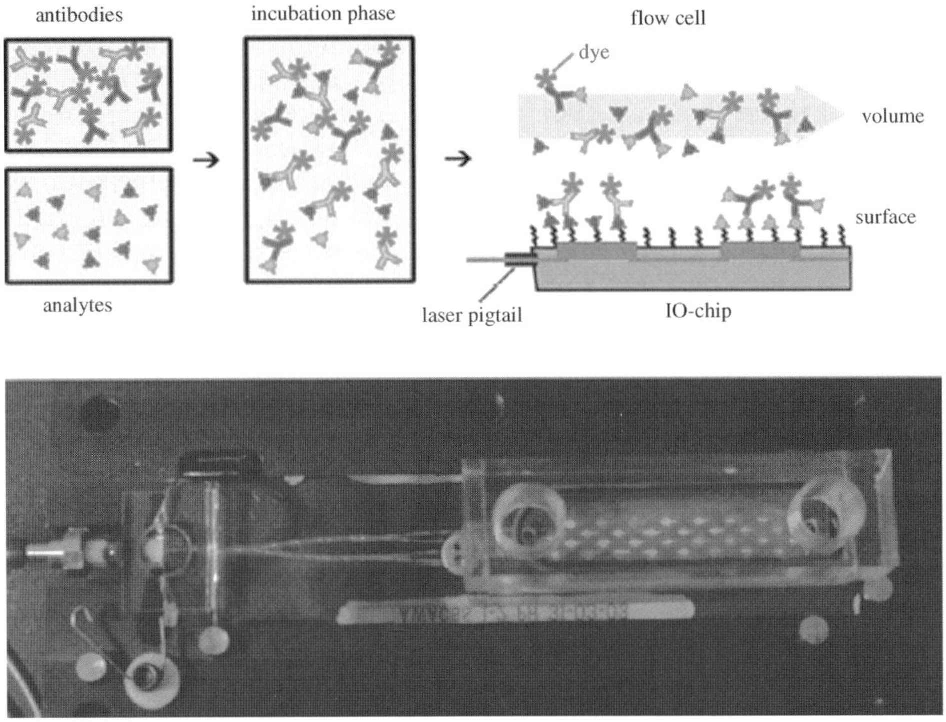

Fig. 2.8.6. The AWACSS instrument employs fluorescence-based detection of the binding of fluorophore-tagged biomolecules to the surface of an optical waveguide chip (upper part of the figure). The design of AWACSS allows for the fast simultaneous measurement of up to 32 different analytes with the IO-transducer chip shown in the lower part of this figure. Reprinted from Ref. [125]. Copyright (2005), with permission from Elsevier.

fibres. The possibility of using this multianalyte biosensor to quantify pharmaceuticals has been demonstrated through the quantification of estrone sulfamethiazole and caffeine in addition to other environmental contaminants. For instance, estrone has been determined in water samples at concentrations ranging between 0.1 and 1.0 μg/L and in sediments at 50–500 ng/g, with coefficient of variations between 3 and 10%.

More recently, other EW immunosensor approaches such as *Grating Couplers* [127–130] or *Mach-Zehnder Interferometers* (MZI) [130–134] have been investigated to make possible direct measurement of small analytes without the use of fluorescent labels. In the *grating coupler* the change produced in the critical angle, as a consequence of the immunoreaction, is measured. The critical angle is the angle that produces total reflection and is very sensitive to the refractive index and thickness at the sensor surface. From this change the N_{eff} (total refractive index due to the effect of the evanescent field) can be calculated. Assuming that n_1 of the waveguide is constant, an optical thickness can be obtained. As an example, an integrated optical grating coupler biosensor for progesterone detection in whole blood samples has been reported [135]. In this case, the modified indirect competitive immunoassay format has been used to detect binding of progesterone antibody. Under these conditions progesterone concentration could be determined in buffer solution and whole blood in a range between 0.005 and 10 ng/mL with a LOD of 3pM. However, experiments were performed for measuring chloramphenicol immobilizing a specific antibody on a grating coupler in order to detect the analyte by optical waveguide lightmode spectroscopy (OWLS) in the range of 10^{-7}–10^{-3} M [136]. Using a *Mach-Zehnder interferometer* the propagating light is splinted in two arms, one of them having the appropriate sensing layer and the other acting as a reference. The evanescent field of the measuring arm collects information regarding the bioreaction, due to the change produced in the refractive index. Consequently, the velocity of the wave in this arm varies. At the end recombination of the waves from both arms allows observation of a constructive or destructive interference, which is related to the extent of the bioreaction that has occurred on the sensing arm.

Surface plasmon Resonance (SPR) is a physical phenomenon that can occur when plane-polarized light hits a metal film under total internal reflection (TIR) conditions (the incoming light is reflected on the interface of a half circular prism). When the prism is coated with a thin film of a noble metal (gold) on the reflection site, the energy of the photon electrical field can interact with the free electron constellations of the gold surface (see Fig. 2.8.7). The incident light photons are absorbed and converted into surface plasmons. A SPR is an evanescent electromagnetic field generated at the surface of a metal conductor

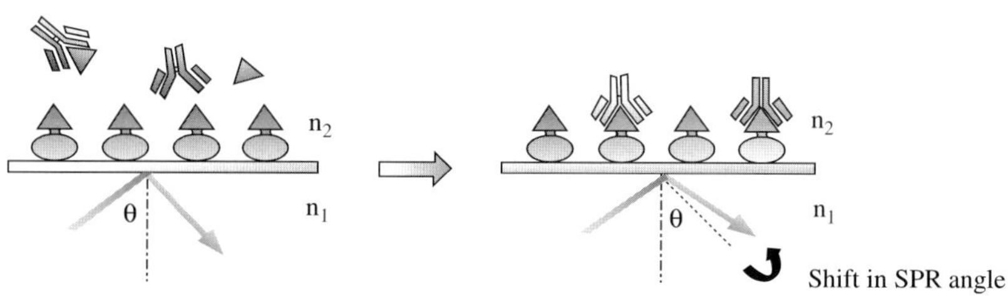

Fig. 2.8.7. Surface plasmon resonance principle (SPR). Reprinted from Ref. [75], with kind permission of Springer Science and Business Media.

(usually Ag or Au) when excited by the impact of light of an appropriate wavelength at a particular angle (θ_p). The absorption of light energy by the surface plasmons during resonance is observed as a sharp minimum in light reflectance when the varying angle of incidence reaches the critical value. The critical angle depends of the wavelength and polarization state of the incident light, but also of the dielectric properties of the medium adjacent to the metal surface and therefore is affected by analytes binding to that surface. This principle allows monitoring of biological interactions. The pioneer of commercial SPR-based biosensing was Pharmacia Biosensor AB, now BIAcore AB. Nowadays, several other brands offer commercial SPR-based sensor devices including IASys from Fisons, IBIS iSPR from IBIS Technologies BV, SensiQ from Nomadics, Plasmoon from BioTul, SPREETA from Sensata (Texas Instruments) and β-SPR sensor from Sensia. In more recent times, surface plasmons produced from an extended flat-film have been exchanged by utilizing the three-dimensional structure of noble metals. Formed by either nanolithography methods or simple colloids covalently attached to the substrate, silver and gold materials have been used to prepare what is termed *localised plasmon resonance sensors (LPR).* The apparent advantages are in the smaller size of the sensor. Each colloid or structure can be prepared to the order of single to tens of nanometers. This allows greater surface area through the three-dimensionality of the structures and thus lower detection limits. Enhanced resonances can be tweaked through the careful choice of material, size and shape of the structures, also contributing to decreases in detection levels. Of the nanolithography-based LPR sensors, the pioneering work of Van Duyne using silver triangles, must be quoted [137–139]. In the colloid-based systems, only model systems measuring large proteins (60–150 kDa) like streptavidin or IgG class molecules has been reported [140–142]. However, a new technique has been described for the quantitative determination of a small steroid molecule called stanozolol (MW 328Da), significantly smaller than model systems. The preliminary features for this assay showed a LOD of 6 μg/L with good reproducibility. This work showed the viability through very simplistic means and setup, when compared to nanolithography, to offer an alternative screening method for this banned anabolic steroid [143].

With the aim to improve performance, there have been some approaches to combine the SPR principle with *grating couplers* and *Mach-Zender interferometers*. Thus, a reported high-sensitivity SPR biosensor based on the *Mach-Zender interferometer design* uses a

Wollaston prism through which the phase quantities of the P and S polarizations are interrogated simultaneously. Since SPR affects only the P polarization, the signal due to the S polarization can be used as the reference. The differential phase between the two polarizations allows eliminating all common-path phase noise while keeping the phase change caused by the SPR effect. A sensitivity limit of 5.5×10^{-8} refractive-index units per 0.01 degrees phase change is reached with this configuration, which has been considered as a significant improvement over previously obtained results when gold was used as the sensor surface [144]. Similarly, a new approach to SPR biosensing based on spectroscopy of multiple surface plasmons on a special multi-diffractive grating coupler has recently been reported. In this approach, the binding event at a surface of the *SPR grating coupler* is simultaneously observed by surface plasmons of different wavelengths, which makes possible the determination of binding-induced changes in the refractive index distribution at the sensor surface [145].

2.8.4.1.3 Piezoelectric transducers

Within mass-sensitive biosensors, acoustic wave biosensors operate on the basis of an oscillating crystal that resonates at a fundamental frequency. The crystal element is coated with a layer containing the biorecognition element designed to interact selectively with the target analyte. A measurable change in the resonance frequency occurs after the binding of the analyte on the sensing surface according to the mass change of the crystal. Most of these biosensors utilize piezoelectric materials as the signal transducers [123,146]. Piezoelectric materials are ideal due to their ability to generate and transmit acoustic waves in a frequency-dependent manner. Acoustic wave biosensors offer label-free, on-line analysis and cost effectiveness combined with ease of use. Disadvantages associated with these sensors include problems with crystal surface regeneration and relatively long-incubation times. Limitations may also be found in the coating or the immobilization of the biorecognition elements on the crystal surface along with the inability to measure small hapten molecules as the mass changes associated are too small to be detected by this method. An emerging group of mass-sensitive biosensors are the so-called cantilever biosensors. Adsorption of biochemical species on a functionalized surface of a microfabricated cantilever can cause surface stress and consequently the cantilever bends. The mass change causes a differential surface stress that can be detected by electric or optical methods [147,148]. Microcantilever biosensors offer various advantages, only small quantities of receptor and analyte are necessary, and limits of detection achieved are often lower than those obtained by classical methods [149]. Chloramphenicol antibiotic was measured in standard solutions with a label-free immunosensor using an anti-chloramphenicol antibody immobilized onto the gold surface of quartz crystal by using 3-mercaptopropionic acid (MPA) and dimethylaminopropyl-ethylcarbodiimide-hydroxysuccinimide ester (EDC-NHS) chemistry. The linear measuring range was found between 5×10^{-6} and 5×10^{-2} M [136].

2.8.4.1.4 Thermometric transducers

Thermometric biosensors exploit the absorption or evolution of heat in biological reactions [150]. This is reflected as a change in the temperature within the reaction medium and is monitored as a change in the resistance of the transducer. The thermal biosensors constructed have been based on direct attachment of the immobilized enzyme or cell to a thermistor. Calorimetric transducers are only useful with enzyme-catalyzed reactions. These reactions exhibit the same enthalpy changes as spontaneous chemical reactions and considerable heat is evolved, ranging from 5 to 100 KJ/mol. Therefore, calorimetric transducers are universally applicable in enzyme sensors.

2.8.4.2 Biorecognition principle

All biosensor technologies and transducing principles described previously require the immobilization of either the receptor or the ligand on the inorganic surface of the transducer. Immobilization of receptor on a solid support has often proven to be difficult due to the loss of functional integrity. Receptors can be covalently attached by chemical cross-linking or non-covalently deposited on the surface. Covalent immobilization may lead to irreversible structural alterations and moreover gives rise to random orientation of the proteins on the surface. Non-covalent immobilization of the receptor can be achieved via adsorption, via incorporation in lipid bilayers, via an affinity tag (i.e. biotin, hexahistidine, etc.). Crucial parameters to consider are the receptor's structural integrity and the orientation of the receptor so that the ligand can be bound without steric restrictions. In addition to the requirement of unchanged receptor affinity and specificity, the receptor should not denature or be released from its support during analysis and the support should demonstrate limited non-specific ligand binding. Moreover, immobilization of the receptor onto a solid support is a key aspect. Enzymes, protein receptors, nucleic acids, antibodies, whole cells and biomimetic receptors are some of the most frequently used biorecognition molecules employed [151].

2.8.4.2.1 Enzymes

Enzyme biosensors are catalytic biosensors which principle relies on the conversion of a non-detectable substrate into an optically or electrochemically detectable product. This process allows the detection of substrates, products, inhibitors and modulators of the catalytic reaction. Combination of different enzymes has been described to extend the range of detectable analytes by converting a non-detectable primary product to a secondary detectable one by the action of a second or third enzyme or to improve sensitivity by increasing amplification. Detection of a particular analyte is thus dependent on its enzymatic transformation or on its capability to act as an inhibitor of an enzyme reaction. Although in nature may exist enzymes able to transform or degrade the pharmaceutical, it is not always possible to have them available on an isolated form, which limits the possibility to develop enzymatic devices [152–155]. Enzyme biosensors have been reported for penicillin V and G using penicillinase [156,157] as biorecognition element. Similarly, enzyme-based amperometric biosensors have been described for the enantioselective analysis of antihypertensive agents [158–160]. S-Enalapril and S-ramipril are angiotensin-converting enzyme (ACE) inhibitors which are used for treatment of hypertension. Owing to the fact that only the S-enantiomer possesses the ACE-inhibiting activity, enantioselective analytical method is necessary in the pharmaceutical industry to discriminate them from the less active R-enantiomers. This system is based on the immobilization of 1-amino acid oxidase in a carbon paste electrode. A sequential injection device allows enantioanalysis of the *S*-enantiomer from the raw materials as well as from their pharmaceutical formulations, with a rate of 75 samples per hour and R.S.D. values better than 0.1% ($n = 10$) [160]. As an example of an enzymatic biosensor, Gustavsson et al. [161] developed a SPR-based biosensor (BIAcore) using as biorecognition element a β-lactam enzyme, achieving LOD of 2.6 μg/kg^{-1} for Penicillin G.

2.8.4.2.2 Protein receptors

As mentioned above, many biochemical processes involve biomolecular recognition events and this fact can be used to develop biochemical assays and consequently also biosensors (see Section 2.8.3). The advantage of this biosensor approach is that the activity of pharmaceuticals has usually been designed to specifically interact with certain protein receptor or process. To overcome denaturation or lost of activity, often transmembrane receptor proteins are successfully immobilized onto the transducer within a lipid bilayer to keep their natural environment. Several receptors have been immobilized onto SPR sensor surfaces, amongst them the β_2-adrenergic receptor. This receptor was incorporated into a solid-supported egg phosphatidylcholine lipid bilayer and followed the binding of full agonists (isoproterenol, epinephrine), a partial agonist (dobutamine), an antagonist (alprenolol) and an inverse agonist (ICI-118,551) to the receptor [162,163]. Seifert et al. [68] used the BIAcore system for the determination of substances with potential estrogenic activity in water using estrogen receptor previously employed in their ELRA (see Section 2.8.3.1). Schmid et al. [164,165] immobilized the purified histidine-tagged serotonin receptor ($5HT_3$-R) via a nickel(II) nitrilotriacetic acid (Ni-NTA) chelating group on SiO_2 of a TIRF sensor. The pharmacological properties of agonists and antagonists towards the receptor were studied by mixing the competing ligands at various concentrations with the fluorescent tracer ligand before measuring binding to the immobilized receptor with TIRF.

2.8.4.2.3 Antibodies

Immunoassay biosensors make use of the specific binding between an antibody and antigen. Main advantages of these kinds of biosensors are the wide range of affinities available expanding thus the number of analytes that can selectively detected. The scope of selectivity of the antibodies is almost unlimited. Additional benefits of using antibodies as sensing elements derive from the possibility to conveniently tailor their affinity and selectivity. Immunosensors can also profit from the monoclonal antibody technology which offers a longer supply period of antibodies with defined chemical and biological properties, and the chance to carefully screen antibodies having the desired characteristics and the feasibility of producing recombinant antibodies in hosts other than mouse at lower cost. Research on the antibody field is still growing and future perspectives also count on the use of small antibody fragments, better defined regarding their chemical structure. This fact would surely help standardization of procedures involved on immunosensor development such as immobilization, stabilization, calibration or storage. Several examples using antibodies as a biorecognition element have been described in the literature and are summarized in Table 2.8.5.

2.8.4.2.4 Nucleic acids

Two main strategies are employed for this type of sensors. A general principle for nucleic acid recognition is base pairing leading to the construction of hybridization devices. Such sensors rely on the immobilization of a short (20–40 mer) synthetic oligomer or single-stranded DNA probe (ssDNA probe) whose sequence is complementary to the sought-after target. Exposure of the sensor to the sample containing the target results in the formation of the hybrid on the surface of the transducer. This strategy has been used for detecting a wide variety of microbial and viral pathogens. As a second approach, these biosensors monitor the interaction of small organic molecules with affinities for the immobilized single (ss) or double (ds) stranded DNA. As an example, an electrochemical technique suitable for the rapid and sensitive screening of the β-blocker atenolol has been proposed based on surface-stabilized bilayer lipid membranes (s-BLMs) composed from egg phosphatidylcholine (PC) and where ss-DNA has been incorporated for the interaction with the analyte [166]. The interactions of atenolol with the DNA modified s-BLMs produced electrochemical ion current increases that reproducibly appeared within a few seconds after the exposure of the membranes to the drug. The detection limit is 1.8 μM (for $S/N = 3$ and for noise levels of 7 nA) and has been applied in the determination of these compounds in pharmaceutical preparations.

2.8.4.2.5 Whole cells

Whole cell (bacteria, fungi, eukaryotic cells or yeast) biosensors examine the effects of the analyte on an intact microorganism. Although purified single molecules are attractive as sensing elements, their preparation can be expensive. In contrast whole living cells may be easily isolated from nature (river water, sediments, soil, activated sludge, etc.). Moreover, these whole living cells are less sensitive to inhibition by other compounds present on the matrix, are more tolerant to variations of the pH or the temperature and seem to have a longer lifetime. Living cells have been used to assess toxicity or to detect a given group of substances. The first approach is based on the fact that living cells provide information on the effects over the living systems. When speaking about pharmaceuticals, it is important to know the overall effect over a living system or organism. These types of sensors rely on measuring the change of certain biological parameters of the organisms exposed to analyte. Whole cell sensors have been developed with a variety of taxonomic groups such as invertebrates, fish, plants, algae and microorganisms; however, the latter group has provided the

TABLE 2.8.5

Reported biosensors for the detection of pharmaceuticals according the transducer classification

	Biorecognition element	Analyte	Sensitivity	Matrix	Time	Ref.	Miscellaneous
Electrochemical							
	E. coli (ATCC 11303)	Tetracycline Quinolone	Posible detection ⩽25 μg/L	Milk samples	120 min	[176]	A miniaturized prototype tested suitable for field determination
	Penicillinase	Penicillin G Ampicillin Amoxicillin	LOD~0.05 μg	Suitable Working buffer		[209]	Capacitive penicillin sensor based on a pH-sensitive electrolyte–insulator–semiconductor EIS sensor (Diffusion barrier)
	Penicillinase enzyme	Penicillin G	LOD~5 μM	Suitable working buffer		[210]	Enzyme field-effect transistor (Enfet) pH-sensitive sensor
	Immobilised Penicillin Binding Protein (PBP)	Penicillin G	(LOD) < mg/kg	Milk	< 7 min	[211]	Assay capable of distinguishing 10, 5, 0 mg/kg Penicillin G levels Assay reproducibility requires improvement
	Penicillinase membrane	Penicillin	0.05 mM	Aqueous solution		[157]	Sensor-based on a microarray electrode coated with pH-Responsive polypyrrole
	Double-stranded calf thymus DNA	Levofloxacin	LOD 25 μg/ml	Urine		[212]	Modified carbon paste electrode with dsDNA for designing a sensitive biosensor for levofloxacin
	Polyphenol oxidase (PPO)	Paracetamol	LR 1.2×10^{-4} to 5.8×10^{-3} M LOD 8.8×10^{-5} M	Pharmaceutical formulations (tablets)	70 s	[213]	*Persea Americana* source of PPO, −0.12 V reduction, chronoamperometry, Vaseline carbon paste (graphite) WE Recovery 97.9–100.7%
	Prussian Blue (PB) - Indium Tin Oxide	Morphine (MO)	LOD 0.1 mM	Buffer, pH 5	60 s	[214]	Oxidation (+0.7 V) of MO by PB mediator. PB acts as artificial peroxidase for MO. Discriminates Codeine (vs. similar) LR 90 μM to 1 mM,Sensitivity of 16.8 mA/cm^2 per mM MO
	Diclofenac selective electrode (SE) with a PVC membrane incorporating cyclodextrin	Diclofenac	LOD 4×10^{-6} M	Buffer & Pharmaceutical formulations (tablets)	Sequential Injection Analysis, < 20 s	[215]	Potentiometric detection based on the formation of hexadecylpyridinium bromide complex with diclofenac 33 samples/hr Recoveries 99–101%

	Nonronite clay	Codeine	LR 0.625–15 μM (R^2, 0.999) LOD 0.15 μM	Buffer, pH 6, & +Urine *Pharmaceutical formulations (tablets)	<60 s	[216]	Square Wave Stripping Voltammetry & amperometry on a clay-modified screen-printed electrode in flow Pre-concentration time, 30 s Pre-concentration potential, −0.6 V Amperometric potential, +1.4 V
	Recombinant *E. coli* (B-Lactamase)	Penicillin G Amoxicillin	5 mM	Fermentation Broth, milk	8 min	[169]	Specificity not determined. Organism immobilized to acetyl cellulose membrane Detection based on change in pH
	Recombinant *E. coli* (B-Lactamase)	Cephalosporins (Cefamandole, Cefotaxime, Cefoperazone)	Cefamandole 0.4 mM	Aqueous Solution	3.5–11 min	[170]	Specificity not determined. Organism immobilised to flat pH electrode through a membrane. Detection based on change in pH
Optical	Commercial antibody against ampicillin	Penicillins	LOD Ampicillin: 33 & 12.5 μg/L Penicillins G < 4 μg/L Penicillins M < 30 μg/L	Milk		[217]	BIAcore biosensor based on SPR. Sample pre-treatment to increase the sensitivity No cross-reactions with cephalosporins Percentages of CR with 9 penicillins were often higher in milk than in buffer
	Monoclonal Antibodies against Digoxigenin	Benzylpenicillin	LOD 4 μg/kg	Milk	40 min (>80 samples a day)	[218]	Surface plasmon resonance biosensor. Use of the penicillin-binding protein PBP 2x*
	Monoclonal antibody (MAb) against Sulfamethazine (21C7)	Sulfamethazine	LOD Sulfamethazine: 10 ng/mL	Chicken Serum	7 min	[219]	Optical biosensor (Biacore Q) used to develop a rapid biosensor immunoassay (BIA) Can detect eight Sulfonamides at comparable sensitivities (LODs 7–20 ng/mL).
	Polyclonal Antibodies	Sulfamethazine (SMZ) Sulfadiazine (SDZ)	LOD SMZ: 0.015–0.029 μg/mL SDZ: 0.028–0.052 μg/mL	Porcine Bile	Up to 650 bile samples per day	[220]	Prototype multi-channel SPR biosensor The instrument allows simultaneous analysis of eight samples for a single or multi analyte analysis
	A binding protein of Sulfonamide derivative (Qflex® Kit)	Sulfonamides	16.9 ng/g	Porcine Muscle	ND	[221]	Generic Sulfonamide SPR Biosensor. No cross-reactivity with inactive acetylated metabolites

TABLE 2.8.5 (*continued*)

Biorecognition element	Analyte	Sensitivity	Matrix	Time	Ref.	Miscellaneous
Thre type of Anti-Sulfonamide Antibodies (Mab 21C7, Qflex and M.3.4.)	17 Sulfonamides	LOD (Mab 21C7): 7–1000 ng/mL (Qflex): 15–340 ng/mL (Mutant M.3.4): 4–82 ng/mL	Chicken Serum	8–10 min	[222]	Comparison of multi-sulfonamide biosensor immunoassays in an optical BIACORE Biosensor (Mab 21C7) was sensitive for the *N*4-acetyl SA Qflex Kit detected 5 sulfonamides registered for application in poultry in Netherlands within the narrowest measurement range
Antibodies	Sulfamethazine	1.7 μg/kg	Raw milk	8–30 min	[223]	Inhibition assay based on antibiotic-immobilised sensor chip and SPR detection Tested against six sulfonamides
Antibodies	Sulfamethazine	1 μg/kg^{-1}	Raw milk	8–30 min	[224]	Inhibition assay based on antibiotic-immobilised sensor chip and SPR detection Good correlation with HPLC method
Antibodies	Sulfamethazine (SMZ)	0.5 μg/kg	Raw milk	5 min	[156]	New inhibition assay based on indirect immobilisation of SMZ to a sensor (SPR) Good HPCL correlation except in 10 μg/kg levels
Antibodies	Sulfadiazine (SDZ)	0.02 μg/mL	Pig bile	2 min	[225]	Inhibition assay based on SDZ immobilized sensor chip and SPR detection
Antibody Streptomycin	Streptomycin Dihydrostreptomycin	15–70 μg/kg Depend on the matrix	Whole Cows' milk, honey, pig kidney and pig muscle		[226]	Qflex™ Kit SPR-based biosensor system They immovilized the Ab in the sensor chip
Polyclonal antibody	Tylosin	(CCβ) 2.5 μg/kg	Honey	10 s	[227]	BiacoreQ SPR-based biosensor system They immovilized the Ab in the sensor chip
Antibody Chloramphenicol	chloramphenicol chloramphenicol glucuronide	(CCα) 0.005–0.04 μg/kg (CCβ) 0.02–0.07 μg/kg^{-1}	poultry muscle, honey, prawn and cows' milk	180–480 s	[228]	BiacoreQ SPR-based biosensor system They immovilized the Ab in the sensor chip

							(HBMChip)
	E. coli JM-109, pQE60-EGFP	Tetracycline Chloramphenicol	Working range 0–5 0–30≠g/L	Buffer	2 h	[229]	Measured of Luminiscence of enhanced green fluorescent protein
	Policlonal antibody (Rabbit)	Chloramphenicol		Buffer		[136]	Anti-chloramphenicol antibody immobilization optical waveguide lightmode spectroscopy (OWLS)
	Liposomes (POPC) SPR sensor	Naproxen, Ketoprofen (16 in total)	RSD <1.5% LR 15.6–500 μM LOD~10 μM	Buffer	120 s	[230]	Sensor regenerates very well (0.4% RSD)
	Tyrosinase and 3 mM Besthorn's hydrazone (MBTH)	Paracetamol	LR 10×10^{-6}–1.4×10^{-3} M	Pharmaceutical formulations (tablets)	<60 s	[231]	UV/Vis at 443 nm Scan rate up to 1000 nm/min
	$Ru(bpy)_2(phen)^{2+}$	Codeine	LOD 0.1 μM	*Buffer & Pharmaceutical formulations		[232]	Electrochemiluminescence (ECL) in both batch & flow modes with Pt interdigitated microelectrodes. +0.75 V vs pseudo ref. in flow mode
	$Ru(bpy)_3^{2+}$	Codeine	LR 100 μM to 2 mM	Buffer & Pharmaceutical formulations (tablets)		[233]	Flow injection analysis mode with Pt electrodes at 750 mV
	Antibodies	Estrone	IC50 = 0.60, 0.53 and 0.56 μg/L	Bidestilled water, river water and groundwater		[124]	RIANA system
	Antibodies	Estrone Sulfathiazole	LOD 0.007 μg/L LOD 0.018 μg/L	Buffer		[125,126]	AWACSS system
	Estrogen receptor	Estradiol	$K_D = 2.33 \times 10^{-10}$ M	Buffer		[68]	The ER binds estradiol-BSA with similar binding characteristic for free estradiol, as it can be showed with the K_D
	Estrogen receptor	Estradiol Estriol	IC50 12.5 Nm IC50 12.5 nM	Buffer		[234]	BIAcore biosensor
Mass sensitive							
	Anti-CAP antibody	Chloramphenicol				[235]	Anti-chloramphenicol antibody immobilization piezoelectric quartz crystal microbalance measurement (QCM) On Develop
	Polyclonal Antibody (Rabbit)	Chloramphenicol	LOD 10^{-5} M	Buffer	10 min	[136]	Anti-chloramphenicol antibody immobilization piezoelectric quartz crystal microbalance measurement (QCM)

most rapid, direct and sensitive approaches. Sensors based on the use of microorganisms are often named *microbial biosensors* and are usually based on direct or indirect measurement of bioprocesses such as the transformation rate of the carbon, nitrogen or sulfur, enzyme activity, growth, mortality, oxygen consumption or luminescence. Cells with a given type of receptor can be considered as sensors for agonists, although biosensors based on measuring enzyme inhibition have also been described. The reason because bacteria cells have been often used as sensing elements in biosensor field is that they can be genetically engineered to respond to specific substances [167]. This strategy has opened up new specificities and sensor possibilities. However, specific biosensors have also been reported based on the use of the genes responsible of resistance mechanisms. Microorganisms have evolved a variety of mechanisms that allow them to survive and grow in contaminated environments. The resistance implies the ability of the cell to exclude the toxic by the membrane, sequestration of metals, chemical modification to a less toxic form, etc. The genes responsible of these resistance mechanisms are organized in operons, usually found on plasmids carried by the resistant bacteria. In many cases, these plasmids confer resistance to one or more toxic substances. Because of the specificity of these regulation mechanisms, the promoters and regulatory genes can be used to construct promoter-reporter-gene fusion for specific biosensors. Non-specific biosensors have also been developed based on the heat-shock or stress-response. Exposure to heat, toxic compounds or heavy metals may induce the expression of stress-response genes linked to stress promoters. Induction of bioluminescent proteins or enzymes that can be detected electrochemically using an electrode or a chemiluminescent substrate allows development of biosensing devices. However, some limitations inherent to microbial sensors are the need of longer response time than enzymes do, and the fact that selectivity is more difficult to accomplish than with single enzymes, due to the variety of metabolic processes occurring on a living cell.

A microbial biosensor for the antihypertensive agent enalapril maleate has been reported based on the use of *Bacillus subitilis* cells. The biosensor measures the acceleration of respiration during specific metabolic pathways of this drug. It has been applied, with good results, for determination of the active ingredient in the pharmaceutical tablet formulations [168]. Microbial cells can also be the recognition element of a biosensor configuration to monitor environmental chemical contaminants. The bacteria used is genetically modified to produce a recombinant organism (i.e. *E. coli*) which can exhibits different traits such as expression of cellular degradative enzymes, specific binding proteins or a reporter enzyme. Several examples of this type of biosensors are described to detect β-lactam antibiotics using a degradative enzyme (β-lactamase) as the recognition element. Penicillin and amoxicillin can be detected in milk, in 8 min, by means of the immobilization of the organism to acetyl cellulose membrane and measuring change in pH. The sensitivity described is around 5 mM [169]. Again, the change of extracellular pH observed in the transduction part of another sensor is also used as a signal for analysing some cephalosporins (cefamandole, cefotaxime and cefoperazone) in aqueous solution. Time required is around 10 min with range of sensitivity 0.1–1 mM depending of the compound [170].

2.8.4.2.6 Biomimetic receptors

Biomimetic biosensors employ non-biological receptors that mimic behaviour of natural biomolecules. That is the case of the molecularly imprinted polymers (MIPs) [171]. As an example, a voltammetric sensor for the β-blocker albuterol was investigated where the microfabrication techniques have been combined with molecular imprinting to construct on-chip devices using photoirradiation of cross-linkable polymers [172]. In this case, a selective MIP was coated as a thin film onto the gold working electrode on chip and the analyte was directly quantified by differential pulse voltammetric measurements. There have been also some examples that use other types of synthetic receptors for the determination of pharmaceuticals. Thus, a potentiometric quasi-array detection system consisting in seven polyvinyl chloride (PVC) based liquid membrane electrodes has been developed to screen libraries of β-adreneraic and β-blocking chiral drugs. Five calix[6]arene derivatives and one modified calyx[4]resorcinarene were used as neutral ionophores to compose mentioned set of PVC-based electrodes [173]. Similarly, five PVC matrix membrane sensors based on the use of the ion-association complexes of the β-blocker cations with tungstophosphate anion as electroactive materials have been developed for the analysis of β-blockers (atenolol, bisoprolol, metoprolol, propranolol and timolol) in some pharmaceutical preparations [174]. Sensors reveal fast, stable and near-Nernstian response in the interval between 10^{-2} and 2×10^{-7} M for different β-blockers and also over a wide range pH values. Many inorganic and organic cations as well as drug excipients and diluents normally used in drug formulations do not interfere.

ACKNOWLEDGMENTS

This work has been supported by the Ministry of Science and Technology (Contract numbers AGL2005-07700-C06-01 and TEC2004-0121-E) and by the European Community (IST-2003-508774). The AMR group is a recognized Grup de Recerca by the Generalitat de Catalunya and has support from the Departament d'Universitats, Recerca i Societat de la Informació la Generalitat de Catalunya (expedient 2005SGR 00207).

REFERENCES

1 K. Kummerer, *Chemosphere*, 45 (2001) 957–969.
2 B. Halling-Sorensen, S.N. Nielsen, P.F. Lanzky, F. Ingerslev, H.C. Holten-Lutzhoft and S.E. Jorgensen, *Chemosphere*, 36 (1998) 357–393.
3 T. Heberer, *Toxicol. Lett.*, 131 (2002) 5–17.
4 O.A. Jones, N. Voulvoulis and J.N. Lester, *Environ. Technol.*, 22 (2001) 1383–1394.
5 R. Andreozzi, M. Raffaele and P. Nicklas, *Chemosphere*, 50 (2003) 1319–1330.
6 C. Daughton and T. Ternes, *Environ. Health Persp.*, 107 (1999) 907–937.
7 C. Daughton and T. Jones-Leep, *ACS Symp. Ser.*, 791 (2001) 39–54.
8 J. Koschorreck, C. Koch and I. Ronnefahrt, *Toxicol. Lett.*, 131 (2002) 117–124.
9 VICH, in, 2000 European Agency for the Evaluation of Medicinal Products, London, UK.
10 G.G. Ying, R.S. Kookana and Y.-J. Ru, *Environ. Int.*, 28 (2002) 545–551.
11 T.A. Ternes, *Water Res*, 32 (1998) 3245–3260.
12 D.R. Dietrich, S.F. Webb and T. Petry, *Toxicol. Lett.*, 131 (2002) 1–3.
13 F.M. Christensen, *Regul. Toxicol. Pharmacol.*, 28 (1998) 212–221.
14 A.R. Mire-Sluis, R. Gaines Das and A. Padilla, *J. Immunol. Methods*, 216 (1998) 103–116.
15 T. Zemlo, R. Rothrock, N. Karle and R. Street, *The Science Advisory Board*.
16 EPA, http://www.epa.gov/region01/oarm/testmeth.pdf (2003).
17 I.E. Tothill and A.P.F. Turner, *TrAC Trend Anal. Chem.*, 15 (1996) 178–188.
18 M. Farre, R. Brix and D. Barcelo, *TrAC Trend Anal. Chem.*, 24 (2005) 532–545.
19 L. da Silveira Ev and E.E.S. Schapoval, *J. Pharmaceut. Biomed. Anal.*, 27 (2002) 91–96.
20 J.A. Bernabeu, M.A. Camacho, M.E. Gil-Alegre, V. Ruz and A.I. Torres-Suarez, *J. Pharmaceut. Biomed. Anal.*, 21 (1999) 347–353.

21 J. Nouws, H. van Egmond, I. Smulders, G. Loeffen, J. Schouten and H. Stegeman, *Int. Dairy J.*, 9 (1999) 85–90.

22 H.R.N. Salgado, C.C.G.O. Lopes and M.B.B. Lucchesi, *J. Pharmaceut. Biomed. Anal.*, 40 (2006) 443–446.

23 M.J.e. Souza, C.F. Bittencourt and P.d.S.e.S Filho, *Int. J. Pharm.*, 271 (2004) 287–291.

24 A. Montero, R.L. Althaus, A. Molina, I. Berruga and M.P. Molina, *Small Ruminant Res*, 57 (2005) 229–237.

25 M.P. Molina, R.L. Althaus, A. Molina and N. Fernandez, *Int. Dairy J.*, 13 (2003) 821–826.

26 L. Okerman, K.D. Wasch and J.V. Hoof, *Analyst*, 123 (1998) 2361–2365.

27 L.H. Hansen, F. Aarestrup and S.J. Sorensen, *Vet. Microbiol.*, 87 (2002) 51–57.

28 M.T. Korpela, J.S. Kurittu, J.T. Karvinen and M.T. Karp, *Anal. Chem.*, 70 (1998) 4457–4462.

29 T. Pellinen, G. Bylund, M. Virta, A. Niemi and M. Karp, *J. Agric. Food Chem.*, 50 (2002) 4812–4815.

30 T. Pellinen, T. Huovinen and M. Karp, *Anal. Biochem.*, 330 (2004) 52–57.

31 A.M. Soto, J.M. Calabro, N.V. Prechtl, A.Y. Yau, E.F. Orlando, A. Daxenberger, A.S. Kolok, L.J. Guillette, B. le Bizec, I.G. Lange and C. Sonnenschein, *Environ. Health Persp.*, 112 (2004) 346.

32 E.J. Routledge and J.P. Sumpter, *Environ. Toxicol. Chem.*, 15 (1996) 241–248.

33 A.C. Gutleb, I.A.T.M. Meerts, J.H. Bergsma, M. Schriks and A.J. Murk, *Environ. Toxicol. Pharmacol.*, 19 (2005) 231–238.

34 M. Fukuzawa, J. Satoh, S. Ohta, K. Takahashi, S. Miyaguchi, X.L. Qiang, Y. Sakata, T. Nakazawa, Y. Takizawa and T. Toyota, *Immunopharmacology*, 48 (2000) 65–74.

35 L.B. Tan, J.G. Burniston, W.A. Clark, Y. Ng and D.F. Goldspink, *J. Cardiovasc. Pharmacol.*, 41 (2003) 518–525.

36 L.A.A. de Jong, D.R.A. Uges, J.P. Franke and R. Bischoff, *J. Chromatogr. B*, 829 (2005) 1–25.

37 L. Aaltonen and M. Scheinin, *Acta Pharmacol. Toxicol. (Copenh)*, 50 (1982) 206.

38 P. Hunt, J.M. Husson and J.P. Raynaud, *J. Pharm. Pharmacol.*, 31 (1979) 448.

39 K. Ensing, D.A. Bloemhof, W.G.i.t. Hout, J.v.d. Lende and R.A.d. Zeeuw, *Pharm. Res.*, 5 (1988) 283.

40 I. Creese and S.H. Snyder, *Nature*, 270 (1977) 180.

41 W. Sadee, D.C. Perry, J.S. Rosenbaum and A. Herz, *Eur. J. Pharmacol.*, 81 (1982) 431.

42 P. Odou and V. Cánovas Gustavo, *Clin. Drug Invest.*, 19 (2000) 283–292.

43 G.G. Belz, *Blood Pressure*, 10 (2001) 13–18.

44 G.G. Belz, R. Butzer, S. Kober, C. Mang and E. Mutschler, *Clin. Pharmacol. Ther.*, 66 (1999) 367–373.
45 N. Bosworth and P. Towers, *Nature*, 341 (1989) 167.
46 S. Udenfriend, L. Gerber and N. Nelson, *Anal. Biochem.*, 161 (1987) 494.
47 R.T. McCabe, B.R.d. Costa, R.L. Miller, R.H. Havunjian, K.C. Rice and P. Skolnick, *J. FASEB*, 4 (1990) 2934–2940.
48 T. Takeuchi, T. Nishikawa, R. Matsukawa and J. Matsui, *Anal. Chem.*, 67 (1995) 2655–2658.
49 B. Ilien, C. Franchet, P. Bernard, S. Morisset, C.O. Weill, J.J. Bourguignon, M. Hibert and J.L. Galzi, *J. Neurochem.*, 85 (2003) 768–778.
50 H.S.J. Hang, D. Li, Y. Liao, D. Lian, Y. Xiao and H. Xue, *J. Biol. Chem.* 275 (2000) 18818; *J. Biol. Chem.* 275 (2000) 18818–18823.
51 K.D. Pfleger and K.A. Eidne, *Pituitary*, 6 (2003) 141–151.
52 E.G. Milligan, *J. Pharm. Sci.*, 21 (2004) 397–405.
53 S. Angers, A. Salahpour, E. Joly, S. Hilairet, D. Chelsky, M. Dennis and M. Bouvier, *Proc. Natl. Acad. Sci. USA*, 97 (2000) 3684–3689.
54 N. Boute, R. Jockers and T. Issad, *Trends Pharmacol. Sci.*, 23 (2002) 351–354.
55 T. Issad, N. Boute and K. Pernet, *Biochem. Pharmacol.*, 64 (2002) 813–817.
56 S. Kohler, S. Belkin and R.D. Schmid, *Fresenius' J. Anal. Chem.*, 366 (2000) 769–779.
57 M. Allen, J. Reeves and G. Mellor, *J. Biomol. Screen.*, 5 (2000) 63–69.
58 A.P. Tairi, R. Hovius, H. Pick, H. Blasey, A. Bernard, A. Surprenant, K. Lundstrom and H. Vogel, *Biochemistry*, 37 (1998) 15850–15864.
59 J. Mellentin-Michelotti, L.T. Evangelista, E.E. Swartzman, S.J. Miraglia, W.E. Werner and P.M. Yuan, *Anal. Biochem.*, 272 (1999) 182–190.
60 E.F. Ullman, H. Kirakossian, A.C. Switchenko, J. Ishkanian, M. Ericson, C.A. Wartchow, M. Pirio, J. Pease, B.R. Irvin, S. Singh, R. Singh, R. Patel, A. Dafforn, D. Davalian, C. Skold, N. Kurn and D.B. Wagner, *Clin. Chem.*, 42 (1996) 1518–1526.
61 N. Rouleau, S. Turcotte, M.H. Mondou, P. Roby and R. Bosse, *J. Biomol. Screen.*, 8 (2003) 191–197.
62 P.C. Simons, S.M. Biggs, A. Waller, T. Foutz, D.F. Cimino, Q. Guo, R.R. Neubig, W.J. Tang, E.R. Prossnitz and L.A. Sklar, *J. Biol. Chem.*, 279 (2004) 13514–13521.
63 A. Waller, P. Simons, E.R. Prossnitz, B.S. Edwards and L.A. Sklar, *Comb. Chem. High T. Scr.*, 6 (2003) 389–397.
64 M.-L. Scippo, C.V.D. Weerdt, P. Willemsen, J.-M. Francois, F. Rentier-Delrue, M. Muller, J.A. Martial and G. Maghuin-Rogister, *Anal. Chim. Acta*, 473 (2002) 135–141.
65 E.R.S. Bauer, N. Bitsch, H. Brunn, H. Sauerwein and H.H.D. Meyer, *Chemosphere*, 46 (2002) 1107–1115.

66 E.R.S. Bauer, H.H.D. Meyer, P. Stahlschmidt-Allner and H. Sauerwein, *Analyst*, 123 (1998) 2485–2487.

67 S.C. Maness, D.P. McDonnell and K.W. Gaido, *Toxicol. Appl. Pharmacol.*, 151 (1998) 135–142.

68 M. Seifert, S. Haindl and B. Hock, *Anal. Chim. Acta*, 386 (1999) 191–199.

69 W. Li, M. Seifert, Y. Xu and B. Hock, *Environ. Int.*, 30 (2004) 329–335.

70 M. Seifert, *Anal. Bioanal. Chem.*, 378 (2004) 684–687.

71 W. Li, M. Seiffert, Y. Xu and E. Hock, *Fresen. Environ. Bull.*, 15 (2006) 74–79.

72 UCB-Bioproducts, http://www.ucbfooddx.com/appben/betastar.htm#test%20kit%20sensitivities.

73 M. Rosner, J. Grassman and R. Haas, *Environ. Health Persp.*, 94 (1991) 131–134.

74 M.G. Weller, *Fresen. J. Anal. Chem.*, 366 (2000) 635–645.

75 M.-C. Estévez, H. Font, M. Nichkova, J.-P. Salvador, B. Varela, F. Sánchez-Baeza and M.-P. Marco, Immunochemical determination of pharmaceuticals and personal care products as emerging pollutants. In: D. Barceló (Ed.), *Emerging Organic Pollutants in Waste Waters and Sludge*, Springer-Verlag, Berlin Heidelberg, 2005, pp. 181–244.

76 M.-C. Estévez, H. Font, M. Nichkova, J.-P. Salvador, B. Varela, F. Sánchez-Baeza and M.-P. Marco, Immunochemical determination of industrial emerging pollutants. In: D. Barceló (Ed.), *Emerging Organic Pollutants in Waste Waters and Sludge*, Springer-Verlag, Berlin Heidelberg, 2005, pp. 119–180.

77 A. Oubiña, B. Ballesteros, P. Bou Carrasco, R. Galve, J. Gascón, F. Iglesias, N. Sanvicens and M.P. Marco, Immunoassays for environmental analysis. In: D. Barcelo (Ed.), *Sample Handling and Trace Analysis of Pollutants: Techniques, Applications and Quality Assurance*, Elsevier Science B.V, Amsterdam, 1999, pp. 287–339.

78 M.-C. Hennion and D. Barcelo, *Anal. Chim. Acta*, 362 (1998) 3–34.

79 G.S. Nunes, I.A. Toscano and D. Barcelo, *TrAC Trend Anal. Chem.*, 17 (1998) 79–87.

80 T.J. Campbell and K.M. Williams, *British J. Clin. Pharmacol.*, 52 (2001) 21S–34S.

81 K. Matalka, T. Arafat, M. Hamad and A. Jehanli, *Fundam. Clin. Pharmacol.*, 16 (2002) 237–244.

82 K. Matalka, T. El-Thaher, M. Saleem, T. Arafat, A. Jehanli and A. Badwan, *J. Clin. Lab. Anal.*, 15 (2001) 47–53.

83 J. Cooper, P. Delahaut, T.L. Fodey and C.T. Elliott, *Analyst*, 129 (2004) 169–174.

84 E.R. Campagnolo, K.R. Johnson, A. Karpati, C.S. Rubin, D.W. Kolpin, M.T. Meyer, J.E. Esteban, R.W. Currier, K. Smith, K.M. Thu and M. McGeehin, *Sci. Total Environ.*, 299 (2002) 89–95.

85 R. Verheijen and W. Haasnoot, *Analyst*, 123 (1998) 2437.

86 Biopharm AG. RIDASCREEN (IA kit). http://www.r-biopharm.com
87 R. Draisci, F.d. Quadri, L. Achene, G. Volpe, L. Palleschia and G. Palleschib, *Analyst*, 126 (2001) 1942–1946.
88 S. Bucknallt, J. Silverlightt, N. Coldhamt, L. Thorne and R. Jackmant, *Food Addit. Contam.*, 20 (2003) 221–228.
89 H. Watanabe, A. Satake, Y. Kido and A. Tsuji, *Analyst*, 127 (2002) 98–103.
90 E. VanCoillie, J. DeBlock and W. Reybroeck, *J. Agric. Food Chem.*, 52 (2004) 4975–4978.
91 J. Duan and Z. Yuan, *J. Agric. Food Chem.*, 49 (2001) 1087–1089.
92 E. Schneider, E. Usleber, R. Dietrich, E. Maertlbauer and G. Terplan, Rapid detection of chloramphenicol in raw milk by enzyme-linked immunofiltration and dipstick assay. In: N. Haagsma and A. Ruiter (Eds.), *Proceedings of the EuroResidues II Conference ''Residues of Veterinary Drugs in Food''*, Veldhoven, Netherlands, 1993, pp. 627–631.
93 C.-H. Huang and D.L. Sedlak, *Environ. Toxicol. Chem.*, 20 (2001) 133–139.
94 G.W. Aherne, J. English and V. Marks, *Ecotoxicol. Environ. Saf.*, 9 (1985) 79–83.
95 P. Onnerfjord, S. Eremin, J. Emneus and G. Markovarga, *J. Immunol. Methods*, 213 (1998) 31–39.
96 A. Reiter, J. Hake, C. Meissner, J. Rohwer, H.J. Friedrich and M. Oehmichen, *Foren. Sci. Int.*, 119 (2001) 248–253.
97 S.A. Eremin, I.A. Ryabova, J.N. Yakovleva, E.V. Yazynina, A.V. Zherdev and B.B. Dzantiev, *Anal. Chim. Acta*, 468 (2002) 229–236.
98 E. Benito-Peña, M.C. Moreno-Bondi, G. Orellan, A. Maquieira and A.v. Amerongen, *J. Agric. Food Chem.*, 53 (2005) 6635–6642.
99 M. Höferl, S. Krist and G. Buchbauer, *Archiv der Pharmazie*, 338 (2005) 493–497.
100 S. Rodriguez-Mozaz, M.-P. Marco, M.J. Lopez de Alda and D. Barceló, *Anal. Bioanal. Chem.*, 378 (2004) 588–598.
101 M.-P. Marco and D. Barcelo, Immunosensors for environmental analysis. In: D. Barceló (Ed.), *Sample Handling and Trace Analysis of Pollutants. Techniques, Applications and Quality Assurance*, Elsevier, Amsterdam, The Netherlands, 2000, pp. 1075–1103.
102 S. Girousi, *Sci. Pap. U. Pardubice, Ser. A: Fac. Chem. Technol*, 8 (2002) 5–17.
103 K. Durick and P. Negulescu, *Biosens. Bioelectron.*, 16 (2001) 587–592.
104 D.J. Monk and D.R. Walt, *Anal. Bioanal. Chem.*, 379 (2004) 931–945.
105 J.P. Hart, A. Crew, E. Crouch, K.C. Honeychurch and R.M. Pemberton, *Anal. Lett.*, 37 (2004) 789–830.
106 K.K. Jain, *Expert Rev. Mol. Diagn.*, 3 (2003) 153–161.
107 I.M. Christie, P.H. Treloar, Z.B. Koochaki, P.M. Vadgama and G.N. Smith, *Anal. Chim. Acta*, 257 (1992) 21–28.

108 Y. Xu, H.B. Halsall and W.R. Heineman, *J. Pharma. Biomed. Anal.*, 7 (1989) 1301–1311.
109 R.Q. Thompson, G.C. Barone III, H.B. Halsall and W.R. Heineman, *Anal. Biochem.*, 192 (1991) 90–95.
110 I. Rosen and J. Rishpon, *J. Electroanal. Chem.*, 258 (1989) 27–39.
111 O. Niwa, Y. Xu, H.B. Halsall and W.R. Heineman, *Anal. Chem.*, 65 (1993) 1559–1563.
112 B. Lu, M. Smyth, J. Quinn, D. Bogan and R. O'Kennedy, *Electroanalysis*, 8 (1996) 619–622.
113 D.J. Pritchard, H. Morgan and J.M. Cooper, *Anal. Chim. Acta*, 310 (1995) 251–256.
114 T. Kaláb and P. Skládal, *Anal. Chim. Acta*, 304 (1995) 361–368.
115 B. Deasy, E. Dempsey, M. Smyth, D. Egan, D. Bogan and R. O'Kennedy, *Anal. Chim. Acta*, 294 (1994) 291.
116 B.B. Dzantiev, A.V. Zherdev, M.F. Yulaev, R.A. Sitdikov, N.M. Dmitrieva and I.Y. Moreva, *Biosens. Bioelectron.*, 11 (1996) 179–185.
117 D. Ivnitski and J. Rishpon, *Biosens. Bioelectron.*, 11 (1996) 409–417.
118 D.V. Brown and M.E. Meyerhoff, *Biosens. Bioelectron.*, 6 (1991) 615.
119 S. Joseph, J.F. Rusling, Y.M. Lvov, T. Friedberg and U. Fuhr, *Biochem. Pharmacol.*, 65 (2003) 1817–1826.
120 A.J. Marshall, D.S. Young, J. Blyth, S. Kabilan and C.R. Lowe, *Anal. Chem.*, 76 (2004) 1518–1523.
121 E. Katz and I. Willner, *Electroanalysis*, 15 (2003) 913–947.
122 K. Yagiuda, A. Hemmi, S. Ito, Y. Asano, Y. Fushinuki, C.-Y. Chen and I. Karube, *Biosens. Bioelectron.*, 11 (1996) 703–707.
123 A.F. Collings and F. Caruso, *Rep. Prog. Phys.*, 60 (1997) 1397–1445.
124 S. Rodriguez-Mozaz, S. Reder, M. Lopez de Alda, G. Gauglitz and D. Barcelo, *Biosens. Bioelectron.*, 19 (2004) 633–640.
125 J. Tschmelak, G. Proll, J. Riedt, J. Kaiser, P. Kraemmer, L. Barzaga, J.S. Wilkinson, P. Hua, J.P. Hole and R. Nudd, *Biosens. Bioelectron.*, 20 (2005) 1499–1508.
126 J. Tschmelak, G. Proll, J. Riedt, J. Kaiser, P. Kraemmer, L. Barzaga, J.S. Wilkinson, P. Hua, J.P. Hole and R. Nudd, *Biosens. Bioelectron.*, 20 (2005) 1509–1519.
127 F.F. Bier and R.D. Schmid, *Biosens. Bioelectron.*, 9 (1994) 125–130.
128 F.F. Bier, R. Jockers and R.D. Schmid, *Analyst*, 119 (1994) 437–441.
129 A. Akkoyun and U. Bilitewski, *Biosens. Bioelectron.*, 17 (2002) 655–664.
130 G. Gauglitz, *Anal. Bioanal. Chem.*, 381 (2005) 141–155.
131 S.H. Hsu and Y.T. Huang, *J. Lightwave Technol.*, 23 (2005) 4200–4206.
132 S.H. Hsu and Y.T. Huang, *Opt. Lett.*, 30 (2005) 2897–2899.
133 F. Prieto, B. Sepulveda, A. Calle, A. Llobera, C. Dominguez and L.M. Lechuga, *Sens. Actuat. B-Chem.*, 92 (2003) 151–158.
134 B.J. Luff, J.S. Wilkinson, J. Piehler, U. Hollenbach, J. Ingenhoff and N. Fabricius, *J. Lightwave Technol.*, 16 (1998) 583–592.

135 E. Ehrentreich-Forster, F.W. Scheller and F.F. Bier, *Biosens. Bioelectron.*, 18 (2003) 375–380.
136 N. Adanyi, M. Varadi, N. Kim and I. Szendro, *Curr. Appl. Phys.*, 6 (2006) 279–286.
137 A.J. Haes and R.P. Van Duyne, *J. Am. Chem. Soc.*, 124 (2002) 10596–10604.
138 A.D. McFarland and R.P. Van Duyne, *Nano Letts*, 3 (2003) 1057–1062.
139 A.J. Haes, W. Paige Hall, L. Chang, W.L. Klein and R.P. Van Duyne, *Nano Letts*, 4 (2004) 1029–1034.
140 G. Raschke, S. Kowarick, T. Franzl, C. Sönnichsen, T.A. Klar, J. Feldmann, A. Nichtl and K. Kurzinger, *Nano Letts*, 3 (2003) 935–938.
141 L.A. Lyon, M.D. Musick, P.C. Smith, B.D. Reiss, D.J. Peña and M.J. Natan, *Sens. Actuat. B*, 54 (1999) 118–124.
142 F. Frederix, J.M. Friedt, K.H. Choi, W. Laureyn, A. Campitelli, D. Mondelaers, G. Maes and G. Borghs, *Anal. Chem.*, 75 (2003) 6894–6900.
143 M.P. Kreuzer, R. Quidant, G. Badenes and M.-P. Marco, *Biosens. Bioelectron.*, 21 (2006) 1345–1349.
144 N. Kinrot, *J. Lightwave Technol.*, 22 (2004) 2296–2301.
145 P. Adam, J. Dostalek and J. Homola, *Sensor. Actuat. B-Chem.*, 113 (2006) 774–781.
146 P. Leonard, S. Hearty, J. Brennan, L. Dunne, J. Quinn, T. Chakraborty and R. O'Kennedy, *Enzy. Microb. Technol.*, 32 (2003) 3–13.
147 J. Tamayo, M. Alvarez and L.M. Lechuga, *Sens. Actuat. B.*, 89 (2003) 33–39.
148 A.M. Moulin, S.J. O'Shea and M.E. Welland, *Ultramicroscopy*, 82 (2000) 23–31.
149 H.P. Lang, M.K. Baller, R. Berger, C. Gerber, J.K. Gimzewski, F.M. Battiston, P. Fornaro, J.P. Ramseyer, E. Meyer and H.J. Güntherodt, *Anal. Chim. Acta*, 393 (1999) 59–65.
150 K. Ramanathan and B. Danielsson, *Biosens. Bioelectron.*, 16 (2001) 417–423.
151 D.J. Monk and D.R. Walt, *Anal. Bioanal. Chem.*, 379 (2004) 931–945.
152 M.D. Rubianes and G.A. Rivas, *Electroanalysis*, 17 (2005) 73–78.
153 M. Trojanowicz, *Electroanalysis*, 14 (2002) 1311–1328.
154 M. Albareda-Sirvent, A. Merkoci and S. Alegret, *Sensor. Actuat. B-Chem.*, 69 (2000) 153–163.
155 P.U. Abel, T. von Woedtke, B. Schulz, T. Bergann and A. Schwock, *J. Mol. Catal. B-Enzym.*, 7 (1999) 93–100.
156 M.R.S. Fuh, L.W. Burgess and G.D. Christian, *Anal. Chem.*, 60 (1988) 433–435.
157 M. Nishizawa, T. Matsue and I. Uchida, *Anal. Chem.*, 64 (1992) 2642–2644.
158 R.I. Stefan, H.Y. Aboul-Enein and G.L. Radu, *Prep. Biochem. Biotechnol.*, 28 (1998) 305–312.

159 R.I. Stefan, J.K.F. van Staden and H.Y. Aboul-Enein, *Electroanalysis*, 11 (1999) 1233–1235.
160 R.I. Stefan, J.F. van Staden, C. Bala and H.Y. Aboul-Enein, *J. Pharmaceut. Biomed. Anal.*, 36 (2004) 889–892.
161 E. Gustavsson, P. Bjurling and A. Sternesjo, *Anal. Chim. Acta*, 468 (2002) 153–159.
162 R.J. Whelan, T. Wohland, L. Neumann, B. Huang, B.K. Kobilka and R.N. Zare, *Anal. Chem.*, 74 (2002) 4570.
163 S. Devanathan, Z. Yao, Z. Salamon, B. Kobilka and G. Tollin, *Biochemistry*, 43 (2004) 3280–3288.
164 R. Hovius, E.L. Schmid, A.P. Tairi, H. Blasey, A.R. Bernard, K. Lundstrom and H. Vogel, *J. Recept. Signal. Transduct. Res.*, 19 (1999) 533–545.
165 E.L. Schmid, A.P. Tairi, R. Hovius and H. Vogel, *Anal. Chem.*, 70 (1998) 1331–1338.
166 D.P. Nikolelis, S.S.E. Petropoulou and M.V. Mitrokotsa, *Bioelectrochemistry*, 58 (2002) 107–112.
167 S. Ramanathan, M. Ensor and S. Daunert, *Trends Biotech*, 15 (1997) 500–506.
168 S. Fleschin, C. Bala, A.A. Bunaciu, A. Panait and H.Y. Aboul-Enein, *Prep. Biochem. Biotechnol.*, 28 (1998) 261–269.
169 R.A. Lovett, *Inside Lab. Manage.*, 2 (1998) 12.
170 J.L. García, C.J. Nuñez, E.G. González, J. Osuna, X. Soberón and E. Galindo, *Appl. Biochem. Biotechnol.*, 73 (1998) 243.
171 K. Haupt, Molecularly imprinted polymers as recognition elements in sensors. In: V.M. Mirsky (Ed.), *Ultrathin Electrochemical Chemo- and Biosensors*, Springer-Verlag, Germany, 2004.
172 H.C. Huang, C.I. Lin, A.K. Joseph and Y.D. Lee, *J. Chromatogr. A*, 1027 (2004) 263–268.
173 G. Bazylak, L.J. Nagels and H.J. Geise, *Combinat. Chem. High T. Scr.*, 7 (2004) 345–359.
174 S.S.M. Hassan, M.M. Abou-Sekkina, M.A. El-Ries and A.A. Wassel, *J. Pharmaceut. Biomed. Anal.*, 32 (2003) 175–180.
175 L.H. Hansen and S.J. Sorensen, *FEMS Microbiol. Lett.*, 190 (2000) 273–278.
176 G.E. Pellegrini, G. Carpico and E. Coni, *Anal. Chim. Acta*, 520 (2004) 13–18.
177 M. Farre and D. Barcelo, *TrAC Trend Anal. Chem.*, 22 (2003) 299–310.
178 M. Cleuvers, *Toxicol. Lett.*, 142 (2003) 185–194.
179 B. Ferrari, N. Paxeus, R.L. Giudice, A. Pollio and J. Garric, *Ecotoxicol. Environ. Safe.*, 55 (2003) 359–370.
180 M. Isidori, M. Lavorgna, A. Nardelli, A. Parrella, L. Previtera and M. Rubino, *Sci. Total Environ.*, 348 (2005) 93–101.
181 P. Cliquet, E. Cox, W. Haasnoot, E. Schacht and B.M. Goddeeris, *Anal. Chim. Acta*, 494 (2003) 21–28.

182 M.T. Muldoon, S.A. Buckley, S.S. Deshpande, C.K. Holtzapple, R.C. Beier and L.H. Stanker, *J. Agric. Food Chem.*, 48 (2000) 545.
183 W. Heering, E. Usleber, R. Dietrich and E. Martlbauer, *Analyst*, 123 (1998) 2759.
184 B. Lang, E. Maertlbauer and G. Terplan, *Arch. Lebensmittelhyg*, 43 (1992) 77.
185 M. Matsumoto, K. Tsunematsu, A. Tsuji and Y. Kido, *Anal. Chim. Acta*, 346 (1997) 207–213.
186 Z. Yuan, J. Duan, S. Fan and K. Kong, *Food Agric. Immunol*, 13 (2001) 199–204.
187 C.K. Holtzapple, S.A. Buckley and L.H. Stanker, *J. Agric. Food Chem.*, 45 (1997) 1984–1990.
188 R.C. Beier, L.C. Creemer, R.L. Ziprin and D.J. Nisbet, *J. Agric. Food Chem.*, 53 (2005) 9679–9688.
189 R.C. Yao and D.F. Mahoney, *Appl. Environ. Microbiol.*, 55 (1989) 1507–1511.
190 L. Lynas, D. Currie, C.T. Elliott, J.D.G. McEvoy and S.A. Hewitt, *Analyst*, 123 (1998) 2773.
191 C. Van de Water, *J. Assoc. Off. Anal. Chem.*, 73 (1990) 534–540.
192 H. Keukens, *J. AOAC Int.*, 75 (1992) 245–257.
193 S. Impens, W. Reybroeck, J. Vercammen, D. Courtheyn, S. Ooghe, K. De Wasch, W. Smedts and H. De Brabander, *Anal. Chim. Acta*, 483 (2003) 153–163.
194 P.N. Rao and I.B. Taraporewala, *Steroids*, 57 (1992) 154–161.
195 H.W. Hagedorn, R. Schulz and G. Jaeschke, *Biomed. Chromatogr.*, 8 (1994) 63–68.
196 N.J. Barrett, A.T. Kicman, S. Bansal and D.A. Cowan, *Pharm. Sci.*, 1 (1995) 87–89.
197 R.J. McConnell, S.P. Fitzgerald and J.V. Lamont, Food Saf. Qual. Assur.: Appl. Immunoassay Syst., *Proc.*, 1st (1992) 245–250.
198 W.N. Sawaya, K.P. Lone, A. Husain, B. Dashti and S. Al-Zenki, *Food Chem*, 63 (1998) 563–569.
199 P.W. Tang, D.L. Crone, C.S. Chu, H.K. Yeung and K. Chan, *Anal. Chim. Acta*, 275 (1993) 139–146.
200 F. Valentini, D. Compagnone, A. Gentili and G. Palleschi, *Analyst (Cambridge, United Kingdom)*, 127 (2002) 1333–1337.
201 M.S. Preti, S. Lodi, P. Busacchi, M. Filicori and C. Flamigni, *Steroids*, 43 (1984) 469–479.
202 B. Saha and C. Das, *J. Immunoassay*, 12 (1991) 391–412.
203 A. Basu and T.G. Shrivastav, *J. Immunoassay*, 21 (2000) 39–50.
204 M.L. Rodriguez, I. McConnell, J. Lamont, J. Campbell and S.P. FitzGerald, *Analyst (Cambridge, United Kingdom)*, 119 (1994) 2631–2634.
205 L.M. Ribeiro Neto, H.S. Spinosa and M.C. Salvadori, *J. Anal. Toxicol.*, 21 (1997) 393–396.

206 M.-L. Scippo, C. Argiris, C.c. VanÂ DeÂ Weerdt, M. Muller, P. Willemsen, J. Martial and G. Maghuin-Rogister, *Anal. Bioanal. Chem.*, 378 (2004) 664–669.
207 J. Legler, C. van den Brink, A. Brouwer, A. Murk, P. van der Saag, A. Vethaak and B. van der Burg, *Toxicol. Sci.*, 48 (1999) 55–66.
208 H. Fang, W. Tong, W.S. Branham, C.L. Moland, S.L. Dial, H. Hong, Q. Xie, R. Perkins, W. Owens and D.M. Sheehan, *Chem. Res. Toxicol.*, 16 (2003) 1338–1358.
209 A. Poghossian, M. Thust, M.J. Schöning, M. Müller-Veggian, P. Kordos and H. Lüth, *Sensor. Actuat. B*, 68 (2000) 260–265.
210 A. Poghossian, M.J. Schöning, P. Schroth, A. Simonis and H. Lüth, *Sensor. Actuat. B*, 76 (2001) 519–526.
211 S.J. Setford, R.M. Van Es, Y.J. Blankwater and S. Kroger, *Anal. Chim. Acta*, 398 (1999) 13–22.
212 A. Radi, M.A. El Ries and S. Kandil, *Anal. Chim. Acta*, 495 (2003) 61–67.
213 O. Fatibello-Filho, K.O. Lupetti and I.C. Vieira, *Talanta*, 55 (2001) 685–692.
214 K.-C. Ho, C.-Y. Chen, H.-C. Hsu, L.-C. Chen, S.-C. Shiesh and X.-Z. Lin, *Biosens. Bioelectron.*, 20 (2004) 3–8.
215 M. Shamsipur, F. Jalali and S. Ershad, *J. Pharmaceut. Biomed. Anal.*, 37 (2005) 943–947.
216 Y. Shih, J.-M. Zen and H.-H. Yang, *J. Pharmaceut. Biomed. Anal.*, 29 (2002) 827–833.
217 V. Gaudin, J. Fontaine and P. Maris, *Anal. Chim. Acta*, 436 (2001) 191–198.
218 G. Cacciatore, M. Petz, S. Rachid, R. Hakenbeck and A.A. Bergwerff, *Anal. Chim. Acta*, 520 (2004) 105–115.
219 W. Haasnoot, M. Bienenmann-Ploum and F. Kohen, *Anal. Chim. Acta*, 483 (2003) 171–180.
220 C. Situ, S.R.H. Crooks, A.G. Baxter, J. Ferguson and C.T. Elliott, *Anal. Chim. Acta*, 473 (2002) 143–149.
221 T. McGrath, A. Baxter, J. Ferguson, S. Haughey and P. Bjurling, *Anal. Chim. Acta*, 529 (2005) 123–127.
222 M. Bienenmann-Ploum, T. Korpimaki, W. Haasnoot and F. Kohen, *Anal. Chim. Acta*, 529 (2005) 115–122.
223 V. Gaudin and M.L. Pavy, *J. AOAC Int.*, 82 (1999) 1316.
224 A. Sternesjö, C. Mellgren and L. Björck, *Immunoassays for Residue Analysis (Food Safety)*, ACS, Washington, DC, 1996.
225 C.T. Elliott, G.A. Baxter, S.R.H. Crooks and W.J. McCaughey, *Food Agric. Immunol.*, 11 (1999) 19.
226 J.P. Ferguson, G.A. Baxter, J.D.G. McEvoy, S. Stead, E. Rawlingsc and M. Sharmanc, *Analyst*, 127 (2002) 951–956.
227 M. Caldow, S.L. Stead, J. Day, M. Sharman, C. Situ and C. Elliott, *J. Agric. Food Chem.*, 53 (2005) 7367–7370.

228 J. Ferguson, A. Baxter, P. Young, G. Kennedy, C. Elliott, S. Weigel, R. Gatermann, H. Ashwin, S. Stead and M. Sharman, *Anal. Chim. Acta*, 529 (2005) 109–113.
229 D.O. Fesenko, T.V. Nasedkina, D.V. Prokopenko and A.D. Mirzabekov, *Biosens. Bioelectron.*, 20 (2005) 1860–1865.
230 C.L. Baird, E.S. Courtenay and D.G. Myszka, *Anal. Biochem.*, 310 (2002) 93–99.
231 E. Valero, P. Carrion, R. Varon and F. Garcia-Carmona, *Anal. Biochem.*, 318 (2003) 187–195.
232 P.E. Michel, G.C. Fiaccabrino, N.F. de Rooij and M. Koudelka-Hep, *Anal. Chim. Acta*, 392 (1999) 95–103.
233 E. L'Hostis, P.E. Michel, G.C. Fiaccabrino, D.J. Strike, N.F. de Rooij and M. Koudelka-Hep, *Sensor. Actuat. B-Chem.*, 64 (2000) 156–162.
234 M. Usami, K. Mitsunaga and Y. Ohno, *J. Steroid Biochem. Mol. Biol.*, 81 (2002) 47–55.
235 I.-S. Park, D.-K. Kim, N. Adanyi, M. Varadi and N. Kim, *Biosens. Bioelectron.*, 19 (2004) 667–674.

CHAPTER 3: FATE AND OCCURRENCE OF PHARMACEUTICALS IN THE WATER CYCLE

Chapter 3.1

Occurrence of pharmaceuticals in the aqueous environment

Kevin V. Thomas and Katherine Langford

3.1.1 INTRODUCTION

The occurrence of pharmaceutical substances in the aquatic environment serves as a timely reminder that not only those substances traditionally targeted, or those that occur on priority lists for monitoring programmes, contaminate the aquatic environment. Pharmaceutical substances are used by man in quantities similar to those of many pesticides. It is therefore hardly surprising that once the analytical instrumentation was established to accurately and specifically analyse for pharmaceuticals in complex environmental matrices they have been detected.

The occurrence of pharmaceutical substances in the aquatic environment was first reported in the UK in 1984 [1]. Clofibric acid, the metabolite of an antihyperlipidaemic drug, and the antibiotics erythromycin and tetracylcin were qualitatively determined in surface waters from the UK. Further research in the 1990s has led to further occurrence data being published. In the 1990s, increased focus began to be placed on substances other than those traditionally placed on priority lists of substances for monitoring, primarily through increased attention on the occurrence of estrogenic chemicals in wastewater treatment works (WTW) effluent resulting in feminisation of fish [2,3] and the advancement in analytical instrumentation. Many thousands of different pharmaceutical substances are used, whilst use patterns vary from country to country, even city to city. It would be too costly to analyse for all these substances, therefore a prioritisation step is required. This typically involves an assessment of the data available on the sources, volumes used, fate, and transformation and toxicity to aquatic organisms. The prioritisation procedure attempts to establish which pharmaceutical substances have the potential to pose the greatest hazard and risk to the aquatic environment. Often the priority lists

Comprehensive Analytical Chemistry 50
Petrović and Barceló (Eds)
Volume 50 ISSN: 0166-526X DOI: 10.1016/S0166-526X(07)50010-3

of substances selected for establishing occurrence data take into account the availability or potential availability of accurate and specific methods of analysis.

Currently, there are a number of priority list of substances that are used to direct the efforts of national and international monitoring programmes. In Europe, those of the European Union and Oslo and Paris Commission (OSPAR) have greatest influence and do contain pharmaceutical substances (Table 3.1.1). In addition to these lists, national prioritisation procedures have also taken place and prioritised pharmaceutical substances based on the potential risk that they are perceived to pose to the aquatic environment. In the UK, 12 pharmaceutical substances were prioritised for targeted monitoring based upon their predicted environmental concentration (PEC), predicted no-effect concentration (PNEC) and PBT (persistence, bioaccumulation and toxic) properties [4]. Elsewhere, prioritisation has been conducted in Sweden [5] and the United States [6]. Although they offer a good start in focusing effort, these lists should be viewed with caution since they are based upon the acute, principally lethal, ecotoxicological test data and may therefore not include those substances that may be exerting effects following chronic exposure.

These prioritisation procedures have been instrumental in deciding which pharmaceuticals are targeted and therefore the occurrence data available are heavily influenced by such an approach. Of course, other compounds have been selected in a less formal manner and serve to complement the data available on the occurrence of pharmaceutical substances in the aquatic environment. What is important is that the occurrence data are used not only to confirm the presence of a substance in the aquatic environment, but is used in combination with relevant ecotoxicological test data to allow the refinement of risk assessments. It is therefore important that occurrence data are not assessed in isolation, and that some sort of risk assessment is performed to establish whether the concentrations at which the particular substances occur are at a level that may pose a threat to aquatic organisms.

3.1.2 SURFACE WATER

3.1.2.1 Marine

Marine surface waters and direct discharges are one area where there remains a paucity of data concerning the occurrence of pharmaceutical

TABLE 3.1.1

Examples of prioritised pharmaceutical substances

Substance	CAS	Priority list
Clotrimazole	23593-75-1	OSPAR list of substances for priority action (2005)
Chlorpromazine	50-53-3	OSPAR list of substances of possible concern
Chloroquine bis(phosphate)	50-63-5	
Chloroquine	54-05-7	
Prochloroperazine	58-38-8	
Fluphenazine	69-23-8	
Fluphenazine dihydrochloride	146-56-5	
10H-phenothiazine, 10-[3-(4-methyl-1-piperazinyl)propyl]-2-(trifluoromethyl)-, dihydrochloride	440-17-5	
Trifluperidol	749-13-3	
1,2-Ethanedisulfonic acid, compd. with 2-chloro-10-[3-(4-methyl-1-piperazinyl)propyl]-10H-phenothiazine (1:1)	1257-78-9	
10(9H)-acridinepropanamine, *N*,*N*,9,9-tetramethyl-, [R-(R*,R*)]-2,3-dihydroxybutanedioate (1:1)	3759-07-7	
	4394-00-7	
Dimetacrine	4757-55-5	
Noclofolan	10331-57-4	
Miconazole nitrate	22832-87-7	
Timiperone	57648-21-2	
Closantel	57808-65-8	
Midazolam	59467-70-8	
Diammonium N-ethylheptadecafluoro-N-[2-(phosphonatooxy)ethyl]octanesulphonamidate	67969-69-1	
Trimethoprim	738-70-5	UK Environment Agency prioritisation
Diclofenac	15307-86-5	
Sulfamethoxazole	723-46-6	
Paracetamol	103-90-2	
Mefenamic acid	61-68-7	
Ibuprofen	15687-27-1	
Erythromycin	114-07-8	
Dextropropoxyphene	469-62-5	
Lofepramine	23047-25-8	
Tamoxifen	10540-29-1	

substances. It may have been expected that dilution would result in only low concentrations of pharmaceutical substances occurring in marine surface waters; however, the five studies published in Europe have demonstrated that certain pharmaceutical substances are occurring in the marine environment from estuaries to the open sea (Table 3.1.2). In

TABLE 3.1.2

Examples of pharmaceuticals that have been shown to occur in the marine surface waters

Substance	CAS	Structure	Concentration range (ng/L)	Median	Location	Reference
Clofibric acid	882-09-7	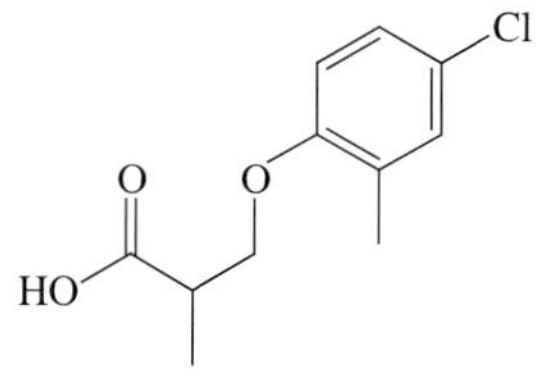	<20–111	20	Selected UK estuaries	[9,10]
Clotrimazole	23593-75-1	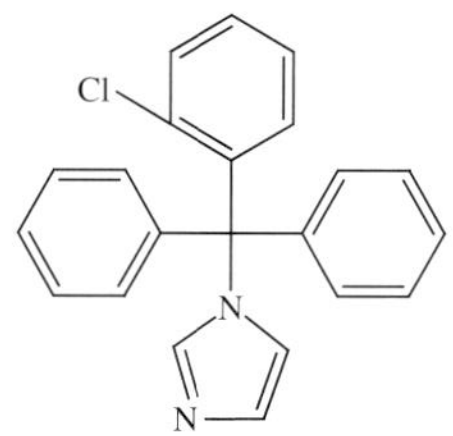	0.5–19 <1–34	 16	North Sea Selected UK estuaries	[7,11] [9,10]
Dextropropoxyphene	469-62-5	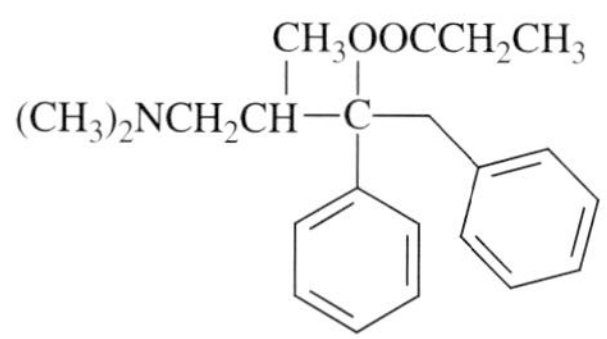	<8–98	8	Selected UK estuaries	[9,10]
Diclofenac	15307-86-5	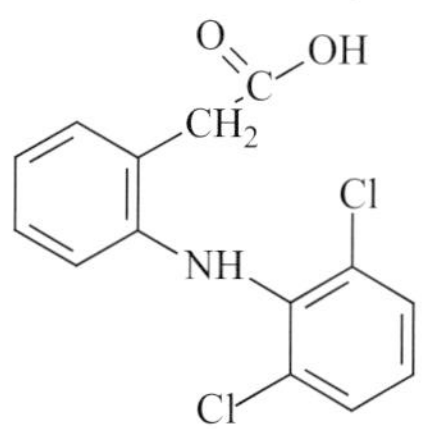	<8–195	8	Selected UK estuaries	[9,10]

Erythromycin	114-07-8	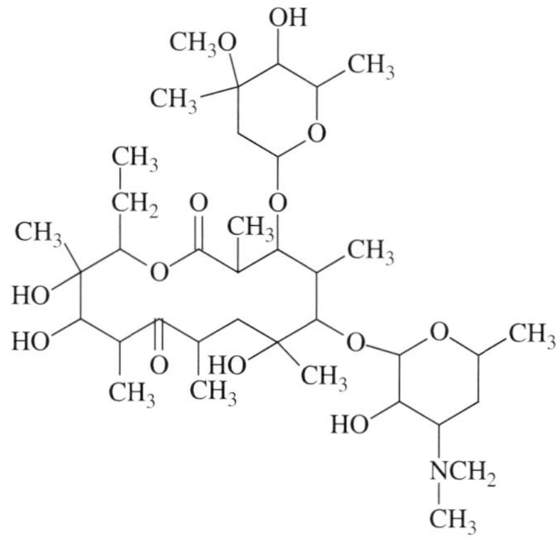	<4–70	6	UK estuaries	[9,10]
Ibuprofen	15687-27-1	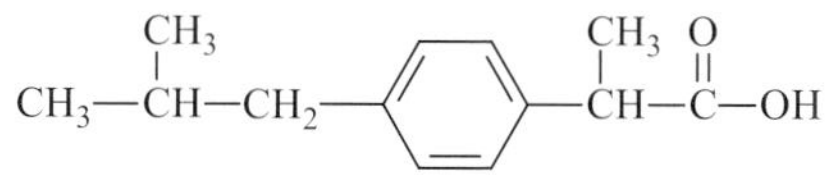	<8–2370	247	Selected UK estuaries	[9,10]
			n.d.–0.7	—	Tromsø Sound, NO	[11]
Ibuprofen-CX		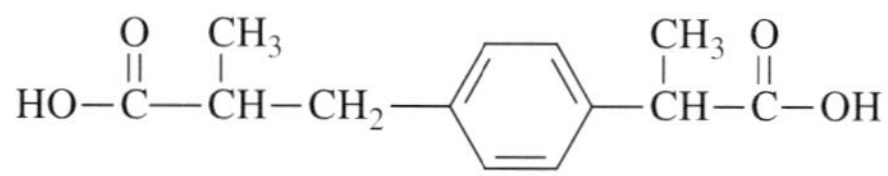	n.d.–7	—	Tromsø Sound, NO	[11]
Ibuprofen-OH		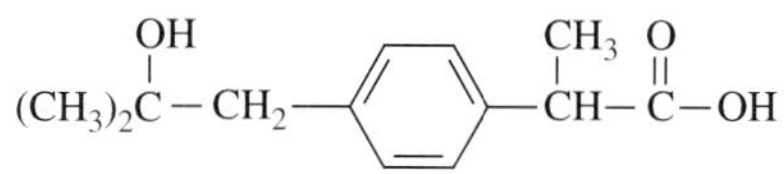	n.d.–1.5	—	Tromsø Sound, NO	[11]
Lofepramine			<4	—	Selected UK estuaries	[9,10]
Mefenamic acid	61-68-7	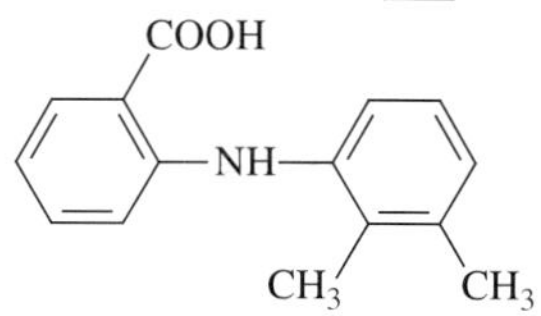	<2–196	20	Selected UK estuaries	[9,10]

TABLE 3.1.2 (*Continued*)

Substance	CAS	Structure	Concentration range (ng/L)	Median	Location	Reference
Paracetamol	103-90-2	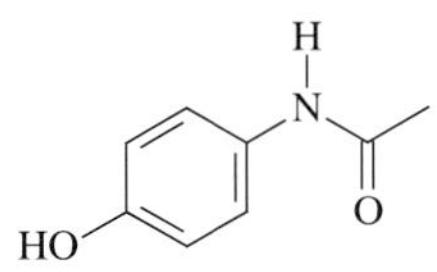	<20	—	Selected UK estuaries	[9,10]
Propranolol	525-66-6	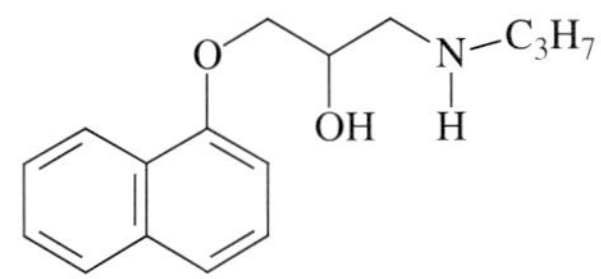	<4–107	36	Selected UK estuaries	[9,10]
Sulfamethoxazole	723-46-6	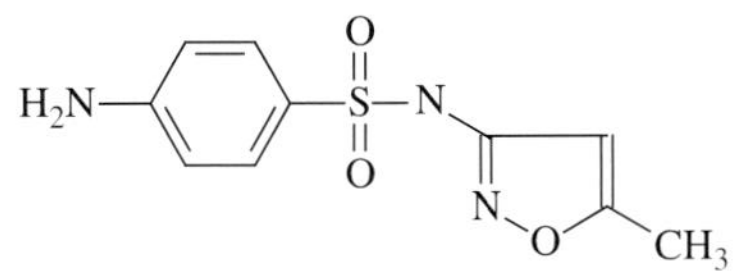	<20	—	Selected UK estuaries	[9,10]
acetyl-sulfamethoxazole	21312-10-7		<20		Selected UK estuaries	[9,10]
Tamoxifen	10540-29-1	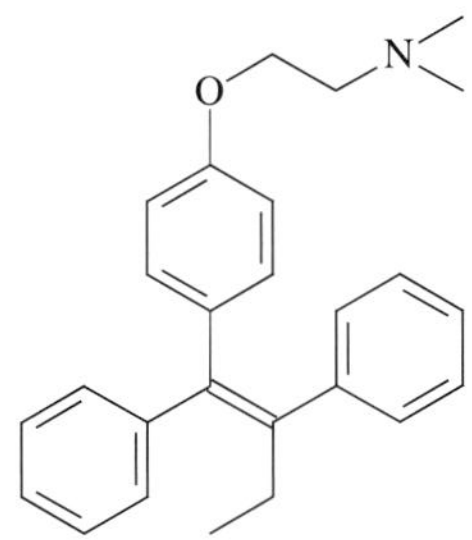	<4–212	25	Selected UK estuaries	[9,10]

Tricolsan	3380-34-5	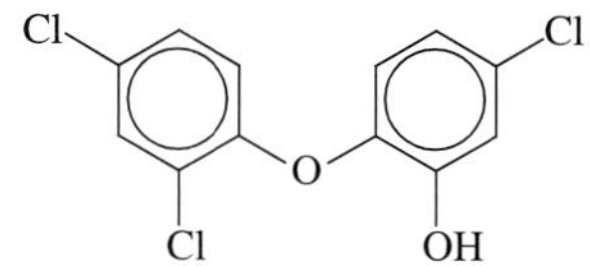	n.d.	—	Tromsø Sound, NO	[11]
Trimethoprim	738-70-5	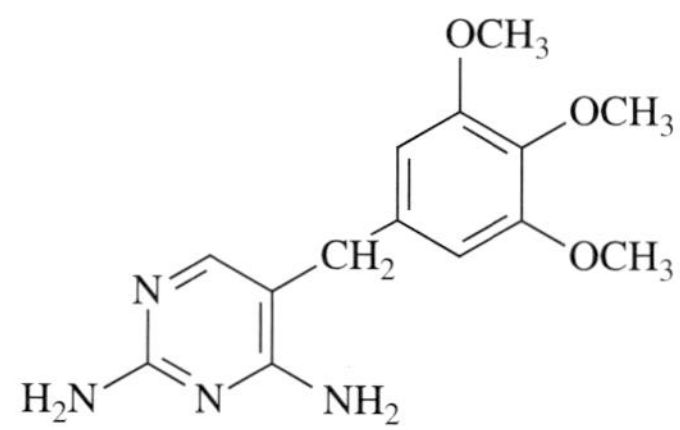	<4–569	8	Selected UK estuaries	[9,10]

1998, the occurrence of clofibric acid was reported in the North Sea at concentrations of between 1 and 2 ng/L, whilst further data were added in 2002 that showed concentrations as high as 19 ng/L [7,8]. All other studies have focused on the occurrence of pharmaceuticals in estuaries or coastal areas (Table 3.1.2). Thomas and Hilton [9] reported the occurrence of the OSPAR priority substance clotrimazole and a number of other selected pharmaceuticals prioritised by the UK Environment Agency in a number of UK estuaries: Tyne, Tees, Belfast Lough, Thames and Mersey. Roberts and Thomas [10] determined the occurrence of the same substances, however, focused on the inputs into the lower Tyne catchment in an attempt to establish whether WTW effluent was the only source. Work in Norway has also shown that pharmaceutical substances occur in the artic environment [11].

Of the 17 samples listed in Table 3.1.2, ibuprofen has been detected at the highest maximal and median concentration. Ibuprofen, like many pharmaceutical compounds, primarily enters the aquatic environment through WTW effluent discharges that contain ibuprofen due to the large amounts used as a prescription and an 'over-the-counter' anti-inflammatory and painkiller in many countries. Another contributory factor is that ibuprofen has also been reported as relatively persistent in aquatic systems (half-life 50 days; [12]), although studies by both Ternes [12] and Roberts and Thomas [10] report that certain WTW remove up to 90% of the ibuprofen entering the works. Ibuprofen is known to metabolise to hydroxy-ibuprofen, carboxy-ibuprofen, their glucuronide derivatives, ibuprofen-glucuronide and carboxy-hydratropic acid. Considering the high concentrations of ibuprofen being detected it is not surprising to also find elevated concentrations of these metabolites (Table 3.1.2).

Two of the most interesting pharmaceutical substances found to occur in marine surface waters are clotrimazole and tamoxifen. Clotrimazole, a topical anti-fungal agent, is included in the OSPAR Commission's list of substances for priority action, which has the consequence that clotrimazole is subject to a cessation target of all discharges, emissions and losses by 2020 [13]. The concentrations of clotrimazole found in UK estuaries and coastal areas are between <1 and 34 ng/L, with a median concentration of 16 ng/L, which is not as high as many other pharmaceuticals. However, clotrimazole is frequently detected in marine surface water that receives high inputs of wastewater effluents. OSPAR has prepared a background document on the environmental properties of clotrimazole and predict an environmental concentration (PEC) in marine surface waters of 15.5 ng/L [13].

This PEC seems in line with published occurrence data and although very few acute or chronic toxicity data are available for the effects of clotrimazole on marine aquatic species, a PEC/PNEC risk assessment assumes that there is no environmental risk.

Tamoxifen is a known estrogen receptor antagonist and is used in the treatment of breast cancer. Tamoxifen has been reported to occur in selected samples collected from UK estuaries and in particular the Tyne estuary in the North East at concentrations of up to 210 ng/L. For tamoxifen, an acute predicted no-effect concentration (PNEC) of 200 ng/L has been reported [14]. The generation of simple risk quotients through using $PEC/PNEC_A$ suggests that there is a risk of acute environmental effects on certain aquatic organisms at this concentration, however since tamoxifen is known to bind to the estrogen receptor [15], the chronic non-lethal effects tamoxifen may be potentially having, should be of greater concern and warrant further investigation.

3.1.2.2 Freshwater

Greater amounts of data are available for freshwater surface waters (Table 3.1.3). The number of pharmaceutical compounds that have been targeted and the number of locations samples are significantly greater (Table 3.1.3). Much of these data are for North America and Europe. For example, of 18 antibiotics targeted in a German study of river waters [16], a degradation product of erythromycin was detected in the highest concentration (maximum 1.7 μg/L), whilst four other compounds were also detected at lower concentrations. These are the same five antibiotics the authors detected in WTW effluent. In another study of German rivers and ponds, pharmaceuticals were detected in all but two samples [17]. In this study, clofibric acid, diclofenac, ibuprofen, ketoprofen, indomethacin, fenoprofen and the main metabolites of ibuprofen, 2-[4-(2-hydroxy-2-methylpropyl)phenyl] propionic acid (hydroxy-ibuprofen) and 2-[4-(2-carboxypropyl)phenyl]propionic acid (carboxy-ibuprofen), were targeted.

Ibuprofen has also been detected in several lakes and rivers in Switzerland at concentrations up to 7.8 ng/L but in this study its metabolites were not observed [26]. The concentrations of ibuprofen in this study were low in comparison to studies in other countries [4,10,12] where concentrations of over 2000 ng/L have been reported. A median concentration of 297 ng/L measured in the river Tyne in the UK supports other work where ibuprofen has been detected in receiving waters with a WTW effluent source [4,29].

TABLE 3.1.3

Examples of pharmaceuticals that have been shown to occur in the freshwater surface waters

Substance	CAS	Concentration (μg/L)	Median (μg/L)	Location	Reference
Carbamazepine	298-46-4		0.185	River water, Canada	[18]
			0.23	River water, Germany	[19]
Clofibric acid	882-09-7		0.066	Selected rivers, Germany	[12]
		n.d.–0.051		River Rhine, Germany	[20]
			n.d.	Selected German rivers	[21]
		0.0032–0.0076		River Elbe, Germany	[22]
		0.0193–0.0435		Selected Austrian rivers	[23]
		n.d.–0.07		Selected rivers and ponds, Germany	[17]
			0.059	River water, Canada	[18]
Diclofenac	15307-86-5		0.15	Selected rivers, Germany	[12]
		0.015–0.304		River Rhine, Germany	[20]
			0.046	Selected German rivers	[21]
		n.d.–0.035		3 rivers in Finland	[24]
		0.031–0.067		River Elbe, Germany	[22]
		<0.02–0.568	0	Selected rivers, UK	[4]
		0.0283–0.392		Selected Austrian rivers	[23]
		0.0011–15.033		Selected rivers and ponds, Germany	[17]
			0.026	River water, Canada	[18]
		<0.01–1.022	<0.01	Selected rivers, UK	[4]
17β-estradiol	50-28-2		n.d.	Selected rivers, Germany	[25]
Gemfibrozil	25812-30-0		0.052	Selected rivers, Germany	[12]
			0.066	River water, Canada	[18]
Ibuprofen	15687-27-1	n.d.–0.0078		Selected rivers and Lakes, Switzerland	[26]
			0.07	Selected rivers, Germany	[12]
		0.144–2.37	0.297	River Tyne, UK	[10]
		n.d.–0.041		River Rhine, Germany	[20]
		0.0051–0.032		River Elbe, Germany	[22]
			nd	Selected German rivers	[21]
		n.d.–0.014		3 rivers in Finland	[24]
		<0.002–5.044	0.826	Selected rivers, UK	[4]
		n.d.–0.063		Selected rivers and ponds, Germany	[17]
		n.d.–0.201		Selected surface waters, Italy	[27]
			0.141	River water, Canada	[18]

TABLE 3.1.3 (*continued*)

Substance	CAS	Concentration (μg/L)	Median (μg/L)	Location	Reference
Naproxen	22204-53-1	n.d.–0.045		3 rivers in Finland	[24]
		n.d.–0.022		Selected surface waters, Italy	[27]
			0.207	River water, Canada	[18]
Sulphamethizole	144-82-1	n.d.		Selected rivers, Germany	[28]
Sulphadiazine	68-35-9	n.d.–0.007		Selected rivers, Germany	[28]
Sulphamethoxazole	723-46-6	n.d.		Selected rivers, Germany	[28]
		<0.05	<0.05	Selected rivers, UK	[4]
Tamoxifen	10540-29-1	0.027–0.212	0.053	River Tyne, UK	[10]
		<0.01	<0.01	Selected rivers, UK	[4]
Triclosan	3380-34-5	n.d.–0.0041		River Elbe, Germany	[22]

Dilution effects are an important consideration when measuring the concentration of pharmaceutical compounds in rivers and streams. Lipid regulators, bezafibrate and gemfibrozil, demonstrated a 5–10 times dilution in rivers receiving WTW effluent compared to the effluent discharged [12]. Elsewhere, elevated concentrations of drugs detected in a small tributary receiving a large contribution of effluent were rapidly diluted to near detection limits when they flowed into a larger volume river [18] and most pharmaceuticals were only detected in freshwater sites receiving WTW effluent. However, in the low flow system of a smaller river, virtually no dilution was shown to occur. The hydrology of the receiving water therefore plays an important role in the dilution of any pharmaceutical substances that may be present and is specific to a given location.

In addition to ibuprofen, other pharmaceutical metabolites have been detected in receiving waters. Clofibrate was not detected in rivers and streams in a German study whereas its metabolite, clofibric acid was present in the ng/L range [12].

In Slovenia, naproxen and diclofenac have been detected in river samples [30]. Typically, WTW effluents are the source of much of the human pharmaceuticals detected in the aquatic environment, however this study showed that concentrations of these two substances were 4–5 times higher in samples collected downstream of a pharmaceuticals factory than in other samples. In Germany, river water samples were screened for eight pharmaceuticals and their metabolites [17]. In this study the rivers that received an input form WTW all showed higher concentrations than those receiving no known WTW effluent.

Uptake of pharmaceuticals by aquatic organisms is an important consequence of elevated pharmaceutical concentrations in receiving waters such as effluent dominated rivers and streams. Fish in US streams have demonstrated uptake of the antidepressants, fluoxetine, norfluoxetine, sertraline and desmethylsertraline [31].

3.1.3 GROUNDWATER

Since the mid-1990s there have been reports of the occurrence of pharmaceutical substances in groundwater (Table 3.1.4). Groundwater can become contaminated from a number of sources, for example, historic contamination from sites of production, runoff from agricultural land,

TABLE 3.1.4

Examples of pharmaceuticals that have been shown to occur in groundwaters

Substance	CAS	Concentration (μg/L)	Location	Reference
Clofibric acid	882-09-7	0.07–7.3	Groundwater wells, Germany	[33]
Diclofenac	15307-86-5	n.d.–0.38	Groundwater wells, Germany	[35]
Erythromycin	114-07-8	n.d.	Selected groundwater, Germany	[16]
Gemfibrozil	25812-30-0	n.d.–0.34	Groundwater wells, Germany	[35]
Ibuprofen	15687-27-1	n.d.–0.2	Groundwater wells, Germany	[35]
Salicylic acid		n.d.–1.225	Groundwater wells, Germany	[35]
Sulphamethizole	144-82-1	n.d.–0.33	Selected groundwater,	[34]
Sulphadiazine	68-35-9	n.d.–0.48	Selected groundwater,	[34]
Sulphamethoxazole	723-46-6	n.d.–0.47	Selected groundwater, Germany	[16]
Trimethoprim	738-70-5	n.d.	Selected groundwater, Germany	[16]

landfill and wastewater effluent [32–34]. The disposal of industrial waste in a landfill site in Denmark has been shown to be the source of pharmaceutical compounds in leachate-contaminated groundwater adjacent to the site [34]. A number of sulphonamide antibiotics (sulphadiazine, sulphamethiozole) were present in groundwater samples collected at concentrations of up to 0.5 μg/L (Table 3.1.4). The pharmaceutical contamination of groundwater around the city of Berlin in Germany has been extensively investigated [32,33]. Heberer [33] and his co-workers have reported the occurrence of clofibric acid, phenazone, propylphenazone, carbemazapine, diclofenac, ibuprofen and fenofibrate in Berlin groundwater with WTW effluent contamination being identified as the source via surface water. In another study conducted in Germany, a number of antibiotic compounds were identified as present in groundwater with application of animal slurry to fields being the likely source due to runoff [16]. Sulfamethoxazole was determined at a maximum concentration of 0.47 μg/L along with sulfamethazine which was detected at a maximum concentration of 0.16 μg/L. However, these two compounds were only detected in 2 of the 59 samples collected, whilst another 16 targeted antibiotics were below the detection limits of the methods used indicating that the load of antibiotics from livestock treatment to groundwater was small.

3.1.4 WASTEWATER

There is much evidence that WTW effluent is a considerable source of pharmaceuticals to receiving waters (Table 3.1.5). Dilution effects are frequently observed downstream of effluent discharge points with increases in measured pharmaceutical concentrations after additional point sources. An example of this can be seen in Fig. 3.1.1, where the pattern of pharmaceuticals determined downstream resembles that of the WTW effluent.

The introduction of pharmaceuticals to the aquatic environment via wastewater treatment and effluent discharge is dependent on drug use, pharmokinetic and physicochemical properties of the drug and on the water treatment processes involved [46]. Aspirin, for example, is readily metabolised to carbon dioxide and water so is unlikely to be detected in effluent [47]. More hydrophobic compounds such as mefenamic acid will sorb to the sludge and may undergo degradation whereas, more hydrophilic compounds, frequently metabolites, such as clofibric acid, will remain in the aqueous phase and be detected in effluents [48]

TABLE 3.1.5

Examples of pharmaceuticals that have been shown to occur in wastewater effluents

Substance	CAS	Concentration (μg/L)	Median (μg/L)	Location	Reference
Carbamazepine	298-46-4	0.3–1.2	0.87	Selected WTW, Europe	[36,37]
			0.7	Selected Canadian WTW	[38]
			1.65	WTW Germany	[19]
Clofibric acid	882-09-7		0.36	Selected WTW, Germany	[12]
		n.d.–0.68	0.23	Selected WTW, Europe	[36,37]
		0.007–0.180		Selected German WTW	[20]
			n.d.	Selected Canadian WTW	[38]
Clotrimazole	23593-75-1	0.01–0.033		1 WTW, UK	[10]
Diclofenac	15307-86-5	0.038–0.489		Selected WTW, Germany	[20]
		n.d.–5.45	0.47	Selected WTW, Europe	[36,37]
			0.81	Selected WTW, Germany	[12]
		0.261–0.598		1 WTW, UK	[10]
		n.d.–2.34	0.424	Selected WTW, UK	[4]
		0.012–0.56		Selected Greek WTW	[39]
			n.d.	Selected Canadian WTW	[38]
		n.d.–0.282		Selected rivers, Slovenia	[30]
Erythromycin	114-07-8	0.145–0.290		River Tyne, UK	[10]
		n.d.–1.842	n.d.	Selected WTW, UK	[4]
Erythromycin-$H_2$0	114-07-8	n.d.–0.838	0.08	Selected WTW, Canada	[40]
17β-estradiol	50-28-2		0.001	Selected WTW, Germany	[25]
			0.009	Selected WTW, Canada	[25]
		n.d.–0.0017	0.00045	Selected WTW, Italy	[41]
Fenoprofen	31879-05-7	n.d.		Selected WTW, Germany	[42]
Fluoxetine	54910-89-3	n.d.		1 WTW USA	[43]
		n.d.–0.099 (mean)		Selected WTW, Canada	[38]
Gemfibrozil	25812-30-0	n.d.–190		Selected WTW, Germany	[20]
			0.4	Selected WTW, Germany	[12]
		0.81–4.76	0.84	Selected WTW, Europe	[36]
			1.3		[38]

TABLE 3.1.5 (*continued*)

Substance	CAS	Concentration (μg/L)	Median (μg/L)	Location	Reference
				Selected Canadian WTW	
Galaxolide	1222-05-5	0.49–0.6		1 WTW, Spain	[44]
Ibuprofen	15687-27-1	0.017–0.139		Selected WTW, Germany	[20]
		0.05–7.11		Selected WTW, Europe	[36]
		n.d.		1 WTW, USA	[43]
		0.002–0.081		3 WTW, Switzerland	[26]
			0.37	Selected Germany	[12]
		1.9–4.2	2.972	1 WTW, UK	[10]
		0.91–2.1		1 WTW Spain	[44]
		0.99–3.3		Selected WTW, Switzerland	[26]
		<0.02–27.25	3.08	Selected WTW, UK	[4]
		0.0012–0.095		Selected surface waters, Italy	[27]
			4.0	Selected Canadian WTW	[38]
Mefenamic acid	61-68-7	0.290–0.396		1 WTW, UK	[10]
Naproxen	222204-53-1	0.8–2.6		1 WTW, Spain	[44]
		n.d.–5.22	1.12	Selected WTW, Europe	[36]
		0.0001–0.0022		Selected surface waters, Italy	[27]
			12.5	Selected Canadian WTW	[38]
		n.d.–0.313		Selected rivers, Slovenia	[30]
Paracetamol	103-90-2	n.d.		6 WTW, UK	[4,10]
Propranolol	525-66-6		0.17	Selected WTW, Germany	[12]
		0.01–0.09	0.01	Selected WTW, Europe	[36]
		0.016–0.284	0.076	Selected WTW, UK	[4]
Propylphenazone			0.095	Selected WTW, Germany	[45]
Salicylic acid	69-72-7		3.6	Selected Canadian WTW	[38]
Sulphamethizole	144-82-1	n.d.–0.006		Selected WTW, Germany	[28]
		0.25		1 WTW, Spain	[44]
Sulphadiazine	68-35-9	0.026–0.081		Selected WTW, Germany	[28]
		n.d.–0.019	0.019	Selected WTW, Canada	[40]
Sulphamethoxazole	723-46-6	0.3–1.5		Selected WTW, Germany	[28]
		n.d.–0.09	0.05	Selected WTW, Europe	[36]
		<0.05–0.132	<0.05	Selected WTW, UK	[4]

TABLE 3.1.5 (*continued*)

Substance	CAS	Concentration (µg/L)	Median (µg/L)	Location	Reference
		n.d.–0.871	0.243	Selected WTW, Canada	[40]
Tamoxifen	10540-29-1	146–369		1 WTW, UK	[10]
		n.d.–0.042	<0.001	1 WTW, UK	[4]
Trimethoprim	738-70-5	218–322		1 WTW, UK	[10]
		n.d.–1.288	0.07	1 WTW, UK	[4]

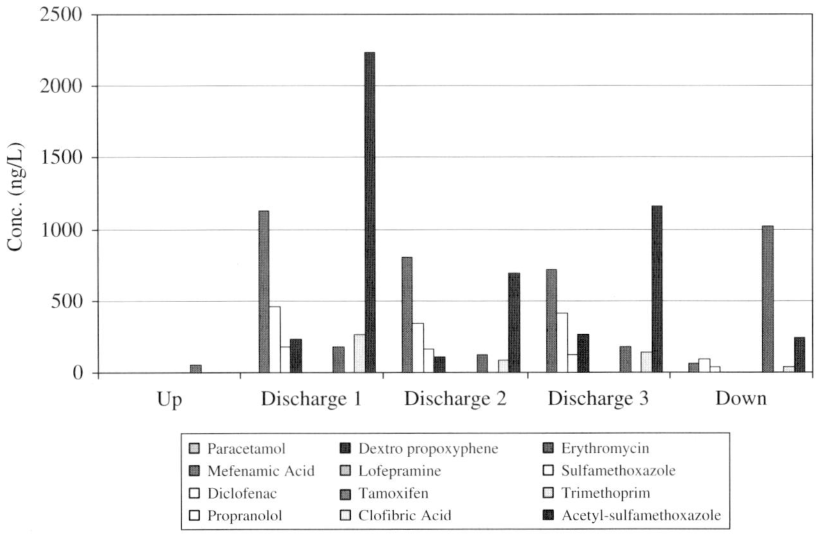

Fig. 3.1.1. Pattern of selected pharmaceuticals in effluent and receiving water from the UK [14].

(see Chapter 4). A large variety of drug residues have been detected in WTW secondary effluent, such as antiphlogistics, lipid regulators and beta blockers, and the concentration of these pharmaceuticals often exceeds several nanograms per litre.

Some pharmaceuticals are frequently detected at concentrations greater than 1 µg/L such as acetylsalicylic acid, bezafibrate, clofibric acid, diclofenac and ibuprofen [12,42] although most are detected at concentrations less than 1 µg/L. The data collated in Table 3.1.4 indicate that ibuprofen and naproxen are detected at the highest concentrations.

Penicillins are unlikely to be detected in effluent as they are easily susceptible to hydrolysis therefore are easily removed during wastewater treatment. Tetracyclines are unlikely to be detected due to their high-metabolic rate and the formation of stable complexes with calcium ions [49].

In a German study looking solely at antibiotics [16], 5 out of 18 measured compounds in effluent were detected, with the maximum concentrations ranging from 0.56 μg/L for clarithromycin to 6 μg/L for a erythromycin degradation product, roxithromycin, sulfamethoxazole and trimethaprion occurred at concentrations between these. The development of antibiotic resistant bacteria in WTW is an important consideration when assessing their fate during biological treatment.

Some of the active ingredients used in important antibiotics, such as ciprofloxacin, ofloxacin and metronidazole, have demonstrated limited biodegradation potential during biological treatment [50]. Cytostatic agents are a less important group of compounds in terms of the quantities used but in terms of their potential environmental impact they remain significant. Most of their active ingredients have proved to have low biodegradability so are expected to pass unchanged through WTW via effluent unless removed by adsorption.

One of the most commonly monitored groups of pharmaceuticals in wastewater treatment are female steroid hormones, in particular 17β-estradiol and 17α-ethinylestradiol which have been reported in raw sewage and effluents. There are large variations in the concentrations of 17β-estradiol and 17α-ethinylestradiol reported in different WTW compartments and this is due to the complex nature of treatment processes and of sampling procedures. Concentrations below the detection limits [42,51] up to 42 ng/L [25] and 62 ng/L [42] have been recorded for 17β-ethinylestradiol. Estrogens are mainly excreted from the body in their conjugated form and deconjugation is likely to occur during biological treatment which has resulted with free estrogen concentrations in effluent occasionally being higher than those found in influent such as in the analysis of estrone [41].

Deconjugation of conjugated metabolites of other pharmaceuticals, such as popranolol and erythromycin [10], can also occur during treatment resulting in the discharge of parent compounds in WTW effluent at higher concentrations than detected in raw influents [12,35,52]. However, sometimes pharmaceutical metabolites, such as clofibric acid, the metabolite of three lipid-regulating agents are more stable than parent compounds and are detected more frequently than the parent compound. Several metabolites of carbamazepine have also been detected in WTW effluent [53]. Fluoxetine is rapidly metabolised to the active compound, norfluoxetine, in the body [54]; however, it is fluoxetine that has been detected rather than its metabolite in effluent samples [18].

During a 3-stage pilot treatment process with denitrification, activated sludge treatment and final settling ibuprofen demonstrated 40% removal whereas only 5% of clofibric acid and diclofenac were removed [55] which is supported by other studies at municipal treatment plants [12]. An example of the changes in pharmaceutical concentration through a modern WTW can be seen in Table 3.1.6. This shows the change in median pharmaceutical concentration following the collection of residence time adjusted 24 h composites from a WTW in the UK [10]. The changes shown are following, primary screening, trickling filer, activated sludge and ultra violet (UV) treatment (Fig. 3.1.2).

3.1.5 DRINKING WATER

In summer 2004, the Observer newspaper in the UK incorrectly reported the alarming news that prozac (fluoxetine) had been detected in UK drinking water [56]. Alarming as this headline may sound, there have indeed been reports of the occurrence of pharmaceuticals in water intended for human consumption (Table 3.1.7). Again, occurrence alone may not be a problem since the doses may be well below those required to exert any effect. Potable water treatment (Chapter 4.3) is also important since the presence of any contaminant in source water does not mean that it will be present in potable water supplies. It is therefore the effectiveness of any treatment process, if present, that is key to the presence of pharmaceutical compounds in drinking water.

TABLE 3.1.6

Percentage change of selected pharmaceuticals through a UK WTW [10]

Compound	% change (±%)
Clofibric acid	−91
Clotrimazole	−41
Dextropropoxyphene	91
Diclofenac	−61
Erythromycin	96
Ibuprofen	−87
Mefenamic acid	46
Paracetamol	−100
Propranolol	250
Tamoxifen	40
Trimethoprim	5

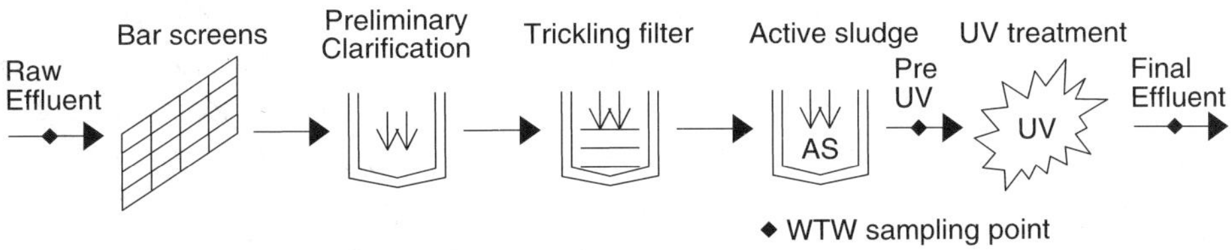

Fig. 3.1.2. Schematic of a modern UK WTW [10].

TABLE 3.1.7

Examples of pharmaceuticals that have been shown to occur in drinking water

Substance	CAS	Highest concentration (ng/L)	Location	Reference
Carbamazapine	298-46-4	258	USA	[61–63]
		20	Berlin	
		<25	The Netherlands	
Dehydronifedipine	67035-22-7	4	USA	[62]
Clofibric Acid	882-09-7	165	Berlin, Germany	[58,59,61,63]
		25–100	The Netherlands	
Primidone	125-33-7	15	Berlin, Germany	[61,64]
Phenadazone	60-80-0	400	Berlin, Germany	[64]
Propylphenazone	479-92-5	120	Berlin, Germany	[61]
1-acetyl-1-methyl-2-dimethyl-oxamoyl-2-phenylhydrazide (AMDOPH)		900	Berlin, Germany	[64]
Diclofenac	15307-86-5	<10	Berlin, Germany	[61]
Acetyl(salicyclic acid)	50-78-2	25–100	The Netherlands	[63]
Sulfamethoxazole	723-46-6			
		<25	The Netherlands	[63]

In the early 1990s, clofibric acid, the pharmacologically active metabolite of blood lipid-regulating drugs used in human medical care, was detected in ground- and drinking water samples collected in Berlin, Germany [57–60] (Table 3.1.7). This initial discovery was due to the structural similarity between clofibric acid and the herbicide mecoprop. In Berlin, concentrations of up to 165 ng/L have been reported, whilst concentrations of between 25 and 100 ng/L have been reported in drinking water collected from the Netherlands (Table 3.1.7). Finding clofibric acid in the drinking water of Berlin prompted further investigation into the groundwater wells and how pharmaceutical substances were entering them [33]. Additional work focused on whether other pharmaceutical substances were occurring in Berlin groundwater and present in drinking water following treatment [35]. In addition to clofibric acid, carbamazapine, primidone, phenadazone, propylphenadazone and dicolfenac have been detected in samples of Berlin drinking water (Table 3.1.7). Elsewhere in Germany, 17α-ethynylestradiol has been detected at <1 ng/L concentrations, whilst also in Berlin, phenazone drugs and their metabolites (e.g. 1-acetyl-1-methyl-2-dimethyl-oxamoyl-2-phenylhydrazide, AMDOPH) have been detected in drinking water samples at concentrations of up to 900 ng/L (Table 3.1.7). Outside of Germany, carbemazapine has also been detected in drinking water samples collected from the US and the Netherlands. In the US, dehydronifedipine has also been reported to occur in drinking water samples, whilst in the Netherlands the occurrence of acetyl(salicyclic acid), the widely used non-prescription analgesic commonly known as aspirin, and the antibiotic, sulfamethoxazole, have also been reported (Table 3.1.7).

The data summarised in Table 3.1.7 show that pharmaceutical substances have been reported in drinking water collected from certain countries. The source of these substances has typically been sewage, with the pharmaceutical substances persisting in the drinking water following treatment.

3.1.6 SUMMARY

There are occurrence data for a wide range of pharmaceutical compounds, whilst in certain countries (e.g. US and Germany) national monitoring surveys have been performed. These data show that many of the targeted pharmaceuticals occur in aqueous samples and that wastewater and treated wastewater is the principle source of human pharmaceutical compounds to the environment.

The occurrence of elevated concentrations of pharmaceutical compounds in the environment is of concern, however occurrence alone does not indicate that they are causing any harmful effects or whether they may have the potential to harm the environment. The data available and that generated in the future must be considered in light of available acute and chronic biological effects data, whilst where no data are available we should employ a precautionary approach.

REFERENCES

1 B. Crathorne, M. Fielding, C.P. Steel and C. Watts, *Environ. Sci. Technol.*, 18 (1984) 797.

2 C. Desbrow, E.J. Routledge, G.C. Brighty, J.P. Sumpter and M. Waldock, *Environ. Sci. Technol.*, 32 (1998) 1549.
3 K.V. Thomas, M. Hurst, P. Matthiessen and M. Waldock, *Environ. Toxicol. Chem.*, 20 (2001) 2165.
4 D. Ashton, M. Hilton and K.V. Thomas, *Sci. Total Environ.*, 333 (2004) 167.
5 Swedish Medical Products Agency, Environmental effects of pharmaceuticals, cosmetics and hygiene products, 2004.
6 D.W. Kolpin, E.T. Furlong, M.T. Meyer, E.M. Thurman, S.D. Zaugg, L.B. Barber and H.T. Buxton, *Environ. Sci. Technol.*, 36 (2002) 1202.
7 H.-R. Buser, M.D. Muller and N. Theobald, *Environ. Sci. Technol.*, 32 (1998) 188.
8 S. Weigel, J. Kuhlmann and H. Huhnerfuss, *Sci. Total Environ.*, 295 (2002) 131.
9 K.V. Thomas and M. Hilton, *Mar. Poll. Bull.*, 49 (2004) 436.
10 P.H. Roberts and K.V. Thomas, *Sci. Total Environ.*, 356 (2006) 143.
11 S. Weigel, U. Berger, E. Jensen, R. Kallenborn, H. Thoresen and H. Huhnerfuss, *Chemosphere*, 56 (2004) 583.
12 T.A. Ternes, *Water Res.*, 32 (1998) 3245.
13 OSPAR, *SPDS 30/3/7-E.* London, UK, 2003.
14 M. Hilton, K.V. Thomas, D. Ashton, *R&D Technical Report P6-012/06/TR.* Environment Agency, 2003.
15 C.M. Taylor, B. Blanchard and D.T. Zava, *Cancer Res.*, 44 (1984) 1409.
16 R. Hirsch, T.A. Ternes, K. Haberer and K.-L. Kratz, *Sci. Total Environ.*, 225 (1999) 109.
17 U. Jux, R.M. Baginski, H.-G. Arnold, M. Kronke and P.N. Seng, *Int. J. Hygiene Environ. Health*, 205 (2002) 393.
18 C.D. Metcalfe, X.-S. Miao, B. Koenig and J. Struger, *Environ. Toxicol. Chem.*, 22 (2003) 2881.
19 S. Zuehlke, U. Duennbier and T. Heberer, *Anal. Chem.*, 76 (2004) 6548.
20 M. Stumpf, T.A. Ternes, K. Haberer, P. Seel and W. Baumann, *Vom Wasser*, 86 (1996) 291.
21 F. Sacher, E. Lochow, D. Bethmann and H.-J. Brauch, *Vom Wasser*, 90 (1998) 233.
22 S. Weigel, R. Kallenborn and H. Huhnerfuss, *J. Chromatogr. A*, 1023 (2004) 183.
23 W. Ahrer, E. Scherwenk and W. Buchberger, *J. Chromatogr. A*, 910 (2001) 69.
24 N. Lindqvist, T. Tuhkanen and L. Kronberg, *Water Res.*, 39 (2005) 2219.
25 T.A. Ternes, M. Stumpf, J. Mueller, K. Haberer, R.D. Wilken and M. Servos, *Sci. Total Environ.*, 225 (1999) 81.
26 H.-R. Buser, T. Poiger and M.D. Muller, *Environ. Sci. Technol.*, 33 (1999) 2529.
27 S. Marchese, D. Perret, A. Gentili, R. Curini and F. Pastori, *Chromatographia*, 58 (2003) 263.

28 C. Hartig, T. Storm and M. Jekel, *J. Chromatogr. A*, 854 (1999) 163.
29 M. Hilton and K.V. Thomas, *J. Chromatogr. A*, 1015 (2003) 129.
30 T. Kosjek, E. Heath and A. Krbavcic, *Environ. Int.*, 31 (2005) 679.
31 B.W. Brooks, C.K. Chambliss, J.K. Stanley, A. Ramirez, K.E. Banks, R.D. Johnson and R.J. Lewis, *Environ. Toxicol. Chem.*, 24 (2005) 464.
32 T. Heberer, *Toxicol. Lett.*, 131 (2002) 5.
33 T. Heberer, U. Dunnbier, C. Reilich and H.-J. Stan, *Fresen. Environ. Bull.*, 6 (1997) 438.
34 J.V. Holm, K. Rugge, P.L. Bjerg and T.H. Christensen, *Environ. Sci. Technol.*, 29 (1995) 1415.
35 T. Heberer, *J. Hydrol.*, 266 (2002) 175.
36 R. Andreozzi, M. Raffaele and P. Nicklas, *Chemosphere*, 50 (2003) 1319.
37 B. Ferrari, N. Paxeus, R.L. Giudice, A. Pollio and J. Garric, *Ecotoxicol. Environ. Safety*, 55 (2003) 359.
38 C.D. Metcalfe, B. Koenig, D.T. Bennie, M. Servos, T.A. Ternes and R. Hirsch, *Environ. Toxicol. Chem.*, 22 (2003) 2872.
39 V. Koutsouba, T. Haberer, B. Fuhrmann, K. Schmidt-Baumler, D. Tsipi and A. Hiskia, *Chemosphere*, 51 (2003) 69.
40 X.-S. Miao, F. Bishay, M. Chen and C.D. Metcalfe, *Environ. Sci. Technol.*, 38 (2004) 3533.
41 C. Baronti, R. Curini, G. D'Ascenzo, A. Di Corcia, A. Gentili and R. Samperi, *Environ. Sci. Technol.*, 34 (2000) 5059.
42 M. Stumpf, T. Ternes, K. Haberer and W. Baumann, *Vom Wasser*, 87 (1996) 251.
43 G.R. Boyd, H. Reemtsma and D.A. Grimm, *S. Mitra Sci. Total Environ.*, 311 (2003) 135.
44 M. Carballa, F. Omil, J.M. Lema, M. Llompart, C. Garcia-Jares, I. Rodriguez, M. Gomez and T. Ternes, *Water Res.*, 38 (2004) 2918.
45 T.A. Ternes, M. Bonerz and T. Schmidt, *J. Chromatogr. A*, 938 (2001) 175.
46 C.G. Daughton and T.A. Ternes, *Environ. Health Persp.*, 107 (1999).
47 M.L. Richardson and J.M. Bowron, *J. Pharm. Pharmacol.*, 37 (1985) 1.
48 B. Halling-Sørensen, S.N. Nielsen, P.F. Lanzky, F. Ingerslev, H.C.H. Lutzhøft and S.E. Jørgensen, *Chemosphere*, 36 (1998) 357.
49 Kummerer, K. (Ed.) Emission and biodegradability of pharmaceuticals, contrst media, disinfectants and AOX from hospitals. *Pharmaceuticals in the environment*, Springer, Berlin, Germany, 2001, p. 29.
50 K. Kummerer, A. Al-Ahmad and V. Mersch-Sundermann, *Chemosphere*, 40 (2000) 701.
51 G.W. Aherne and R. Briggs, *J. Pharm. Pharmacol.*, 41 (1989) 735.
52 X.-S. Miao, B. Koenig and C.D. Metcalfe, *J. Chromatogr. A*, 952 (2002) 139.
53 X.-S. Maio and C.D. Metcalfe, *Anal. Chem.*, 75 (2003) 3731.
54 C. Hiemke and S. Hartter, *Pharmacol. Therapy*, 85 (2000) 11.

55 C. Zwiener, T.J. Gremm and F.H. Frimmel, Pharmaceutical residues in the aquatic environment and their significance for drinking water production. In: K. Kummerer (Ed.), *Pharmaceuticals in the Environment*, Springer, Berlin, Germany, 2001, p. 81.
56 M. Townsend, *Observer, UK*, 2004, 8 August.
57 T. Heberer and H.J. Stan, *Vom Wasser*, 86 (1996) 19.
58 T. Heberer and H.J. Stan, *Int. J. Environ. Anal. Chem.*, 67 (1997) 113.
59 H.J. Stan, T. Heberer and M. Linkerhagner, *Vom Wasser*, 83 (1994) 57.
60 H.J. Stan and M. Linkerhagner, *Vom Vasser*, 79 (1992) 75.
61 T. Heberer, K. Reddersen and A. Mechlinski, *Water Sci. Technol.*, 46 (2002) 81.
62 P.E. Stackelberg, E.T. Furlong, M.T. Meyer, S.D. Zaugg, A.K. Henderson and D.B. Reissman, *Sci. Total Environ.*, 329 (2004) 99.
63 A.A.M. Stolker, W. Niesing, E.A. Hogendoorn, J.F.M. Versteegh, R. Fuchs and U.A.T. Brinkman, *Anal. Bioanal. Chem.*, 378 (2004) 955.
64 K. Reddersen, T. Heberer and U. Dunnbier, *Chemosphere*, 49 (2002) 539.

Chapter 3.2

Transformation of pharmaceuticals in the environment: Photolysis and other abiotic processes

William A. Arnold and Kristopher McNeill

3.2.1 INTRODUCTION

Once released into the environment via the discharge of treated or untreated wastewater, pharmaceuticals are subject to the same potential transport and degradation processes as other organic contaminants. Pharmaceuticals differ in several respects to "traditional" environmental organic pollutants (e.g., solvents, polychlorinated biphenyls, (poly)aromatic hydrocarbons, herbicides, and insecticides). A major difference is that pharmaceutical compounds, in general, have passed through a human or animal digestive tract and, for human drugs, possibly through a wastewater treatment system. Two consequences of this pre-exposure to biochemical metabolism are that many drugs will enter the aquatic environment in a modified form and those that are unaltered share a resistance to biochemical transformation. This allows certain inferences to be drawn regarding the importance of various abiotic transformation processes for pharmaceutical compounds in the aquatic environment.

For a pharmaceutical compound to enter the aquatic environment unaltered after having passed through a human or animal digestive system, it must be resistant to acid- and enzyme-promoted hydrolysis reactions. The synthetic prostaglandin misoprostol (Cytotec) and aspirin (Fig. 3.2.1) are two examples of ester-containing drugs that are rapidly hydrolyzed in the stomach. Consequently, aspirin metabolites salicylic acid, *o*-hydroxyhippuric acid, and gentisic acid, are the forms detected in sewage influent [1].

Similarly, the compound must be resistant to other enzyme-catalyzed transformation reactions if it is to be released to the

Comprehensive Analytical Chemistry 50
Petrović and Barceló (Eds)
Volume 50 ISSN: 0166-526X DOI: 10.1016/S0166-526X(07)50011-5

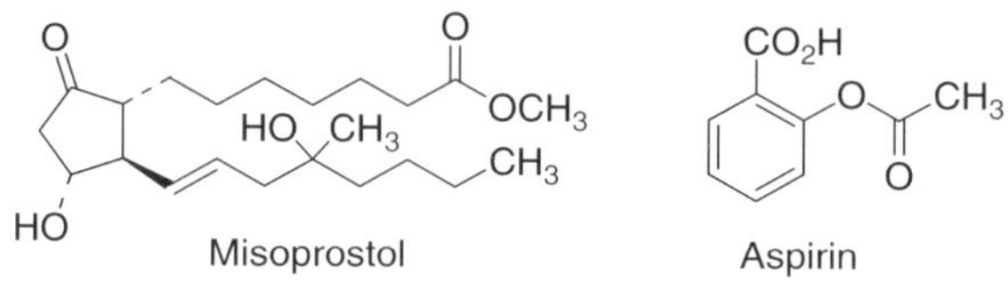

Fig. 3.2.1. Structures of misoprostol and aspirin, pharmaceuticals that are rapidly hydrolyzed in the stomach, and thus not expected to be released into the environment as the parent compound.

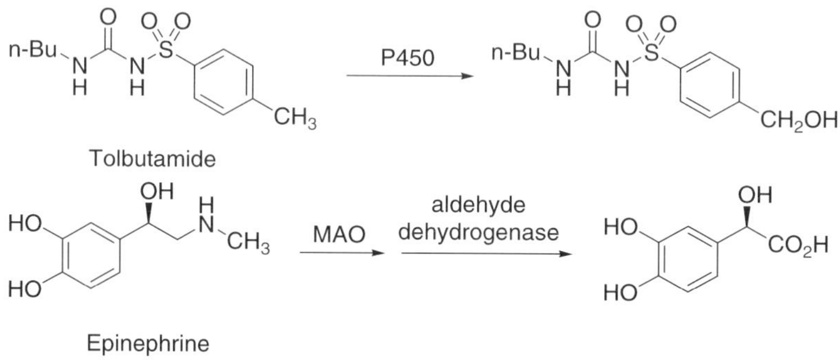

Fig. 3.2.2. Metabolic oxidation processes for tolbutamide and epinephrine, which lead to additional possibilities for pharmaceutical species being released into the environment.

environment in its original form. Two of the most important metabolic enzyme systems are cytochrome P450 enzymes located in the liver and the widely distributed monoamine oxidase (MAO) enzymes. While both enzyme systems oxidize drugs, they differ in their substrate scope and product distribution. Cytochrome P450 enzymes hydroxylate C–H bonds, while MAO enzymes oxidatively deaminate primary and secondary amines. Two examples of enzyme-mediated oxidations are shown in Fig. 3.2.2 for tolbutamide, a drug used to treat diabetes, and epinephrine [2,3].

It has been estimated that over half of all drugs are metabolized by cytochrome P450 [4]. In some cases, metabolic processing of a pharmaceutical compound is important to its activity. For example, the product of P450 metabolism of the blockbuster antihistamine loratidine (Claritin, Alavert) is the active form, desloratidine (Clarinex, Aerius), which is now marketed as an antihistamine in its own right (Fig. 3.2.3).

Another type of metabolic modification is conjugate formation, in which the parent drug or its metabolite is covalently bound to a small organic fragment. Typical conjugates are glucuronide, sulfate, acyl, methyl, and glutathione adducts. Formation of these conjugates can

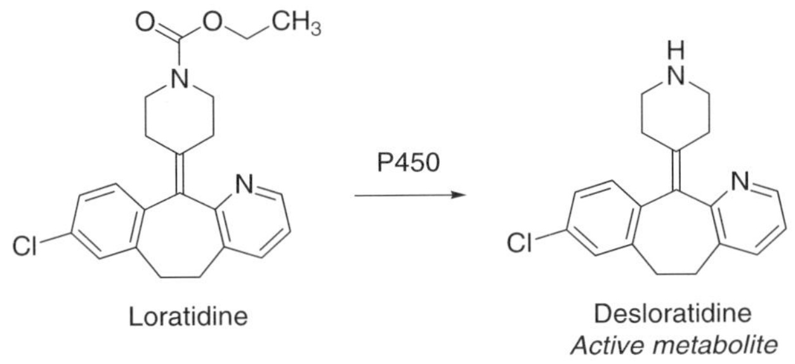

Fig. 3.2.3. Oxidation of loratidine leads to the active form of the drug. Thus, the abiotic fate of both compounds needs study.

lead to higher water solubility, which makes elimination more facile, or decreased toxicity by capping reactive groups. An example of conjugate formation is in the metabolism of moxifloxacin (Fig. 3.2.4), a fluoroquinolone antimicrobial, which forms both *N*-sulfate (M1) and acyl glucuronide (M2) conjugates [5]. Modifications of this type can complicate environmental monitoring because they are potentially reversible. Fluctuations in carbamazepine and gemfibrozil concentrations in a field study in Sweden were attributed to release of the parent drugs from their glucuronide metabolites [6].

Overall, studies of the transformations of pharmaceutical compounds in aquatic systems must consider the ingested form of the drug, active metabolites produced via hydrolysis or enzymatic reaction, and drug conjugates. Most studies to date, however, have focused on the parent form of the drug or known, active metabolites. Overall, the passage of pharmaceutical compounds through the digestive tract and their relatively long-residence time in aqueous environments within the wastewater treatment process indicates that hydrolysis reactions are unlikely to be important in the aquatic fate of most pharmaceutical compounds that reach the environment.

Pharmaceutical compounds generally also contain acidic or basic functional groups, such as carboxylic acids, phenols, and amines. The pK_a value thus becomes important in predicting environmental transport and transformation. For example, several non-steroidal anti-inflammatory drugs (NSAIDs) have carboxylic acid functionalities with pK_a values much less than 7. These compounds are thus likely to remain in the solution phase (i.e., low organic-water partitioning values), and removal by sorption to/settling of particles may be limited. Similarly, the antiepileptic/antidepressant carbamazepine, which contains an amine moiety, is well known to not be removed by any

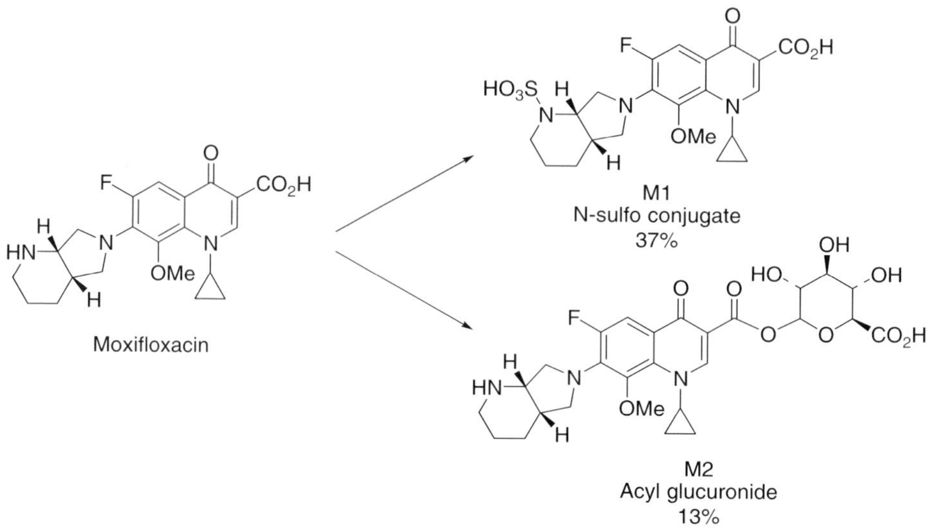

Fig. 3.2.4. Conjugate formation of moxifloxacin leading to additional species for which abiotic fate should be evaluated, particularly considering that degradation of the conjugate may lead to regeneration of the active drug in the environment.

standard wastewater treatment unit operations (over 90% of influent carbamazepine levels are generally detected in effluents, and it is one of the most commonly detected drugs in the environment) [1]. Other compounds have multiple pH-sensitive functional groups. For example, the tetracycline antibiotics are reported to have three or four pK_a values [7,8]. This leads to the possibility of protonated/positive, neutral (or zwitterionic), and deprotonated/negative forms of a drug being present depending on the pH of the specific water body. In fact, many antibiotics and antimicrobials have a pK_a at ~7.4 that is key for uptake and effectiveness. Thus, pH will be important in evaluating environmental processing of the compounds, even though they are not subject to hydrolytic reactions. The speciation of the drug will influence its partitioning behavior, as well as its light-absorbing properties.

As for biological degradation reactions, most human pharmaceuticals have passed through the biochemical wastewater treatment environment, suggesting that they are either not biodegradable or not present at sufficient concentrations to be used as a substrate by bacteria or other microorganisms in these systems. Some pharmaceutical compounds are inherently biodegradable (e.g., ibuprofen [9]), and a high percentage of such compounds is removed in wastewater treatment processes [1]. The biodegradability of antibiotics under wastewater treatment conditions has been found to vary widely, with some degraded rapidly [10] and others to a limited extent or not at all [11,12]. Field evidence has shown that biochemical oxygen demand and the degradation of selected compounds are comparable in streams [13], suggesting that biological transformation in the aquatic environment may occur.

The focus of this chapter, however, is abiotic processes that lead to the transformation of pharmaceutical compounds in environmental systems. Because pharmaceutical compounds or their degradates emitted in wastewater effluent are expected to be resistant to hydrolytic processes, direct and indirect photolysis may be the only relevant abiotic loss processes in sunlit aquatic systems. Given the rich variety of functional groups in pharmaceuticals known to be subject to direct and indirect photolysis (e.g., conjugated aromatics, nitro-compounds, furans, and phenols), a diverse set of photochemical processes are expected. Oxidative losses via reaction with mineral and humic materials will also occur in sediments or soils.

Abiotic transformations, such as oxidations via reactions with disinfectants (e.g., chlorine, ozone) or in advanced oxidation processes (i.e., reactions with hydroxyl radical), are also important in water and wastewater treatment systems (see Chapters 4.2 and 4.3). Additionally, sorption (to sediments, soils, or wastewater solids) is also an important abiotic removal pathway in natural and engineered systems (Chapter 4.1), but does not result in transformation of the parent compound. Abiotic reactions in engineered systems and non-transformative loss processes are discussed in other chapters of this book.

3.2.2 PHOTOLYSIS IN THE ENVIRONMENT

Two types of photolysis processes occur in aquatic systems: direct photolysis and indirect photolysis (Fig. 3.2.5). In direct photolysis, the target contaminant (in this case, the pharmaceutical compound) absorbs a solar photon. In an indirect photolysis mechanism, the target does not need to or is unable to absorb light because another chromophore in the system such as dissolved organic matter acts as a sensitizing species.

Direct photolysis of pharmaceuticals is initiated by photon absorption, a fact that is well known to pharmacists who dispense medication in amber bottles and advise patients to stay out of the sun when taking

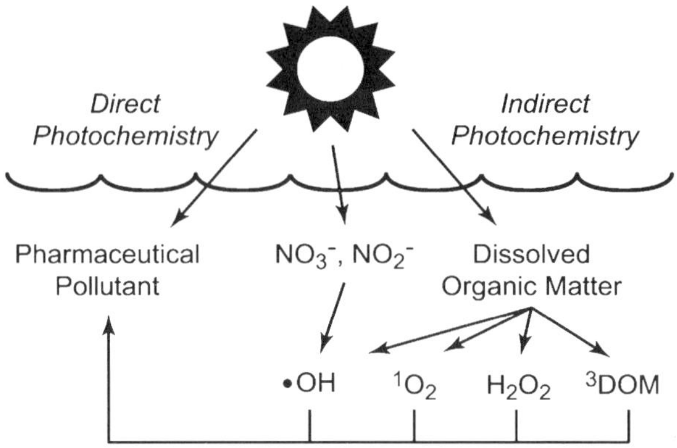

Fig. 3.2.5. Direct and indirect photolysis in aquatic systems.

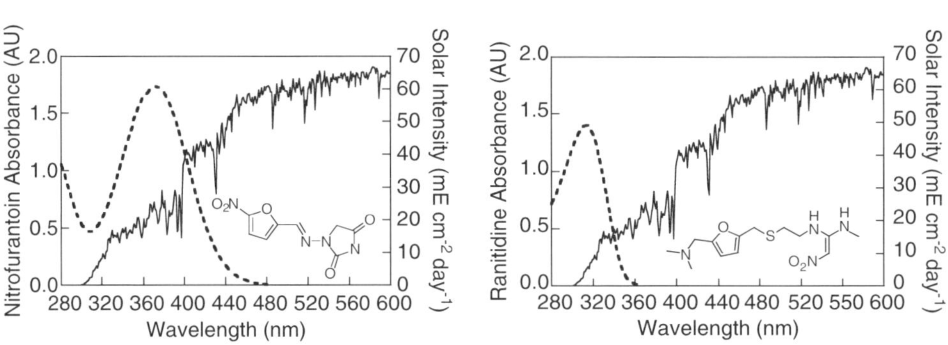

Fig. 3.2.6. Spectral overlap of sunlight emission and nitrofurantoin absorbance and ranitidine absorbance (mE is equal to millieinsteins). Nitrofurantoin and ranitidine absorption indicated by a dashed line and sunlight emission is indicated by a solid line. Spectra are based on the data from Refs. [14,15]. Sunlight emission was calculated using SMARTS [67] for Minneapolis, MN in July at noon.

certain medications. The phototoxicity of some drugs has led to an active area of research on this topic that complements the study of the environmental photochemistry of pharmaceuticals (see Section 3.2.3).

The rate of photon absorption is determined by the action spectrum, or the product of the compound's absorbance spectrum and the solar spectrum. Examples of the overlap between light absorbance and the solar spectrum for nitrofurantoin and ranitidine are shown in Fig. 3.2.6. Note that both of these compounds contain similar chromophores, a furan and a nitro group, and that nitrofurantoin, which has these groups in conjugation, has larger overlap with the solar spectrum due to the longer wavelength absorption feature.

The excited state of a molecule is short-lived and may undergo various physical or chemical relaxation processes. Physical relaxation processes, such as vibrational energy loss, energy transfer to another species, or emission of a photon lead to the regeneration of the parent compound. Only those processes that lead to chemical changes in the parent compound lead to a decrease in the concentration of the species being photolyzed. Such transformations may include fragmentation, isomerization/intramolecular rearrangement, H-abstraction, dimerization/polymerization, and electron transfer. The fraction of chemical transformation events per photon absorbed is defined as the quantum yield (Φ) for that process. This value may range from 0 to 1, but values between 0.0001 and 0.1 are common for compounds that are photodegraded on reasonable time scales (half-lives of minutes to days). The first-order rate constant for the transformation is given by Eq. (3.2.1).

$$k_{\mathrm{obs}} = 2.303 \int_{\lambda} (\Phi_{\lambda} \varepsilon_{\lambda} I_{\lambda}) \mathrm{d}\lambda \qquad (3.2.1)$$

where k_{obs}, the rate constant for direct photolysis, is equal to the product of the quantum yield, the molar absorptivity of the compound (ε), and the solar irradiance (I), integrated over all wavelengths (λ).

The photolysis rate is a function of both the rate of light absorption (i.e., the action spectrum) and the quantum yield. Note that a compound with a large spectral overlap with sunlight and a small quantum yield may be more persistent than a compound with a small spectral overlap and a large quantum yield. In the example shown in Fig. 3.2.6, ranitidine has a smaller rate of absorption than nitrofurantoin, yet both compounds degrade at similar rates due to the fact that ranitidine has a higher quantum yield [14,15]. Another example is the anti-inflammatory medication diclofenac, which absorbs very little light with wavelengths greater than 300 nm (leading to a small spectral overlap) compared with the structurally similar mefenamic acid (which absorbs wavelengths up to 400 nm), both shown in Fig. 3.2.7. The quantum yield for diclofenac is ~0.1 [16–18] while that for mefenamic acid is 0.00015 [19]. While mefenamic acid absorbs 100 times more sunlight, it is 1000 times less efficient at being transformed, resulting in a direct photolysis half-life for mefenamic acid that is ~10 times greater than that of diclofenac.

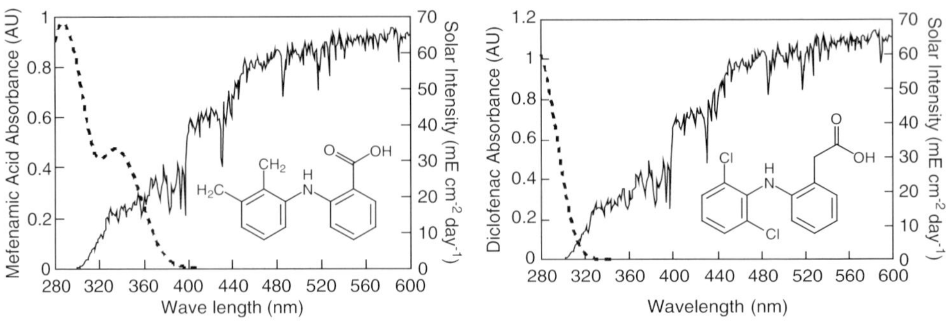

Fig. 3.2.7. Spectral overlap of sunlight emission and diclofenac absorbance and mefenamic acid absorbance (mE is equal to millieinsteins). Diclofenac and mefenamic acid absorption indicated by a dashed line and sunlight emission is indicated by a solid line. Spectra are based on the data from Ref. [19]. Sunlight emission was calculated using SMARTS [67] for Minneapolis, MN in July at noon.

All compounds, whether or not they absorb solar photons, are potentially subject to indirect photolysis. In an indirect photochemical mechanism, a sensitizer absorbs light and subsequently reacts directly with the substrate or produces a reactive intermediate that reacts with the substrate. The principal light-absorbing species in indirect photolysis is the dissolved organic matter (DOM) present in natural waters. Photoexcitation of DOM leads to the production of a variety of photochemically produced reactive intermediates (PPRIs) including the reactive oxygen species hydroxyl radicals (HO•), singlet oxygen ($^1O_2$), peroxy radicals (ROO•), and superoxide (O_2^{-}•) as shown in Fig. 3.2.5. Other PPRIs, such as triplet (excited) DOM and hydrated electrons, are also produced and can react with pharmaceutical pollutants. It should also be mentioned that while DOM is the main sensitizing species in natural waters, other light-absorbing species may also generate PPRIs, such as nitrate and nitrite that produce hydroxyl radicals in sunlight.

Reaction rates with PPRIs are dictated by the product of their steady-state concentration and their bimolecular reaction rate constant. The specificity of the PPRI (i.e., the chemical functional groups it will react with) varies widely. For example, hydroxyl radical is a non-specific oxidant that reacts with most organic compounds at diffusion controlled rates, either by hydrogen atom abstraction from sp^3 hybridized C–H bonds or addition to C–C double bonds. Steady-state concentrations of hydroxyl radical in sunlit waters range from 10^{-18} to 10^{-15} mol/L [20], leading to a disappearance rate of approximately

10^{-5}–$10^{-8}\,s^{-1}$ (half-life of ~1–1000 days) for even the most unreactive organic pollutants. At the other end of the selectivity spectrum is singlet oxygen, which only reacts with specific functional groups such as electron rich olefins, phenolates, and sulfides. The fact that it is present in higher steady-state concentrations (10^{-12}–10^{-14} mol/L, [20]) leads to reaction with singlet oxygen being the dominant photochemical loss mechanism for substrates containing these high-reactivity functional groups.

A good example is cimetidine, a photostable compound in pure water under sunlight irradiation that is rapidly degraded photochemically in DOM-containing waters. The PPRI responsible for its degradation is singlet oxygen, which reacts rapidly with cimetidine's imidazole ring [15]. Other PPRIs do not appear to be important due to the fact that the total reaction rate constant for cimetidine loss matches the calculated reaction rate constant based on the steady-state concentration of singlet oxygen and the bimolecular reaction rate constant.

Overall, the total rate constant for loss via photolysis will be the sum of the first-order direct photolysis rate constant and the second-order rate constants for reactions with PPRIs multiplied by the steady-state concentration of the appropriate PPRI species as shown in Eq. (3.2.2).

$$k_{\text{obs}} = 2.303 \int_{\lambda} (\Phi_{\lambda}\varepsilon_{\lambda}I_{\lambda})\mathrm{d}\lambda + \sum_{j} k_{i,j}[\mathrm{PPRI}_j] \qquad (3.2.2)$$

In this expression, the first term represents the rate of direct photolysis (Eq. 3.2.1). The second term represents the sum of all indirect photolysis pathways, which are each the product of the second-order rate constant for reaction of the species of interest with a PPRI, $k_{i,j}$, and the concentration of the respective PPRI. Depending on the relative importance of these two terms, compounds may react solely via direct photolysis, solely via indirect photolysis, or by a combination of the two. Thus, a continuum of photolysis reactivity is expected, as shown in Fig. 3.2.8 for several compounds for which the environmental photolysis has been studied. The exact ratio of direct to indirect processes will be a function of the quantum yield, the light-absorbing properties of the compound, the magnitude of the rate constants for the reactions with the PPRIs, and the PPRI steady-state concentrations. The rate of light absorption by the compound and the PPRI steady-state concentration will be a function of time of day (light intensity) as well as environmental conditions.

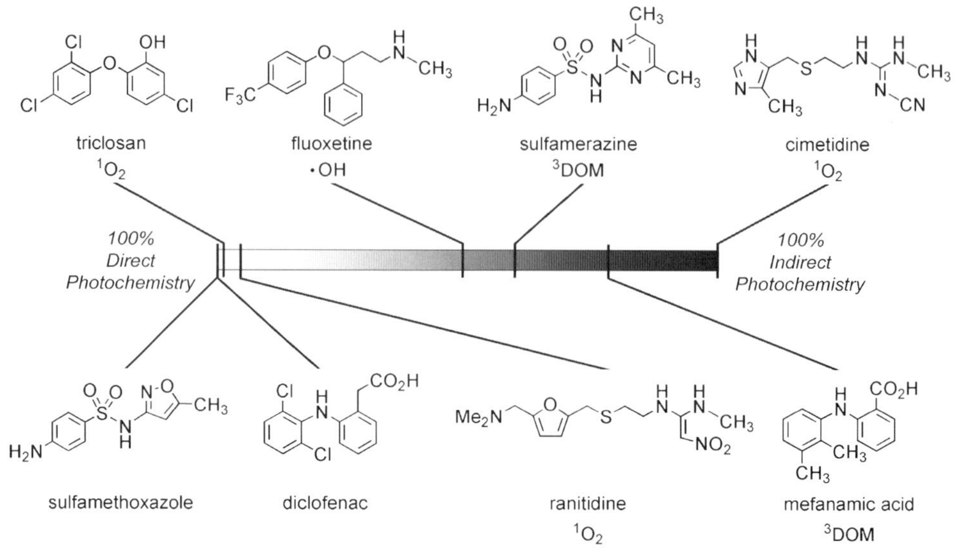

Fig. 3.2.8. Continuum of photoreactivity with respect to direct and indirect processes. The contribution of direct photolysis decreases from left to right. The PPRI listed is the species that contributes most to indirect photolysis for the particular compound. Photolysis rates/mechanisms are compiled from Refs. [15,17,19,31,34,36,38].

3.2.3 PHOTOSTABILITY AND PHOTOTOXICITY OF PHARMACEUTICAL COMPOUNDS

The photochemistry of pharmaceutical compounds has long been studied, but generally not in the context of a transformation process in aquatic systems. As recently reviewed by Boreen et al. [21], dozens of investigations have focused on the photostability of pharmaceutical compounds. In some cases, these studies have used conditions relevant to environmental conditions (i.e., an aqueous solution of the compound irradiated with wavelengths of light >300 nm). For these studies, the findings are applicable to environmental systems/predictive models if the quantum yield is reported (additional information, such as reactor geometry and specific solution conditions may also be required). Even if the information cannot be directly applied to environmental systems, such studies provide guidance for environmental photochemistry studies in that a screening has already been performed that reveals whether a compound is susceptible to direct photolysis.

Photoxicity studies may also provide insight into susceptibility to photochemical processes. The phototoxic/photoallergic reaction to NSAIDs is due to photoexcitation of these species in the blood stream [22]. Miskoski et al. [23] found that the photodegradation of tetracyclines in biological media could be attributed to reactions with singlet oxygen. Additionally, tetracyclines may serve as photosensitizers, for superoxide, hydrogen peroxide, and hydroxyl radical have been observed in illuminated solutions of several tetracyclines and their derivatives [24]. Such processes may give rise to additional products if the PPRIs generated react with the pharmaceutical serving as the sensitizer, although this is unlikely in natural waters.

3.2.4 DIRECT AND INDIRECT PHOTOLYSIS OF PHARMACEUTICAL COMPOUNDS IN AQUATIC SYSTEMS

3.2.4.1 Direct photolysis

Not surprisingly, studies to date have demonstrated that compound structure (which dictates UV-visible light absorbance) has a dramatic effect on the rate of direct photolysis. The two drugs rantitidine and cimetidine, both sold over the counter as antacids have quite similar chemical structures (Fig. 3.2.9). As shown in Fig. 3.2.6, ranitidine absorbs light above 290 nm and reacts rapidly via direct photolysis [15]. Cimetidine, however, does not absorb light above 290 nm and thus is only transformed via indirect photolysis (singlet oxygenation) [15]. Thus, compounds within a therapeutic class and/or with similar chemical structures may react at very different rates or via different processes.

For determination of rates of direct photolysis, the two fundamental parameters necessary are the molar absorptivity of the compound as a function of wavelength (ε_λ, which is easily measured with a UV/visible spectrophotometer) and the quantum yield. The quantum yield may be determined as a function of wavelength, but, in general, an overall quantum yield is reported. Table 3.2.1 provides a list of pharmaceutical

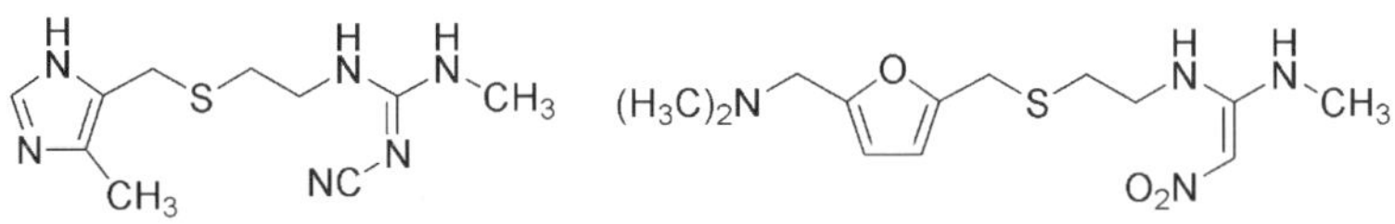

Fig. 3.2.9. Chemical structures of cimetidine and ranitidine.

TABLE 3.2.1

Environmental direct photolysis studies of pharmaceuticals

	CAS number	Quantum yield (Φ)[a]	Reference
Non-steroidal anti-inflammatory drugs			
Diclofenac	15207-86-5	0.094	[18]
		0.037	[16]
		0.13	[17]
Ibuprofen	15687-27-1	nm	[18,68]
Ketoprofen	22071-15-4	nm	[68]
Mefenamic acid	61-68-7	0.00015	[19]
Naproxen	22204-53-1	0.036	[18]
		0.026	[68]
Immunosuppressant/ anti-histmatic drugs			
Cimetidine	51481-61-9	nm	[15]
Ranitidine	66357-35-5	0.0053	[15]
Glucocorticosteroids			
Prednisone	53-03-2	nr	[49]
Psychotropic agents			
Carbamazepine	298-46-4	0.0000477	[16]
		nr	[69]
Fluoxetine	54910-89-3	0.000042	[34]
Paroxetine	61869-08-7	0.000271-0.000377	[35]
Andrenergics			
Propranolol	525-66-6	0.00222	[10]
		0.0052	[68]
Serum lipid-reducing agents			
Clofibric acid	882-09-7	0.002	[10]
		0.00553	[16]
		nr	[69]
Fenofibric acid	42017-89-0	nr	[26]
Hormones			
Estradiol	50-28-2	0.0048	[68]
Estriol	50-27-1	0.0048	[68]
Estrone	53-16-7	0.0296	[68]
Ethinyl estradiol	57-63-6	0.0048	[68]
Dermatologicals			
Triclosan	3380-34-5	0.02–0.93[b]	[33]
		0.31 (313 nm, pH 11)	[66]
		0.12 (sunlight, pH 8)	[31]

TABLE 3.2.1 *(continued)*

	CAS number	Quantum yield (Φ)[a]	Reference
Antibacterials: Sulfa drugs[c]			
Sulfamethoxazole	723-46-6	0.09 (anionic)	[36]
		0.00429	[16]
Sulfisoxazole	127-69-5	0.07 (anionic)	[36]
Sulfamethizole	144-82-1	0.05 (anionic)	[36]
Sulfathiazole	72-14-0	0.07 (neutral)	[36]
		0.40 (anionic)	
Sulfamethazine	57-68-1	0.0003 (neutral)	[38]
		0.005 (anionic)	
Sulfamerazine	127-79-7	0.00023 (neutral)	[38]
		0.003 (anionic)	
Sulfadiazine	68-35-9	0.0012 (anionic)	[38]
		nr	[70]
Sulfachloropyridazine	80-32-0	0.0023 (anionic)	[38]
Sulfadimethoxine	122-11-2	0.00004 (anionic)	[38]
		nr	[70]
Antibacterials: Aromatic derivatives			
Chlortetracycline	57-62-5	nr	[40]
Oxytetracycline	79-57-2	nr	[70]
Tetracycline	60-54-8	nr	[39]
β-lactam antibiotics			
Amoxicillin	26787-788-0	0.00447–0.00597	[10]
Macrolide antibiotics			
Lincomycin	154-21-2	0.00011–0.00013	[45]
Antibacterials: Heterocyclic drugs			
Flumequine	42835-25-6	nr	[70]
Furazolidone	67-45-8	nr	[70]
Ofloxacin	82419-36-1	0.0000779	[16]
Oxolinic acid	14698-29-4	nr	[70]
Trimethoprim	738-70-5	nr	[70]
Diagnostic agents			
Iomeprol	78649-41-9	nr	[69]
Iopromide	73334-07-3	0.00109	[27]

nr, not reported; nm, not measured because indirect photolysis was deemed to be the dominant process.
[a]Quantum yields are given when calculated by the cited works.
[b]Quantum yield is a function of pH and irradiation wavelength.
[c]Quantum yields for sulfa drugs are pH dependent. The quantum yield reported is for the dominant protonation state(s) expected at pH 7.

compounds for which direct photolysis has been studied in the context of environmental photochemistry. Quantum yields for additional compounds are also available in the photostability/phototoxicity literature (see review of Boreen et al. [21]). If indirect photolysis reactions have been studied for the compounds listed, they are presented in Section 3.2.4.2. Table 3.2.1 is organized into therapeutic categories [25].

Pharmaceuticals that absorb solar photons are subject to direct photolysis. As described above, the rate of loss is a product of the light absorption rate and the quantum yield. Thus, a large quantum yield does not guarantee fast disappearance nor is the reverse necessarily true. Compounds that do not absorb solar photons (ibuprofen, ketoprofen, and cimetidine) will not react via this pathway. Additionally, some compounds have both a small spectral overlap integral and a small quantum yield. In this case, direct photolysis is most certainly unimportant. This combination is likely true for the fibrate drugs bezafibrate, gemfibrozil, and fenofibrate, as $<10\%$ loss of these compounds was observed over 200 h of irradiation [26]. Overall, direct photolysis half-lives may be a short as minutes and as long as tens of days. Note that while the direct photolysis of nearly 40 compounds has been studied to evaluate their environmental persistence (not all have reported the quantum yield, however), this is still only a small subset of the total number of pharmaceutical compounds and derivatives released. Additional work is necessary to determine quantum yields for all compounds of concern (if the relevant values are not available in other photochemistry literature) so that inclusion of direct photolysis in environmental fate models is possible.

As compound structure influences the rate of reaction, structure also dictates the mechanism by which the direct photolysis proceeds. For example, compounds with aromatic halogens tend to react via carbon–halogen bond cleavage. The X-ray contrast agent iopromide loses iodide upon photolysis [27]. Diclofenac (non-steroidal anti-inflammatory) and triclosan (antimicrobial), a chlorinated diphenylamine and a chlorinated diphenyl ether respectively, react via similar, yet disparate pathways as depicted in Fig. 3.2.10. Diclofenac products arise from decarboxylation and by loss of chlorine and formation of an inter-ring carbon–carbon bond to form a carbazole [28,29]. Subsequent products are then formed via further degradation of these initial two intermediates. Triclosan also gives rise to a variety of products resulting from removal of a chlorine atom (and replacement by either –OH or –H) [30]. Other products, including 2,4-dichlorophenol, a US EPA priority pollutant, arise from cleavage of the ether bond [31]. Lastly, several

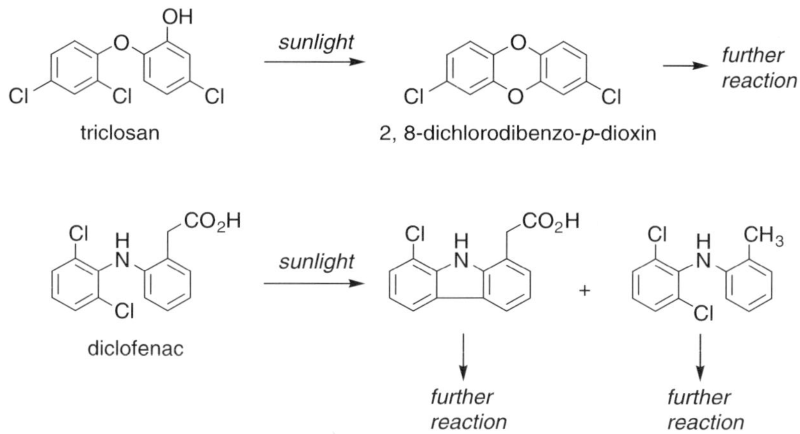

Fig. 3.2.10. Selected photolysis pathways for triclosan and diclofenac [29,31–33,48]. Note that the yield of the dioxin from triclosan is ~5%.

studies have demonstrated that an intramolecular attack of the hydroxyl displacing the chlorine of the neighboring ring leads to the production of 2,8-dichlorodibenzo-*p*-dioxin [31–33].

Not surprisingly, drugs from the same therapeutic class may react via similar pathways. Products of fluoxetine and paroxetine photolysis appear to arise from initial cleavage of an ether bond [34,35]. Sulfa drugs with 5-membered heterocyclic groups react via cleavage of the sulfur–nitrogen bond to give rise to sulfanilic acid [36] after an initial isomerization [37]. The sulfa drugs with 6-membered heterocyclic groups, however, react via an SO_2 extrusion mechanism [38]. Tetracyclines have a particularly complex product suite with fragmentation [39] and oxygenation products being observed [40].

3.2.4.2 Indirect photolysis

The simplest experiment to determine whether indirect photochemistry is important in the photochemical fate of a given pharmaceutical is to compare the photodegradation rates in pure water and a sample containing natural water components. Acceleration of the degradation in the presence of these components, such as dissolved organic matter or nitrate ion, has been taken as *prima facie* evidence for the involvement of PPRIs. For these experiments, two types of approaches have been taken that employ either a representative natural water sample or synthetic field water. The synthetic systems have the advantage of being able to test the relative importance of various components, which can help to identify the important PPRIs in the system. A powerful implementation of this principle to pharmaceuticals and pesticides employs photolysis in an ensemble of synthetic field water samples followed by deconvolution of the contributing processes by multivariate analysis [41,42].

The role of specific PPRIs can be ruled out based on quenching experiments and kinetic arguments. For example, the addition of isopropyl alcohol quenches oxygen-based radicals, such as ROO• and •OH, which undergo hydrogen atom abstraction reactions with the alcohol's weak methine C–H bond. The absence of quenching of the photolysis in the presence of a large excess of isopropyl alcohol is good evidence against such radicals. Similarly, the concentrations of PPRIs can often be estimated using chemical probe methods [43]. Combining this information with the measured bimolecular reaction rate constant between the pharmaceutical of interest and a suspected PPRI allows one to judge whether such a process is kinetically competent.

Through experiments of this type, to date PPRIs have been established as being important in the photochemical fate of a small selection of pharmaceutical compounds. As more studies are conducted in this area, this group is expected to expand given the variety of PPRIs and their collective scope of reactivity.

A role for hydroxyl radical has been proposed in the photodegradation of carbamazepine [16,44], clofibric acid [16,18], ofloxacin [16], ibuprofen [18], fluoxetine [34], and lincomycin [45]. Hydroxyl radical reacts with high-rate constants (10^9–$10^{10}\,M^{-1}\,s^{-1}$) with most organic compounds, but has low steady-state concentrations in sunlit aquatic systems (10^{-18}–10^{-15} M). These two facts combine to give environmental half-lives due to reaction with hydroxyl radical from ones to thousands of days. Thus, compounds that are thought to be degraded predominantly via hydroxyl radicals in sunlit waters are typically unreactive with respect to direct photolysis or other degradation pathways.

Carbamazepine and clofibric acid are two compounds that fit this archetype. Both compounds were found to have conservative behavior in a seasonal field study in Lake Greifensee, Switzerland meaning that no removal process, including photodegradation, was found to compete with the physical flushing rate ($t_{1/2} = 120$ days) [46]. In a separate study, carbamazepine and clofibric acid were estimated to have environmental half-lives due to direct photolysis of 100–400 days and 25–100 days, respectively [16]. The contribution of natural water constituents was found to be complicated for carbamazepine. The presence of 10 mg/L nitrate ion reduced the half-life by a factor of two, but 5 mg/L humic acid increased the half-life by a factor of four [16]. Addition of both nitrate and humic acid decreased the half-life of clofibric acid [16]. In another study of clofibric acid, which found that the photodegradation rate was enhanced by a factor of two in Mississippi River water and that the enhancement was removed in the presence of isopropyl alcohol, confirmed the importance of radical species [18]. Based on kinetic arguments, this study concluded that hydroxyl radical could only account for a portion of the enhancement and that other radicals were likely involved.

The study of fluoxetine photodegradation is notable because it is the rare case in which product identification has been used to bolster the case for involvement of hydroxyl radical [34]. Four products were identified through LC–MS–MS experiments (Fig. 3.2.11). Three of these products were identified in the slow direct photolysis reaction (compounds I, II, and III in Fig. 3.2.11). Under hydroxyl radical-producing conditions, a new product (IV) was formed and two of the direct photolysis products (I and II) were found to be formed faster. The new product, IV, the result of aromatic ring hydroxylation, is reasonably direct evidence for hydroxyl radical involvement.

Recent studies have implicated reaction with singlet oxygen as an important component of the photochemical fate of two histamine H_2-receptor antagonists, drugs used to combat gastric acidity (Fig. 3.2.9). The more dramatic example of the two is cimetidine, which is highly

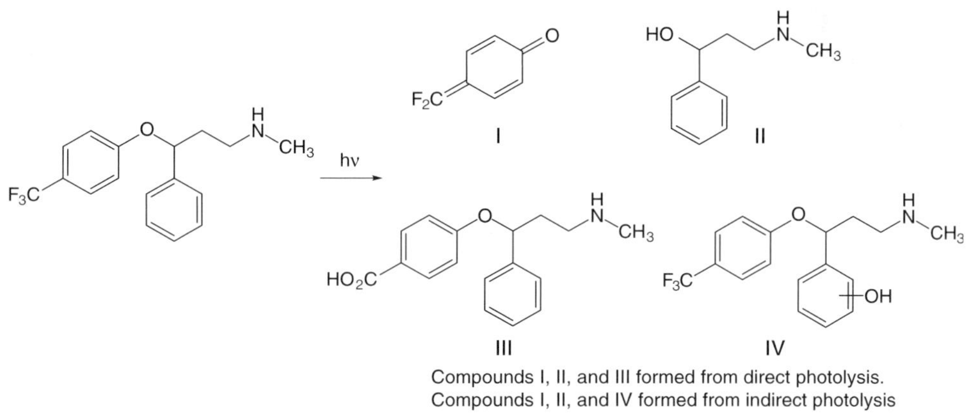

Fig. 3.2.11. Products identified in the photochemical degradation of fluoxetine [34]. Hydroxyl radical-mediated processes are hypothesized for products I, II, and IV.

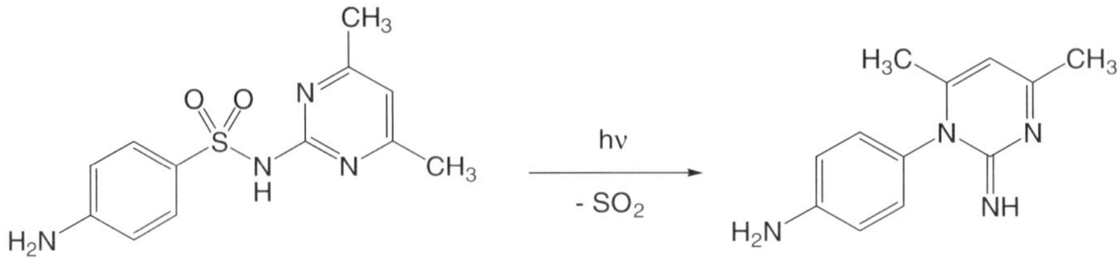

Fig. 3.2.12. Reaction of sulfamethazine via SO_2 extrusion. The process is enhanced by triplet-dissolved organic matter [38].

photoreactive in DOM-containing water and essentially photostable in pure water [15]. Reaction with $^1O_2$ is the key degradation process, accounting for all of the cimetidine loss. The photodegradation of ranitidine is enhanced by 10% in Mississippi River water (compared to the direct photolysis observed in distilled water) due to reaction of $^1O_2$ with the furan and sulfide moieties [15]. The nitro-containing acetamidine functional group is responsible for the significant direct photochemical component of ranitidine.

The involvement of triplet-excited states of dissolved organic matter (^{3}DOM) has been shown in the photodegradation of mefenamic acid [19] and a set of sulfa drugs [38]. The photodegradation rate of mefenamic acid is doubled in Mississippi River water, and the role of ^{3}DOM was established through the addition and removal of the quenchers isoprene and oxygen, respectively. Similarly, sulfa drugs with six-membered heterocyclic substituents were shown to have photodegradation enhancements in natural water (Lake Josephine, Minnesota, USA) versus deionized water. The enhancements were shown to become more pronounced in the absence of oxygen and to be quenched by added isoprene. This stands in contrast to a prior study of sulfa drugs with five-membered heterocyclic substituents for which indirect photochemistry was found not to be important [36]. For the six-membered heterocycle-substituted sulfa drugs, both the direct and indirect photolysis pathways led to SO_2 extrusion to yield novel products (Fig. 3.2.12).

3.2.5 ROLE OF pH

Many pharmaceuticals have weak acid or weak base functional groups as part of their structure that lead to thermodynamically accessible ionized and unionized forms at physiological pH values. The ionized forms aid aqueous solubility, while the non-ionized forms can diffuse more easily through lipophilic membranes. Consequently, numerous pharmaceuticals have environmentally relevant pK_a values, with ionized and non-ionized species being present under natural conditions. Each of these species must be considered to fully anticipate the pharmaceutical's aquatic chemical behavior. This situation is exemplified by a set of sulfa drugs, whose photochemistry is strongly modulated by their speciation.

In Table 3.2.2, there are four sulfa drugs, sulfamethoxazole, sulfisoxazole, sulfamethizole, and sulfathiazole, each have pK_a values ranging from 5.0 to 7.2 and each undergo direct photochemical degradation [36]. These compounds have distinct direct photochemical degradation rates for the non-ionized and anionic forms. It is striking that no trend is observed with half of the sulfa drugs degrading more quickly and half degrading more slowly in the anionic form. The photodegradation rate is a product of the light absorption rate and the quantum efficiency of the degradation process, both of which change as the

TABLE 3.2.2

Structures, pK_a values, and calculated direct photochemical rate constants for selected sulfa drugs containing five-membered heterocyclic substituents.

Compound	Structure	pK_a	k_{direct}[a] ($10^5\ s^{-1}$)	
			SH	S^-
Sulfamethoxazole	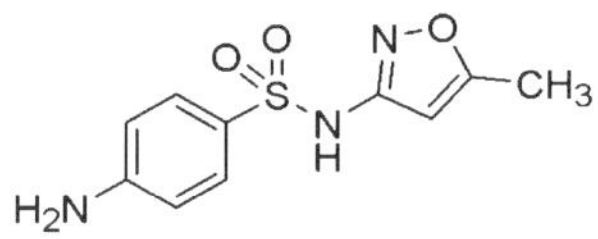	5.7	6	0.8
Sulfisoxazole	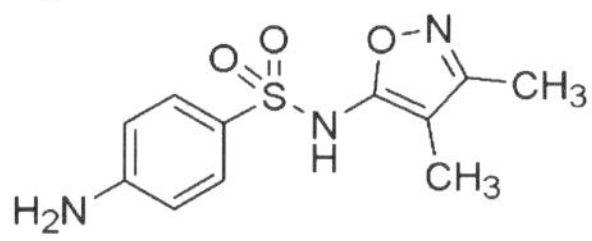	5.0	7	2.1
Sulfamethizole	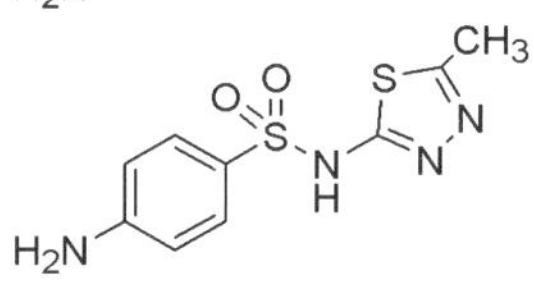	5.3	⩽0.3	1.3
Sulfathiazole	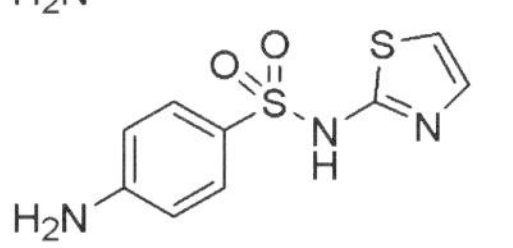	7.2	3.1	14

Source: Adapted from Boreen et al. [36].
[a]Calculated for 45°N latitude, noon, mid-spring sunlight.

compound becomes deprotonated. Kwon and Armbrust have also noted pH effects on direct photolysis quantum yield [35].

The effect of speciation of photochemical reaction rate applies not only to direct processes, but also to indirect processes. For example, it is well documented that singlet oxygen reacts quickly with phenolate ions, but not with phenols [47].

3.2.6 IMPORTANCE OF PRODUCT IDENTIFICATION

As described above, photolysis reactions often lead to multiple reaction products produced by various competing/parallel pathways. Thus, identification of photolysis reaction products is laborious and usually requires isolation of the compounds and/or analysis with sophisticated mass spectrometers. The studies that have gone through the effort of identifying reaction products (e.g., [29,31,33–40,48–50]) have provided several examples that demonstrate why this effort is of ultimate importance in determining the impacts of pharmaceutical compounds in aquatic systems.

For example, Della Greca et al. [49] demonstrated that while prednisone had low acute and chronic toxicity to aquatic producers and primary consumers, the photoproducts had elevated toxicity. In some cases chronic toxicity values were two orders of magnitude smaller (i.e., 100 times more potent) than that of prednisone. Toxicity effects of antibiotics to algae have also been measured, and some of the observed toxicity may be attributed to photoproducts, as the testing was performed under illumination [51]. Toxicity may also be reduced upon photolysis as was demonstrated for iopromide [27], and no toxic effects were observed for the diuretic furosemide or its photoproducts [52].

Residual antibiotic activity of degradation products is also of concern. Some dehydrated products of tetracyclines retain some potency [53]. Additionally, photodegradation products of the fluoroquinolone antibiotic compound ofloxacin also retain antimicrobial activity [54]. These findings may be exceptions, but further study is necessary to determine the potency of reaction products. Because many drugs are site specific, it is commonly assumed that antibacterial function may be lost upon minor structural changes. In support of this hypothesis, Wammer et al. [55] found that antibacterial activity was lost upon photolysis of triclosan and a variety of sulfa drugs, including the sulfa drugs that were degraded via SO_2 extrusion where the photoproducts have similar overall structure to the parent compound. The fact that antimicrobial activity is reduced or eliminated, however, does not mean that the products are not of environmental concern. The products may still be acutely or chronically toxic to aquatic organisms. Additionally, products may also be produced that are of toxic concern to higher organisms, as is the case for triclosan, where photolysis results in production of a dioxin and 2,4-dichlorophenol [31].

3.2.7 OTHER ABIOTIC TRANSFORMATIONS

In soils, sediments, and groundwaters, the absence of light precludes photolysis as an abiotic loss process. Other abiotic processes, however, may lead to pharmaceutical transformation in these situations. For example, it has been demonstrated that manganese oxide minerals are capable of mediating the oxidation of variety of pharmaceuticals and personal care products including the antibacterials triclosan and chlorophene [56], carbadox and olaquindox [57], and fluoroquinolones [58]. For each of these compounds, a specific functional group is present that is oxidized by the mineral. A phenol group (triclosan, chlorophene), an *N*-oxide (carbadox, olaquindox), or an amine (fluoroquinolones, specifically, the piperazine moiety) is the site of reaction, as shown in Fig. 3.2.13. Reaction products retain the core structure of the drug, and thus it will be important to evaluate the antibacterial activity of these reaction products.

Another potential reaction leading to removal of pharmaceuticals containing amine groups is coupling with organic matter. A variety of

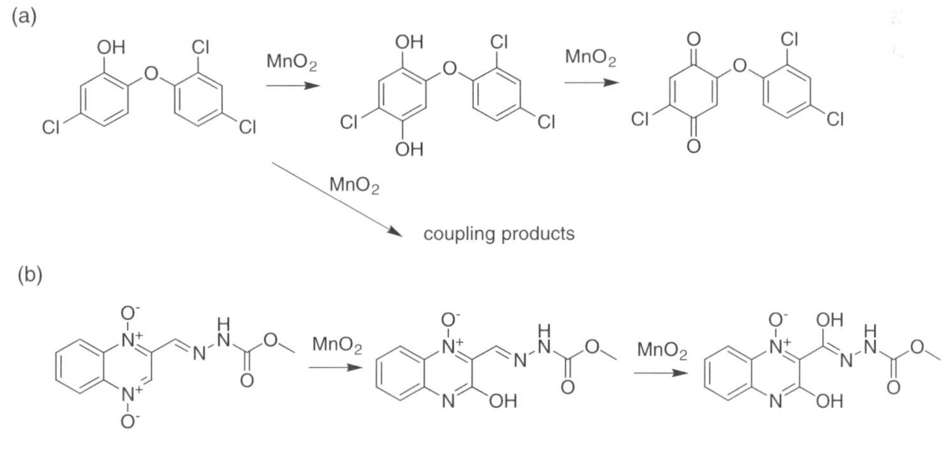

Fig. 3.2.13. Oxidation products of triclosan (a) and carbadox (b) produced via reactions with manganese oxides. Adapted from Refs. [56,57].

sulfa drugs have been shown to cross-couple with model humic constituents [59]. Such coupling would decrease the availability and slow transport of these compounds. Binding to sediment was also a major loss mechanism for paracetamol that contains an acetamide moiety [60].

An additional interesting case arises with the tetracyclines. These compounds have been detected in wastewater/environmental matrices [61,62], yet are known to undergo (pH dependent) isomerizations, epimerizations (i.e., change in chirality), and hydrolysis reactions [63]. The reaction products are also subject to these same reactions, some of which lead to reformation of the parent compound. Also of note is that some of the reaction products also show antibacterial activity and are able to inactivate tetracycline resistant bacteria [53], again indicating that transformation of the parent compound does not necessarily eliminate ecological threats/pressures.

3.2.8 FIELD STUDIES

Studies conducted to date under field conditions indicate that photolysis will be an important loss process for pharmaceuticals in surface waters. Lam et al. [50] dosed a mixture of eight pharmaceutical compounds (acetaminophen, atorvastatin, caffeine, carbamazepine, levofloxacin, sertraline, sulfamethoxazole, and trimethoprim) into pond water in outdoor microcosms meant to simulate shallow lake water. It was determined that photolysis was the major loss process in the system and that hydrolysis and biodegradation were not important processes over the time scale of the experiment. In field mesocosms, the photolysis of enrofloxacin led to the formation of ciprofloxacin, but over the period of exposure (30 days) no effects of the drugs on the bacterial community were observed [64]. Photolysis was also identified as one of the loss processes for four tetracycline antibiotics in field microcosms [65]. Finally, concentration profiles of triclosan in the epilimnion of Lake Greifensee, Switzerland could only be accurately predicted if photolysis was included as a loss process [66]. Similar results were found for diclofenac [17,28,46], and phototransformation was also suggested as a loss process for naproxen and ketoprofen in surface waters [46].

ACKNOWLEDGMENTS

The efforts of W.A.A. and K.M. focused on the photochemistry of pharmaceutical compounds have been supported by the National Institutes for Water Resources/USGS National Water Quality Competitive Grants Program, the National Research Initiative of the USDA Cooperative state Research, Education and Extension Service (grant number 2003-35107-13830), the Camille and Henry Dreyfus Foundation, and the University of Minnesota Center for Urban and Regional Affairs. The comments and aid of Anne Boreen, Kristine Wammer, Betsy Edhlund, and Jeffrey Werner are also appreciated.

REFERENCES

1 T.A. Ternes, *Water Res.*, 32 (1998) 3245.
2 R.C. Thomas and G.J. Ikeda, *J. Med. Chem.*, 9 (1966) 507.
3 H. Blashko, *Pharmacol. Rev.*, 6 (1954) 23.
4 F.P. Guengerich, *AAPS J.*, 8 (2006) E101.
5 H. Stass and D. Kubitza, *J. Antimicrob. Chemoth.*, 43 (1999) 83.
6 D. Bendz, *J. Hazard. Mater.*, 122 (2005) 195.
7 C.R. Stephens, K. Murai, K.J. Brunings and R.B. Woodward, *J. Am. Chem. Soc.*, 78 (1956) 4155.
8 H.A. Duarte, S. Carvalho, E.B. Paniago and A.M. Simas, *J. Pharm. Sci.*, 88 (1999) 111.
9 M.L. Richardson and J.M. Bowron, *J. Pharm. Pharmacol.*, 37 (1985) 1.
10 R. Andreozzi, V. Caprio, C. Ciniglia, M. De Champdore, R. Lo Giudice, R. Marotta and E. Zuccato, *Environ. Sci. Technol.*, 38 (2004) 6832.
11 K. Kummerer, A. Al-Ahmad and V. Mersch-Sundermann, *Chemosphere*, 40 (2000) 701.
12 T. Junker, R. Alexy, T. Knacker and K. Kummerer, *Environ. Sci. Technol.*, 40 (2006) 318.
13 D. Morrall, D. McAvoy, B. Schatowitz, J. Inauen, M. Jacob, A. Hauk and W. Eckhoff, *Chemosphere*, 54 (2004) 653.
14 B.L. Edhlund, W.A. Arnold and K. McNeill, *Environ. Sci. Technol.*, 40 (2006) 5422.
15 D.E. Latch, B.L. Stender, J.L. Packer, W.A. Arnold and K. McNeill, *Environ. Sci. Technol.*, 37 (2003) 3342.
16 R. Andreozzi, M. Raffaele and P. Nicklas, *Chemosphere*, 50 (2003) 1319.
17 H.-R. Buser, T. Poiger and M.D. Mueller, *Environ. Sci. Technol.*, 32 (1998) 3449.
18 J.L. Packer, J.J. Werner, D.E. Latch, K. McNeill and W.A. Arnold, *Aquat. Sci.*, 65 (2003) 342.
19 J.J. Werner, K. McNeill and W.A. Arnold, *Chemosphere*, 58 (2005) 1339.
20 T. Mill, *Chemosphere*, 38 (1999) 1379.
21 A.L. Boreen, W.A. Arnold and K. McNeill, *Aquat. Sci.*, 65 (2003) 320.
22 G. Condorelli, G. De Guidi, S. Giuffrida and L.L. Costanzo, *Coord. Chem. Rev.*, 125 (1993) 115.

23 S. Miskoski, E. Sanchez, M. Garavano, M. Lopes, A.T. Soltermann and N.A. Garcia, *J. Photochem. Photobiol. B*, 43 (1998) 164.
24 J.P. Martin Jr. and K.L. Colina, *J. Bacteriol.*, 169 (1987) 2516.
25 A. Albini and E. Fasani, *Spec. Pub. Royal Soc. Chem.*, 225 (1998) 1.
26 M. Cermola, M. DellaGreca, M.R. Iesce, L. Previtera, M. Rubino, F. Temussi and M. Brigante, *Environ. Chem. Lett.*, 3 (2005) 43.
27 T. Steger-Hartmann, R. Laenge, H. Schweinfurth, M. Tschampel and I. Rehmann, *Water Res.*, 36 (2002) 266.
28 T. Poiger, H.-R. Buser and M.D. Muller, *Environ. Toxicol. Chem.*, 20 (2001) 256.
29 A. Ageuera, L.A. Perez Estrada, I. Ferrer, E.M. Thurman, S. Malato and A.R. Fernandez-Alba, *J. Mass Spectrom.*, 40 (2005) 908.
30 I. Ferrer, M. Mezcua, M.J. Gomez, E.M. Thurman, A. Aguera, M.D. Hernando and A.R. Fernandez-Alba, *Rapid Commun. Mass Spectrom.*, 18 (2004) 443.
31 D.E. Latch, J.L. Packer, W.A. Arnold and K. McNeill, *Environ. Toxicol. Chem.*, 24 (2005) 517.
32 A. Kanetoshi, H. Ogawa, E. Katsura, H. Kaneshima and T. Miura, *J. Chromatogr.*, 454 (1988) 145.
33 D.E. Latch, J.L. Packer, W.A. Arnold and K. McNeill, *J. Photochem. Photobiol. A*, 158 (2003) 63.
34 M.W. Lam, C.J. Young and S.A. Mabury, *Environ. Sci. Technol.*, 39 (2005) 513.
35 J.-W. Kwon and K.L. Armburst, *Environ. Toxicol. Chem.*, 23 (2004) 1394.
36 A.L. Boreen, W.A. Arnold and K. McNeill, *Environ. Sci. Technol.*, 38 (2004) 3933.
37 M.W. Lam and S.A. Mabury, *Aquat. Sci.*, 67 (2005) 177.
38 A.L. Boreen, W.A. Arnold and K. McNeill, *Environ. Sci. Technol.*, 39 (2005) 3630.
39 H. Oka, Y. Ikai, N. Kawamura, M. Yamada, K. Harada, S. Ito and M. Suzuki, *J. Agric. Food Chem.*, 37 (1989) 226.
40 P. Eichhorn and D.S. Aga, *Anal. Chem.*, 76 (2004) 6002.
41 S.S. Walse, S.L. Morgan, L. Kong and J.L. Ferry, *Environ. Sci. Technol.*, 38 (2004) 3908.
42 M.W. Lam, K. Tantuco and S.A. Mabury, *Environ. Sci. Technol.*, 37 (2003) 899.
43 N.V. Blough and R.G. Zepp. In: C.S. Foote, J.S. Valentine, A. Greenberg and S.F. Liebman (Eds.), *Active Oxygen in Chemistry*, Chapman and Hall, New York, 1995, p. 280.
44 R. Andreozzi, R. Marotta, G. Pinto and A. Pollio, *Water Res.*, 36 (2002) 2869.
45 R. Andreozzi, M. Canterino, R. Lo Giudice, R. Marotta, G. Pinto and A. Pollio, *Water Res.*, 40 (2006) 630.

46 C. Tixier, H.P. Singer, S. Oellers and S.R. Mueller, *Environ. Sci. Technol.*, 37 (2003) 1061.
47 P.G. Tratnyek and J. Hoigne, *Environ. Sci. Technol.*, 25 (1991) 1596.
48 T. Poiger, F.G. Kari and W. Giger, *Environ. Sci. Technol.*, 33 (1999) 533.
49 M. DellaGreca, A. Fiorentino, M.R. Iesce, M. Isidori, A. Nardelli, L. Previtera and F. Temussi, *Environ. Toxicol. Chem.*, 22 (2003) 534.
50 M.W. Lam, C.J. Young, R.A. Brain, D.J. Johnson, M.A. Hanson, C.J. Wilson, S.M. Richards, K.R. Solomon and S.A. Mabury, *Environ. Toxicol. Chem.*, 23 (2004) 1431.
51 B. Halling-Sorensen, *Chemosphere*, 40 (2000) 731.
52 M. Isidori, A. Nardelli, A. Parrella, L. Pascarella and L. Previtera, *Chemosphere*, 63 (2006) 785.
53 B. Halling-Sorensen, G. Sengelov and J. Tjornelund, *Arch. Environ. Contam. Toxicol.*, 42 (2002) 263.
54 J. Sunderland, C.M. Tobin, L.O. White, A.P. MacGowan and A.J. Hedges, *Drugs*, 58 (1999) 171.
55 K.H. Wammer, T.M. LaPara, K. McNeill, W.A. Arnold and D.L. Swackhamer, *Environ. Toxicol. Chem.*, 25 (2006) 1480.
56 H. Zhang and C.-H. Huang, *Environ. Sci. Technol.*, 37 (2003) 2421.
57 H. Zhang and C.-H. Huang, *Environ. Sci. Technol.*, 39 (2005) 593.
58 H. Zhang and C.-H. Huang, *Environ. Sci. Technol.*, 39 (2005) 4474.
59 H.M. Bialk, A.J. Simpson and J.A. Pedersen, *Environ. Sci. Technol.*, 39 (2005) 4463.
60 D. Loeffler, J. Roembke, M. Meller and T.A. Ternes, *Environ. Sci. Technol.*, 39 (2005) 5209.
61 S. Yang, J. Cha and K. Carlson, *J. Chromatogr. A*, 1097 (2005) 40.
62 D.W. Kolpin, E.T. Furlong, M.T. Meyer, E.M. Thurman, S.D. Zaugg, L.B. Barber and H.T. Buxton, *Environ. Sci. Technol.*, 36 (2002) 1202.
63 T. Soeborg, F. Ingerslev and B. Halling-Sorensen, *Chemosphere*, 57 (2004) 1515.
64 C.W. Knapp, L.A. Cardoza, J.N. Hawes, E.M.H. Wellington, C.K. Larive and D.W. Graham, *Environ. Sci. Technol.*, 39 (2005) 9140.
65 H. Sanderson, F. Ingerslev, R.A. Brain, B. Halling-Sorensen, J.K. Bestari, C.J. Wilson, D.J. Johnson and K.R. Solomon, *Chemosphere*, 60 (2005) 619.
66 C. Tixier, H.P. Singer, S. Canonica and S.R. Mueller, *Environ. Sci. Technol.*, 36 (2002) 3482.
67 C.A. Gueymard, Florida Solar Energy Center, Technical Report No. FSEC-PF-270-95, Cocoa, Florida, 1995.
68 A.Y.-C. Lin and M. Reinhard, *Environ. Toxicol. Chem.*, 24 (2005) 1303.
69 T.E. Doll and F.H. Frimmel, *Chemosphere*, 52 (2003) 1757.
70 B.T. Lunestad, O.B. Samuelsen, S. Fjelde and A. Ervik, *Aquaculture*, 134 (1995) 217.

Chapter 3.3

Ecotoxicity of pharmaceuticals

Oliver A.H. Jones, Nikolaos Voulvoulis and John N. Lester

3.3.1 INTRODUCTION

During the last three decades, research on the impact of chemical pollution has focused almost exclusively on the conventional "priority" contaminants (i.e. heavy metals). Today, these compounds are less relevant for many first world countries since emissions have been substantially reduced through the adoption of appropriate legal measures and the elimination of many of the dominant pollution sources. The focus has consequently switched to compounds that are present in lower concentrations but which nevertheless may have the ability to cause harm.

The contamination of surface waters by pharmaceutical chemicals has raised concern among environmental scientists in recent years because of their potential to induce adverse biological effects, especially in aquatic environments. While all drugs must, by law, receive considerable pharmacological and clinical testing during development, there has been a paucity of research on their environmental behaviour. Therefore, any deleterious effects that have been observed for most pharmaceuticals have only been seen under laboratory conditions. As a result, the potential ecological effects associated with the presence of these compounds in the environment have been largely ignored and, while there are an increasing number of papers on this topic, their toxicity to organisms is not well documented in the scientific literature. However, the growing, worldwide importance of reducing potential impacts on water supplies has ensured that this issue has been steadily gaining attention in recent years within both the academic community and the general public. For example, the UK Natural Environment Research Council (NERC) strategy for "Science and a Sustainable Future" highlights the importance of the protection of water resources and

Comprehensive Analytical Chemistry 50
Petrović and Barceló (Eds)
Volume 50 ISSN: 0166-526X DOI: 10.1016/S0166-526X(07)50012-7

emphasises the need for a better understanding of the sources, pathways and transfer rates of environmental contaminants.

The term "pharmaceutical" (from the Latin *pharmaceuticus* and the Greek *pharmakeutikos*) may be defined as a chemical used for diagnosis, treatment (cure/mitigation), alteration or prevention of disease, health condition or structure/function of the body [1]. The term "medicine" (as in prescription-only medicine) is sometimes used to distinguish therapeutic drugs from recreational and other drugs (such as opiates) that are used illegally. Thus, terms commonly used in the literature such as pharmaceutically active compounds (PhACs), and pharmaceuticals and personal care products (PPCPs) are somewhat general, catch all, terms for an extremely broad group of compounds with wide ranging physical and chemical characteristics.

The scale of the potential problem is therefore very large, for example there are more than 3000 individual pharmaceutical substances currently licensed for use in the UK alone [2]. Clearly, it is not feasible to test the environmental toxicity of all the compounds that might be found in the environment. Therefore, some form of selection process is needed to narrow down interest to those compounds likely to do most harm, either through their sheer volume of use (e.g. painkillers) or their potential for toxicity (e.g. anti-cancer drugs).

Pollution by pharmaceuticals can occur in concentrations of parts per billion (ppb), or parts per trillion (ppt) (where 1 ppt equates to 1 ng/L). Previously these levels have been considered too low to cause any detrimental environmental effects. However, although these concentrations are indeed very low, many chemicals have been shown to have effects on aquatic life at similar concentrations, for example tributyltin (TBT) [3]. Therefore, it is likely some pharmaceuticals also have the potential to cause detrimental environmental effects, even at these very low levels [4].

Numerous studies have been undertaken to evaluate the risk of other potentially harmful chemicals. Many of which have, eventually, either been banned completely, or had their use severely restricted (the case of TBT being a good example [5]). However, compared with other pollutants, the sources of pharmaceuticals are likely to be much more difficult to control and, due to their importance to human health and the economy, it is highly unlikely that they will be replaced or banned. Therefore, it is evident that the evaluation of the risks posed by environmental contamination by drug compounds is of high priority, especially in view of the lack of current knowledge.

Pharmaceutical dose response relationships are usually very well documented in their intended targets, be they human or animal.

However, there is uncertainty on the range of nontarget species that have been tested, with factors such as dosage, timing of exposure and response, pharmacokinetic action and the presence (or absence) of other chemicals or stressors [6]. Often the type or magnitude of effects at lower doses is predicted from higher doses. This is further complicated by the observation that non-target effects can vary for a given species among drugs of the same therapeutic class, as well among different species of the same genus for the same drug. This means that the approach of assessing the ecological risk of a substance on a class-by-class, or genus-by-genus, basis may be untenable and extrapolations of effects from higher concentrations may not necessarily have any relevance to what may happen at lower concentrations [7].

The range of animals used in tests of pharmaceutical toxicity is narrow, with a large proportion of mammalian tests undertaken on rodents, and others limited to algae, *Daphnia*, and/or fish [8]. This increases the uncertainty where environmental impacts on other species are of concern. In addition intra species variation and other factors, such as gender, may result in different sensitivity and dose responses for the same tests [9]. The critical exposure and timing of any measurements of response may also cause significant variation in the results of any dose response assessments. Additionally, it is likely that the developmental stage of organisms is more sensitive to drugs than the more mature stage. However, the critical time in development may not be clearly defined in different organisms [10].

Because of such difficulties, dose response in wild animals has often been extrapolated from that observed for common laboratory animals [11]. Even though, the demonstrated differences in response between species usually make this inadvisable. These sorts of tests have, in the past, failed to predict many unexpected toxic effects such as egg-shell thinning in birds after exposure to DDT, or the extreme toxicity of organotin compounds to mollusks [12]. More recently, Griffin et al. demonstrated that in terms of both renal tissue and urine composition, wild mammals differed significantly from laboratory rats [13]. In particular, the wild animals were found to have higher concentrations of triglycerides in their kidneys compared with rats. The authors postulated that this may have important toxicological consequences, since many environmental contaminants are highly lipophilic and hence may bioaccumulate in wild animals to a greater extent that predicted from studies on laboratory animals. If data from laboratory animals cannot be extended to wild species, this would have profound implications for environmental toxicology and possibly related environmental policy and/or legislation.

One method to predict how well laboratory tests predict results on wild organisms could be Receiver Operation Characteristic (ROC) analysis. This has been successfully used in the biomedical and military fields for several decades [14]. It allows for the analysis of the accuracy of a diagnostic test, in terms of both its sensitivity (probability of correctly identifying positive cases) and its specificity (probability of correctly excluding negative cases). The ROC itself can be represented as a graph, by plotting the fraction of true positives against the fraction of false positives predicted from a particular test.

The underlying assumption of ROC analysis is that a diagnostic variable (e.g. ELISA test values) is used to discriminate between two mutually exclusive states of tested organisms. It has recently been applied, with some success, to assess how well in vitro bioassays can predict estrogenicity in vivo [14]. It could therefore potentially be a useful tool for measuring how successfully the results of in vitro assays conducted in the laboratory predict the effects of a particular pollutant (or mixture of pollutants) in the wider environment [14].

There are, however, known mammalian effects for certain drug classes. For example, various antiepileptic drugs, e.g. phonation, valproate and carbamazepine (the latter of which is frequently identified in the environment) are becoming more recognised as potential human neuroteratogens [15].

An interesting characteristic of many of the chemicals involved in this type of pollution is that they do not necessarily need to be persistent to cause negative effects. This is because their high transformation and removal rates can be offset by their continuous introduction into the environment. Organisms can thus be exposed to low doses throughout their life, rather than over a limited time as in laboratory tests. This is one reason why there is an increasingly widespread consensus that this kind of contamination may require legislative action sooner rather than later [16]. Unlike many other potential pollutants, however, there are at present no consent standards on concentrations for most of pharmaceuticals that can be discharged to the environment [17].

Aside from their potential continual introduction into the environment, some pharmaceuticals are actually quite resistant to biodegradation, since this is necessary to give the compound time to perform the required pharmacological action in the body. This resilience to degradation means that these compounds may be concentrated in animal tissue. Bioconcentration (uptake via the surrounding phase) and biomagnification (uptake via food) are important factors influencing the extent of bioaccumulation [18]. Bioconcentration of estrone has been demonstrated in *Daphnia magna* feeding on the alga *Chorella vulgaris* [19] and Schwaiger et al. [20] reported that the bioconcentration factor of diclofenac in rainbow trout (*Oncorhynchus mykiss*) were 12-2732 in the liver, 5-971 in the kidney, 3-763 in the gills and 0.3–69 in the muscle, depending on the initial exposure concentrations. Histopathological examinations of the diclofenac-exposed fish revealed changes in the kidney and the gills. The lowest observed effect concentration (LOEC) at which both renal lesions and alterations of the gills occurred was 5 μg/L. In contrast, there were no observable changes in the liver, gastro-intestinal tract or spleen in either exposed fish or control individuals. This LOEC level is in the range of concentrations of diclofenac reported in wastewater effluent. However, diclofenac is known to breakdown quite quickly in the environment, especially as a result of photodegradation [21]. Environmental concentrations of the drug are therefore much lower than those found in wastewater and the potential risk is therefore also likely to be lower.

Bioaccumulation itself is an important factor for risk assessments as the increase in tissue concentration may induce adverse affects on biota and also increase the exposure of predators via dietary intake in the food chain [19]. At present, few studies have investigated the bioaccumulation of drugs or their potential for effects in the higher trophic levels [22]. One exception is that of diclofenac, accumulating in the food of vultures (which is discussed in more detail later [23]). Bioaccumulation of fluoxetine, sertraline and the selective serotonin reuptake inhibitor metabolites norfluoxetine and desmethylsertraline has also been reported in fish [24]. Trophic transfer of oestrogens has been demonstrated by fish feeding on *Artemia* [25] and one study has even shown that the water flea (*Moina macrocopa*) can be used as a novel carrier of the antibiotic norfloxacin to fish [26]. Bioaccumulation of oestrogens has also been reported [18] but as yet there have been a very few corresponding studies on pharmaceuticals.

Of course, the intake of any type of exogenous compound in sufficient quantity, by any species, may interfere with the regulation of metabolic systems and induce adverse, or even fatal, effects. The main problem is arriving at the definition of sufficient quantity. This is not often clearly defined and may vary substantially between species, as well as individuals [27]. Owing to the low (but continuous) levels of pharmaceuticals introduced into the environment, acute toxic effects are unlikely (though not impossible) and hence testing exclusively for them in the first instance is seen as unsatisfactory by many researchers

[28,29]. Instead chronic effects occurring over a period of time are more probable but studies of this type may take several years to prove the issue conclusively. In the remainder of this chapter, laboratory based, acute and chronic toxicity data, as well as studies concerned with the effects of pharmaceuticals on a variety of different organisms are examined.

It should be noted that this work is intended to be a brief summery and general introduction to the area for the non-specialist reader. In general, very little is known about the possible counterparts of human target receptors of pharmaceuticals in invertebrates but for those interested in exploring the topic in more detail, a good synopsis of the modes of actions of different pharmaceutical classes on humans and the potential for effects on similar target receptors and biomolecules in lower organisms is given in Fent et al. [30] and hence is not repeated here.

3.3.2 ACUTE EFFECTS

Since most pharmaceuticals were designed to affect mammalian physiology it is not known what effects they could have on other types of organisms. A major stumbling block to this type of research is that pharmaceuticals were generally never designed to have any intended effects on wildlife. However, as noted by Daughton and Ternes, even if the mode of action is known, this incorrectly assumes that other modes of action are non-existent or minimal [31]. Knowledge as to what types of effects to look for is therefore limited and information on their environmental toxicity is generally insufficient. This situation is not helped by current regulatory guidance, which only requires pharmaceuticals to undergo standard acute toxicity tests (this is often only required for algae, *Daphnia* and fish) unless there is good reason to believe the compound may bioaccumulate [2]. Indeed, this aspect has been highlighted by a number of researchers who have considered that when assessing the environmental risks of pharmaceuticals chronic effect testing should be used because conventional acute toxicity tests (e.g. the Ames test) may be inappropriate [32] and that perhaps alternatives, such as the "Green Screen" could be used instead [33].

The hazards posed to wildlife from pharmaceuticals are not well known [34]. However acute toxic effects seem doubtful, unless the organism is exposed to an extremely high concentration of a drug (or drugs) which is unlikely to happen short of a spill at a manufacturing plant. Nevertheless, it is impossible to rule out acute effects entirely without further testing since certain species may be particularly susceptible to certain classes of drugs.

A current case in point is the decline in vulture populations in the Indian subcontinent. The species most seriously affected are the oriental white-backed vulture (*Gyps bengalensis*), the long-billed vulture (*G. indicus*) and the slender-billed vulture (*G. tenuirostris*). Populations of *G. Bengalensis* have declined by more than 95% in the last 10 years making them now critically endangered. Populations of G. *indicus*, *G. tenuirostris* and related vulture species have also suffered catastrophic losses. The sheer scale of the decline of these populations has no parallel in birds since the disappearance of peregrine falcons (*Falco peregrinus*) and other predatory birds in the 1960s due to exposure to DDT [30].

At first thought to be the result of particularly virulent pathogen, the cause was eventually tracked down to the use of the diclofenac (a painkiller and anti-inflammatory) to treat lameness and mastitis (inflammation of the udder) in cattle [23]. Vultures feeding on the unburied carcasses of animals previously treated with diclofenac would also ingest the drug. Although not exposed to particularly large doses of the compound, they proved particularly susceptible to it, suffering renal failure, visceral gout (the accumulation of uric acid throughout the body cavity following kidney malfunction) and eventually death, in a comparatively short period of time. This is the first known case of a pharmaceutical causing major ecological damage over a large geographic area and threatening species with extinction. There are also human health risks associated with this problem. Without vultures to dispose of them, the number of unattended animal carcasses in India has increased. This has in turn lead to an increase in the numbers of feral dogs, which can now exploit this previously unavailable resource. These animals can be dangerous and spread diseases such as rabies. In March 2005, the Indian Government announced its support for a ban on the veterinary use of diclofenac and its replacement with meloxicam (another NSAID) which has been found to be much less toxic to vultures [35] and from 12 August 2006 the production and importation of veterinary diclofenac has no longer been permitted in India. The neighbouring country of Nepal has also recently deregistered the drug, preventing its manufacture and import. Both countries are promoting the use of meloxicam as a safe alternative.

Many other drugs have unexpected effects on non-target organisms. For instance, acetaminophen (paracetamol) has been shown to be useful in controlling the brown tree snake (*Boiga irregularis*) in Guam, at dose of 40 mg of the active compound [36,37] and has also been shown to inhibit oestrogen-induced vitellogenin production in isolated trout liver cells. Approximately 50% inhibition achieved with 0.05 mM acetaminophen, while using 0.3 mM acetaminophen inhibited secreted vitellogenin levels educed to undetectable levels [38]. The drug may therefore alter other oestrogen-regulated processes. At low mg/L concentrations the beta blocker propranolol can affect the growth and reproduction of Japanese medaka (*Oryias latipes*), a small freshwater fish [39] and may also cause germinal vesicle breakdown (GVBD) of full-grown folliculated oocytes of the catfish (*Clarias batrachus*), cultured in vitro, in a dose-dependent manner [40]. Certain anti-depressants (including fluoxetine) have been show to effect spawning in shellfish at concentrations in the μg/L range [41,42].

Brooks et al. [43] studied the waterborne and sediment toxicity of fluoxetine to several species. Average LC_{50} values for the Daphnia species *Ceriodaphnia dubia* and *Daphnia magna*, and the fathead minnow (*Pimephales promelas*) were 234 μg/L, 820 μg/L, and 705 μg/L, respectively. Growth of the algae *Pseudokirchneriella subcapitata* and *C. dubia* fecundity were decreased by fluoxetine treatments of 14 μg/L and 223 μg/L, respectively. The survival of *O. latipes* survival was not affected by fluoxetine exposure up to a concentration of 8.9 μg/L. While an LC_{50} of 15.2 mg/kg was estimated for the non-biting midge *Chironomus tentans*. Survival of the amphipod *Hyalella azteca* was not affected up to 43 mg/kg fluoxetine sediment exposure. The lowest observed effect concentrations for *C. tentans* and *H. azteca* were 1.3 and 5.6 mg/kg, respectively. This indicates fluoxetine, and possibly related compounds, can have effects on organisms living on/in the sediment and the water column at relatively low concentrations.

Cleuvers [44] evaluated the ecotoxicological potential of ten prescription drugs against aquatic organisms using the cladoceran *D. magna*, the chlorophyte *Desmodesmus subspicatus* and the macrophyte *Lemna minor*. The endpoints were taken to be immobilisation for *Daphnia* and inhibition of the average growth rate for *D. subspicatus* and *L. minor*. For most of the substances, toxicities were moderate, with values of EC_{50} in the range from 10 to 100 mg/L, or even substantially higher. *L. minor* was the most sensitive test species for the majority of all compounds tested. Tests with combinations of various pharmaceuticals revealed greater effects than those expected from the individually measured effects. Clofibric acid and carbamazepine were found to act by a non-specific mode of action (non-polar narcosis). In *Daphnia*, the combination effect of these substances was found to follow the concept of concentration addition, while in the algae test the concept of independent action could be used to calculate the toxicity of the mixture. The anti-inflammatory drugs, diclofenac and ibuprofen, have also been found to act by non-polar narcosis and to follow the concept of concentration addition in the algae test as well as in the *Daphnia* test. However, the measured toxicities of the tested pharmaceuticals in this study indicated that acute toxic effects arising from the presence of these substances in the aquatic environment are very unlikely to occur.

Halling-Sørensen (2000) demonstrated that *Microcystis aeruginosa* (a freshwater cyanobacteria) was approximately two to three orders of magnitude more sensitive to antibiotics than the green algae *Selenastrum capricornutu*. This is unsurprising considering that cyanobacteria are closely related to the pathogenic bacteria antibiotics were originally intended to affect. Having not been routinely exposed to these compounds in the past they have much lower resistance to them. Indeed, dosing with antibiotics is often used to remove unwanted cyanobacteria from fish tanks. Since cyanobacteria occupy the lower trophic levels within foodwebs, changes in their population could have an indirect but significant effect on the rest of the freshwater community. In addition, research on other compounds, such as organotins, has demonstrated that some xenobiotics can have an effect on the hormonal systems of aquatic organisms at concentrations of only a few nanograms per litre [5]. Recently published results also clearly demonstrated effects on the endocrine systems of the fish exposed to sewage effluent due to synthetic contraceptives present in the low ng/L range [45].

Henry et al. [46] examined the acute toxicity of five selective serotonin reuptake inhibitors (SSRIs)—fluoxetine, fluvoxamine, paroxetine, citalopram and sertraline—in the daphnid *C. dubia*. For each SSRI, the 48-h median lethal concentration (LC_{50}) was determined in three static tests with neonate *C. dubia*. These ranged from 0.12 to 3.90 mg/L and the order of toxicity of the compounds was (lowest to highest): citalopram, fluvoxamine, paroxetine, fluoxetine, sertraline.

Jones et al. [28] assessed the 25 most used pharmaceuticals in the National Health Service (NHS) in England in 2000 for their environmental effects. Predicted environmental concentrations (PECs) for the aquatic environment were calculated using conservative assumptions and all PECs exceeded the 1 ng/L cut-off value that requires a phase II assessment to be conducted according to EU guidelines [47]. The calculation of predicted no-effect concentration (PNEC) based on aquatic toxicity data from the literature was possible for eleven of the pharmaceuticals. PNECs were predicted with ECOSAR (Ecological Structure Activity Relationships) for twelve of the remaining fourteen but no data were available for two of the compounds. The PEC/PNEC ratio exceeded one for paracetamol, amoxycillin, oxytetracycline and mefenamic acid. No toxicity data was available for the terrestrial environment and no assessment was made. Similar work was previously performed by Stuer Lauridsen et al. [48] for the 25 most used pharmaceuticals in the primary health sector in Denmark. Here all PECs exceeded 1 ng/L. Measured concentrations were in general within a factor of 2–5 of PECs and ranged from approximately 0.5 to 3 μg/L for nine of the pharmaceuticals tested. The calculation of PNECs based on aquatic ecotoxicity data was possible for six of the pharmaceuticals. The PEC/PNEC ratio exceeded one for ibuprofen, acetylsalicylic acid, and paracetamol while for oestrogens the PEC/PNEC ratio approached one when non-standard tests were used. The ratio was below one for oestrogens (standard test), diazepam and digoxin. Again, for the terrestrial compartment, toxicity data were not available, and so no assessment was carried out.

ECOSAR is a computer program that estimates the toxicity of chemicals to aquatic organisms, such as fish (both in fresh and saltwater), invertebrates and algae in the absence of test data. It does this by using structure–activity relationships (SARs) to predict the aquatic toxicity of chemicals based on the similarity of their molecular structures to other compounds for which the aquatic toxicity is known.

To date, over 150 SARs have been developed for more than 50 chemical classes. The SARs contained within the program (many of which have been validated) are based on test data and express the correlations between a compound's physical and chemical properties and its aquatic toxicity. Using the measured aquatic toxicity and K_{ow} values, regression equations can be developed for a class of chemicals. Toxicity values may then be calculated by inserting the estimated K_{ow} into the regression equation and correcting the resultant value for the molecular weight of the compound. This is a technique routinely used by the US Environmental Protection agency (EPA) to estimate the aquatic toxicity of chemicals being reviewed in response to pre-manufacture notices mandated by Section 5 of the US Toxic Substances Control Act (TSCA). These sorts of methods are helpful in estimating potential toxicity or the behaviour of a compound in the environment but they cannot replace in vivo or in vitro assays [30]. They would also be unlikely to pick up effects such as the extreme toxicity of diclofenac to vultures discussed earlier.

Jos et al. [49] evaluated the toxicity of carbamazepine, using six ecotoxicological model systems with 18 endpoints evaluated at different exposure time periods. These included the immobilisation of *D. magna*, bioluminescence inhibition in the bacterium *Vibrio fischeri*, growth inhibition of the alga *Chlorella vulgaris*, and micronuclei induction and root growth inhibition in the plant *Allium cepa*. Cell morphology, neutral red uptake, total protein content, MTS metabolisation, lactate dehydrogenase leakage and activity, and glucose-6-phosphate dehydrogenase activity were studied in the salmonid fish cell line RTG-2. The total protein content, LDH activity, neutral red uptake, and MTT metabolisation in Vero monkey kidney cells were also investigated. The most sensitive system to carbamazepine was the Vero cell line, followed by *C. vulgaris*, *V. fischeri*, *D. magna*, *A. cepa*, and RTG-2 cells. EC_{50} values were found to be much higher than the concentrations previously reported in aquatic systems although chronic and synergistic effects with other chemicals could not be excluded.

Lalumera et al. [50] studied the occurrence and effects of flumequine and oxytetracycline in sediments sampled from two trout farms and three sea-bass farms and in their surrounding environments in Italy. Flumequine was found to have the highest toxicity in a bioluminescence assay with EC_{50} values varying within the range of 12–15 mg/L, while the EC_{50} values for oxytetracycline were in the range of 121–139 mg/L.

Nunes et al. [51] evaluated the acute toxicity of three therapeutic agents (diazepam, clofibrate and clofibric acid) and a widely used detergent (sodium dodecyl sulfate, SDS) to three aquatic species from distinct trophic levels, namely the mosquito fish *Gambusia holbrooki*, the crustacean *Artemia parthenogenetica* and the marine algae *Tetraselmis chuii*. The toxicity ranking for the compounds (as determined by the 50% LC_{50} values for the two animal species and 50% inhibitory concentration (IC_{50}) for the alga) was found to be, clofibric acid > SDS > diazepam > clofibrate for *G. holbrooki*, Clofibric acid > clofibrate > SDS > diazepam for *A. parthenogenetica* and Clofibric acid > clofibrate > SDS > diazepam for *T. chuii*. Although both LC and IC_{50} values were in the mg/L range in all cases, the three test organisms showed distinct thresholds of toxicity for the compounds studied. These differences show that the inherent differences of organisms used in toxicity testing must be considered when evaluating the results of such tests and their implications for wild organisms and aquatic ecosystems as a whole.

A previous study by Nunes et al. [52] reported on acute toxic effects of the same three compounds to *A. parthenogenetica*. It focused specifically on oxidative stress parameters, namely levels of, total and selenium-dependent glutathione-peroxidase (GPx), glutathione reductase (GRed), total superoxide dismutase (SOD) and glutathione-S-transferases (GSTs). The effects of the substances on lipid peroxidation thiobarbituric acid reactive substances, (TBARS) and soluble cholinesterases (ChE) were also investigated.

Diazepam was found to cause a significant inhibition of ChE (LOEC = 7.04 mg/L) and total GPx activities. SDS exposure resulted in a decrease in the activity of both ChE (LOEC = 8.46 mg/L) and GRed (LOEC = 4.08 mg/L). Both clofibrate and clofibric acid induced significant decreases in Se-dependent GPx, with LOEC values of 176.34 and 3.09 mg/L, respectively. Clofibrate also caused a slight increase of TBARS content in *A. parthenogenetica* homogenates. These results indicate exposure to all the tested compounds induced changes in the cellular redox status of *A. parthenogenetica*. Diazepam was also shown to have the capability of interfering with neurotransmission through the inhibition of ChE. However, as with many similar studies reported here, all these effects were only observed when the organism was exposed to concentrations of the test substances in the mg/L range. As previously stated, levels this high are unlikely to occur in the environment and such results would seem to indicate the relative harmlessness of the drug to the aquatic compartment.

Wollenberger et al. [53] investigated the acute toxicity of nine antibiotics used both therapeutically and as growth promoters in intensive farming on *D. magna*. The effects of the antibiotics, metronidazole, olaquindox, oxolinic acid, oxytetracycline, streptomycin, sulfadiazine, tetracycline, tiamulin and tylosin, were tested in accordance to the ISO (1989) and OECD (1996) standard procedures. The acute toxicities ($48\,h^{-1}$ EC_{50} value, mg/L) in decreasing order were oxolinic acid (4.6), tiamulin (40), sulfadiazine (221), streptomycin (487), tylosin (680) and oxytetracycline (~1000). NOECs were 340 mg/L for tetracycline and 1000 mg/L for metronidazole and olaquindox. Toxic effects on reproduction generally occurred at concentrations one order of magnitude below the acute toxic levels. Similar results were reported by Isidori et al. [54].

Of particular importance are pharmaceutical compounds that affect the nervous or endocrine systems because effects on aquatic organisms are possible at the low concentrations found in the environment. Selective SSRIs are drugs used to treat clinical depression in humans, and have been detected in low concentrations in surface waters. The acute and chronic toxicity of five SSRIs (fluoxetine, fluvoxamine, paroxetine, citalopram and sertraline) to the daphnid *C. dubia* were evaluated by Henry et al. [46]. For each SSRI, the 48-h median lethal concentration (LC_{50}) was determined in three static tests with neonate *C. dubia*. The 48-h LC_{50} for the SSRIs ranged from 0.12 to 3.90 mg/L and the order of toxicity of the compounds was found to be (lowest to highest): citalopram, fluvoxamine, paroxetine, fluoxetine, sertraline. Again the results of this study indicate that SSRIs can impact survival and reproduction of *C. dubia*; but only at concentrations that are considerably higher than those expected in the environment.

Other seemingly innocuous compounds may also affect organisms. For instance, the COX-inhibitor indomethacin has been shown to promote egg-shell thinning in birds at doses of 50–100 mg (comparable to the effects reported as consequences of environmental contamination with DDT) [55]. The ability of dissected polyps of the cnidarian *Hydra vulgaris* to regenerate their hypostome, tentacles and foot was inhibited by diazepam, digoxin and amlodipine at a concentration of only 10 μg/L [56].

Plants have also been shown to beadversely affected by drugs in soil or sewage sludge used to amend soil or through irrigation with contaminated wastewater [57]. Growth rate, nitrogen fixation, heterocyst frequency and bioaccumulation have been investigated and shown to be adversely affected if an appropriate concentration is reached [57–59]. If livestock graze, or are fed, on plants that have absorbed pharmaceuticals there is the possibility that lipophilic substances may be retained in their body tissue and/or milk opening up a potential route to the human food chain. However, the bioavailability of these compounds is greatly dependent on their sorption kinetics as well as the organic matter and pH of the soil [60].

Veterinary medicines may persist in soil [61] and also have the potential to run off to surface waters or leach to groundwaters, while substances used in aquaculture may be released directly to surface water. Possible exposure routes include:

Animals that have accumulated veterinary medicines in their tissues through the food chain.

- Crops that have accumulated veterinary medicines from soils or manure.
- Fish exposed to treatments used in aquaculture to treat disease or promote growth.
- Ground and surface waters that have become contaminated with veterinary medicines.

Animal drugs not only often have a much more direct route to the environment, they can also be much more toxic. For instance because of historical and measurable impacts on both the environment and human health, a number of groups (primarily sheep dip chemicals, fish farm medications and anthelmintics) are known to be of concern [62].

There is little data available in the public domain on the environmental fate, behaviour and effects of other generic groups of animal drugs and so potential impacts are less well understood [63]. Generating such data is important since animal medications may have unintended effects on non-target organisms. For instance, ivermectin (a broad spectrum anti-parasitic drug) may be used as a feed additive to alleviate sea lice infestation of farmed salmon. However, it has been shown to be toxic to two sediment-dwelling organisms, the amphipod, *Corophium volutator* and the starfish, *Asterias rubens*. Additionally, an initial assessment of the potential risk to the marine environment from sediment-associated ivermectin indicated that there may be significant risk to infaunal polychaetes in sediment immediately below and around the fish cages where medicated feed was applied [64]. Other members of the ivermectin group may also have toxic environmental effects to a variety of organisms at ng/L–μg/L concentrations [65–67].

Derksen in CSTEE [68] compiled a list of the most toxic groups of pharmaceutical compounds, a revised version of this is presented in Table 3.3.1.

TABLE 3.3.1

Toxicity of seven major groups of human drugs to the aquatic environment (after CSTEE [68])

Substances	Extremely toxic ($EC_{50} <$ 0.1 mg/L)	Very toxic ($EC_{50} <$ 0.1–1 mg/L)	Toxic (EC_{50} = 1–10 mg/L)	Harmful (EC_{50} = 10-100 mg/L)	Non-toxic ($EC_{50} >$ 100 mg/L)
Analgesics			D	D, E	
Antibiotics	A	B			
Antidepressants		D			
Anti-epileptics			C		D, E
Cardiovascular drugs		D			
Cytostatics		A		D, E	
X-ray contrast media					A, B, D, E
Most sensitive taxonomic groups					
A-Microorganisms	B-Algae	C-*Cnidaria*	D-*Crustacea*	E-Fish	

Source: Reprinted with permission from Ref. [28] © 2002 Elsevier.

3.3.3 CHRONIC EFFECTS

Owing to their low but persistent occurrence pharmaceuticals will most likely have chronic, rather than acute toxic, effects since many aquatic species will be continuously exposed to pollutants over long periods of time, or even over their entire life. Unfortunately, these effects may not become apparent for many years and therefore the evaluation of the chronic effects of pharmaceuticals is important. Unfortunately, studies of this type are extremely limited in the scientific literature and this is a major hindrance to satisfactory risk appraisal of pharmaceuticals.

Wollenberger et al. [53] evaluated the chronic toxicity (EC_{50} values, mg/L) of the antibiotics, metronidazole, olaquindox, oxolinic acid, oxytetracycline, streptomycin, sulfadiazine, tetracycline, tiamulin and tylosin, in the *D. magna* reproduction test. The toxicity, in decreasing order was tiamulin (5.4), sulfadiazine (13.7), tetracycline (44.8) and oxytetracycline (46.2). The NOECs (mg/L) obtained in the reproduction test with oxolinic acid, streptomycin, tylosin and metronidazole were 0.38, 32, 45 and 250 mg/L, respectively. The observed toxicity of oxolinic acid to *D. magna* indicates that this substance (commonly used feed additive in fish farms) has the potential to cause adverse effects on the aquatic environment.

Henry et al. [46] investigated the chronic toxicity of five SSRIs. Chronic (8 per day) tests were conducted to determine no-observable-effect concentrations and the lowest-observable-effect concentrations for reproduction endpoints. The SSRIs negatively affected *C. dubia* reproduction by reducing the number of neonates per female, and for some SSRIs, by reducing the number of broods per female. For sertraline, the most toxic SSRI, the LOEC for the number of neonates per female was 0.045 mg/L and the NOEC was 0.009 mg/L. Therefore, although the tested SSRIs can impact the survival and reproduction of *C. dubia*; it is only liable to occur at concentrations that are considerably higher than those expected in the environment.

Nunes et al. [69] investigated both acute and chronic effects of clofibrate and clofibric acid on the enzymes acetylcholinesterase (AChE), lactate dehydrogenase (LDH) and catalase (CAT) of the mosquitofish (*Gambusia holbrooki*). AChE, commonly used as a biomarker of neurotoxicity, was determined in the entire head of the fish. LDH, an important enzyme of anaerobic metabolism, was quantified in dorsal muscle, and CAT, which has previously been used as an indicative parameter of peroxisome proliferation, was determined in the liver. Alterations in body and liver weight were also determined through the calculation of the final ratios of body weight/initial body weight, liver weight/final body weight, liver weight/gills weight and liver weight/head weight.

Acute exposure of *G. holbrooki* to both clofibrate and clofibric acid was found to induce a decrease in liver CAT activity, an increase in muscle LDH activity, while no effects were observed on AChE activity. However, chronic exposure did not alter significantly the enzymatic activities, suggesting reduced or null effects over these pathways, relative to effects reported in other species. No effects were observed for the calculated ratios, except a significant weight reduction for males chronically exposed to clofibrate.

Sanderson et al. [70] reviewed the ecotoxicological data available for risk assessment of ~4500 compounds and their adjuvants. They ranked 2986 different pharmaceutical compounds from 51 classes relative to their hazards towards algae, daphnids and fish using the Estimation Program Interface (EPI) Suite program. Cardiovascular, gastrointestinal, antiviral, anxiolytic sedatives hypnotics and antipsychotics, corticosteroid and thyroid pharmaceuticals were predicted to be the most hazardous therapeutic classes while the overall relative order of susceptibility was estimated to be daphnids > fish > algae.

Quantitative structure–activity relationships (QSARs) that attempt to correlate structure with activity using statistical approaches have become a popular alternative to other test methods in recent years [71].

Sanderson et al. [72] tested a large proportion of the pharmaceuticals observed in the environment for toxicological properties using QSARs. The results did not indicate significant acute risks prior to application of assessment factors. Compared with measured effect concentrations the QSAR predictions were more "sensitive" 80% of the time. The long-term effects of subtle and chronic changes, additive or synergistic effects and effects on other endpoints, e.g. reproduction, behaviour, metabolism, bacterial resistance, etc. are still uncertain. These results indicate that QSARs can be important prioritisation tools for subsequent experimental risk assessment of pharmaceuticals in surface waters, due to the prevalent deficiency of ecotoxicological data. However, they are only models and there are many unknowns with respect to the rates of absorption, biotransformation and elimination of pharmaceuticals in organisms [72]. Nor can the possible effects of other chemical and environmental stressors that may cause additive, or synergistic, effects be taken into account [73]. This uncertainty is not limited to studies on pharmaceuticals and is found in the risk assessment of many chemical compounds and is unlikely to be overcome due to the complex nature of the environment [74].

Sanderson et al. [75] used ECOSAR to scan 2848 pharmaceuticals which were then categorised according to the OECD aquatic toxicity classification system. The qualitative risk assessment ranking relative to probability and potential severity for human and environmental health effects was found to be: antibiotics > sex hormones > cardiovascular drugs > antineoplastics. A relatively large proportion (a third) of all pharmaceuticals were found to be potentially very toxic to aquatic organisms.

The predicted species susceptibility was found to be: daphnid > fish > algae, and the predicted rank order of relative toxicity: sex hormones > cardiovascular > antibiotics > antineoplastics.

Since chronic data are lacking, QSARs, SARs and pharmacodynamic information could be used (through programs such as ECOSAR) to prioritise and steer experimental risk assessments of pharmaceuticals, and potentially, also in the development of new drugs, for optimising efficacy and in minimising environmental hazards of new products. The results from these methods can then be amended as more data become available.

3.3.4 MIXTURE EFFECTS

Another facet of this problem is exposure to mixtures of chemicals. Some compounds, which have no inherent risk on their own, may contribute to risk by increasing the toxicity of others. An example of this is the potential impact of inhibitors/inducers of multi-drug transport (efflux) systems, which serve to minimise the intracellular concentrations of toxicants in compromising aquatic health [76]. Now recognised for enabling a significant portion of the increasing incidence of antimicrobial resistance among bacteria, efflux pumps also play critical role in protecting many different types of cells from xenobiotics. They are a common defensive strategy for aquatic biota, especially in the aquatic environment where organisms suffer continual, life-long exposure to pollutants. [77]. By minimising the intracellular concentrations of harmful compounds they prevent the accumulation of pollutants and so allow many aquatic organisms to survive in contaminated waters which might otherwise prove toxic [78].

Any of a diverse array of chemicals (some of the more potent being verapamil, reserpine and cyclosporine) can inhibit these pumping systems, thereby allowing toxins to cause adverse effects at lower concentrations than normal. For instance, the incidence and severity of developmental abnormalities and deformities observed in embryos and larvae of the mussel *Mytilus edulis* exposed to vinblastine, methyl methanesulfonate (MMS), chloroquine, mitomycin-C, cadmium chloride and colchicine have been shown to be significantly increased when each toxin is added in the presence of 20 μM of verapamil compared to clean seawater [79]. Organisms in less-polluted aquatic environments may be at higher risk to newly introduced toxicants because of their lower-induced levels of efflux pumps [31]. There is also concern that broad-spectrum antiseptics such as triclosan may promote widespread antibiotic resistance simply by inducing bacteria to produce more efflux pumps. However, these compounds have yet to be the targets of any published, environmental surveys [80].

This issue is further complicated by the fact that exposure to only one drug/toxicant at a time is most likely a rare event [81]. In the aquatic environment, most organisms are continually exposed to a range of toxic substances with possibly only slight temporal and spatial variations in concentration levels [82]. Recent work is beginning to demonstrate the significance of exposure to mixtures of chemical (and non-chemical) stressors at low concentrations and this raises the question of whether additive effects might occur or whether synergy could magnify the effects of certain pharmaceuticals under study [48,83].

Arnold et al. [84] reported that combinations of two weak environmental estrogens, such as dieldrin, endosulfan or toxaphene, could cause an estrogenic effect even if each compound was present at a concentration below its no effect value. This report was later withdrawn because neither the authors nor other workers were able to repeat the results [85]. However, the episode prompted others to re-examine the potential effects of mixtures of organic toxicants.

Mixtures of pesticides can increase the occurrence of various abnormalities in frogs (such as malformed legs) [86] and may also interfere with nitrogen fixation in some plants [87]. Similarly, a mixture of ibuprofen, prozac and ciprofloxacin has been shown to be harmful to plankton, aquatic plants and fish at concentrations of 10–200 times lower than the standard human dose [88]. A mixture of ciprofloxacin (an antibiotic), triclosan (an antimicrobial) and tergitol 10 (a surfactant) was shown to have a significant effect on algal biomass yields, it may therefore influence both the structure and the function of algal communities in streams receiving wastewater effluent [89].

Triclosan may also be weakly androgenic at the μg/L range [90]. Mixtures of quinolones (a group of synthetic antibiotics used in aquaculture) have been shown to be toxic to the bacterium *Vibrio fischeri* [91]. Similar results have been reported for mixtures of estrogens [92] as well as estrogen mimicking compounds such as alkyl phenols [93]. These studies highlight the limitations in the use of NOECs as "safe" concentrations [94] as well as the traditional focus on the effects of single agents, since the additive effects of a mixture is unlikely to simply be the arithmetic sum of the effects of its individual components [95].

Since hazards may differ from those that were anticipated, unexpected effects of individual or groups of drugs cannot be ruled out [96]. Nobody knows what the combination of effects of these and other compounds on aquatic life might be and there is increasing justifiable concern that there could be a threat to aquatic fauna and the food webs they support [97,98]. An effect on one organism at a low-trophic level could have more dramatic effects at higher trophic levels and it may be difficult to disentangle cause and effect in such interrelated systems [44]. Therefore, risk assessments that ignore the possibility of the additive effects of pharmaceutical chemicals (e.g. antibiotics) will almost certainly lead to significant underestimation of risk. A more valid approach to risk assessment may therefore be consider the concentration of compounds within a therapeutic group that shares a similar mode of action such as antibiotics or painkillers [28].

It is unknown what effects if any exposure to repeated doses of a mixture of sub-therapeutic amounts of drug compounds could have on human health. Most likely they will be of little or no consequence in healthy adults. Effects may be more pronounced in young or elderly who may have a reduced ability to remove toxic compounds from their bodies and there is also the possibility that they may interact with other medications that an individual may be taking. For instance, ibuprofen has been demonstrated to interfere with the cardio protective properties of aspirin [99] while caffeine can enhance the effects of certain analgesics [100].

How to tease apart the effects of complex mixtures of pollutants on environmental health, is of course, another question entirely. The effects of environmental stressors on flora and fauna can, be followed, and possibly better understood, by tracking changes in the low-molecular weight chemicals they produce. Indeed, it can be argued that the cellular metabolic status is the most functional measure of the cell's phenotype [101].

The analysis of metabolites is known as metabolomics (it is also, somewhat confusingly, labelled as metabonomics, metabolic profiling, metabolic fingerprinting and metabolic footprinting amongst others). Since it is the study of all the low-molecular weight chemicals in an organism, such as sugars, organic acids, amino acids and nucleotides, the technique can be used to sort through hundreds of molecules to separate out a dozen or so that can serve as the signature of a particular problem [102].

An organism's "metabolome" is its full complement of metabolites, in the same way that its genome is all the genetic information it contains [103]. Metabolomics is ideal for studying the impact of stressors, such as pollution, on environmental species, not least because, unlike genes or proteins, metabolites are conserved across species [104]. Since the biochemical consequences of mutations, changes in the environment and exposure to pollutants can be observed directly, it may also help to demonstrate how pollutants interact.

Metabolites may range in concentration to the order $\sim 10^9$, have mass ranges of the order of ~1500 amu and polarity ranges of $\sim 10^{20}$ [105]. Unsurprisingly, no analytical approach can provide universal coverage of the metabolome, and therefore multiple analytical techniques are needed for complete analysis of a metabolome. The primary analytical technique for metabolomic studies has for many years been nuclear magnetic resonance spectroscopy (NMR). This has a number of advantages in that it requires minimal sample preparation and is fast and robust technique, which allows a wide range of small molecule metabolite to be measured simultaneously. The disadvantage is a lack of sensitivity. For this reason many metabolomic-based studies now also use gas and/or liquid chromatography mass spectrometry, which have the advantage of greatly enhanced sensitivity compared to NMR but with a trade-off of increased sample preparation time.

There are drawbacks to using metabolites as markers for pollution. Not least that their concentrations are heavily influenced by a variety of factors not directly related to pollutant exposure (such as diet or disease) [106]. However, metabolomics can give researchers a more comprehensive look at the complex changes under way in hundreds of molecules, as an adverse effect (or effects) of pollutant exposure, or disease, begin to develop [107]. A useful review of the use of metabolomics in toxicology is given in Robertson [108].

3.3.5 IMPACTS OF PHARMACEUTICALS IN THE ENVIRONMENT

An often-cited concern regarding pharmaceuticals in the environment has been the spread of drug resistant pathogens [109]. Increased antimicrobial action from bacteria has also been observed at fish farms both where antibiotics are added directly to the water as medicated feed (which is often not eaten) and where livestock manure is added to fish ponds as fertiliser to stimulate the growth of photosynthetic organisms to provide food for the fish. If the manure comes from animals treated with antibiotics (or other drugs) these compounds can be transferred to the ponds. In one study, the resistance to ciprofloxacin among bacteria in ponds treated in this way increased from less than 5% to 80%. In one case, resistance was observed for oxytetracycline and sulfamethoxazole in 100% of the sampled population [110].

The continued spread of resistance of bacterial pathogens to the many compounds presently used to control infections is a phenomenon which may be assisted by repeated doses at the low concentrations found in the environment [111]. Wastewater treatment plants have been predicted to be hotspots for horizontal gene transfer and selection of antibiotic resistance genes among aquatic bacteria in wastewater [112]. There has therefore been concern that antibiotics found in sewage effluent may cause increased resistance amongst natural bacterial populations. Many antibiotic resistant isolates of microorganisms can be found in the environment and, although the subject remains controversial, the significant increase in the number of bacterial strains resistant to multiple antibiotics has often been attributed to the irrational use of antibiotics and the increase in discharges to wastewater. The three well-established mechanisms of gene transfer (i.e. conjugation, transduction and transformation) are believed to occur in the aquatic environment [113]. As a result, streams and rivers could provide a source and a reservoir of resistant genes as well as a medium for their spread.

Although one study has shown that bacteria isolated from treated sewage and digested sludge were generally not significantly more resistant to antibiotics than isolates from raw sewage [114], many others have shown the opposite [112,113]. Therefore, the occurrence of antibiotics in sewage effluent and receiving waters as well as their potential effects on exposed microbial populations is of great interest [115]. It is unclear, however, which is the most important source of dug-resistant organisms. The excretion of such organisms by humans and animals that have undergone treatment and the subsequent transfer of plasmids in the environment, particularly wastewater treatment plants, or the induction of resistance by extremely low concentrations of antibiotics exerting selective pressure in microbial populations. Antibiotics also have the potential to affect the microbial community in sewage systems and the inhibition of wastewater bacteria has the potential to seriously affect organic matter degradation as well as processes such as nitrification and de-nitrification [116].

Grabow and van Zyl [117] studied coliforms resistant to ampicillin, chloramphenicol, streptomycin, kanamycin and tetracycline. Their findings suggest that conventional sewage treatment has a limited effect on the incidence of drug resistance in bacteria. Similar results were determined by Bell [118] in bacteria isolated from domestic sewage before and after treatment in an aerobic lagoon in Canada. Although this study has shown that bacteria isolated from treated sewage and digested sludge were generally not significantly more resistant to antibiotics than isolates from raw sewage, others have shown the opposite [112,113,116,119,120]. Some authors have claimed that antibiotics in hospital wastewater may by itself be sufficient to induce resistance but this is not a widely supported view.

Certain antibiotics may also have a toxic effect. For instance, Hartmann et al. [121] identified fluoroquinolone antibiotics as the main source of genotoxicity in hospital wastewater using a bacterial short-term genotoxicity assay, based on a umuC:lacZ fusion gene (umuC assay). The ratio of theoretical mean wastewater concentrations (derived from consumption data) and lowest observable effect concentrations of selected pharmaceuticals were used to calculate umuC induction probabilities. The fluoroquinolone antibiotics ciproxin and noroxin exhibited the highest induction probabilities and exceeded all other investigated drugs by at least one order of magnitude in significance. Antineoplastic drugs, originally thought to be the main effecters, were found to be of marginal significance using this technique. These findings were further supported by investigation of urine samples from hospital patients with the umuC assay. The determination of ciprofloxacin in hospital wastewater revealed concentrations from 3 to 87 μg/L and ciprofloxacin concentrations and umuC induction factors in 16 hospital wastewater samples exhibited a log-linear correlation. The authors suggest that the previously measured umuC genotoxicity in the wastewater of the hospital under investigation is caused mainly by fluoroquinolone antibiotics, especially by ciprofloxacin. However, follow-up work by Hartmann et al. [122] suggested that this could also be due to the presence of additional mutagen that were not identified at the time. Therefore, the occurrence of antibiotics in sewage effluent and receiving waters as well as their potential effects on exposed microbial populations is of interest and concern.

Effects other than the inducement of antibiotic resistance are also possible especially given the fact that many drugs have more than one use and/or unexpected applications. For instance, the analgesics ibuprofen and flurbiprofen have been shown to have antibacterial and antimycotic properties [123,124] and ibuprofen may also interfere with the cardio-protective effects of aspirin at therapeutic doses in patients with established cardiovascular disease [99]. Antimycotic activity has also been observed during degradation of beta-lactam antibiotics [125] and the new antibiotic fosmidomycin also shows promise as an anti-malarial agent (at doses of 1–2 g^{-1} every 8 h) [126].

Hospital wastewater is often assumed to be the most toxic to aquatic life and there are indeed several studies in which genotoxic activity of hospital wastewater has been confirmed. Gartiser et al. [127] demonstrated the genotoxicity of some hospital effluents with the chromosome aberration test (hamster cell line V79). However, the authors could not attribute the observed genotoxic effects to a specific substance or group of substances. Guiliani et al. [128] found that out of over 800 hospital effluent samples from a large cancer hospital 13% were genotoxic in the umuC assay [62]. Genotoxic samples were detected throughout a 24-h period with the morning hours showing the highest activity. Of the toxic wastewater samples 96% showed genotoxic potential without detectable cytotoxic effects. The authors considered that anti-neoplastic agents were the possible causative agents however they concluded that there was no obvious pollution hazard attributable to the waste because no genotoxic activity was detected in the influx of the sewage treatment plant (STP) receiving the wastewater of the hospital.

Some workers have tried to identify the causal agents of genotoxicity activity in hospital wastewater. Steger-Hartmann et al. [129] investigated the effects of cyclophosphamide in the umuC assay. No genotoxic effects were found at concentrations as high as 1 g/L. This was in agreement with the SOS chromotest in which Hellmér and Bolcsfoldi did not detect a genotoxic effect of cyclophosphamide at concentrations of up to 4.6 g/L [130]. Hartmann et al. [121] found evidence to suggest that one-single class of antibiotic drug, the fluoroquinolone antibiotics (e.g. ciprofloxacin) were responsible for the genotoxic activity for a specific hospital under investigation.

X-ray contrast media, which biodegrade very slowly and may contribute to the adsorbable organic halogen (AOX) load, were investigated by Steger-Hartmann et al. [131] who studied the widely used compound iopromide. The authors calculated the surface water PEC to be 2 μg/L. This was then compared with the predicted no-effect concentration as derived from a range of ecotoxicity tests. In short-term toxicity tests with bacteria (*V. fisheri, Pseudomonas putida*), algae (*Scenedesmus subspicatus*), crustaceans (*D. magna*) and fish (*Danio rerio*, *Leuciscus idus*) no toxic effects were detected at the highest tested concentration of 10 g/L. In a chronic toxicity test with *D. magna*, no effect was observed at the highest tested concentration of 1 g/L. Using an assessment factor of 100, the ratio between the PEC and the PNEC was calculated to be less than 0.0002. The authors concluded that this low value indicates that no environmental risk would be expected as a result of the release of iopromide into the aquatic environment.

Clearly, when evaluating pharmaceuticals, the health benefits to humans must take precedence over any environmental problems. Therefore, the normal methods for the control of pollutants, for example by restricting or banning their use if a problem is found, is neither appropriate nor desirable in this case. Instead it may be better to try to regulate the pathways by which pharmaceuticals enter the environment, perhaps through the labelling of medicinal products. Given the enormous importance of the pharmaceutical industry both to human health and the economy, any increased controls could have significant economic and social ramifications.

Notwithstanding the above, the poorly characterised processes involved with pharmaceuticals in the environment (occurrence, transport, etc.) warrant a more precautionary view on their possible environmental fate and effects. A large amount of research remains to be completed before a thorough understanding of this subject is available, at present the available scientific knowledge is less than that needed to fully assess the risks these compounds pose to the environment. Future work will need to focus on more detailed ecotoxicity testing, using a wide range of aquatic organisms as well as how these compounds are sorbed, transferred and biodegraded during sewage and water treatment plants, and the environment. This will provide a better understanding of how they may affect both the aquatic and terrestrial environments and indicate possible remediation strategies. Currently, antibacterial resistance probably represents the most significant human health hazard, and natural and synthetic sex hormones are potentially the largest hazard to non-target organisms (by acting as endocrine modulators in wildlife).

3.3.6 ENVIRONMENTAL RISK ASSESSMENT OF PHARMACEUTICALS

The information available on the ecotoxicology of pharmaceutical compounds is weak and traditional endpoints in current test regimes often do not capture the typical mode of action of these biologically active substances. A full risk assessment is therefore difficult because of the paucity of relevant data [17]. Regulations associated with drugs are generally overseen by human health agencies, which usually have limited experience in environmental issues and until recently pharmaceuticals were not seen as potentially toxic substances. Therefore, unlike many other anthropogenic contaminants, they have not been subjected to detailed research regarding their possible, environmental effects.

Ecological/environmental risk assessment (ERA) is the practice of determining the nature and likelihood of the effects of human actions on animals, plants and the environment [132]. It provides a framework for eventual risk management and typically involves three tiers: problem formulation or hazard assessment (initial planning and information gathering), effects and exposure assessment (data gathering and analysis) and risk characterisation (assimilation and integration) [133]. ERA differs from human health risk assessment (HHRA) since it must consider a very large number of genera, rather than a single specie (e.g. *Homo sapiens*). Furthermore, whereas an HHRA aims to protect individuals, an ERA is more concerned with populations as well as ecosystem processes and functions [134]. Risk management necessitates knowledge of hazards incurred as well as the quantification of the exposure which the organism or systems subject to those hazards would be likely to incur. At present this is missing with regard to the majority of pharmaceutical compounds [135,136]. The process of assessing a particular drug's potential impact on the environment varies across countries. Two of the more developed testing methods are those of the EU and the USA.

Pharmaceuticals in the environment have been controlled in the USA since 1977 with the US Food and Drug Administration (FDA) taking responsibility for their control under the auspices of the National Environmental Policy Act of 1969. Regulation occurs through the environmental review process for any new drug application (NDA) submitted to the FDA. In the late 1980s, additional data were required from pharmaceutical companies and more extensive information was provided in environmental risk assessments that accompanied NDAs. An evaluation and review of the data submitted from the late 1980s through the mid-1990s led the FDA to revise the regulations in 1995. The process led to a tightening of regulations and revised environmental assessment requirements. These minimised environmental risk assessment data included in NDAs since it was believed a lot of the required information was not relevant to assessing the drug's potential impact [137]. Environmental assessments of veterinary pharmaceuticals have also been required by the US FDA since 1980 [62].

In contrast, the development of specific ecological risk assessment for pharmaceuticals did not begin in earnest in Europe until in the early 1990s. This was initially based on Directive (65/65/EEC) and later refined in (93/39/EEC). The process was significantly advanced by the "Discussion Paper on Environmental Risk Assessment of Non-Genetically Modified Organism (Non-GMO) Containing Medicinal Products for Human Use" [138]. Most recently, the European Agency for the Evaluation of Medicinal Products "Note for Guidance" on the published in July 2003 and updated in 2005 is now under consultation [139,140]. All of these approaches are based on general assumptions about the fate of these compounds in the environment, and hence have been subjected to criticism [141].

Currently, the environmental assessment procedure for new drugs in the USA is a two-stage process. Firstly, the manufacturer is required to estimate the expected introductory concentration (EIC) entering the environment based on total fifth year production estimates. If the EIC of a drug, or any of its active metabolites, at the point of entry (e.g. sewage effluent) in the aquatic environment is shown to be less than 1 μg/L (1 ppb) the drug is considered to be acceptable and is given environmental "category exclusion" status. In this case, no further environmental risk assessment is needed and no monitoring is conducted to confirm the environmental concentration after a drug is marketed.

If the EIC is calculated to be over 1 μg/L then a formal environmental assessment has to be conducted, this will include data on environmental fate and a tiered set of ecotoxicity tests. The base set usually includes effects on microbial respiration and acute toxicity to at least one algal, invertebrate and fish species. Chronic testing need only be considered under certain circumstances, for example if the drug has the potential to bioaccumulate [137].

The procedure is not as developed in Europe as it is in the USA. From 1995, any company applying for registration of a new drug has had to demonstrate that it will not have an impact on the environment through the submission of environmental risk assessments to accompany marketing authorisation approval. Like the FDA system, it defines a cut-off limit for a detailed risk assessment [2]. In this case, a manufacture is required to calculate the PEC. If this is <0.01 μg/L no ERA is obligatory (unless there are indications that effects may occur below that concentration). If the PEC >0.01 μg/L, then the PEC/PNEC ratio has to be calculated. If the PEC/PNEC is <1, no risk is anticipated and no further tests are conducted. If the PEC/PNEC is >1 then phase II risk assessment will be required which again includes further tests on algae, *Daphnia* and fish, to specified OECD guidelines [2].

The main criticisms of both guidelines include the difficulty in obtaining PEC/EIC (since little to no data on national or regional use are available for most products). The level of the threshold PEC/EIC that triggers an assessment, the lack of consideration for the terrestrial compartment and the specificity of the toxicity screening and the bias towards acute rather than chronic toxicity assessment [28]. In addition, while it is evident that the guidelines outlined above could be used to also assess existing medications (which they do not apply to at present) there is no legal requirement to do so. It is also not to be expected that potential adverse environmental effects will block the admittance of new drugs at the expense of public health benefits. The procedures therefore still give rise to extensive discussions between regulators, environmental agencies and industry, owing to their different points of view. This demonstrates the need for the revision of the regulatory guidelines within a more thorough and effective assessment framework.

It is evident therefore that the evaluation of the risks posed by environmental exposure to drug compounds deserves a high priority especially in view of the lack of current knowledge [142]. Since it is impractical to fully assess the risks of every medicine and application authorised for use one approach may be to develop a prioritisation scheme to identify those substances that might pose a risk to human health and which warrant further study [63,143,144].

3.3.7 DISCUSSION

There are numerous concerns regarding the hazards of pharmaceutical compounds in the environment and it is frequently recommended that more research should be done in this area [109]. In this chapter, some of the suggestions made that are relevant to the undertaking of ecological risk assessments for the pharmaceuticals have been discussed. Overall, a more diverse range of animal species with defined genders and physiological status should be assessed using in vitro tests (with detailed correlation to in vivo tests). This technique is of benefit since in vitro studies not only allow the analysis of specific biochemical interactions in both animal and plant cells but also mean that a large number of compounds, or mixtures of compounds, can be screened rapidly. This additionally allows for a significant reduction in the number of animal experiments necessary. However, to facilitate accurate exposure assessments, speciation, bioaccumulation and bioavailability of pharmaceuticals in the environment should also be examined.

Pollution from pharmaceuticals poses several, difficult challenges to policymakers, not least because the potential risks are poorly defined, and interventions likely to be costly and/or only marginally effective. In the United Kingdom both pharmaceutical manufacturing plants, and STPs are subject to regulation under the Environmental Protection Act 1990, and the Water Resources Act 1991 [145,146]. This legislation could be used by the EA to set limits, if deemed appropriate, on the quantities of pharmaceuticals being released into the environment. Powers within the Water Industry Act 1991 also enable sewage undertakers to set limits on substances in trade discharges to sewer networks. In addition, future regulation such as the European water framework directive (directive 2000/60/EC) will also substantially influence the environmental regulation of chemicals [147].

One area of particular interest is the potential for pharmaceuticals to re-enter the human body. When surface waters are used as sources of drinking water, abstraction points may often be downstream of effluent discharge points and groundwater sources have also been found to be contaminated with pharmaceutical compounds [148]. In densely populated urban areas with high-municipal wastewater discharges and low-surface water flows there is a potential risk of drinking water contamination by polar organic compounds [149]. While many countries employ advanced technologies such as ozonation, reverse osmosis and granular activated carbon (GAC) for drinking water treatment, some persistent and highly stable compounds have been shown to be unaffected by such processes and to consequently appear in drinking water [150]. Examples include the anti-epileptic drug carbamazepine, and the lipid-regulating compound gemfibrozil which have both been found at the ng/L level in German and Canadian drinking water samples [151,152].

Drinking water is a direct route to the human body and for any drug compounds that may be present [153]. Other pathways such as ingestion (eating crops irrigated with effluent or grown on sewage sludge amended soil) or bodily interaction (bathing or showering in waters containing effluent) can also place the body in contact with pharmaceutical compounds. While technologies such as granular-activated carbon, membrane filtration and ozone treatment can remove the majority of drug compounds from potable waters, some compounds are resistant to even this type of treatment [154] and information concerning their occurrence, (eco)toxic risk, and fate is virtually non-existent. This makes carrying out accurate risk assessments on the potential health impacts to humans of ingestion of a mixture of compounds at many times less than the therapeutic dose difficult.

Therefore, whilst exposure to pharmaceuticals via potable water supplies and associated risks is likely to be relatively minor, the increasing demands on the worlds freshwater supplies is likely to lead to greater dependency on indirect and direct water reuse and the potential for adverse effects from this should not be overlooked [155]. For instance, if drugs were to find their way into drinking water, then potential health concerns would need to focus not only on individual and possible mixture effects over an extended period of time (approximately 80 years or more). Their possible interactions with other medications (or even illegal drug substances) which people may also be taking must also be considered [156].

It is difficult to extrapolate laboratory-based acute toxicity data to the lower concentrations and routes of exposure encountered in the environment. There also remain a wide range of issues relating to the occurrence of potential effects that requires further investigation before the environmental significance of this problem can be fully evaluated. Advances in analytical chemistry have driven this area of research (pharmaceutical pollution at these levels was not routinely detectable even ten years ago and hence was not considered a threat). Nevertheless, the development of analytical methods is still an essential part of improving uncertainty, and methods for the determination of drugs in solid phases, such as sediments, would also be very useful. The main challenge may lie in the separation and identification of drugs from the plethora of other chemicals that may be present in large quantities in waters and sediments [157]. It is an issue of increasing international importance that could potentially have long-term effects on population levels and species diversity. While new policy strategies designed to reduce the quantities of drugs that eventually enter watercourses are being introduced or considered in many parts of the world [158] these will not affect the large number of medicinal compounds already in use. Only the European Unions proposed REACH (Research Evaluation and Assessment of Chemicals) regulations take this unique radical step.

Owing to the beneficial health effects and economic importance of pharmaceuticals and since their use is expected to grow with the completion of the human genome project and the increasing age of the population, it may be better for pollution control efforts to focus more on reduction, minimisation and elimination at source. Other policies could include the development of clearer labelling on medicinal products and better guidelines for the disposal of pharmaceutical compounds by patients and medical professionals. This approach would have the potential benefit of improved consumer health (by minimizing the intake of active substances) as well as reduced healthcare spending [92].

3.3.8 CONCLUSIONS

1) Pharmaceuticals are a part of life for modern society, being important both in improving human health and welfare, and the economy. However, they are also increasingly being found in the natural environment (notably the aquatic compartment) throughout the world. Since they are designed to elicit a biological effect there are justifiable concerns over their potential effects on flora and fauna.
2) It is unlikely that most pharmaceutical compounds are present in the environment at high enough concentrations to cause significant harm. However, at sufficient concentrations they have been observed to induce effects in both animals and plants and it is possible they may have other effects which have not yet been observed due to the differences in biology of the organisms exposed (as the collapse of vulture populations in the Indian subcontinent shows). There is also the possibility of chronic, long-term effects and it would therefore seem unwise to conclude that these compounds are having no effect at all until there is more proof.
3) The effects induced by pharmaceutical compounds are likely to be dependant on the dose, exposure route and timing, pharmacokinetic mechanisms and the physiological status of the target organism. Life cycle studies linking different habitats and exposure routes of organisms at various life stages to a mixture of substances may be important to assess the effects of drugs on individual species.
4) Sewage treatment works are likely to be the most significant source of human medicinal compounds to surface waters while the application of contaminated livestock manure may also contribute a high load of veterinary drugs to the aqueous phase after run-off events. Sewage is a continuous, point source, while runoff from agriculture is diffuse and concentrations are dependent on the application rate and run-off parameters.

5) Laboratory data on the toxicity of compounds gathered during product development may be able to provide useful information for risk assessment but more data are needed about the ecotoxicological effects of medicines in the wider environment. There will also be a need to adjust any risk assessment to the specific environmental compartment, organism and endpoint of the drug in question and to take into account as many potential effects of the compound(s) in question as possible. More investigation is also needed into potential long-term ecotoxicological effects.
6) While there is no evidence that most pharmaceuticals pose a human health risk their presence is of concern. This is especially important in potential indirect and direct water re-use situations and this should be taken into account in new policies on drug compounds.
7) The potentially large social and economic impacts of regulating drugs means the need for reviewing the testing methods for these products is all the more important. Before any changes in policy are considered it is recommended that environmental monitoring for the more commonly used pharmaceuticals or groups of pharmaceuticals should be undertaken as a priority. Detailed toxicity testing utilizing a wide variety of test organisms and compounds should then be performed in the light of observed concentrations.
8) Efforts to remove pharmaceuticals (or indeed any pollutant) from wastewater would not be without environmental cost (e.g. energy usage and associated release of CO_2) and so should be seen and assessed in the context of the total environmental impact.

REFERENCES

1 H.R. Buser, T. Poiger and M.D. Muller, *Environ. Sci. Technol.*, 33 (1999) 2529–2535.
2 N.J. Ayscough, J. Fawell, G. Franklin and W. Young, Review of Human Pharmaceuticals in the Environment (P390), Environment Agency of England and Wales, Bristol, UK, 2000, p. 106.
3 J.P. Sumpter, *Toxicol. Lett.*, 103 (1998) 337.
4 O.A.H. Jones, N. Voulvoulis and J.N. Lester, *Bull. World Health Organ.*, 81 (2003) 768–769.
5 N. Voulvoulis, M.D. Scrimshaw and J.N. Lester, *Appl. Organomet. Chem.*, 13 (1999) 135–143.

6 J.C. O'Connor, J.C. Cook, M.S. Marty, L.G. Davis, A.M. Kaplan and E.W. Carney, *Crit. Rev. Toxicol.*, 32 (2002) 521.
7 G. Carbonell, C. Ramos, M.V. Pablos, J.A. Ortiz and J.V. Tarazona, *Sci. Total Environ.*, 247 (2000) 107.
8 K.P. Henschel, A. Wenzel, M. Diedrich and A. Fliedner, *Regul. Toxicol. Pharmacol.*, 25 (1997) 220.
9 K.M. Lai, M.D. Scrimshaw and J.N. Lester, *Crit. Rev. Toxicol.*, 32 (2002) 113.
10 J. Miyamoto and W. Klein, *Pure Appl. Chem.*, 70 (1998) 1829.
11 D.E. Kime, *Sci. Total Environ.*, 225 (1999) 3.
12 D.G. Crosby, *Environmental Toxicology and Chemistry*, Oxford University Press, New York, 1998, p. 336.
13 J.L. Griffin, L.A. Walker, S. Garrod, E. Holmes, R.F. Shore and J.K. Nicholson, *Comp. Biochem. Physiol. B*, 127 (2000) 357.
14 O.V. Martin, M.D. Scrimshaw, J.N. Lester and K.M. Lai, *Environ. Sci. Technol.*, 39 (2005) 5349.
15 J.W. Olney, N.B. Farber, D.F. Wozniak, V. Jevtovic-Todorovic and C. Ikonomidou, *Environ. Health Perspect.*, 108 (2000) 383.
16 M. Petrović, S. Gonzalez and D. Barcelo, *Trends Anal. Chem.*, 22 (2003) 685.
17 F.M. Christensen, *Regul. Toxicol. Pharmacol.*, 28 (1998) 212–221.
18 K.M. Lai, M.D. Scrimshaw and J.N. Lester, *Appl. Environ. Microbiol.*, 68 (2002) 859.
19 R.L. Gomes, H.E. Deacon, K.M. Lai, J.W. Birkett, M.D. Scrimshaw and J.N. Lester, *Environ. Toxicol. Chem.*, 23 (2004) 105.
20 J. Schwaiger, H. Ferling, U. Mallow, H. Wintermayr and R.D. Negele, *Aquatic. Toxicol.*, 68 (2004) 141.
21 T. Poiger, H.R. Buser and M.D. Muller, *Environ. Toxicol. Chem.*, 20 (2001) 256.
22 K.M. Lai, M.D. Scrimshaw and J.N. Lester, *Sci. Total Environ.*, 289 (2002) 159.
23 J.L. Oaks, M. Gilbert, M.Z. Virani, R.T. Watson, C.U. Meteyer, B.A. Rideout, H.L. Shivaprasad, S. Ahmed, M. Jamshed, I. Chaudhry, M. Arshad, S. Mahmood, A. Ali and A.A. Khan, *Nature*, 427 (2004) 630.
24 B.W. Brooks, C.K. Chambliss, J.K. Stanley, A. Ramirez, K.E. Banks, R.D. Johnson and R.J. Lewis, *Environ. Toxicol. Chem.*, 4 (2005) 464–469.
25 L. Migliore, C. Civitareale, G. Brambilla and G.D. di Delupis, *Water Res.*, 31 (1997) 1801.
26 R. Wiwattanapatapee, N. Padoongsombat, T. Choochom, S. Tang and A. Chaimongkol, *J. Controlled Release*, 83 (2002) 23.
27 H.P. Rang, M. Dale and J.M. Ritter, *Pharmacology*, Churchill Livingstone, St. Louis, USA, 1999, p. 832.
28 O.A.H. Jones, N. Voulvoulis and J.N. Lester, *Water Res.*, 36 (2002) 5013.

29 B. Halling-Sørensen, S.N. Nors-Nielsen, P.F. Lanzky, F. Ingerslev, H.C. Holten Lutzhoft and S.E. Jorgensen, *Chemosphere*, 36 (1998) 357.
30 K. Fent, A.A. Weston and D. Caminada, *Aquat. Toxicol.*, 76 (2006) 122.
31 C.G. Daughton and T.A. Ternes, *Environ. Health Perspect.*, 107 (1999) 907.
32 B. Halling-Sørensen, H.-C. Holten Lützhøft, H.R. Andersen and F. Ingerslev, *J. Antimicrob. Chemother.*, 46 (2000) 53.
33 A.M. Knight, P.O. Keenan, N.J. Goddard, P.R. Fielden and R.M. Walmsley, *J. Environ. Manage.*, 71 (2003) 71.
34 R. Andreozzi, R. Marotta, G. Pinto and A. Pollio, *Water Res.*, 36 (2002) 2869.
35 R.E. Green, M.A. Taggart, D. Das, D.J. Pain, K.C. Sashi, A.A. Cunningham and R. Cuthbert, *J. Appl. Ecol.*, 43 (2006) 949.
36 P.J. Savarie, J.A. Shivik, G.C. White, J.C. Hurley and L. Clark, *J. Wildlife Manage.*, 65 (2001) 356–365.
37 J.J. Johnston, P.J. Savarie, T.M. Primus, J.D. Eisemann, J.C. Hurley and D.J. Kohler, *Environ. Sci. Technol.*, 36 (2002) 3827–3833.
38 M.R. Miller, E. Wentz and S. Ong, *Toxicol. Sci.*, 48 (1999) 30–37.
39 D.B. Huggett, B.W. Brooks, B. Peterson, C.M. Foran and D. Schlenk, *Archiv. Environ. Contam. Toxicol.*, 43 (2002) 229.
40 S. Haider and S. Baqri, *Comp. Biochem. Physiol.*, 126 (2000) 517.
41 P.P. Fong, P.T. Huminski and L.M. D'urso, *J. Exp. Zool.*, 280 (1998) 260–280.
42 P.P. Fong, *Biol. Bull.*, 194 (1998) 143.
43 B.W. Brooks, P.K. Turner, J.K. Stanley, J.J. Weston, E.A. Glidewell, C.M. Foran, M. Slattery, T.W. La Point and D.B. Huggett, *Chemosphere*, 52 (2003) 135.
44 M. Cleuvers, *Toxicol. Lett.*, 142 (2003) 185.
45 D.G.J. Larsson, M. Asolfsson-Erici, J. Parkkonen, M. Petterson, A.H. Berg, P.-E. Olsson and L. Forlin, *Aquat. Toxicol.*, 45 (1999) 91.
46 T.B. Henry, J.-W. Kwon, K.L. Armbrust and M.C. Black, *Environ. Toxicol. Chem.*, 23 (2004) 2229.
47 EU, Assessment of potential risks to the environment posed by medicinal products for human use, excluding products containing genetically modified organisms. III/5504/94 Draft 4, 1994.
48 F. Stuer-Lauridsen, M. Birkved, L.P. Hansen, H.C.H. Lützhoft and B. Halling-Sørensen, *Chemosphere*, 40 (2000) 783.
49 A. Jos, G. Repetto, J.C. Rios, M.J. Hazen, M.L. Molero, A. del Peso, M. Salguero, P. Fernandez-Freire, J.M. Perez-Martin and A. Camean, *Toxicol. in Vitro*, 17 (2003) 525.
50 G.M. Lalumera, D. Calamari, P. Galli, S. Castiglioni, G. Crosa and R. Fanelli, *Chemosphere*, 54 (2004) 661.
51 B. Nunes, F. Carvalho and L. Guilhermino, *Chemosphere*, 62 (2006) 581.
52 B. Nunes, F. Carvalho and L. Guilhermino, *Chemosphere*, 61 (2005) 413.

53 L. Wollenberger, B. Halling-Sørensen and K.O. Kusk, *Chemosphere*, 40 (2000) 723.
54 M. Isidori, M. Lavorgna, A. Nardelli, L. Pascarella and A. Parrella, *Sci. Total Environ.*, 346 (2005) 87.
55 C.E. Lundholm, *Comp. Biochem. Physiol.*, 118 (1997) 113.
56 D. Pascoe, W. Karntanut and C.T. Muller, *Chemosphere*, 51 (2003) 521.
57 L. Migliore, S. Cozzolino and M. Fiori, *Chemosphere*, 52 (2003) 1233.
58 C. Forni, A. Cascone, M. Fiori and L. Migliore, *Water Res.*, 36 (2002) 3398.
59 L. Migliore, S. Cozzolino and M. Fiori, *Chemosphere*, 40 (2000) 741.
60 P.K. Jjemba, *Agr. Ecosyst. Environ.*, 93 (2002) 267.
61 J.V. Tarazona and M.M. Vega, *Toxicology*, 181–182 (2002) 187.
62 A.B.A. Boxall, L.A. Fogg, P. Kay, P.A. Blackwel1, E.J. Pemberton and A. Croxford, *Toxicol. Lett.*, 142 (2003) 207.
63 A.B.A. Boxall, L. Fogg, P.A. Blackwell, P. Kay and E.J. Pemberton, *Review of veterinary medicines in the environment*, P6-012/8 (2002).
64 I.M. Davies, P.A. Gillibrand, J.G. McHenery and G.H. Rae, *Aquaculture*, 163 (1998) 29.
65 B.A. Halley, W.J.A. VandenHeuvel and P.G. Wislocki, *Vet. Parasitol.*, 48 (1993) 109.
66 R.A. Bloom, I. Matheson and C. John, *Vet. Parasitol.*, 48 (1993) 281.
67 R.J. Nessel, D.H. Wallace, T.A. Wehner, W.E. Tait and L. Gomez, *Chemosphere*, 18 (1989) 1531.
68 CSTEE, Opinion on "Draft discussion paper on environmental risk assessment of non-genetically modified organism (non-gmo) containing medicinal products for human use," 2001.
69 B. Nunes, F. Carvalho and L. Guilhermino, *Chemosphere*, 57 (2004) 1581.
70 H. Sanderson, D.J. Johnson, T. Reitsma, R.A. Brain, C.J. Wilson and K.R. Solomon, *Regul. Toxicol. Pharmacol.*, 39 (2004) 158.
71 R. Perkins, H. Fang, W. Tong and W.J. Welsh, *Environ. Toxicol. Chem.*, 22 (2003) 1666.
72 H. Sanderson, D.J. Johnson, C.J. Wilson, R.A. Brain and K.R. Solomon, *Toxicol. Lett.*, 144 (2003) 383.
73 J.U. Skaare, H.J. Larsen, E. Lie, A. Bernhoft, A.E. Derocher, R. Norstrom, E. Ropstad, N.F. Lunn and O. Wiig, *Toxicology*, 164 (2002) 193.
74 B.I. Escher and J.L.M. Hermens, *Environ. Sci. Technol.*, 36 (2002) 4201.
75 H. Sanderson, R.A. Brain, D.J. Johnson, C.J. Wilson and K.R. Solomon, *Toxicology*, 203 (2004) 27.
76 B. Kurelec, *Crit. Rev. Toxicol.*, 22 (1992) 23.
77 S.M. Bard, *Aquat. Toxicol.*, 48 (2000) 357.
78 M. Bamdad, P. Brousseau and F. Denizeau, *FEBS Lett.*, 456 (1999) 389.
79 I. McFadzen, N. Eufemia, C. Heath, D. Epel, M. Moore and D. Lowe, *Mar. Environ. Res.*, 50 (2000) 319.

80 D. Epel and T. Smital, Multidrug-multixenobiotic transporters and their significance with respect to environmental levels of pharmaceuticals and personal care products. In: C.G. Daughton and T.L. Jones-Lepp (Ed.), *Pharmaceuticals and Personal Care Products in the Environment: Scientific and Regulatory Issues*, American Chemical Society, Washington, DC, 2001, p. 244.
81 C.R. Tyler, S. Jobling and J.P. Sumpter, *Crit. Rev. Toxicol.*, 28 (1998) 319.
82 D. Schowanek and S. Webb, *Toxicol. Lett.*, 131 (2002) 39.
83 S. Broderius and M. Kahl, *Aquat. Toxicol.*, 6 (1985) 307.
84 S.F. Arnold, D.M. Klotz, B.M. Collins, P.M. Vonier, L.J. Guillette and J.A. McLachlan, *Science*, 272 (1996) 1489.
85 J.A. McLachlan, *Science*, 277 (1997) 459.
86 R. Renner, *Environ. Sci. Technol.*, 37 (2003) 52A.
87 J.E. Fox, M. Starcevic, K.Y. Kow, M.E. Burow and J.E. McLachlan, *Nature*, 413 (2001) 128.
88 R. Renner, *Environ. Sci. Technol.*, 36 (2002) 268A.
89 B.A. Wilson, V.H. Smith, F. deNoyelles and C.K. Larive, *Environ. Sci. Technol.*, 37 (2003) 1713.
90 C.M. Foran, E.R. Bennett and W.H. Benson, *Mar. Environ. Res.*, 50 (2000) 153.
91 T. Backhaus, M. Scholze and L.H. Grimme, *Aquat. Toxicol.*, 49 (2000) 49.
92 C.G. Daughton, *Lancet*, 360 (2002) 1035.
93 E. Silva, N. Rajapakse and A. Kortenkamp, *Environ. Sci. Technol.*, 36 (2002) 1751.
94 M. Crane and M.C. Newman, *Environ. Toxicol. Chem.*, 19 (2000) 516.
95 A. Kortenkamp and R. Altenburger, *Sci. Total Environ.*, 221 (1998) 59.
96 C.-Y. Chen and C.-L. Lu, *Sci. Total. Environ.*, 289 (2002) 123.
97 E. Voutsas, K. Magoulas and D. Tassios, *Chemosphere*, 48 (2002) 645.
98 B. Ferrari, N. Paxeus, R.L. Giudice, A. Pollio and J. Garric, *Ecotoxicol. Environ. Saf.*, 55 (2003) 359.
99 T.M. Macdonald and L. Wei, *Lancet*, 361 (2003) 573.
100 I.J. Buerge, T. Poiger, M.D. Muller and H.-R. Buser, *Environ. Sci. Technol.*, 37 (2003) 691.
101 J.K. Nicholson, J. Connelly, J.C. Lindon and E. Holmes, *Nat. Rev. Drug Discov.*, 1 (2002) 153.
102 J.L. Griffin, *Curr. Opin. Chem. Biol.*, 7 (2003) 648–654.
103 S.G. Oliver, *Philos. Trans. R. Soc. B*, 357 (2002) 17.
104 M.R. Viant, J.G. Bundy, C.A. Pincetich, J.S. de Ropp and R.S. Tjeerdema, *Metabolomics*, 1 (2005) 149.
105 H.J. Atherton, N.J. Bailey, W. Zhang, J. Taylor, H. Major, J. Shockcor, K. Clarke and J.L. Griffin, *Physiol. Genomics*, 27 (2006) 178.
106 M.R. Viant, E.S. Rosenblum and R.S. Tjeerdema, *Environ. Sci. Technol.*, 37 (2003) 4982.

107 M.R. Pears, J.D. Cooper, H.M. Mitchison, R.J. Mortishire-Smith, D.A. Pearce and J.L. Griffin, *J. Biol. Chem.*, 280 (2005) 42508.
108 D.G. Robertson, *Toxicol. Sci.*, 85 (2005) 809.
109 O.A.H. Jones, N. Voulvoulis and J.N. Lester, *Environ. Technol.*, 22 (2001) 1383.
110 A. Petersen, J.S. Andersen, T. Kaewmak, T. Somsiri and A. Dalsgaard, *Appl. Environ. Microbiol.*, 68 (2002) 6036.
111 S.E. Jorgensen and B. Halling-Sørensen, *Chemosphere*, 40 (2000) 691–699.
112 T. Schwartz, W. Kohnen, B. Jansen and U. Obst, *FEMS Microbiol. Ecol.*, 1470 (2002) 1.
113 M.D. Cooke, *New Zealand J. Mar. Fresh. Res.*, 10 (1983) 717.
114 L. Guardabassi, M. Danilo, L.-F. Fong and A. Dalsgaard, *Water Res.*, 36 (2002) 1955.
115 K. Kümmerer, *Pharmaceuticals in the Environment: Sources, Fate, Effects and Risks*, Springer-Verlag, Berlin, 2001, p. 265.
116 F.F. Reinthaler, J. Posch, G. Feierl, G. Wust, D. Haas, G. Ruckenbauer, F. Mascher and E. Marth, *Water Res.*, 37 (2003) 1685.
117 W.O.K. Grabow, M. Van Zyl and O.W. Prozesky, *Water Res.*, 10 (1976) 717.
118 R.B. Bell, *Can. J. Microbiol.*, 24 (1978) 886.
119 M. Goni-Urriza, M. Capdepuy, C. Arpin, N. Raymond, P. Caumette and C. Quentin, *Appl. Environ. Microbiol.*, 66 (2000) 125.
120 J. Lin, P.T. Biyela and T. Puckree, Antibiotic resistance profiles of environmental isolates from Mhlathuze river, KwaZulu-Natal (RSA), *Water SA*, 2004, p. 23.
121 A. Hartmann, A.C. Alder, T. Koller and R.M. Widmer, *Environ. Toxicol. Chem.*, 17 (1998) 377.
122 A. Hartmann, E.M. Golet, S. Gartiser, A.C. Alder, T. Koller and R.M. Widmer, *Arch. Environ. Contam. Toxicol.*, 36 (1999) 115.
123 K.T. Elvers and S.J.L. Wright, *Lett. Appl. Microbiol.*, 20 (1995) 82.
124 B. Chowdhury, M. Adak and S.K. Bose, *Lett. Appl. Microbiol.*, 37 (2003) 158.
125 A.K. Sanyal, B. Chowdhury and A.B. Banerjee, *Lett. Appl. Microbiol.*, 14 (1992) 221.
126 M.A. Missinou, S. Borrmann, A. Schindler, S. Issifou, A.A. Adegnika, P.-B. Matsiegui, R. Binder, B. Lell, J. Wiesner, T. Baranek, H. Jomaa and P.G. Kremsner, *Lancet*, 360 (2002) 1941.
127 R. Gartiser, L. Brinker, A. Uhl, R. Willmun, K. Kummerer and F. Dachner, *Industrieabwasser*, 9 (1994) 1618.
128 F. Giuliani, T. Koler, F.E. Wurgler and R.M. Widmer, *Mutat. Res.*, 368 (1996) 49.
129 T. Steger-Hartmann, K. Kummerer and A. Hartmann, *Ecotoxicol. Environ. Saf.*, 36 (1997) 174.

130 L. Hellmér and G. Bolcsfoldi, *Mutat. Res.*, 272 (1992) 145.

131 T. Steger-Hartmann, R. Lange and H. Schweinfurth, *Ecotoxicol. Environ. Saf.*, 42 (1999) 274.

132 D.J. Paustenbach, *Human and Ecological Risk Assessment: Theory and Practice*, Wiley-Interscience, Chichester, 2002, p. 1556.

133 T.H. Hutchinson and D.B. Pickford, *Toxicology*, 181–182 (2002) 383.

134 L.J. Schulman, E.V. Sargent, B.D. Naumann, E.C. Faria, D.G. Dolan and J.P. Wargo, *Human Ecol. Risk Assess.*, 8 (2002) 657.

135 D.R. Dietrich, S.F. Webb and T. Petry, *Toxicol. Lett.*, 131 (2002) 1.

136 J.P. Seiler, *Toxicol. Lett.*, 131 (2002) 105.

137 FDA, Guidance for industry for the submission of an environmental assessment in human drug applications and supplements, 1995, p. 39.

138 EMEA, Draft CPMP discussion paper on environmental risk assessment of non-genetically modified organisms (non-GMO) containing medicinal products for human use. CPMP/SWP/4447/00 draft corr., EMEA, London, UK, 2001.

139 EMEA, Note for guidance on environmental risk assessment of medicinal products for human use. CPMP/SWP/4447/00, The European Agency for the Evaluation of Medicinal Products, London, UK, 2003.

140 EMEA, Note for guidance on environmental risk assessment of medicinal products for human use. CMPC/SWP/4447/draft, The European Agency for the Evaluation of Medicinal Products, London, UK, 2005.

141 J.P. Bound and N. Voulvoulis, *Chemosphere*, 56 (2004) 1143.

142 A.P. van Wezel and T. Jager, *Chemosphere*, 47 (2002) 1113–1128.

143 R. Lange and D. Dietrich, *Toxicol. Lett.*, 131 (2002) 97.

144 A.B.A. Boxall, D. Oakes, P. Ripley and C.D. Watts, *Chemosphere*, 40 (2000) 775.

145 M.J. Hilton, K.V. Thomas and D. Ashton, Targeted monitoring program for pharmaceuticals in the aquatic environment, R&D Technical Report P6-012/06/TR, 2003, p. 57.

146 S. Ball, S. Bell and D. McGillivray, *Environmental Law*, Blackstone Press, London, UK, 2000, p. 776.

147 G. Kallis and D. Butler, *Water Policy*, 3 (2001) 125.

148 F. Sacher, F.T. Lange, H.J. Brauch and I. Blankenhorn, *J. Chromatogr. A*, 938 (2001) 199.

149 K.L. Froese and K.A.M. Bodo, *Water Environ. Res.*, 71 (1999) 1119.

150 T. Heberer, K. Schmidt-Baumler and H.J. Stan, *Acta Hydroch. Hydrob.*, 26 (1998) 272.

151 T. Heberer, K. Reddersen and A. Mechlinski, *Water Sci. Technol.*, 46 (2002) 81.

152 R. Tauber, Quantitative analysis of pharmaceuticals in drinking water from ten Canadian cities, Enviro-Test Laboratories/Xenos Division, Ontario, 2003, p. 6.

153 S. Webb, T. Ternes, M. Gibert and K. Olejniczak, *Toxicol. Lett.*, 142 (2003) 157.
154 T.A. Ternes, M. Meisenheimer, D. McDowell, F. Sacher, H.-J. Brauch, B. Haist-Gulde, G. Preuss, U. Wilme and N. Zulei-Seibert, *Environ. Sci. Technol.*, 36 (2002) 3855.
155 T. Heberer, *J. Hydrol.*, 266 (2002) 175.
156 T. Heberer, U. Dunnbier, C. Reilich and H.J. Stan, *Fresen. Environ. Bull.*, 6 (1997) 438.
157 K. Langford, M.D. Scrimshaw, J.W. Birkett and N. Lester J, *Water Res.*, 39 (2005) 870.
158 T. Heberer, *Toxicol. Lett.*, 131 (2002) 5.

CHAPTER 4: REMOVAL OF PHARMACEUTICALS IN WASTEWATER AND DRINKING WATER TREATMENTS

Chapter 4.1

Removal of pharmaceutical residues during wastewater treatment

Jörg E. Drewes

4.1.1 INTRODUCTION

Some of the most representative pharmaceutical residues found in wastewater treatment plants are antibiotics, blood lipid regulators, anti-inflammatories, antiepileptics, tranquillizers, X-ray contrast agents, and contraceptives [1–6]. Until very recently, relatively little was known about the efficiency with which pharmaceutical residues are removed during wastewater and biosolid treatment processes. Conventional wastewater treatment facilities are not specifically designed to remove pharmaceutical residues, and the degrees with which they are removed vary from nearly complete to very little [7,8].

There are a number of operational factors that are likely to influence the biological removal of pharmaceutical residues in activated sludge systems such as biochemical oxygen demand (BOD_5), suspended solids (SS) loading, hydraulic residence time (HRT), solids retention time (SRT), food–microorganism ratio (F/M ratio), mixed liquor-suspended solids (MLSS), pH, and temperature. The SRT is related to the growth rate of microorganisms. High SRTs allow the enrichment of slowly growing bacteria leading to the establishment of a more diverse biocoenosis with broader physiological capabilities. Frequently, however, these operational details are lacking in studies reported in the literature on the fate and transport of pharmaceutical residues during wastewater treatment.

Removal of organic micropollutants in wastewater unit processes is determined by their biodegradability and physicochemical properties, most notable their water solubility, hydrophobicity, and tendency to volatilize. Physicochemical properties of pharmaceutical residues are summarized in the appendix. These properties will influence whether a

Comprehensive Analytical Chemistry 50
Petrović and Barceló (Eds)
Volume 50 ISSN: 0166-526X DOI: 10.1016/S0166-526X(07)50013-9

compound will remain in the aqueous phase (like many of the acidic, neutral, and basic pharmaceuticals) or interact with solid particles, such as estrogens or certain antibiotics, which have a higher potential to be adsorbed to sewage sludges. Findings from multiple studies as presented in this chapter have demonstrated that adsorption onto suspended solids, aerobic, and anaerobic biodegradation, chemical (abiotic) degradation (via processes such as hydrolysis), and volatilization are the primary removal mechanisms for pharmaceutical residues in wastewater. Sorption and volatilization are physical processes and their relevance for specific contaminants can be predicted using physicochemical property information. Hydrophobic contaminants may partition onto primary or secondary sludge solids and the tendency to accumulate in sludge solids can be assessed using the octanol–water partition coefficient (K_{ow}). Rogers [9] proposed the following guide to assess the sorption potential of organic contaminants:

$\log K_{ow} < 2.5$	Low sorption potential
$2.5 < \log K_{ow} < 4.0$	medium sorption potential
$\log K_{ow} > 4.0$	High sorption potential

During wastewater treatment many contaminants partition onto solids as a consequence of their hydrophobic nature resulting in enrichment in biosolids at concentrations several orders of magnitude higher than in the raw wastewater.

Volatilization losses of organic compounds during wastewater treatment can be estimated using the following empirically defined categories based on Henry's Law constant (H_c) and K_{ow} [9]:

$H_c > 1 \times 10^{-4}$ and $H_c/K_{ow} > 1 \times 10^{-9}$	High-volatilization potential
$H_c < 1 \times 10^{-4}$ and $H_c/K_{ow} < 1 \times 10^{-9}$	Low-volatilization potential

Biodegradation of pharmaceutical residues might occur during secondary treatment, which involves both aerobic (trickling filters, activated sludge treatment) and anaerobic (sludge digestion) processes. Although the mechanism of degradation of the bulk organic matter of wastewater during aerobic and anaerobic processes is well understood [10], the effects of such processes on pharmaceutical residues occurring at the parts-per-trillion (ppt) level have received relatively little focused study.

An exact determination of biodegradation rates is extremely difficult due to the large number of mostly unknown products [11]. No systematic and comprehensive work has described the dimensions of pharmaceutical residues issues in wastewater treatment, including origins, distributions, fate, and transport. Beside these uncertainties regarding available information, various approaches have been proposed to estimate pharmaceutical residue concentrations in sewage. Some of these studies have focused on closed systems such as hospitals [12,13]. Other studies have utilized prescription rate data in combination with per-capita sewage volume [14–16] or combined prescription data, physicochemical information and mass balances approaches [11,17] to predict the fate of pharmaceutical residues through conventional wastewater unit operations. Up to now, concentration predictions derived from these studies for secondary-treated effluent qualities can only be considered as illustrative due to the lack and uncertainties of input data and limitations in biodegradation modeling. However, findings of these studies can assist in highlighting priorities for further research into the fate and transport of pharmaceutical residues during wastewater treatment.

4.1.1.1 Acidic, neutral, and basic pharmaceutical residues

Most of the acidic, neutral, and basic pharmaceutical residues come either from domestic sources or from hospital or industrial discharges and contribute to municipal sewage. Previous studies have demonstrated that removal of these pharmaceuticals in municipal wastewater treatment plants is incomplete [1], with efficiencies varying between 60 and 90 percent. The removal efficiency of pharmaceutical residues is influenced by their physicochemical properties as well as operational conditions of the secondary treatment process. Given their physicochemical properties (see Appendix), volatilization in the sewer or during wastewater treatment is negligible for the majority of pharmaceutical residues. The removal is dependent upon adsorption of sewage sludge and degradation/transformation processes during biological treatment.

4.1.1.2 Antibiotics

Among the wide variety of pharmaceutical residues, antibiotics assume special significance due to their extensive use in human therapy, veterinary medicine, and as husbandry growth promoters. Antibiotics can be classified into sulfonamides, macrolides, tetracyclines, fluoroquinolones, and others (Table 4.1.1). Sulfonamides have become the most widely used class of antibiotics in the world. Antibiotics used to

TABLE 4.1.1

Antibiotics commonly present in municipal wastewater

Sulfonamides	Macrolides	Tetracyclines	Fluoroquinolones	Others
Sulamethizole	Tylosin	Tetracycline	Norfloxacin	Lincomycin
Sulfathiazole	Roxithromycin	Oxytetracycline	Enrofloxacin	Trimethoprim
Sulfamerazine		Erythromycin–H_2O	Chlortetracylcine	Ciprofloxacin
Sulfamethazine	Clarithromycin	Doxycycline	Sarafloxacin	
Sulfochloropyridazine			Ofloxacin	
Sulfamethoxazole			Oxolinic acid	
Sulfadimethoxine			Pipemidic acid	
Sulfacetamide				
Sulfadiazine				
Sulfaguanidine				
Sulfamethoxypyridazine				
Sulfamoxole				
Sulfapyridine				
Sulfaquinoxaline				
Sulfisomidin				
Sulfisoxazole				

Source: Adapted from [42] and [43].

treat humans in hospitals or by prescription are ultimately released into domestic sewage. Members of the lactam class of antibiotics, including penicillins and cephalosporins, readily undergo hydrolysis [18] and were never detected in concentrations above 20 ng/L in wastewater effluents as reported by Hirsch et al. [19]. Studies focusing exclusively on wastewater treatment systems regarding fate and transport of antibiotics are, however, limited.

4.1.1.3 X-ray contrast agents

Contrast agents are used to get detailed images of soft tissues in X-ray radiography. Among them is the group of iodinated organics, which are widely used. Most of these contrast media are derivatives of 2,4,6-triiodobenzoic acid possessing polar carboxylic and hydroxy moieties in their side chains. The iodinated X-ray contrast agents can be divided into ionic agents containing a free carboxulic moiety (diatrizoate, iothalamic acid, and ioxithalamic acid) and nonionic agents in which all carboxylic moieties are amide derivatives (iopamidol, iopromide, and iomeprol) [20]. Iodinated X-ray contrast media are biologically inert and metabolically stable during their passage through the body. Within a day, they are excreted unmetabolized almost quantitatively. Contrast agents are designed to be highly hydrophilic and therefore are not expected to accumulate in organisms or biosolids.

4.1.1.4 Steroid hormones

The synthetic steroid hormone 17α-ethinylestradiol (EE2) is the main estrogenic component of the combined oral contraceptive pill, which contains between 30 and 50 μg of EE2 per pill [21]. It is important to note that steroid hormones, especially 17α-ethinylestradiol and 17β-estradiol, can be present in wastewater effluents at concentration above the levels that are shown to cause endocrine effects in fish [8,22,23]. The results of multiple studies have demonstrated that adsorption onto suspended solids, aerobic and anaerobic degradation, and hydrolysis are the primary removal mechanisms for steroid hormones in wastewater treatment processes [24–26]. Photolytic decay has also been suggested as a removal mechanism for steroid hormones during exposure times exceeding 24 h [26,27], but this mechanism has less relevance for conventional wastewater treatment.

4.1.2 REMOVAL DURING PRIMARY TREATMENT

Acidic drugs, such as ibuprofen and naproxen, the antibiotic sulfamethoxazole, and the contrast agent iopromide with very low solid–liquid partition coefficients, exhibited no removal during primary treatment [28]. Tauxe-Wuersch et al. [29], however, reported a 32 percent removal of ibuprofen during primary treatment. Of the acidic drugs, only diclofenac exhibited higher sorption characteristics in controlled experiments using wastewater sludges leading to a 5–15 percent partitioning to particulate matter present in raw wastewater [11]. Acidic drugs, such as fenofibric acid, ibuprofen or clofobric acid, as well as neutral pharmaceuticals, such as phenazone and propyphenazone, did not sorb to primary sludges in a study reported by Ternes et al. [11]. During mechanical treatment, Golet et al. [30] reported that the fluoroquinolones, ciprofloxacin and norfloxacin, were removed by 35 and 28 percent, respectively, due to sorption onto suspended solids and recirculated excess sludge. A study conducted by Nasu et al. [31] suggested little to no elimination of estrogens during primary treatment. Sorption onto primary sludges was also not observed for iopromide [11]. Diclofenac and certain antibiotics are characterized by higher log K_{ow} values exceeding 4, which might indicate a tendency to sorb to suspended matter.

4.1.3 REMOVAL DURING SECONDARY TREATMENT

4.1.3.1 Acidic, neutral, and basic pharmaceutical residues

Influent and effluent concentrations reported by various studies are summarized in Table 4.1.2. Moehle and Metzger [32] conducted controlled batch experiments simulating activated sludge treatment and observed an initial loss in concentrations of fortified wastewater after 15 min of exposure to activated sludge. This removal of acidic and neutral drug residues (e.g., diclofenac, propyphenazone, carbamazepine, and primidone) was attributed to initial adsorption to the sludge although these compounds span a wide range of hydrophobicities (log K_{ow}), which would not suggest a high tendency to adsorb onto biosolids. Likely, the adsorption observed in these experiments was not in equilibrium and Ternes et al. [11] reported no appreciable sorption of carbamazepine onto biosolids in their controlled experiments. Sorption experiments conducted in the latter study suggested no significant role of sorption for the removal of acidic (fenofibric acid, ibuprofen, and clofibric acid) and neutral drugs (phenazone, propyphenazone, and glibenclamide) [11].

Kreuzinger et al. [33] investigated highly loaded activated sludge plants with an SRT of one day or less and observed no removal of select pharmaceutical residues (i.e., ibuprofen, diclofenac, bezafibrate). During activated sludge treatment, ibuprofen and naproxen were removed by 60–70 percent and 40–55 percent, respectively [28]. A similar removal of ibuprofen and naproxen as well as other acidic drugs was reported by Stumpf et al. [34] for an activated sludge system in Brazil and for activated sludge system in Switzerland [29]. Vieno et al. [35] observed an ibuprofen removal exceeding 98 percent in an oxidation ditch system in Finland. Clara et al. [36] reported no removal of ibuprofen in a full-scale facility with short SRT (2 days) providing no nitrification, but a removal of 98 percent in a denitrifying facility with a SRT of 48 days. Findings of this study allowed deriving a critical SRT of 10 days for complete removal of ibuprofen in activated sludge systems. Buser et al. [37] reported the efficient removal of ibuprofen exceeding 98 percent in activated sludge facilities with longer SRTs. Additional findings derived from controlled laboratory studies revealed a residence time of wastewater in excess of 6 h for complete removal of ibuprofen. Complete removal of ibuprofen and the lipid regulator bezafibrate was also reported for an activated sludge facility employing an SRT of more than 50 days [36]. Ternes [1] reported a removal of 83 percent for

TABLE 4.1.2

Removal of acidic, neutral, and basic pharmaceutical residues during secondary treatment processes

Category	Compound	Treatment process	HRT (hours)	SRT (days)	MLSS (mg/L)	Influent concentration (ng/L)	Effluent concentration (ng/L)	Reference
Acidic pharmaceuticals	Ibuprofen	Nitrification/denitrification	24	N/A	2100	2600–5700	910–2100	[28]
	Ibuprofen	Trickling filter	N/A	N/A	N/A	330	260	[34]
	Ibuprofen	Partially nitrifying	N/A	N/A	N/A	330	80	[34]
	Ibuprofen	No nitrification	N/A	2	4000	2300	2400	[36]
		Nitrification/denitrification	N/A	48	3100	1200	24	[36]
	Ibuprofen	Nitrification	N/A	N/A	N/A	990–3300	2	[37]
	Ibuprofen	Oxidation ditch w/chem. P-removal	36	20	N/A	23,400	40	[35]
	Naproxen	Nitrification/denitrification	24	N/A	2100	1800–4600	800–2600	[28]
	Naproxen	Trickling filter	N/A	N/A	N/A	600	520	[34]
	Naproxen	Partially nitrifying	N/A	N/A	N/A	600	120	[34]
	Naproxen	Oxidation ditch w/chem. P-removal	36	20	N/A	8600	420	[34]
	Gemfibrozil	Trickling filter	N/A	N/A	N/A	300	250	[34]
	Gemfibrozil	Partially nitrifying	N/A	N/A	N/A	300	170	[34]
	Diclofenac	Trickling filter	N/A	N/A	N/A	790	720	[34]
	Diclofenac	Partially nitrifying	N/A	N/A	N/A	790	200	[34]
	Diclofenac	No nitrification	N/A	2	N/A	1400	1300	[36]
	Diclofenac	Nitrification/denitrification	N/A	48	N/A	905	780	[36]
	Diclofenac	Oxidation ditch w/chem. P-removal	36	20	N/A	460	400	[35]
	Ketoprofen	Trickling filter	N/A	N/A	N/A	520	260	[34]
	Ketoprofen	Partially nitrifying	N/A	N/A	N/A	520	180	[34]
	Ketoprofen	Oxidation ditch w/chem. P-removal	36	20	N/A	2900	230	[35]
	Fenofibric acid	Trickling filter	N/A	N/A	N/A	420	400	[34]
	Fenofibric acid	Partially nitrifying	N/A	N/A	N/A	420	240	[34]
	Bezafibrate	Trickling filter	N/A	N/A	N/A	1180	950	[34]
	Bezafibrate	Partially nitrifying	N/A	N/A	N/A	1180	590	[34]
	Bezafibrate	Oxidation ditch w/chem. P-removal	36	20	N/A	460	140	[35]
	Clofibric acid	Trickling filter	N/A	N/A	N/A	1000	850	[34]
	Clofibric acid	Partially nitrifying	N/A	N/A	N/A	1000	650	[34]
Neutral and basic pharmaceuticals	Carbamazepine	No nitrification	N/A	2	4000	670	690	[36]
	Carbamazepine	Nitrification/denitrification	N/A	48	3100	325	465	[36]

bezafibrate and 90 percent in an activated sludge plant in Germany. In studies investigating activated sludge treatment, another commonly occurring lipid regulator, gemfibrozil, was also removed by 69 percent [1] and 46 percent [34]. The same acidic pharmaceuticals were less efficiently removed during trickling filter treatment [34].

In the study conducted by Clara et al. [36] and Strenn et al. [38], contradictory results were obtained for diclofenac where a significant removal was observed in some facilities, but in other wastewater treatment plants at comparable SRTs no or only slight removal was obtained. Similar contradictory results are documented in the literature for diclofenac. Buser et al. [37] and Heberer [4] reported no significant removal of diclofenac during wastewater treatment, two studies [1,39] reported an elimination of diclofenac in excess of 70 percent and one study [36] listed a removal between 40 and 60 percent, respectively. Vieno et al. [35] reported a 13 percent removal of diclofenac in an oxidation ditch in Finland. Clofibric acid was not degraded in full-scale activated sludge systems as reported in two studies [4,29]. However, the results of Ternes [1] and Stumpf et al. [34] revealed a removal of 15 percent during trickling filter treatment, 34 percent during activated sludge treatment, and 51 percent in an activated sludge system using ferric chloride. The reasons for these discrepancies require further study.

Clara et al. [36] evaluated several full-scale facilities in Austria and reported that the antiepileptic drug carbamazepine was not removed in any of the facilities, which confirms previous findings reported by others [4,33,40]. Only Ternes [1] reported a 7 percent removal of carbamazepine in an activated sludge plant in Germany. Carbamazepine does not adsorb to sewage sludge and usually effluent concentrations vary within the same range as the influent concentrations [40].

Dokianakis et al. [41] investigated the potential inhibition of nitrifiers due to the presence of pharmaceutical residues. Neutral and acidic drugs, such as propranolol, diclofenac, carbamazepine, and clofibrate, caused no measurable inhibition on nitrite-oxidizing bacteria.

4.1.3.2 Antibiotics

Karthikeyan and Meyer [42] studied the removal of four different classes of antibiotics, namely sulfonamides, macrolides, tetracyclines, and fluoroquinolones, in two activated sludge treatment facilities in Wisconsin. A total of six antibiotic compounds were detected in influent samples, including two sulfonamides (sulfamethazine and sulfamethoxazole), one tetracycline (tetracycline), fluoroquinolone (ciprofloxacin), macrolide (erythromycin–H_2O), and trimethoprim. Table 4.1.3 summarizes influent and effluent concentrations of antibiotics during wastewater treatment. Erythromycin is frequently not detected in its original form but as the degradation product with an apparent loss of one molecule of water (erythromycin–H_2O). The degradation product, which forms in aqueous solutions at pH < 7.0, no longer exhibits antibiotic properties [19]. Trimethoprim and tetracycline, the latter a broad-spectrum antibiotic, were present in every wastewater sample with higher concentrations observed in the fall as compared to the summer. Higher occurrence of trimethoprim coincided with elevated concentrations of sulfamethoxazole. First-line treatment of bacterial sinusitis, which is prevalent during the fall season, involves a combination of these two antibiotics [42]. However, in the early summer samples, sulfonamides were either near the detection limit or not detected. Sulfamethoxazole and sulfapyridine were also detected in all secondary effluents in a study from Canada [43]. Sulfamethoxazole has also been frequently detected in wastewater effluents in Germany with a median concentration of 400 ng/L [19]. Tetracyclines have complexing properties, which can easily bind to calcium and similar ions, thus forming stable complexes, which can bound to suspended matter or sewage sludge. These properties might explain why tetracyclines were not detected in concentrations above 50 ng/L in treated secondary effluents in Germany [19]. Tetracyline and doxycycline were, however, detected in Canadian wastewater effluents at median concentration of 151 and 38 ng/L, respectively, employing rather short HRTs and SRTs and no nutrient removal [43]. Erythromycin–H_2O concentrations in the secondary effluents in a study reported from the US varied between < 50 and 300 ng/L [42]. Erythromycin was detected during a field study in Canada in eight out of eight final effluent samples with a median concentration of 80 ng/L, while clarithromycin and roxithromycin were detected in six of the eight facilities with median concentrations of 87 and 8 ng/L, respectively [43]. In a study reported by McArdell et al. [44], only three macrolides, clarithromycin, roxithromycin, and erythromycin–H_2O, were present in secondary effluents in Switzerland with clarithromycin being the most abundant macrolide with concentrations varying between 57 and 328 ng/L. Roxithromycin and erythromycin–H_2O were detected in every wastewater treatment plant effluent in a German study at median concentrations of 680 and 2500 ng/L [19]. In treated effluent samples in a study in Switzerland, ciprofloxacin and norfloxacin were detected at concentrations of 45–108 ng/L and

TABLE 4.1.3

Removal of antibiotics during secondary treatment processes

Category	Compound	Treatment process	HRT (hours)	SRT (days)	MLSS (mg/L)	Influent concentration (ng/L)	Effluent concentration (ng/L)	Reference
Antibiotics	Sulfamethoxazole	Nitrification/denitrification	24	N/A	2100	600	250	[28]
	Sulfamethoxazole	Nitrification	N/A	N/A	N/A	<50–1250	50	[42]
	Sulfamethoxazole	Nitrification	N/A	N/A	N/A	80–170	210	[42]
	Sulfamethazine	Nitrification	N/A	N/A	N/A	110	<50	[42]
	Sulfamethazine	Nitrification	N/A	N/A	N/A	210	<50	[42]
	Tetracycline	Nitrification	N/A	N/A	N/A	510–790	50–160	[42]
	Tetracycline	Nitrification	N/A	N/A	N/A	240–270	<50–70	[42]
	Ciprofloxacin	Nitrification	N/A	N/A	N/A	<50–210	<50–60	[42]
	Ciprofloxacin	Nitrification	N/A	N/A	N/A	<50	<50	[42]
	Ciprofloxacin	Nitrification	N/A	11	N/A	331 ± 53	95 ± 15	[30]
	Norfloxacin	Nitrification	N/A	11	N/A	383 ± 61	69 ± 15	[30]
	Erythromycin–H_2O	Nitrification	N/A	N/A	N/A	<50–480	<50–270	[42]
	Erythromycin–H_2O	Nitrification	N/A	N/A	N/A	430–1200	90–300	[42]
	Trimethoprim	Nitrification	N/A	N/A	N/A	210–1100	70–550	[42]
	Trimethoprim	Nitrification	N/A	N/A	N/A	140–580	<50	[42]

48–120 ng/L, respectively [30]. Norflaxin was detected at similar concentrations in a study from Canada, but ciprofloxacin was detected at slightly higher concentrations (median concentration 118 ng/L) [43]. The differences in concentrations observed in secondary effluents in the US and Canada as compared to Germany and Switzerland might reflect differences in prescription practices in different countries.

The activated sludge processes investigated by Karthikeyan and Meyer [42] were capable of reducing the concentration of antibiotics, with a percent reduction of 100 percent for sulfamethazine, 53–100 percent for sulfamethoxazole, 70–100 percent for tetracycline, 70 percent for ciprofloxacin, 44–80 percent for erythromycin–H_2O, and 50–100 percent for trimethoprim, respectively. A sulfonamide and a macrolide were investigated in full-scale facilities by Kreuzinger et al. [33], who reported removal percentages of 33–62 percent for sulfamethoxazole and 0–61 percent for roxithromycin, respectively. Sorption to sewage sludge has been suggested as the primary removal mechanism for fluoroquinolones (ciprofloxacin and norfloxacin) during secondary wastewater treatment [30]. In a field study at a full-scale activated sludge facility in Switzerland, Golet et al. [30] attributed a 53–64 percent reduction of ciprofloxacin and norfloxacin concentrations to sorption processes during activated sludge treatment. McArdell et al. [44] confirmed the partial removal of macrolides during activated sludge treatment.

The presence of ofloxacin and sulfamethoxazole at the milligram per liter range exhibited a significant inhibition of nitrite-oxidizing bacteria in controlled batch experiments [41].

4.1.3.3 X-ray contrast agents

Six iodinated X-ray contrast media were present in German municipal influents and effluents with maximum concentrations exceeding 3000 ng/L and median concentrations varying between 250 and 750 ng/L [20]. It is noteworthy that different X-ray contrast agents were predominantly present in samples from different facilities, which is likely attributed to application practice of hospitals and radiological centers located in the service area of a wastewater plant. Occurrence and removal results for X-ray media from different studies are summarized in Table 4.1.4. Testing iopromide degradation in a laboratory-scale test system using activated sludge resulted in an elimination of more than 80 percent after a lag period of 31 days and the formation of a stable metabolite [45]. No significant removal of iopromide was

TABLE 4.1.4

Removal of X-ray contrast agents during secondary treatment processes

Category	Compound	Treatment process	HRT (hours)	SRT (days)	MLSS (mg/L)	Influent concentration (ng/L)	Effluent concentration (ng/L)	Reference
X-ray contrast agents	Iopromide	Nitrification/denitrification	24	N/A	2100	6000–7000	9300	[28]
	Iopromide	Nitrification	N/A	N/A	N/A	7500 ± 1500	8100 ± 1600	[20]
	Iopamidol	Nitrification	N/A	N/A	N/A	4300 ± 900	4700 ± 1000	[20]
	Diatrizoate	Nitrification	N/A	N/A	N/A	3300 ± 700	4100 ± 800	[20]
	Iothalamic acid	Nitrification	N/A	N/A	N/A	180 ± 100	140 ± 100	[20]
	Ioxithalamic acid	Nitrification	N/A	N/A	N/A	170 ± 100	160 ± 100	[20]
	Iomeprol	Nitrification	N/A	N/A	N/A	1600 ± 400	1300 ± 300	[20]

observed during secondary treatment [28] (Table 4.1.4). Iopromide was also not removed in full-scale facilities employing short SRTs (one day or less) but exhibited a certain degree of removal at longer SRTs [33]. Kalsch [46] conducted controlled batch experiments with activated sludges over 54 h and yielded biotransformation of iopromide into two stable metabolites. Under aerobic conditions, neither deiodination nor breakdown of the aromatic ring took place. In the same experiment, diatrizoate remained unaltered in the aqueous solution. Both contrast media and their metabolites are highly hydrophilic and were not adsorbed to activated sludge solids. Dehalogenation or mineralization of iodinated contrast agents during full-scale secondary treatment has not been observed [47].

4.1.3.4 Steroid hormones

A large proportion of 17α-ethinylestradiol ingested is excreted as unmetabolized glucuronide conjugates primarily through urine [48,49]. Steroid hormones, including EE2, are deconjugated quickly through enzymatic hydrolysis in the wastewater collection system or primary treatment process introducing the biologically active form of EE2 into wastewater [50–52]. Studies that assess both influent and effluent steroid hormone concentrations are still rare. However, based upon studies reported from Europe and North America (Table 4.1.5), EE2 usually occurs in domestic sewage at concentrations ranging from not detectable (<0.1 ng/L) to less than 10 ng/L [7,8,21,23,40,50,51,53,54,55]. Facilities employing nitrogen removal (nitrifying or nitrifying/denitrifying) can achieve effluent concentrations consistently below 1 ng/L. Ternes et al. [7] examined EE2 removal at a trickling filter plant in Brazil and reported a less efficient removal as compared to an activated sludge system receiving the same influent. This finding is supported by a study in Sweden, which reported that trickling filter facilities did not remove a significant amount of estrogenic compounds [56].

During a pilot-scale study simulating an activated sludge process, Onda et al. [57] was unable to establish strong correlations between estrogen removal and BOD loading or other operational conditions. By measuring mass fluxes of EE2 at a full-scale nitrifying/denitrifying facility, Andersen et al. [51] were able to quantify that about 90 percent of the EE2 was eliminated through aerobic degradation. The sorbed load of EE2 onto the excess and digested sludge at this facility was lower than 6 percent of the inlet load suggesting little removal of EE2

TABLE 4.1.5

Removal of steroid hormones during secondary treatment processes

Category	Compound	Treatment process	HRT (hours)	SRT (days)	MLSS (mg/L)	Influent concentration (ng/L)	Effluent concentration (ng/L)	Reference
Steroid hormones	17α-ethinylestradiol	Nitrification	14	N/A	N/A	1.8	0.36	[8]
		Nitrification	14	N/A	N/A	0.4	<0.3	[8]
		Nitrification	14	N/A	N/A	4.6	0.35	[8]
		Nitrification	14	N/A	N/A	2.9	0.73	[8]
		Nitrification	14	N/A	N/A	1.7	0.31	[8]
		Nitrification/denitrification	N/A	11–13	2800	6.2–10.1	<1	[51]
		Nitrification, biophosphorus removal	13.7	10.0	1900	1.9	<0.7	[55]
		Nitrification, chem. phosphorus removal	5.5	8.5	2320	9.1	<0.7	[55]
		No nitrification	1.2	1.7	2180	14.4	4.1	[55]
		No nitrification	N/A	2	4000	8	5	[36]
		Nitrification/denitrification	N/A	48	3100	70	< 1	[36]

through adsorption onto suspended solids, which is supported by rather small sorption coefficients for EE2 onto colloidal organic carbon derived from activated sludge as determined by others [11,58,59]. The plant targeted in this study had just been updated to achieve nutrient removal at SRTs of 11–13 days. Prior to the update, the facility operated at SRTs of less than four days and a previous study at the same facility had revealed only minor reductions of estrogens [7,39]. Removal of EE2 during activated sludge treatment between 75 and 90 percent was also observed during full-scale studies of wastewater treatment facilities in Italy, however, no information was provided in this study regarding SRTs or nitrification capabilities of the plants investigated [8]. In a study reported by Drewes et al. [55], a not nitrifying facility with a SRT of less than 1.7 days achieved only a 70 percent removal of EE2 with an effluent concentration of 4.1 ng/L, whereas all nitrifying plants employing longer SRTs exhibited effluent concentrations of less than 0.7 ng/L. Kreuzinger et al. [33] conducted mass balances for EE2 at various full-scale facilities in Austria and reported EE2 removal between 70 and 80 percent for plants with SRTs exceeding nine days. Consistently, higher concentrations of EE2 in not nitrified effluents representing short SRTs were also reported in other studies [21,50] stressing the fact that longer SRTs seem to promote the growth of microorganisms capable of degrading EE2. Longer SRTs can also result in modified sorbent characteristics as suggested by Holbrook et al. [25], which might be beneficial for estrogen mineralization as well. Layton et al. [60] conducted controlled biodegradation experiments with radiolabled EE2 using biosolids from a wastewater treatment plant with short SRT. Removal of EE2 under these conditions was considerably less than the removal of ^{14}C–17β-estradiol, 20 vs. 75 percent, indicating that the ethinyl group might inhibit degradation. Controlled experiments conducted by Shi et al. [49] with nitrifying activated sludge collected from a facility employing longer SRTs demonstrated very similar degradation rate constants for natural hormones and the synthetic hormone EE2. The study also confirmed that a consortium of bacteria rather than an individual specie (such as *Nitrosomonas europaea*) is likely responsible for the biodegradation of estrogens including EE2. Similar observations regarding the decrease of EE2 concentrations in controlled experiments using nitrifying activated sludge was also reported by Vader et al. [61]. Since nitrifying bacteria have a lower growth rate at lower temperatures, prevalent during winter operation, ammonia removal is usually lower during winter seasons. Monitoring studies available today, however, do not suggest that a declining nitrification activity will also result in a less efficient removal of EE2 [8,21,23].

4.1.4 REMOVAL DURING TERTIARY TREATMENT

Findings from Jar tests conducted by Chang et al. [62] over a range of ferric chloride dosages and pH conditions showed that coagulation was ineffective in removal of steroid hormones from secondary effluent. This finding is consistent with results from a bench-scale study using drinking water samples spiked with pharmaceutical residues including EE2 [63]. Of the pharmaceutical residues screened in this study, no compound exhibited a removal of more than 20 percent during alum coagulation and only erythromycin–H_2O showed a removal of 33 percent. Ferric coagulation in this study achieved comparable removals as equivalent alum dosages. Three full-scale wastewater treatment facilities in Sweden employing only chemical precipitation with Al or Fe without biological treatment had no significant reduction in estrogenic activity [56]. A fourth plant employing only lime softening at elevated pH [11.4] and no biological treatment was more effective and removed 73 percent of estrogenic compounds (Table 4.1.6).

Golet et al. [30] reported a minor removal of ciprofloxacin (4 ± 1 percent) and norfloxacin (3 ± 2 percent) during tertiary treatment using flocculation/filtration, probably due to sorption of fluoroquinolones to small particles and precipitates.

4.1.5 REMOVAL DURING DISINFECTION

Limited studies have focused on the removal of estrogens during wastewater disinfection. Lee et al. [64] explored the removal of 17β-estradiol during oxidation with free chlorine at dosages of 1–7 mg/L, commonly applied in drinking water applications. Whereas low-chlorine levels of 1.5 mg/L required more than 36 h for complete

TABLE 4.1.6

Removal of antibiotics during tertiary treatment processes

Category	Compound	Treatment process	Influent concentration (ng/L)	Effluent concentration (ng/L)	Reference
Antibiotics	Ciprofloxacin	Flocculation/filtration	95 ± 15	71 ± 11	[30]
	Norfloxacin	Flocculation/filtration	69 ± 15	51 ± 7	[30]

removal of 17β-estradiol, chlorination for 10 min at a higher dose of free chlorine (7.5 mg/L) achieved a complete removal. Westerhoff et al. [63] conducted controlled chlorination studies with surface water using dosages between 3.5 and 3.8 mg/L and demonstrated a complete removal of steroid hormones within a contact time of 24 h. Under these conditions, other pharmaceutical residues also exhibited a high degree of reactivity with chlorine to concentrations below the limit of detection (acetaminophen, diclofenac, naproxen, oxybenzone, sulfamethoxazole, and triclosan) [63]. Controlled studies conducted by Boyd et al. [65] also confirmed that naproxen readily reacts with free chlorine.

During wastewater disinfection, chlorine doses of 10–20 mg/L are commonly applied with contact times often exceeding 10 min [10]. Drewes et al. [66] collected composite samples prior to and after chlorine disinfection of tertiary-treated effluent at a facility applying chlorine dose of 3.5 mg/L with a contact time of approximately 45 min. Estrogens present in the tertiary effluent were removed below the detection limit (less than 0.4 ng/L). Since a common structural characteristic of all estrogenic chemicals is the presence of a phenolic ring and considering that a phenolic ring is likely to undergo a transformation in an aqueous chlorinated solution, the observations made by Lee et al. [64] and Drewes et al. [66] are likely applicable to EE2 as well.

4.1.6 REMOVAL DURING MEMBRANE TREATMENT

4.1.6.1 Microfiltration/ultrafiltration

The molecular weight of the majority of pharmaceutical residues ranges between 200 and 400 g/mole. Iodinated contrast media can have molecular weights between 700 and 1000 g/mole. Common pore sizes of microfiltration (MF) and ultrafiltration (UF) membranes are well above several thousand Daltons. Therefore, steric exclusion of pharmaceutical residues in MF and UF membranes is not relevant. However, highly hydrophobic pharmaceutical residues can still adsorb onto MF and UF membrane surface and might partition through the membrane into the filtrate. Chang et al. [62] conducted controlled experiments with hollowfiber MF membranes and reported a significant accumulation of estrone on the membrane surface.

4.1.6.2 Nanofiltration/reverse osmosis

Some of the factors affecting the permeation of pharmaceutical residues during reverse osmosis (RO) and nanofiltration (NF) treatment are fairly well understood, such as physical sieving of solutes larger than the molecular cutoff in an NF membrane. Other mechanisms of rejection such as electrostatic exclusion and hydrophobic–hydrophobic interactions between membrane and solute have been studied mainly at the bench-scale [67–71]. The following key solute parameters have been identified by Bellona et al. [72] in a comprehensive literature review to primarily affect solute rejection during NF/RO: molecular weight (MW), molecular size (length and width), acid dissociation constant (pK_a), hydrophobicity/hydrophilicity ($\log K_{ow}$), and diffusion coefficient (D_p). Key membrane properties affecting rejection that were identified include molecular weight cutoff (MWCO), pore size, surface charge (quantified as zeta potential), hydrophobicity/hydrophilicity (measured as contact angle), and membrane surface morphology (measured as roughness).

Findings from laboratory-scale studies indicated that the membrane surface charge of NF and RO membranes determines the rejection of ionic hydrophilic solutes, such as acidic pharmaceutical residues and dissociated antibiotics [73,74]. Speciation of acidic pharmaceuticals as a function of pH may result in a dramatic change of rejection [74,75]. For neutral pharmaceutical species, intrinsic physicochemical properties of the pharmaceuticals can substantially affect their rejection. Polarity (represented by the dipole moment) and hydrophobicity can affect the separation of molecules in NF and RO membranes [74]. In controlled laboratory experiments, hydrophilic and hydrophobic non-ionic solutes, such as neutral pharmaceuticals (carbamazepine, primidone) and 17α-ethinylestradiol, exhibited a high degree of rejection dependent upon the solute size and molecular weight cutoff of the membrane [73]. Increased recovery resulting in an increased concentration gradient across the membrane did not affect rejection of hydrophilic negatively charged and nonionic compounds. Hydrophobic compounds such as estrogens have a tendency to adsorb onto the hydrophobic membrane surfaces. Nghiem et al. [71] reported that steroid hormone rejection decreased rapidly initially, which was caused by the decrease in feed hormone concentration due to adsorption and increased permeate concentration due to diffusion across the membrane skin layer. The decline in hormone rejection with time was much more severe when colloidal fouling took place [76]. During field evaluations of a full-scale RO train reported by Drewes et al. [77], certain pharmaceutical residues and steroid hormones were detected in the microfiltered secondary effluent (i.e., primidone, carbamazepine, salicylic acid, estrone, and 17β-estradiol). None of the compounds were detected in the permeate samples. During field operation of membrane applications, membranes are exposed to organic matter and tend to foul. Owing to foulant precipitation and cake-layer formation, membrane surface characteristics change considerably (in terms of hydrophobicity, surface charge, functionality, and surface morphology), which potentially will affect transport of pharmaceutical residues as compared to virgin (unfouled) membranes. Xu et al. [78] demonstrated in controlled laboratory experiments that the transport of ionic organic micropollutants was hindered as a result of electrostatic exclusion likely due to a more negative membrane surface charge. Membrane fouling is also resulting in an increased adsorption capacity and reduced mass transport through partitioning and diffusion across the membrane. Ng and Elimelech [76] observed higher permeate concentrations of steroid hormones in the presence of colloidal fouling and speculated that the cake layer formed on the membrane surface hinders the back diffusion of the hormone from the membrane surface back to the bulk solution. Consequently, buildup of estradiol on the membrane surface provided a larger concentration gradient for its diffusion across the RO membrane. Therefore, findings of these studies suggest that fouling significantly affects the rejection of organic solutes by NF membranes while it is less important for thin film composite RO membranes.

4.1.7 CONCLUSIONS

In controlled batch experiments, most pharmaceutical residues exhibited low K_d values, indicating negligible sorption onto sewage sludges. These findings suggest that the removal of pharmaceutical residues observed in wastewater treatment plants is mainly the result of biodegradation. In biological systems, such as activated sludge, the solid retention time (SRT) is related to the growth rate of microorganisms. Although not all pharmaceutical residues exhibited a critical SRT, in general an increased biological degradation was observed with increasing SRT. A minimum hydraulic residence time (HRT) is also important for the removal of biodegradable pharmaceutical residues and rain events in areas with combined sewer systems compromised the removal efficiencies of activated sludge systems, likely due to shorter HRTs and washout of certain microorganisms [1,29]. The biological system is also affected by the temperature of the water and less-efficient treatment of pharmaceutical residues was observed during winter seasons in colder climates [35]. In addition to short SRTs, Kreuzinger et al. [33]

concluded that F/M ratios above 0.5 kg BOD_5 (kg TSS/d) are not suitable to remove biodegradable pharmaceutical residues in activated sludge systems.

REFERENCES

1 T.A. Ternes, *Water Res.*, 32 (1998) 3245.
2 C.G. Daughton and T.A. Ternes, *Environ. Health Perspect.*, 107 (1999) 907.
3 J.E. Drewes and L.S. Shore, *ACS Symp. Ser.*, 791 (2001) 206.
4 T. Heberer, *Toxicol. Lett.*, 131 (2002) 1–2, 5.
5 X. Miao, B.G. Koenig and C.D. Metcalfe, *J. Chromatogr. A.*, 952 (2002) 139.
6 S.T. Glassmeyer, E. Furlong, D. Kolpin, J.D. Cahill, S. Zaugg, L. Werner, Meyer and D.D. Kryak, *Environ. Sci. Technol.*, 39 (2005) 5157.
7 T.A. Ternes, M. Stumpf, J. Mueller, K. Haberer, R.D. Wilken and M. Servos, *Sci. Total Environ.*, 225 (1999) 81.
8 C. Baronti, R. Curini, G. D'Ascenzo, A.D. Corcia, A. Gentili and R. Samperi, *Environ. Sci. Technol.*, 34 (2000) 5059.
9 H.R. Rogers, *Sci. Total Environ.*, 185 (1996) 3.
10 G. Tchobanoglous, F.L. Burton and H.D. Stensel, *Wastewater Engineering Treatment and Reuse*, 4th ed., McGraw-Hill, 2003.
11 T.A. Ternes, N. Herrmann, M. Bonerz, T. Knacker, H. Siegriest and A. Joss, *Water Res.*, 38 (2004) 4075.
12 K. Kuemmerer, T. Steger-Hartmann and M. Meyer, *Water Res.*, 31 (1997) 2705.
13 T. Heberer and D. Feldmann, *J. Hazar. Mater.*, 122 (2005) 211.
14 F. Stuer-Lauridsen, M. Birkved, M. Hansen, L.P. Lutzhoft and B. Halling-Sorensen, *Chemosphere*, 40 (2000) 783.
15 C. Huang, J.E. Renew, K.L. Smeby, K.E. Pinkston and D.L. Sedlak, *Water Resour. Update*, 120 (2001) 30.
16 D.L. Sedlak and K.E. Pinkston, *Water Resour. Update*, 120 (2001) 56.
17 S.J. Khan and J.E. Ongerth, *Chemosphere*, 54 (2004) 355.
18 A. AL-Ahmad, F.D. Daschner and K. Kuemmerer, *Arch. Environ. Contam. Toxicol.*, 37 (1999) 158.
19 R. Hirsch, T. Ternes, K. Haberer and K. Kratz, *Sci. Total Environ.*, 225 (1999) 109.
20 T.A. Ternes and R. Hirsch, *Environ. Sci. Technol.*, 34 (2000) 2741.
21 C. Desbrow, E.J. Routledge, G.C. Brighty, J.P. Sumper and M. Walddock, *Environ. Sci. Technol.*, 32(11) (1998) 1549.
22 E.J. Routledge, D. Shehan, C. Desbrow, G.C. Brighty, M. Waldock and J.P. Sumpter, *Environ. Sci. Technol.*, 32 (1998) 1559.

23 A.C. Belfroid, A. Van der Horst, A.D. Vethaak, A.J. Schaefer, G. Rijs, J. Wegener and W.P. Cofino, *Sci. Total Environ.*, 225 (1999) 101.

24 J.W. Birkett and J.L. Lester, *Endocrine Disruptors in Wastewater and Sludge Treatment Processes*, Lewis Publishers, 2003.

25 R.D. Holbrook, J.T. Novak, T.J. Grizzard and N.G. Love, *Environ. Sci. Technol.*, 36 (2000) 4533.

26 J. Mansell, J.E. Drewes and T. Rauch, *Water Sci. Technol.*, 50(2) (2004) 29.

27 J. Gray and D. Sedlak, *Proceedings 3rd International Conference on Pharmaceuticals and Endocrine Disrupting Chemicals in Water*, Minneapolis, MN, March 19–21, 2003. National Ground Water Association.

28 M. Carballa, F. Omil, J. Lema, M. Llompart, C. Garcia-Jares, I. Rodriguez, M. Gomez and T. Ternes, *Water Res.*, 38 (2004) 2918.

29 A. Tauxe-Wuersch, L.F. De Alencastro, D. Grandjean and J. Tarradellas, *Water Res.*, 39 (2005) 1761.

30 E.M. Golet, I. Xifra, H. Siegrist, A.C. Alder and W. Giger, *Environ. Sci. Technol.*, 37 (2003) 3243.

31 M. Nasu, M. Goto, H. Kato, Y. Oshima and A. Tanaka, *Water Sci. Technol.*, 43 (2001) 101.

32 E. Moehle and J.W. Metzger, *ACS Symp. Ser.*, 791 (2001) 192.

33 N. Kreuzinger, M. Clara, B. Strenn and H. Kroiss, *Water Sci. Technol.*, 50 (2004) 149.

34 M. Stumpf, T. Ternes, R. Wilken, S.V. Rodrigues and W. Baumann, *Sci. Total Environ.*, 225 (1999) 135.

35 N.M. Vieno, T. Tuhkanen and L. Kronberg, *Environ. Sci. Technol.*, 39 (2005) 8220.

36 M. Clara, N. Kreuzinger, B. Strenn, O. Gans and H. Kroiss, *Water Res.*, 39 (2005) 97.

37 H.-R. Buser, T. Poiger and D. Mueller, *Environ. Sci. Technol.*, 33 (1999) 2529.

38 B. Strenn, M. Clara, O. Gans and N. Kreuzinge, *Water Sci. Technol.*, 50(5) (2004) 269.

39 T.A. Ternes, P. Kreckel and J. Mueller, *Sci. Total Environ.*, 225 (1999) 91.

40 M. Clara, B. Strenn, M. Ausserleitner and N. Kreuzinger, *Water Sci. Technol.*, 50 (2004) 29.

41 S.N. Dokianakis, M.E. Kornaros and G. Lyberatos, *Water Sci. Technol.*, 50(5) (2004) 341.

42 K.G. Karthikeyan and M.T. Meyer, *Sci. Total Environ.*, 39 (2005) 2660.

43 X. Miao, F. Bishay, M. Chen and C.D. Metcalfe, *Environ. Sci. Technol.*, 38 (2004) 3533.

44 C.S. McArdell, E. Molnar, M. Suter and W. Giger, *Environ. Sci. Technol.*, 37 (2003) 5479.

45 T. Steger-Hartmann, R. Laenge, H. Schweinfurth, M. Tschampel and I. Rehmann, *Water Res.*, 36 (2002) 266.

46 W. Kalsch, *Sci. Total Environ.*, 225 (1999) 143.
47 J.E. Drewes, P. Fox and M. Jekel, *J. Environ. Sci. Health A*, 36 (2001) 1633.
48 J.L. Maggs, S.F. Grimmer, M.l'E. Orme, A. Brekenridge, B. Park and I.T. Gilmore, *Xenobiotica*, 13(13) (1983) 421.
49 J. Shi, S. Fujisawa, S. Nakai and M. Hosomi, *Water Res.*, 38 (2004) 2323.
50 C. Huang and D. Sedlak, *Environ. Toxicol. Chem.*, 20 (2001) 133.
51 H. Andersen, H. Siegrist, B. Halling-Sorensen and T. Ternes, *Environ. Sci. Technol.*, 37 (2003) 4021.
52 G. D'Ascenzo, A. Di Corcia, A. Gentili, R. Mancini, R. Mastropasqua, M. Nazzari and R. Samperi, *Sci. Total Environ.*, 302 (2003) 199.
53 A.C. Johnson, A. Belfroid and A. Di Corcia, *Sci. Total Environ.*, 256 (2000) 163.
54 S. Snyder, D. Villeneuve, E. Snyder and J. Giesy, *Environ. Sci. Technol.*, 35 (2001) 3620.
55 J.E. Drewes, J. Hemming, S. Ladenburger, J. Schauer and W. Sonzogni, *Water Environ. Res.*, 77(1) (2005) 12.
56 A. Svenson, A. Allard and M. Ek, *Water Res.*, 37 (2002) 4433.
57 K. Onda, Y. Nakamura, C. Takatoh, A. Miya and Y. Katsu, *Water Sci. Technol.*, 47 (2003) 109.
58 R.D. Holbrook, N.G. Love and J. T Novak, *Environ. Sci. Technol.*, 38 (2004) 3322.
59 H. Yamamoto and H.M. Liljestrand, *Water Sci. Technol.*, 47(9) (2003) 77.
60 A.C. Layton, B.W. Gregory, J.R. Seward, T.W. Schultz and G.S. Sayler, *Environ. Sci. Technol.*, 34 (2000) 3925.
61 J.S. Vader, C.G. van Ginkel, F.M. Sperling, J. de Jong, W. de Boer, J.S. de Graaf, M. van der Most and P.G. Stokman, *Chemosphere*, 41 (2000) 1239.
62 S. Chang, T.D. White, P. Ong, A.I. Schaefer and A.G. Fane, *J. Environ. Eng.*, 130 (2004) 736.
63 P. Westerhoff, Y. Yoon, S. Snyder and E. Wert, *Environ. Sci. Technol.*, 39 (2005) 6649.
64 B. Lee, M. Kamata, Y. Akatsuka, M. Takeda, K. Ohno, T. Kamei and Y. Magara, *Water Res.*, 38 (2004) 733.
65 G.R. Boyd, S. Zhang and D.A. Grimm, *Water Res.*, 39 (2000) 668.
66 J.E. Drewes, J.D. Hemming, J.J. Schauer and W. Sonzogni, Final Report 01-HHE-20-T, Water Environment Research Foundation (WERF), Alexandria, VA, 2006.
67 Y. Kiso, Y. Nishimura, T. Kitao and K. Nishimura, *J. Membr. Sci.*, 171 (2000) 229.
68 Y. Kiso, T. Kon T. Kitao and K. Nishimura, *J. Membr. Sci.*, 182 (2001) 205.
69 Y. Kiso, Y. Sugiura, T. Kitao and K. Nishimura, *J. Membr. Sci.*, 192 (2001) 1.
70 H. Ozaki and H. Li, *Water Res.*, 36 (2002) 123.

71 L.D. Nghiem, A.I. Schaefer and M. Elimelech, *Environ. Sci. Technol.*, 38 (2004) 1888.
72 C. Bellona, J.D. Drewes, P. Xu and G. Amy, *Water Res.*, 38 (2004) 2795.
73 P. Xu, J.E. Drewes, C. Bellona, G. Amy, T. Kim, M. Adam and T. Heberer, *Water Environ. Res.*, 77 (2005) 40.
74 L.D. Nghiem, A.I. Schaefer and M. Elimelech, *Environ. Sci. Technol.*, 39 (2005) 7698.
75 C. Bellona and J.E. Drewes, *J. Membr. Sci.*, 249 (2005) 227.
76 H.Y. Ng and M. Elimelech, *J. Membr. Sci.*, 244 (2004) 215.
77 J.E. Drewes, C. Bellona, M. Oedekoven, P. Xu, T. Kim and G. Amy, *Environ. Prog.*, 24(4) (2005) 400.
78 P. Xu, J.E. Drewes, T. Kim, C. Bellona and G. Amy, *J. Membr. Sci.*, 279 (2006) 165.

Chapter 4.2

Removal of pharmaceuticals by advanced treatment technologies

Maria Dolores Hernando Guil, Mira Petrović, Jelena Radjenovic, Antonio Rodríguez Fernández-Alba, Amadeo Rodríguez Fernández-Alba and Damia Barceló

4.2.1 INTRODUCTION

Many scientific reports have shown that elimination of pharmaceuticals from municipal sewage treatment plants (STP) is often incomplete and therefore a large number of pharmaceutically active compounds (PhACs) are detected in effluents entering the aquatic environment [1–6]. Owing to their physico-chemical properties (many of them are highly polar compounds) they are able to penetrate through all natural and man–made filtration steps and enter groundwater as well as drinking water [7–9].

The significance of pharmaceuticals as trace environmental pollutants in waterways, and on land to which treated sewage sludge or wastewater has been applied, is largely unknown. Owing to several facts that they deserve special attention: (i) because of continuous introduction via effluents from sewage treatment facilities and from septic systems pharmaceuticals are referred to as "pseudo" persistent contaminants (i.e. high transformation/removal rates are compensated by their continuous introduction into environment) (ii) they are developed with the intention of performing a biological effect, (iii) pharmaceuticals often have the same type of physico-chemical behaviour as other harmful xenobiotics (persistence in order to avoid the substance to be inactive before having a curing effect, and lipophilicity in order to be able to pass membranes), and (iv) pharmaceutical substances are used by man in quantities similar to those of many pesticides.

Numerous studies have been undertaken to evaluate the risk of other potentially harmful chemicals. Many of which have, eventually, either been banned completely, or had their use severely restricted

Comprehensive Analytical Chemistry 50
Petrović and Barceló (Eds)
Volume 50 ISSN: 0166-526X DOI: 10.1016/S0166-526X(07)50014-0

(i.e. tributyltin, alkylphenolic surfactants, penta and octa PBDEs). However, compared with other pollutants, the sources of pharmaceuticals are likely to be much more difficult to control and it is highly unlikely that they will be replaced or banned. Therefore, an efficient treatment of wastewaters is of great importance.

Although, adopted as the best available technology; biological treatment permits only partial removal of a wide range of PhACs, especially polar ones, which are consequently discharge into the aquatic environment. Thus, it has become evident that application of more enhanced technologies may be crucial for the fulfillment of the requirements of an indirect potable reuse of municipal and industrial wastewater. In recent years, new technologies are being studied for wastewater treatment. Among them membrane treatment, using both biological (membrane bioreactors) and non-biological processes (reversed osmosis, ultrafiltration, nanofiltration), and advanced oxidation processes (AOP) are most frequently considered as treatments that may be appropriate to remove trace concentrations of polar emerging contaminants.

Membrane bioreactor (MBR) is a technology that combines biological treatment with membrane filtration and can offer several advantages to conventional wastewater treatment including reduced footprint, consistent and superior effluent water quality. The MBR technology has become technically and economically feasible alternative for water and wastewater treatment, especially because of high-sludge retention time (SRT) achieved within compact reactor volumes (which improves treatment efficiency) and no limitations by settling characteristics of sludge (separation of suspended solids is achieved by membranes). A superior effluent quality not only allows reaching of an improved discharge quality, but also opens perspectives for direct and indirect reuse of industrial and municipal effluents. Although it is generally agreed that MBRs perform better than CAS for biological removal of bulk organic matter, their behaviour with respect to trace pollutants is much less documented. To date, MBR application has been related with industrial wastewater treatments [10], elimination of micro-contaminants (e.g. surfactants) [11–13]. Few works concerning the efficacy of MBR to eliminate specific chemicals such as pharmaceuticals have been published [14–18]. However, many MBR applications have been mainly developed evaluating removal efficiency in function of global parameters such as chemical oxygen demand (COD) or total organic carbon (TOC) [19–21] without an in-depth study of behaviour of trace organics. This is however crucial to understand the full potential of this technology and to assess its applicability for water reuse.

Another promising technology that has been applied for the treatment of wastewaters is based on AOPs. AOPs are defined as processes that generate free radicals, mainly hydroxyl radical ($^{\bullet}OH$) able to destruct water pollutants, via chemical (O_3/OH^-, O_3/H_2O_2; Fe^{2+}/H_2O_2) and photochemical (UV–C/H_2O_2, UV–C/O_3) (UV–A/TiO_2) processes, as well as different radiolytic processes (electron beam irradiation (EBI), γ-irradiation). Efficacy of AOPs to removal refractory organic pollutants in wastewaters is well documented in many studies [22–24] and a growing number of works related with pharmaceuticals is being published [25–28].

This chapter gives an overview of recent literature on efficacy of advanced treatment technologies (MBR and AOPs) applied to removal pharmaceuticals from wastewaters. Pharmaceuticals such as antibiotics, anti-inflammatories, lipid regulating agents, antiepileptics, beta-blockers, X-ray contrast agents and estrogens have been included in this review.

4.2.2 MEMBRANE BIOREACTOR (MBR) TECHNOLOGY

4.2.2.1 General aspects of MBR

The efficacy of MBR system is determined by the combination of filtration technology and biological treatment. Membranes can provide high removals, but dependent upon their properties as well as compounds structure (size, polarity). Generally, MBR process is performed using low-pressure membranes such as microfiltration (MF) or ultrafiltration (UF) that are capable of display a barrier effect based on size exclusion to high-molecular weight solutes. In the case of pharmaceuticals, most of them can range from 150 to 500 Daltons in molecular size, so those associated with particles or colloidal organic matter could be removed using MF or UF membranes. An additional removal by membranes can be expected for hydrophobic compounds since they are sorbed onto the particles that form a deposition layer on the membrane surface, while this effect for hydrophilic compounds is still not very well defined. The application of nanofiltration (NF) systems is also reported showing moderate to good removal for steroid hormones [29,30].

Other properties such as the material of membranes can also determine the removal efficiency of micro-contaminants. But, with the available information concerning removal of pharmaceuticals where the use of various membranes has been explored, the preliminary conclusions point towards no significant effects, at least, in case of ibuprofen in MBR system using membranes of different materials (organic polymers vs. ceramic) [31].

In MBR systems, factors such as high-sludge ages or sludge retention time (SRT) and high-biomass concentration could be also correlated with the improvement of degradation capacity [32,33]. Typical biomass concentrations in MBR systems are 10.000 mg/L to 15.000 mg/L and even as high as 20.000 mg/L, but in certain instances. The high-biomass concentrations requires of high-oxygen concentrations in aerobic membrane system to ensure continuous biosynthesis and cell growth. Thus, high SRTs allow the enrichment of slowly growing bacteria and consequently the establishment of a more diverse biocoenosis with broader capabilities (i.e. nitrification or the capability for certain elimination pathways) [34,35]. Therefore, the chemical structures being microbiologically transformed could broaden with increasing sludge age and the biological degradation can be considered dependent on the SRT.

However, as result of very high-biomass concentrations, the permeate flow can be limited by the physical presence of solids at or near the pores which restricts flow. But, on the other hand, positive influence of the higher SRT increases the adsorption capacity hydrophobicity of micropollutants towards sludge particles. This is expected for compounds that are not easily biodegraded and for hydrophobic compounds (log K_{ow} ~4). For those substances, most of the removal achieved could involve adsorption to the organic-rich solid phase (i.e. estrogenic steroid hormones) [36–38]. Contrary, for polar compounds, removal is achieved due to biodegradation rather than sorption phenomena [38]. For charged compounds that interact with membrane surfaces the removal is better than for less polar or neutral compounds [39].

4.2.2.2 Removal of pharmaceuticals by MBR

Many works have been carried out to determine the optimal operating conditions [10–12,18] and applications of MBR [10–12] for treating industrial wastewaters. In most of them, the efficiency of MBR has been evaluated throughout measurements of global chemical parameters (i.e. COD; biochemical oxygen demand (BOD) or TOC). Functional aspect such as kinetics properties of the MBR process, which are important in understanding the mechanism and controlling the system have been also subject of several studies [40]. Contrary, few works related to the removal efficiency for drugs have been published. Among pharmaceuticals explored with MBR are lipid-regulating drugs, anti-inflammatories, antiepileptics, steroid hormones or X-ray contrast agents [14–18].

TABLE 4.2.1

Removal efficiency for pharmaceuticals by membrane bioreactor (MBR) and conventional activated sludge (CAS) systems

Pharmaceuticals	Removal efficiency			
	CAS (%)	Reference	MBR (%)	Reference
Anti-inflammatories				
Ibuprofen	>95	[1,2,14]	>95	[1,14,15,42]
	83	[42]		
Naproxen	66	[2]	50–71	[15]
	85	[42]	99	[42]
Ketoprofen	52	[42]	50–65	[15]
			92	[42]
Diclofenac	40–60	[14,42]	40–60	[17]
	>69	[1,2]	1–23	[14]
	10–39	[7]	10–50	[1,14,17]
			88	[42]
Mefenamic acid	75	[42]	29	[42]
Propyphenazone	65	[42]	43	[42]
Lipid regulator agents				
Clofibric acid	15–51	[1,2]	<10	[1]
	28	[42]	72	[42]
Bezafibrate	>95	[14]	>95	[1,14,42]
	48	[42]	80–91	[15]
Gemfibrozil	39	[42]	90	[42]
Pravastatin	62	[42]	91	[42]
Beta-blockers				
Atenolol	<10	[42]	66	[42]
Metoprolol	<10	[42]	59	[42]
Antiepileptics				
Carbamazepine	<10	[2,14,42]	<10	[2,14,42]
Steroid hormones				
17α-ethinylestradiol (EE2)	60–70	[14]	60–70	[14]
17β-estradiol (E2)	10–50	[1,2]	>90	[1]
Estrone (E1)	<10	[1,2]	>90	[1]
X-ray contrast media				
Diatrizoate	<10	[1,2]	<10	[1]
Iopromide	<10	[1,2]	10–50	[1]

Table 4.2.1 shows removal efficiencies that have been reported for pharmaceuticals in MBR and CAS systems.

4.2.2.2.1 Acidic, neutral and basic pharmaceuticals

Centre of Competence for Water in Berlin (Germany) conducted a two-year investigation with the objective to assess two MBR configurations designed to achieve enhanced nutrient removal from municipal wastewater and to compare the trace organics elimination with a CAS plant. Configurations included an anaerobic zone ahead of the biological reactor and differed by the position of the anoxic zone: standard pre-denitrification or post-denitrification. The results [41] showed that in contrast to the removal of bulk organics which showed only a slight improvement with MBR the removal rate of most of the pharmaceuticals and steroids was higher in MBR than in CAS. The removal was mainly related to biological mechanisms and improved with increasing SRT, reactor temperature and influent concentration, providing a significant adaptation time (of up to seven months). The maximal elimination rates of 65% for formylaminoantipyrine, 70% for phenazone, 75% for propyphenazone and acetylaminoantipyrin were achieved in MBR as shown in Fig. 4.2.1. However, some compounds such as carbamazepine and metabolite 1-acetyl-1-methyl-2-dimethyloxamyl-2-phenyl hydrazide were very persistent in both systems showing elimination rates of <10%.

Radjenovic et al. [42] studied behaviour of 22 pharmaceuticals belonging to different therapeutic categories (analgesics and anti-inflammatory drugs, lipid regulators, beta-blockers, antibiotics, etc.)

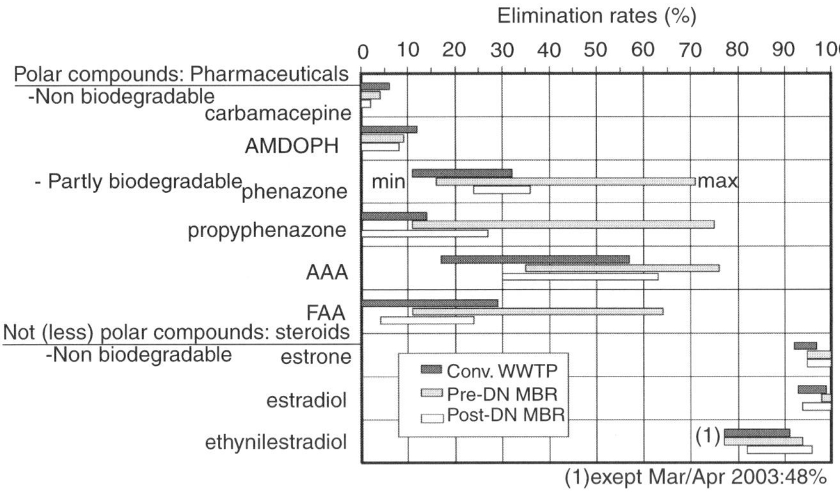

Fig. 4.2.1. Elimination rates of relevant monitored compounds through membrane and conventional activated sludge treatment (minimum–maximum of mean bimonthly values). Reprinted with permission from Ref. [41] © 2005 IWA Publishing.

during treatment of wastewater in a laboratory-scale MBR. The results were compared with their removal in a CAS process in an existing full-scale wastewater treatment facility. Performance of an MBR was monitored during approximately two months to investigate a long-term operational stability of the system and a possible influence of SRT on the removal efficiencies of target compounds. For some of the investigated compounds, the removal efficiencies of both treatments had very similar values (e.g. ibuprofen, naproxen, acetaminophen and paroxetine). They were all removed to a large extent in both of the investigated systems (removal rates were higher than 80%).

However, for a range of compounds removal rates in an MBR were steadier and significantly higher compared to CAS (e.g. diclofenac, ketoprofen, bezafibrate, ranitidine, pravastatin and ofloxacin) as shown in Fig. 4.2.2.

A possible explanation for significantly higher removal rates for diclofenac in a membrane system (average MBR removal efficiency of 87% compared to 50% in CAS [42]) could be higher sludge age of MBR sludge. Another explanation could be higher sorption potential of the MBR sludge, as the organic matter content is higher with respect to the CAS sludge. Literature data on this matter is still very contradictory. While Heberer et al. [6] reported low-removal efficiencies of conventional treatment, Ternes et al. [8] documented a significant elimination of diclofenac (69%). Clara et al. [43] determined removal rates of up to 70% in one STP and no removal in another two conventional STP. In the MBR, no removal occurred during the first sampling campaign when the MBR was operated with an SRT of approximately 10 days. With increasing SRT a partial removal (33–50%) is observed.

Another comparative studies between CAS and MBR systems have evidenced that there is no considerable differences in the efficacy of elimination of studied lipid-regulating drugs (e.g. bezafibrate), and anti-inflammatories (e.g. ibuprofen) [1,14]. Removal rates of more than 95% for ibuprofen and bezafibrate have been reported. Ibuprofen is metabolized to two isomers of hydroxyl-ibuprofen and carboxy-ibuprofen while only one intermediate has been identified in the bezafibrate transformation. 4-Chlorobenzoic acid is the result of the hydrolytic cleavage of the amide bond from bezafibrate [15]. The formation of these intermediates is by microbial degradation process. But, due that both are preserved under microbial process over a long period of time (28 days), other degradation process could be involved in CAS treatments, since ibuprofen and bezafibrate are practically

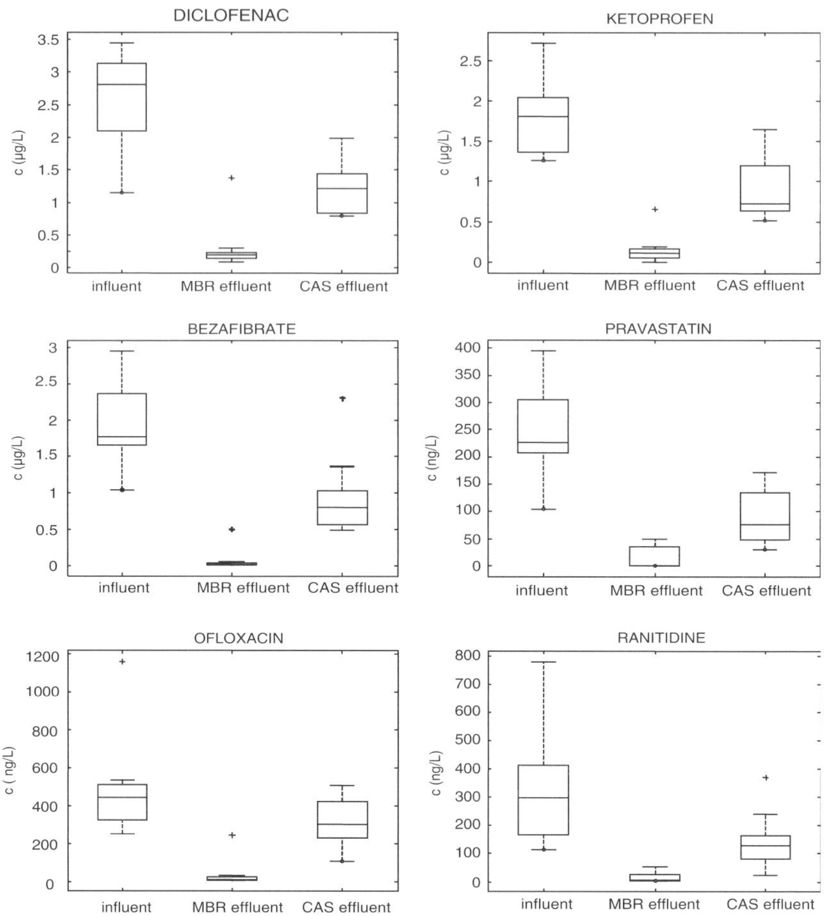

Fig. 4.2.2. Removal of diclofenac, ketoprofen, bezafibrate, pravastatin, ofloxacin and ranitidine by MBR and CAS treatment (Influent and MBR effluent variables have 10 measured values, and variable of CAS effluent has 8 measured values (missing data for two sampling campaigns)). For each variable, the box has lines at the lower quartile (25%), median (50%) and upper quartile (75%) values. The whiskers are the lines extending from each end of the box to show the extent of the data up to 1.5 times the interquartile range (IQR). Outliers are marked with "+" symbols). Modified from Ref. [42].

eliminated (95%). Experiments conducted with the addition of an external carbon source showed that co-metabolic degradation may be the major removal mechanism since these pharmaceuticals could not be utilized as sole source of carbon and energy in microbial degradation processes [15].

Similar behaviour was observed for other anti-inflammatories such as naproxen, which elimination is by co-metabolic degradation (approximately 50%) leading to the formation of desmethylnaproxen that occurs in the initial step of this process [15]. Naproxen is moderately resistant to wastewater treatment and shows an average removal of 66% [1,2]. On the contrary, ketoprofen is completely degraded by microbial transformation during a similar period of time (28 days) and the metabolites appear to be formed along pathway known for biphenyls and related compounds [15].

The antiepileptic drug carbamazepine turned out to be the most persistent pharmaceutical and it passed both through MBR and CAS system untransformed. Poor elimination of this neutral drug has been reported previously by many authors in both CAS and MBR [2,14,38,44,45]. Carbamazepine is not degraded by biological mechanisms and neither is eliminated by processes such as absorption or membrane ultrafiltration. Moreover, glucuronide conjugates of carbamazepine can presumably be cleaved in sewage and thus increase effluent concentrations [8]. Effluent concentrations greater than the ones recorded for the influent could be explained by the presence of input conjugate compounds that are being retransformed during treatment into the original compounds. However, the analysis of conjugates was not performed in none of the studies using MBR, no firm conclusion about their biotransformation could be made, especially because bigger outputs could also be found due to sampling inaccuracy (erroneous hydraulic retention time). For carbamazepine no dependency on the SRT, that in general, influences a higher elimination, has been established [44,46].

Low removal is observed for clofibric acid even at SRT of 25 days (Table 4.2.1) by MBR and by CAS (15–51%) treatment [1,2]. However, it seems plausible that although not all pharmaceuticals are degraded better with increasing SRT, in general, it should be expected that with increasing SRT, the biological degradation of pharmaceuticals will increase too.

Contradictory removal rates for CAS and MBR treatment of a lipid-regulating agent gemfibrozil are reported. While values of biological degradation constants reported by Joss et al. [47] show a significantly higher rate of biodegradation by CAS sludge, the results of Radjenovic et al. [42] show a much better performance of an MBR (89.6% of elimination compared to 38.8% found for CAS system).

High and steady removal rates over the period of two months (>80%) in the MBR were observed for antibiotic ofloxacin, while in the conventional treatment it was eliminated with a wide range of efficiencies [42].

The removal of another antibiotic sulfamethoxazole was found to be very variable in both CAS and MBR systems. According to Drillia et al. [48] its microbial degradation will depend on the presence of easily biodegradable organic matter in wastewater, which is submitted to changes both in MBR and CAS systems. Also a significant amount of sulfamethoxazole enters WWTPs as its human metabolite N_4-acetyl-sulfamethoxazole that can be possibly converted during process back to the original compound.

4.2.2.2.2 Steroid hormones

Publications concerning the behaviour of steroid hormones in MBR systems are relatively scarce. The available data show that high removal rates (>80%) could be achieved for less polar estrogenic steroids (estrone, 17β-estradiol and ethinylestradiol) [41] which represents only a slight improvement since the CAS treatment commonly yields elimination of 75–95% (see Fig. 4.2.1). However, no complete removal was observed, although in some cases (depending on the temperature and SRT) almost 100% removal could be achieved. The comparative study of Clara et al. also showed similar elimination of 17α-ethinylestradiol in MBR and CAS systems, with removal rates of 60–70% [14]. Joss et al. [49] studied removal of estrone, 17β-estradiol and ethinylestradiol in various municipal STP and MBR. The results show comparable removal of all estrogens in the MBR (96% for estrone, >98% for estradiol and >75% for eahinylestradiol) and in three CAS systems (49–99% for estrone, 88–98% for estradiol and 71–94% for ethinylestradiol). According to their study, the sludge originating from an MBR (sludge age 30 days) showed significantly higher activity in batch experiments than the sludge from a CAS with a SRT of 12 days (kinetic values increase by a factor of 2–3).

4.2.2.2.3 X-ray contrast agents

X-ray contrast media are very stable biochemically, and generally resistant to biodegradation. For diatrizoate and iopromide the elimination by CAS treatments is reported to be very low (<10%). The low elimination during CAS treatments is also due to their high polarity (e.g. log K_{ow} of iopromide is −2.33), thus the sorption on the sludge is negligible. In MBR system, it has been determined that degradation of iopromide is low to moderate (10–50%) for SRT of 25 days, while diatrizoate is not removed at all (<10%) [1,2]. By comparing the elimination of iopromide in the biological treatment step of municipal full-scale CAS and pilot MBR Joss et al. [38] measured 40–75% removal in MBR and 45–80%

removal in CAS. However, significant variations in the observed removal and apparent negative dependency on sludge age (lower removal at higher SRTs) lack clear explanation. The authors discussed the representativeness of 24 h composite sampling since the observed incoming daily load of iopromide in medium-sized municipal STP (25,000–50,000 natural inhabitants) is generated by only a small number of patients that receive high doses (daily administered dose per adult patient is 4–60 g). Taking into account dilution in the sewer each patient increases the average influent concentration by approximately 0.1–0.4 μg/L, which was in the same range as the measured influent concentrations.

4.2.3 ADVANCED OXIDATION PROCESSES (AOP)

Most common AOPs applied to remove pharmaceuticals are based on the photocatalytic processes using titanium dioxide (TiO_2) or Fenton's reagent (reaction between iron II salts and hydrogen peroxide, H_2O_2), light (UV), ozone (O_3) and H_2O_2 [1,50–55]. The AOPs are able to completely oxidize recalcitrant compounds rendering less harmful and easily biodegradable components. TiO_2 photocatalysis UV/H_2O_2, O_3 or O_3/H_2O_2 combined treatments are frequently used in the treatment of pharmaceutical wastewaters. The effluents originated from pharmaceutical industry typically show low biodegradability due to high concentration of active substances. In particular, certain antibiotics, anti-tumour agents and analgesics are neither degraded by biological treatment processes nor adsorbed on sewage sludge. The use of AOPs as pre-treatment for improving the biodegradability and efficacy of further treatments is recommended for effluents that contain high concentrations of pollutants [56,57].

Table 4.2.2 summarizes removal efficiencies obtained by two most common AOPs applied for the treatment of pharmaceuticals in wastewaters: photocatalysis and ozonation.

4.2.3.1 Photocatalysis

Application of photocatalysis is extending in the area of wastewater treatment, since the operating conditions are favourable. Complete mineralization can be achieved for many pollutants at ambient temperature and mild conditions of pressure. The possibility to effectively use sunlight or near UV light for irradiation, represent also a major advantage, since should result in considerable economic savings in particular for large-scale operations [58,59]. Photocatalysis performance

TABLE 4.2.2

Removal efficiency for pharmaceuticals by photocatalytic and ozonation processes

Pharmaceuticals	Removal efficiency			
	Photocatalysis (%)	Reference	Ozonation (%)	Reference
Anti-inflammatories				
Ibuprofen	50–90	[1,52,53]		
Diclofenac	>90	[53]	60	[25]
	50–90	[1,53]	>90	[1,74]
Lipid regulator agents				
Bezafibrate	50–90	[1]	60	[1,74]
			10–50	
Clofibric acid	50–90	[1,52]	40	[1,74]
	>90		<10	
Beta-blockers			>90	[25]
Antiepileptics				
Carbamazepine	>90	[52,65]	>90	[73,74]
	50–90	[1]		
Steroid hormones				
17β-estradiol (E2)	>90	[1,61]	>90	[1]
17α-ethinylestradiol (EE2)	>90	[1]	>90	[1]
Estrone (E1)	>90	[1]	80	
X-ray contrast media				
Iomeprol	>90	[52]	>80	[1]
Iopromide	>90	[52]	>80	[1]
			10–50	
Iopam	50–90	[1]		
Iopamidol	50–90	[1]	>80	[1]
			10–50	
Diatrizoate	10–50	[1]	<14	[1]
Antibiotics				
Trimethroprim			>90	[25]
Sulfamethoxazole			>90	[25]
Clarithromycin			>90	[25]
Erythromycin			>90	[25]
Roxithromycin			>90	[25]
Procaine penicillin G			51–79	[71]

can be also more efficient using TiO_2 catalyst. The chemical stability of TiO_2 in aqueous media and in larger range of pH ($0 \leqslant pH \leqslant 14$), its low cost and effectiveness at low concentrations makes that its use is widely extended. TiO_2 photocatalysis should be also efficient with some halogenated compounds that are very toxic for bacteria in biological treatment [60]. However, there are some drawbacks that limit its application on full-scale operation for wastewater treatment. Although photocatalytic processes appear to be much more developed, the treatment of complex effluents need high degree of energy efficiency which is not always feasible. Fenton's reagent is not often applied in photocalysis, since there are many chemicals that are refractory towards it [60] and its use to treat complex effluents is limited. In addition, Fenton's reaction is strongly dependent on pH, with optimal operating conditions at low pH (2.5–4). However, the degree of oxidation in Fenton's reaction could be increased with other advanced oxidation technique such as photocatalysis or ultrasonic irradiation [60].

4.2.3.1.1 Steroid hormones

Most of organic compounds can be oxidized to CO_2 by TiO_2 photocatalytic reaction because of its strong oxidizing power [24,60]; however, the formation of intermediate products is of particular relevance. Species such as 17β-estradiol (E2) has a high-estrogenic activity at very low concentrations, but unexpected increase of the estrogenicity may be due to degradation products. E2 is totally mineralized by TiO_2 photocatalysis and 17β-dihydroxy-1,4-estradien-3-one and testosterone species are generated as intermediate products [1,61]. The mechanisms of E2 degradation are related with the oxidation of the phenol moiety, which plays the main role in the interaction with estrogen receptors (hER) [62]. In this way, estrogenic activity should be almost lost concurrently with the initiation of the photocatalytic degradation. In fact, recent studies demonstrated that estrogenic activity of intermediates of E2 was negligible evaluating the transcriptional estrogenic activity in response to hER in a yeast assay system [54]. On this basis, other natural estrogens (e.g. estrone, estriol) or synthetic (e.g. ethynilestradiol, EE2) that have steroid structures with a phenol moiety similar to E2, could show similar behaviour by TiO_2 photocatalytic processes [1].

In this sense, this technology could be applied to water treatment to effectively remove estrogens without producing active intermediary products [61]. As example, EDCs including estrone (E1) or17α-ethynylestradiol (EE2) have been degraded by photocatalytic processes using TiO_2 thin film under UV light [1,63].

Additional studies have been also focused on predicting the photooxidation of hormones with high-estrogenic activity such as diethylstilbestrol (DES) in water, since might provide a reference for the development of AOP processes in wastewater [64]. Experiments were based on the use Fe(III)–oxalate complexes because they are common composites in natural waters. It was demonstrated that degradation of DES is induced by Fe(III)–oxalate complexes throughout two oxidation pathways. Main intermediates (DES-4-semiquinone and DES-o-catechol) are formed but to achieve mineralization of catechol intermediates further oxidation and decarboxylation should be necessary.

4.2.3.1.2 Antiepileptics

Contrary to biological treatments, photocatalysis assisted by TiO_2 appears to be a promising technology for the removal of carbamazepine. High photoreactivity of TiO_2 materials, in particular P25, has a positive influence making faster the photocatalyst degradation of carbamazepine [52]. But, to date studies performed using this technology point on primary transformation of carbamazepine to several still active intermediates, but not on complete mineralization [52,65]. Among the intermediates (acridine-9-carboxaldehyde, salicylic acid, catechol, anthranilic acid, epoxide species), acridine is the main degradation product which formation occurs under sunlight irradiation and with UV/H_2O_2-induced degradation [65]. Acridine is considered as a chemical that induces mutagenic and carcinogenic activity [66,67], and even at low concentrations possible environmental impact could be derived. Therefore, prolonged oxidation treatments to ensure complete degradation is of importance. Using AOP based on UV/H_2O_2 removal rates for carbamazepine vary from 50 to 90% [1].

4.2.3.1.3 Anti-inflammatories and lipid-regulating drugs

Solar photocatalysis using both photocatalyst TiO_2 and Fenton reagent have been evaluated as alternative for the decomposition of anti-inflammatory drugs, in particular, diclofenac which removal is only partial by biological treatments [53]. Major advantage of application of solar-photocatalysis is the low cost. The use of Fenton reagent is also an attractive option, since the process produces OH radicals in a very simple way and is environmentally safe. Comparison of both treatments has showed that decomposition of diclofenac is completely achieved after around 100 min under photo-Fenton treatment, while with TiO_2 decomposition took around twice as long. However, with photo-Fenton process, there is incompatibility, at first stage because of insolubility of diclofenac at the low pH of the process. But, in spite of that the reaction rate is high and effective removal is achieved [53]. Partial to high elimination is also achieved when UV/H_2O_2 is applied (50–90%) [1].

Similar behaviour is observed in light-induced degradation of clofibric acid using TiO_2 (P25) that is transformed within 40 min, but without fast mineralization and dechlorination [52]. The great number of intermediates (isobutyric, hydroquinone and chlorinated products, such as 4-chlorophenol, show the complexity of the photocatalysis process that suggests the existence of various reductive and oxidative degradation routes and that further prolonged oxidation treatments should be applied for mineralization) [45]. By UV/H_2O_2 process, clofibric acid and bezafibrate are removed in a range of 50–90% [1].

4.2.3.1.4 X-ray contrast agents

Few studies have been reported related with AOPs to treat iodinated contrast agents. In last years, there has been also a growing interest about the occurrence of these compounds in the environment and the need to assess water treatment processes for their elimination. Recent studies have evaluated the removal efficiency of photocatalytic degradation of iomeprol and iopromide applying TiO_2 [52]. A fast transformation of both iodinated contrast media is developed under photocatalytic degradation but complete mineralization is not obtained. As result, organic degradation products as well as formation of iodinated intermediates are formed. For other compounds such as iopam or iopamidol, removal rate was found to vary from 50 to 90% by UV/H_2O_2, while low to medium removal (10–50%) is achieved for diatrizoate [1].

4.2.3.2 Ozonation

Ozonation has already been demonstrated to be an effective process for removing refractory and/or toxic chemicals from water and wastewater [60,68]. Ozone is a very powerful oxidizing agent able to react with most species containing multiple bonds, but not with single functional bonds (e.g. C–C, C–O and O–H) at high rates. During ozonation pollutants can be oxidized either by O_3 directly or by hydroxyl radicals ($^{\bullet}OH$) that are formed as a consequence of O_3 decay [69]. O_3 is a selective oxidant attacking certain functional groups, whereas $^{\bullet}OH$ reacts very fast with a large number of moieties. As a consequence the oxidation of compounds that react fast with O_3 is always more efficient since most $^{\bullet}OH$ is scavenged by the wastewater matrix.

In raw wastewater, O_3 doses ranging from 5 to 15 mg/L led to a complete disappearance of most of the pharmaceuticals; however, it may result only impartial oxidation and therefore could yield biologically still active oxidation products [69]. By treating wastewater with a dissolved organic carbon (DOC) concentration representative for good-quality secondary or tertiary effluent (6.6 to 7.7 mg/L) Huber et al. [69] demonstrated that macrolide and sulphonamide antibiotics, estrogens and the acidic pharmaceuticals such as diclofenac, naproxen and indomethacine were oxidized by more than 90–99% for O_3 doses $\geqslant$2 mg/L. However, DOC seemed to be the water quality parameter that has a stronger influence on the efficiency of the ozonation process, since in case of wastewaters with a higher DOC (23 mg/L) O_3 doses $\geqslant$5 mg/L had to be applied to achieve a comparable result [54]. Combinations with other oxidation techniques such as UV irradiation and H_2O_2 resulted in higher efficiency.

A major limitation of ozonation is that the process efficiency is severely dependent on the efficient gas–liquid mass transfer, which is difficult to achieve due to the low solubility of ozone in the aqueous solutions. New designs of contactors have been developed to improve the efficiency, but applications to large-scale treatments of wastewaters and the high cost of ozone generation hamper its wide application.

4.2.3.2.1 Antibiotics

As mentioned above, ozonation is expected to be an effective tool for the partial oxidation of potentially refractory and/or toxic pharmaceuticals, for instance antibiotics. Variables such as applied ozone dose and reaction pH play a critical role in determining treatment efficiency. The effectiveness of ozonation process at natural pH of waters has been demonstrated achieving high levels of chemical oxygen demand (COD) and aromaticity removals. Addition of hydrogen peroxide (H_2O_2) in the ozonation process have been also demonstrated to provide almost 100% of COD and aromaticity removal, although the O_3/H_2O_2 combination could not show advantage for COD removal kinetics over the direct O_3 application [70].

However, ozonation would be exceedingly costly and time consuming to achieve a substantial degree of degradation, and therefore process optimization needs to be considered prior to detailed kinetic investigation as well as full-scale process implementation. Proper selection of ozone dosage for the destruction of chemicals is very crucial for process optimization and as expected, increasing the applied dose has a profound effect on COD removal. Ozone dosage of 10–15 mg/L and natural pH values has been demonstrated to provide an effective oxidation of antibiotics such as trimethroprim, sulfamethoxazole, clarithromicin, erythromycin or roxithromycin [25]. However, in the case of ozonation for PPG (Procaine Penicillin G) degradation, at economically acceptable doses, only partial COD and TOC removals can be achieved (79 and 51%, respectively) [71].

Ozonation could also be successfully used as a pre-treatment step to improve the biodegradability of wastewaters containing antibiotics, since ozone can leads to the formation of low-molecular weight oxygenated by-products that are more amenable to biodegradation [70]. Experiments performed on synthetic pharmaceutical wastewater containing human and veterinary antibiotics (cephalosporine, penicillin and quinolone group) have evidenced the enhancement of biodegradability [70]. In general, ozonation increases the number of functional groups and the polarity of molecules, so the intended properties of pharmaceuticals or the original mode of action could be modified. For instance, hydroxylated antibiotics should not promote the formation of antibiotic resistant strains.

However, when ozone is used in disinfection treatments, effective reduction of concentration of pathogens is obtained, but as consequence of low-ozone dosage, intermediates more toxic than original antibiotics could be formed [72]. In particular, ozonation experiments on amoxicillin show how ozone attack is on phenolic ring and on the protonated amino group leading to the formation of intermediates. In addition, the low degree of mineralization, even for long treatment times, evidence that application partial oxidation treatments for wastewater containing antibiotics could be inappropriate [73]. In this sense, the identification of intermediates occurring with the ozonation, their fate or ecotoxicity, is of crucial importance and need further studies.

4.2.3.2.2 Anti-inflammatories, lipid-regulating drugs, beta-blockers and antiepileptics

The removal of pharmaceuticals, which use is widely extended such as anti-inflammatories, lipid-regulating drugs or beta-blockers, has been also evaluated by ozonation process. Different results have been also published reporting high removals for diclofenac [1] or beta-blockers [25] and moderate and low removals for bezafibrate and clofibric acid, respectively (Table 4.2.2). Huber et al. [69] detected 70–90% removal of diclofenac by 1 mg/L O_3 and >95% removal by 2 mg/L, independent on the water matrix (CAS, CAS+suspended solids, MBR), while bezafibrate exhibits an intermediate reactivity towards O_3 showing approximately

40% removal under the same conditions. Other tested acidic pharmaceuticals, such as naproxen and indomethacin, were oxidized by more than 90–99% for O_3 doses $\geqslant 2$ mg/L in all effluents.

Contrary to other AOPs or MBR treatments, the ozonation for carbamazepine has been demonstrated to be a suitable treatment option. Its disappearance is attained even at the process conditions usually adopted in drinking water facilities. But, on the other hand, low degree of mineralization is achieved even for long ozonation times and intermediates (oxalic acid, glyoxylic acid, glyoxal, oxamic acid, anthranilic acid and ketomalonic acid) have been identified. However, the preliminary results from toxicity tests performed with algae showed that no toxicity has been observed throughout the different times of ozone exposition [73,74].

4.2.3.2.3 Steroid hormones

For steroid hormones in low DOC matrices such as groundwaters or tertiary-treated effluents, in general ozone and AOPs is expected to be particularly effective. In fact, it has been determined that with the application of O_3 and UV-disinfection, steroid hormones, such as estrone [1], are eliminated to a great extent. Specifically, by applying ozone dose of 5 mg/L over a contact time of 18 min, estrone is reduced to approximately 80% and efficient removal for EE2 and E2 ($>90\%$) is also achieved (Fig. 4.2.3).

4.2.3.2.4 X-ray contrast agents

Ternes et al. [54] used a pilot plant for ozonation and UV-disinfection to treat effluents from a municipal STP (DOC 23 mg/L, COD 30 mg/L). Elimination of X-ray contrast agents by ozonation process was rather effective ($>80\%$) for iodinated derivates such as iopamidol, iopromide or iomeprol by applying 10–15 mg/L ozone dosage (contact time 18 min), although appreciable concentrations were still detected after the treatment (Fig. 4.2.4). At the same dosage, the ionic diatrizoate only exhibited removal efficiencies of not higher than 14%. Using O_3/H_2O_2 combination or O_3/UV-low pressure mercury arc, slightly increased oxidation

---

Fig. 4.2.3. Relative residual concentrations of four compounds (iopromide, roxithromycin, sulfamethoxazole and EE2) that represent the classes of contrast media, macrolides, sulfonamides and estrogens, respectively. The residuals for the effluents CAS (pH = 7), CAS+SS (pH = 7) and MBR (pH = 7.5), measured at the outlet of column 2 (SP-OUT), are plotted versus O_3 dosages. Furthermore, absolute O_3 concentrations measured at the outlet of column 1 (SP-MID) are given. Reprinted with permission from Ref. [69] © 2005 American Chemical Society.

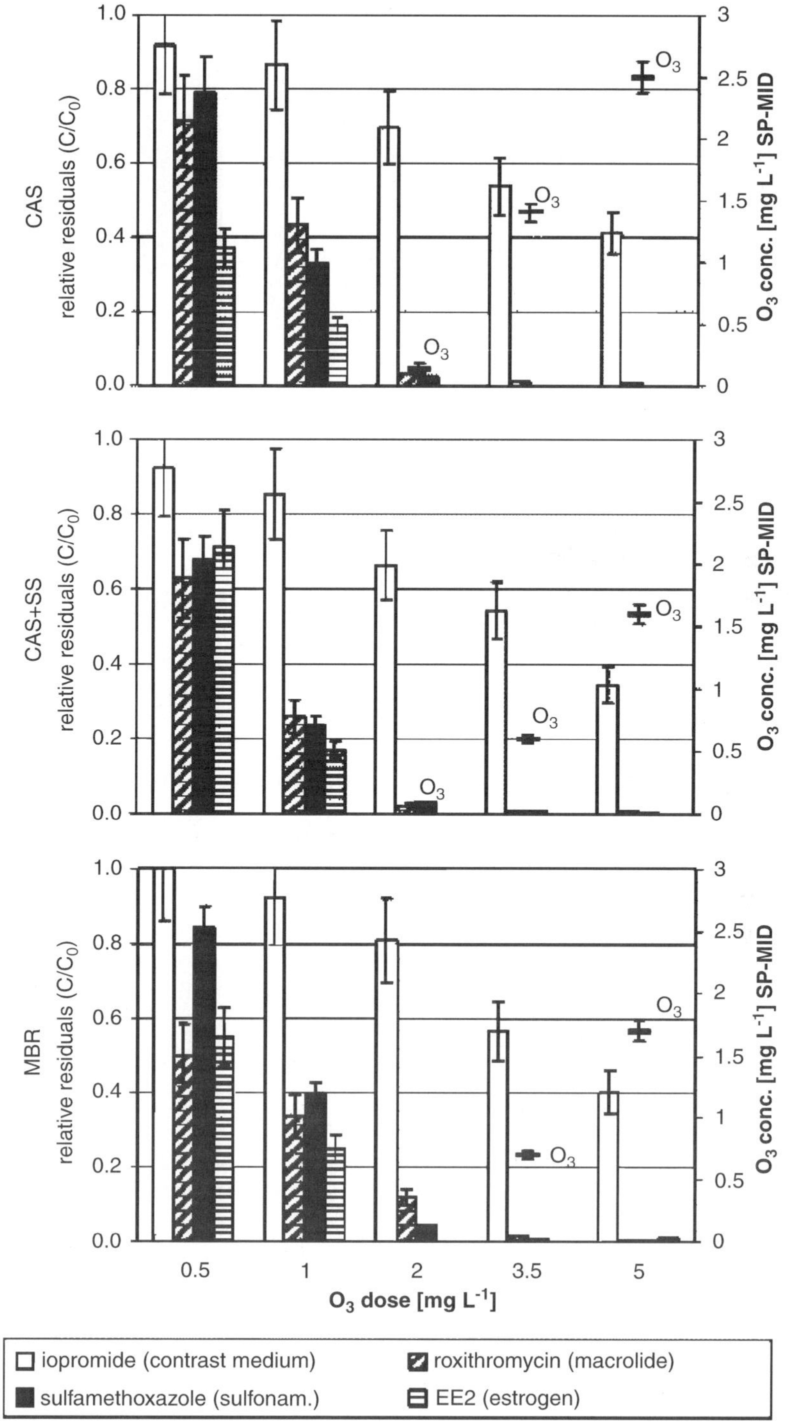
CAS
relative residuals (C/C$_0$)
CAS+SS
relative residuals (C/C$_0$)
MBR
relative residuals (C/C$_0$)
O$_3$ conc. [mg L^{-1}] SP-MID
O$_3$
1.0
0.8
0.6
0.4
0.2
0.0
3
2.5
2
1.5
1
0.5
0
0.5
1
2
3.5
5
O$_3$ dose [mg L^{-1}]
iopromide (contrast medium)
roxithromycin (macrolide)
sulfamethoxazole (sulfonam.)
EE2 (estrogen)

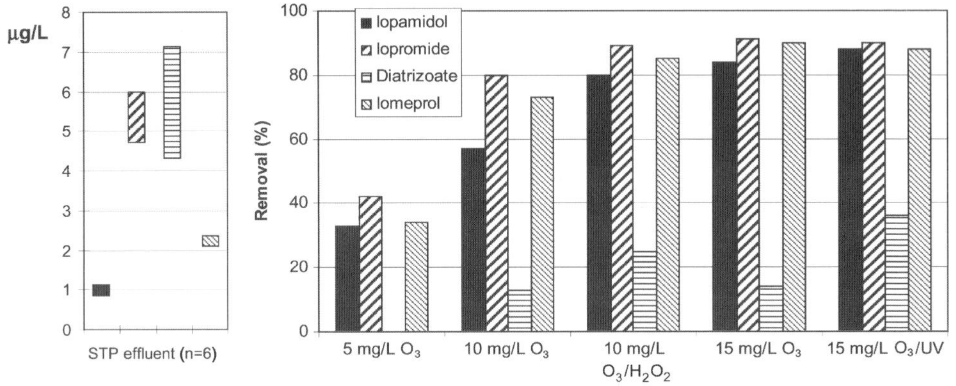

Fig. 4.2.4. Removal of X-ray contract media in the STP effluent by ozonation and AOPs. Data from Ref. [54].

efficiency is obtained in comparison to ozone alone, but as higher ozone doses are required, treatment is unrealistic due to its high cost, an efficient ozone/UV AOP treatment seems not to be feasible.

Huber et al. [69] demonstrated that contrast media do not react with O_3 directly, but with $^{\bullet}OH$. The comparable extent of parent compound oxidation was observed for structurally similar compounds (non-ionic iopamidol, iopromide or iomeprol), while the anionic diatrizoate showed different pattern suggesting a substantially lower reactivity to $^{\bullet}OH$.

4.2.4 CONCLUSIONS

AOPs and MBR are emerging technologies which position in the range of water treatment processes has not been well defined to date, since better performances are not yet required by the current regulations. AOPs and MBR still need of further advances that demonstrate the levels of reliability and full-scale implementation. Evaluation of AOPs and MBR for removal of pharmaceuticals is topic of ongoing scientific research. Current research has demonstrated the potential application of AOPs and MBR in the treatment of pharmaceutical wastewater. But, pharmaceuticals are a wide variety of structures and until now research are based on a limited number of pharmaceuticals. However, aspects related with contamination of pharmaceutical intermediates and improvements of wastewater treatments needs also of more information.

ACKNOWLEDGEMENTS

This work has been developed under the framework of Project EVITA, CTM2004-06265-CO3-01, CTM2004-06265-CO3-03 and Project CE-CSD2006-004, financially supported by Spanish Ministerio de Ciencia y Tecnología. M.D.H. acknowledges the reserach contrat "Contrato de Retorno de Investigadores" from Consejeria de Educación y Ciencia, Junta de Andalucia, Spain. J.R. gratefully acknowledges the I3P Program, co-financied by CSIC and European Social Funds, for a predoctoral grant.

REFERENCES

1 T. Ternes, Assessment of technologies for the removal of pharmaceuticals and personal care products in sewage and drinking water facilities to improve the indirect potable water reuse, POSEIDON EU Project Report, http:// www.eu-poseidon.com.
2 European Commission Report on Pollutants in urban waste water and sewage sludge, Chapter 6: Case studies, http://europa.eu.int/comm./environment/waste/sludge/sludge_pollutants_6.pdf.
3 D. Barceló, Ed., *Emerging Organic Pollutants in Waste Waters and Sludges*, Springer-Verlag, Berlin Heidelberg, 2005, Vol. XIV, p. 280.
4 M. Petrović, M. Sole, M. Lopez de Alda and D. Barcelo, *Environ. Toxicol. Chem.*, 21 (2002) 2146.
5 S. Diaz-Cruz and D. Barceló, Occurrence and analysis of selected pharmaceuticals and metabolites as contaminants present in wastewaters, sludge and sediments. In: D. Barceló (Ed.), *The Handbook of Environmental Chemistry*, Springer-Verlag, Berlin Heidelberg, 2004, Vol. 5, Part I, pp. 227–260.
6 T. Heberer, *Toxicol. Lett.*, 131 (2002) 5.
7 Assessment and removal technologies of antibiotics in waste-waters REMPHARMAWATER EU Project Report, http://www.unina.it/~rmarotta.
8 T.A. Ternes, *Water Res.*, 32 (1998) 3245.
9 C. Zwiener, T.J. Gremm and F.H. Frimmel, Pharmaceutical residues in the environment: Pharmaceutical residues in the aquatic environment and their significance for drinking water production. In: K. Kümmerer (Ed.), *Pharmaceuticals in the Environment: Sources, Fate, Effects and Risks*. Springer-Verlag, Heidelberg, Berlin and New York, 2001, p. 81.
10 K. Brindle and T. Stephenson, *Biotechnol. Bioeng.*, 49 (1996) 601.
11 H. De Wever, S. Van Roy, C. Dotremont, J. Müller and T.P. Knepper, *Water Sci. Technol.*, 50 (2004) 219.

12 H. De Wever, S. Van Roy, J. Vereecken, C. Dotremont, L. Diels, J. Müller and T.P. Knepper, *Proceedings of the European Symposium on Environmental Biotechnology (ESEB)*, 2004, pp. 271–275.
13 S. Weiss and T. Reemtsma, *Proceedings of the Annual Conference of the German Water Chemical Society*, 2004, pp. 226–230.
14 M. Clara, B. Strenn, M. Ausserleitner and N. Kreuzinger, *Water Sci. Technol.*, 50 (2004) 29.
15 J.B. Quintana, S. Weiss and T. Reemtsma, *Water Res.*, 39 (2005) 2654.
16 N. Kreuzinger, M. Clara, B. Strenn and H. Kroiss, *Water Sci. Technol.*, 50 (2004) 149.
17 H. De Wever, J. Vereecken, T. Reemtsma, S. Weiss, M.D. Hernando, S. González, D. Barceló, O. Rörden, J. Miller and T.P. Knepper, *Proceedings of the Annual Meeting of SETAC-Europe*, 2005, TU3-2/1.
18 Removal of Persistent Polar Pollutants through improved treatment of wastewater effluents, P-THREE EU Project, http://www.pthree.de.
19 L. Innocenti, D. Bolzonella, P. Paven and F. Cecchi, *Desalination*, 146 (2002) 467.
20 R. Liu, X. Huang, J. Xi and Y. Qian, *Process Biochem.*, 40 (2005) 3165.
21 C.G. Dosoretz and K.W. Böddeker, *J. Membrance Sci.*, 239 (2004) 81.
22 M.R. Hoffmann, S.T. Martin, W. Choi and D.W. Bahnemann, *Chem. Rev.*, 95 (1995) 69.
23 M.R. Prairie, L.E. Evans, B.M. Stange and S.L. Martinez, *Environ. Sci. Technol.*, 27 (1993) 1776.
24 S. Malato, J. Blanco, A. Vidal and C. Richter, *Appl. Catal.*, 37 (2002) 1.
25 T.A. Ternes, J. Stüber, N. Herrmann, D. McDowell, A. Ried, M. Ried, M. Kampmann and B. Teiser, *Water Res.*, 37 (2003) 1976.
26 M.M. Huber, S. Canonica, G.-Y. Park and U. Von Gunten, *Environ. Sci. Technol.*, 37 (2003) 1016.
27 C. Zwiener and F.H. Frimmel, *Water Res.*, 34 (2000) 1881.
28 M. Sprehe, S.-U. Geissen and A. Vogelpohl, *Water Sci. Technol.*, 44 (2001) 317.
29 T. Wintgens, M. Gallenkemper and T. Melin, *Water Sci. Technol.*, 50 (2004) 1.
30 A.I. Schäfer, L. Nghiem and D. Waite, *Environ. Sci. Technol.*, 37 (2003) 182.
31 G.Y. Park, J.H. Lee, I.S. Kim and J. Cho, *Water Sci. Technol.*, 50 (2004) 239.
32 T.P. Knepper, D. Barceló, K. Lindner, P. Seel, T. Reemtsma, F. Ventura, H. De Wever, E. van der Voet, P. Gehringer and M. Schonerklee, *Water Sci. Technol.*, 50 (2004) 195.
33 P. Peev, M. Schonerklee and H. De Wever, *Water Sci. Technol.*, 50 (2004) 209.
34 J.A. Howell, T.C. Arnot and W. Liu, *Ann NY Acad. Sci.*, 984 (2003) 411.
35 H.Y. Ng and S.W. Hermanowicz, *Water Res.*, 39 (2005) 981.

36 K.M. Lai, K.L. Johnson, M.D. Scrimshaw and J.N. Lester, *Environ. Sci. Technol.*, 34 (2000) 3890.
37 K.I.E. Holthaus, A.C. Johnson, M.D. Jurgens, R.J. Williams, J.J.L. Smith and J.E. Carter, *Environ. Toxicol. Chem.*, 21 (2002) 2526.
38 A. Joss, E. Keller, A.C. Alder, A. Göbel, C.S. McArdell, T. Ternes and H. Siegrist, *Water Res.*, 39 (2005) 3139.
39 P. Berg, G. Hagmeyer and R.P. Gimbel, *Desalination*, 113 (1997) 205.
40 X. Wen, C. Xing and Y. Qian, *Process Biochem.*, 35 (1999) 249.
41 B. Lesjean, R. Gnirss, H. Buisson, S. Keller, A. Tazi-Pain and F. Luck, *Water Sci. Technol.*, 52 (2005) 453.
42 J. Radjenovic, M. Petrović and D. Barcelo, *Anal. Bioanal. Chem.*, 387 (2007) 1365–1377.
43 M. Clara, B. Sternn, O. Gans, E. Martinez, N. Kreuzinger and H. Kroiss, *Water Res.*, 39 (2005) 4797.
44 M. Clara, B. Strenn and N. Kreuzinger, *Water Res.*, 38 (2004) 947.
45 C.D. Metcalfe, B.G. Koenig, D.T. Bennie, M. Servos, T.A. Ternes and R. Hirsch, *Environ. Toxicol. Chem.*, 22 (2003) 2872.
46 B. Strenn, M. Clara, O. Gans and N. Kreuzinger, *Water Sci. Technol.*, 50 (2004) 269.
47 Joss, S. Zabczynski, A. Gobel, B. Hoffmann, D. Loffler, C.S. McArdell, T.A. Ternes, A. Thomsen and H. Siegrist, *Water Res.*, 40 (2006) 1686.
48 P.P. Drillia. J. Hazard, *Mat*, 122 (2005) 259.
49 A. Joss, H. Andersen, T. Ternes, P.R. Richle and H. Siegrist, *Environ. Sci. Technol.*, 38 (2004) 3047.
50 Y. Ohko, D.A. Tryk, K. Hashimoto and A. Fujishima, *J. Phys. Chem. B.*, 102 (1998) 2699.
51 A. Fujishima, T.N. Rao and D.A. Tryk, *Electrochim. Acta*, 45 (2000) 4683.
52 T.E. Doll and F.H. Frimmel, *Water Res.*, 38 (2004) 955.
53 L.A. Pérez-Estrada, M.I. Maldonado, W. Gernjak, A. Agüera, A.R. Fernández-Alba, M.M. Ballesteros and S. Malato, *Catal. Today*, 101 (2005) 219.
54 T.A. Ternes, J. Stüber, N. Herrmann, D. McDowell, A. Ried, M. Kampmann and B. Teiser, *Water Res.*, 37 (2003) 1976.
55 R. Andreozzi, R. Marotta, G. Pinto and A. Pollio, *Water Res.*, 36 (2002) 2869.
56 Enhancing biodegradability and persistent water contaminant removal by advanced oxidation processes (Solar Photocatalysis), FOTODETOX EU Project Report, http://www.psa.es/webesp/projects/fotodetox/index.html.
57 A coupled advanced oxidation—Biological process for recycling of industrial wastewater containing persistent organic contaminants, CADOX EU Project, http://www.psa.es/webeng/projects/cadox/index.html.
58 J.-M. Herrmann, *Catal. Today*, 53 (1999) 115.
59 A.A. Yawalkar, D.S. Bhatkhande, V.G. Pangarkar and A.A.C.M. Becnackers, *J. Chem. Tech. Biot.*, 76 (2001) 363.

60 P.R. Gogate and A.B. Pandit, *Adv. Environ. Res.*, 8 (2004) 501.
61 Y. Ohko, K.-I. Iuchi, C. Niwa, T. Tatsuma, T. Nakashima, T. Iguchi, Y. Kubota and A. Fujishima, *Environ. Sci. Technol.*, 36 (2002) 4175.
62 A.M. Brzozowski, A.C.W. Pike, Z. Dauter Z., R.E. Hubbard, T. Bonn, O. Engstrom, L. Ohman, G.L. Green, J. Gustafsson and M. Carlquist, *Nature*, 389 (1997) 753.
63 T. Tanizaki, K. Kadokami and R. Shinohara, *Bull. Environ. Contam.*, 68 (2002) 732.
64 D. Zhou, F. Wu and N. Deng, *Chemosphere*, 57 (2004) 283.
65 D. Vogna, R. Marotta, R. Andreozzi, A. Napolitano and M. d'Ischia, *Chemosphere*, 54 (2004) 497.
66 R. Shinohara, A. Kido, Y. Okamoto and R. Takeshita, *J. Chromatogr. A*, 256 (1983) 81.
67 S.M. Furst and J.P. Uetrecht, *Biochem. Pharmacol.*, 45 (1993) 1267.
68 S. Chiron, A.R. Fernández-Alba, A. Rodriguez and E. García-Calvo, *Water Res.*, 34 (2000) 366.
69 M.M. Huber, A. Göbel, A. Joss, N. Hermann, D. Löffler, C. McArdell, A. Ried, H. Siegrist, T.A. Ternes and U. Von Gunten, *Environ. Sci. Technol.*, 39 (2005) 4290.
70 I.A. Balcioğlu and M. Ötker, *Chemosphere*, 50 (2003) 85.
71 I. Arslan-Alaton and A.E. Caglayan, *Chemosphere*, 59 (2005) 31.
72 N.J. Ashbolt, *Toxicology*, 198 (2004) 255.
73 R. Andreozi, M. Canterino, R. Marotta and N. Paxeus, *J. Hazard. Mater.*, 122 (2005) 243.
74 F. Sacher, B. Haist-Gulde, H.J. Brauch, G. Preub, U. Wilme, N. Zullei-Seibert, M. Meisenheimer, H. Welsch and T.A. Ternes, Preprints of Extended Abstract 40 (2000) 116. Conference Proceeding, American Chemical Society, Division of Environmental Chemistry 3 (2000) 26.

Chapter 4.3

Removal of pharmaceuticals during drinking water production

Thomas Heberer

4.3.1 INTRODUCTION

Administration of pharmaceuticals in human or animal therapy or administration as food additives in livestock breeding were identified as the main sources for the occurrence of pharmaceutical residues in the environment [1–3]. As comprehensively reported in Chapters 1.1, 3.1 and 4.1, several investigations have shown that residues of pharmaceuticals are often not completely eliminated during conventional wastewater treatment and often also not attenuated in the environment. Under recharge conditions, e.g. when surface water under the influence of sewage effluents is used for groundwater recharge, such residues can also leach into groundwater aquifers. Thus, the occurrence of drug residues in ground- and drinking water samples has already been reported in several studies [4–19] investigating water samples collected from water works using bank filtration or artificial groundwater recharge downstream from municipal sewage treatment plants (STPs). Other sources for the occurrence of pharmaceutical residues in groundwater are the leaching of such residues from manure applied to agricultural areas, reuse of waste water by soil-aquifer treatment (SAT) [4,20,21], leaking sewer lines, spills from landfill leachates [15,22–24] or residues from chemical production plants [25].

In the meantime, residues of more than 100 substances, pharmaceuticals or metabolites from their degradation, have been identified and quantified in sewage or surface water [1,2,26]. The observed concentrations of drug residues in the aquatic environment range from less than 1 ng/L up to more than 10 μg/L. As shown in Table 4.3.1, most of the pharmaceutical residues were found in wastewater effluents or surface waters. The number and the concentrations of pharmaceutical residues are decreasing from sewage to surface, ground and drinking water. Thirty-four compounds have also been detected in samples

Comprehensive Analytical Chemistry 50
Petrović and Barceló (Eds)
Volume 50 ISSN: 0166-526X DOI: 10.1016/S0166-526X(07)50015-2

TABLE 4.3.1

Findings of drug residues in sewage, surface, ground-, and drinking water. Number of compounds reported to occur in different environmental compartments according to their prescription class type. Supplemented as cited in Ref. [1,3]

Analytes (prescription classes)	Number of compounds detected		
	Sewage and surface water	Groundwater	Drinking water
Analgesics/anti-inflammatory drugs+metabolites	26	15	8
Antibiotics	31	3 (+ 5 traces)	–
Antiepileptics	2	2	2
Lipid regulators+metabolites	7	3	3
Beta-blockers	7	1	–
Contrast media+metabolites	8	6	3
Cytostatic drugs	2 (traces)	–	–
Oral contraceptives (EE2 and mestranol)	2 (almost <2/0.5 ng/L)	1 (validity questionable[a])	1 (validity questionable[a])
Other pharmaceuticals	21	4	–
Total	106	34 (39)	16

[a]Validity of positive detects are questionable due to analytical reasons. Results have not been considered as inevitable positive detects.

collected from monitoring or water-supply wells. In drinking water, the presence of 16 pharmaceuticals or their metabolites has been reported. However, these residues almost occur at low ng/L levels in drinking water samples [1,3,15]. Eight out of the sixteen compounds were analgesic drugs and their metabolites, three were blood-lipid regulators including a pharmacologically active metabolite, two were anti-epileptic drugs and three were X-ray contrast media or metabolites.

4.3.2 GROUNDWATER

4.3.2.1 Contamination of groundwater

Polar drug residues can leach into groundwater when surface water under the influence of sewage effluents is used for groundwater recharge. This was the key conclusion of a review article by Verstraeten et al. [6] who compared the results from ten case studies investigating the occurrence of residues of pesticides, pharmaceuticals and other organic contaminants. Especially, some very polar and highly persistent compounds were identified as also being relevant for drinking-water supply. The inclusion of hydroxy- or carboxy-moieties during the degradation of the parent compounds makes the metabolites of such contaminants even more polar. If such metabolites are chemically and biologically persistent, the occurrence of drug residues is becoming even more likely [6]. Besides groundwater recharge areas, findings of drug residues have also been reported for groundwater aquifers contaminated by landfill leachates [15,22,23,27], as a result from spills of production residues [25,28] or when municipal wastewater is reused by SAT [20,21].

First detects of pharmaceutical residues in ground- or drinking water were reported in the mid-1990s [9,13,29]. These and most of the following studies were focused on the monitoring of the occurrence of drug residues and on the development of new, highly sensitive and reliable multi methods for the trace analysis of various pharmaceuticals in all kinds of often very complex matrices.

4.3.2.2 Behavior of drug residues in the groundwater body

Nowadays, research is also focusing on the investigation of natural processes for the attenuation of such residues when they are entering the groundwater body. Cordy et al. [30] identified carbamazepine and sulfamethoxazole as persistent compounds in a soil column study with treated effluent applied at the top of the 2.4 m long, 32.5-cm diameter soil column over 23 days. Scheytt et al. [31,32] investigated the behavior of selected drug residues under varying conditions in sorption and partition studies. Several laboratory studies with fortified solutions of environmentally relevant drug residues were conducted with soil columns under defined saturated or non-saturated conditions [33–36]. For the analgesic compound ibuprofen, a significant removal was observed in laboratory soil-column experiments most likely caused by microbial degradation [35]. Under saturated conditions, no degradation and almost no retardation (retardation factor: $R\mathrm{f} = 1.1$) was observed for clofibric acid [35], a metabolite of blood-lipid regulating drugs and the first drug residue being detected in sewage [37,38] and drinking water [29]. In contrast to clofibric acid, propyphenazone ($R\mathrm{f} = 1.6$) and diclofenac ($Rf = 2.0$) were retarded in the soil columns under saturated conditions. However, no significant degradation was observed for both compounds under the conditions prevailing in these columns [35]. Under unsaturated conditions, Scheytt et al. [36] observed a significant elimination for ibuprofen (54%), propyphenazone (55%) and diclofenac (35%), whereas carbamazepine was not eliminated. Under the non-saturated conditions retardation factors varied between 1.84 for carbamazepine, 2.51 for propyphenazone, 3.00 for ibuprofen and 4.80 for diclofenac [36].

Steroid hormones, such as 17β-estradiol (E2) and 17α-ethinyl estradiol (EE2), have received considerable attention because of their potential for causing endocrine effects in wildlife and humans. The sorption behavior of steroid estrogens has been investigated by several authors [39,40]. E2 has a water solubility of 13 mg/L and a log K_{ow} of 3.94 whereas EE2 has a water solubility of 4.8 mg/L and a log K_{ow} of 4.15 [39]. The sorption coefficient (K_d) of these compounds have been determined by using the ratio of the concentration of the chemical sorbed by the sediment (mg/kg) to the equilibrium solution concentration (mg/L). The mean K_d (L/kg) can also be normalized on the basis of organic carbon content to obtain the so-called K_{oc} value ($K_{oc} = K_d$/fraction of organic carbon). Lai et al. [39] reported modest sorption of estrogenic steroids on a river sediment with sorption coefficients (K_d values) of 51, 36, 21, 52 and 182 L/kg for estrone (E1), E2, estriol (E3), EE2 and mestranol (MeEE2), respectively. Ying et al. [40] measured K_d values of 7.7 ± 3.4 L/kg for E2 and 10.6 ± 5.1 L/kg for EE2 on an aquifer sediment obtained from a groundwater well at Bolivar in South Australia, where aquifer storage and recovery (ASR) is used for the reclamation and reuse of wastewater. These values are relatively low compared with the data reported by Lai et al. [39]. But Ying et al. [40] assumed that this might be due to the low organic carbon and clay contents of this aquifer sediment. The organic carbon normalized sorption coefficients (K_{oc}) were 1540 L/kg and 2120 L/kg for E2 and EE2, respectively, which is relatively low compared to other compounds known as endocrine disrupters such as 4-tert-octylphenol (4-tert-OP) and 4-nonylphenols (4-NP) which have K_{oc} values which are up to 10–200 times higher. Based on the physical properties of the aquifer material in the Bolivar ASR site retardation factors were calculated to be 25 for E2 and 34 for EE2 [40].

Ying et al. [40] also investigated the biodegradation behavior of E2 and EE2 between a sediment-groundwater slurry media (1:1, w/w) and a sediment-effluent slurry media (1:1, w/w) under aerobic and anaerobic conditions at a temperature of 20°C (Figs. 4.3.1 and 4.3.2). Under aerobic conditions, a rapid biodegradation with a DT50 value of around two days was observed for E2 in both incubation media. E2 also degraded slowly under anaerobic conditions whereas EE2 was found to be resistant to biodegradation under both aerobic and anaerobic conditions. Thus, it remained almost unchanged within 70 days under anaerobic conditions in both incubation media [40]. For E2, a slightly better anaerobic degradation in the sediment-effluent media was observed. The authors [40] concluded that this might be due to co-metabolism or increased microbial activity in the effluent media having higher dissolved organic carbon and nutrient contents than the ambient groundwater. Ying et al. [40] also concluded that under

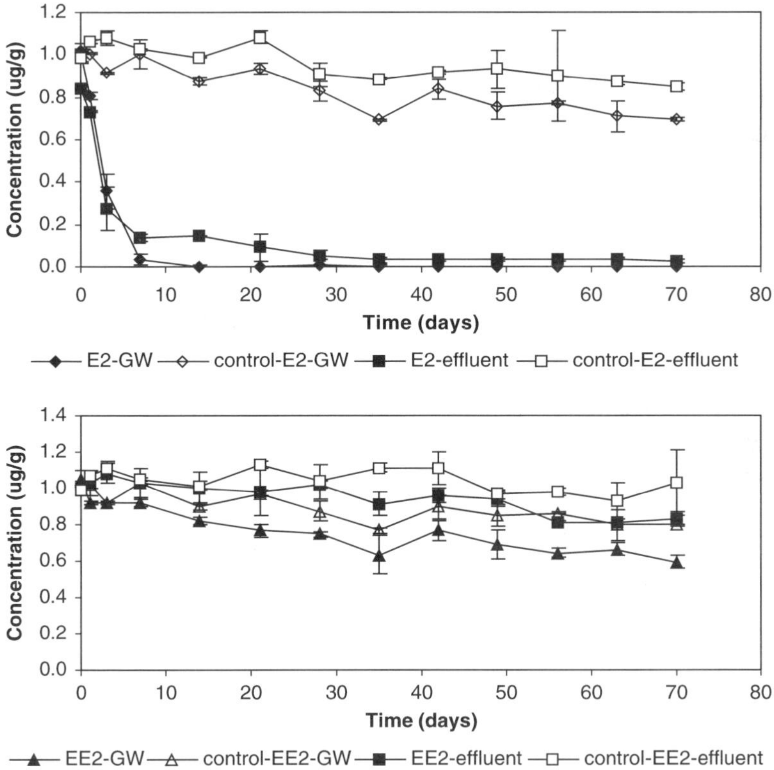

Fig. 4.3.1. Degradation of E2 and EE2 in a sediment–groundwater media (1:1, w/w) and a sediment–effluent media (1:1, w/w) under aerobic conditions. GW indicates the sediment–groundwater media and "effluent" means the sediment–effluent media. Reprinted from Ref. [40] with permission of the National Ground Water Association. Copyright 2004.

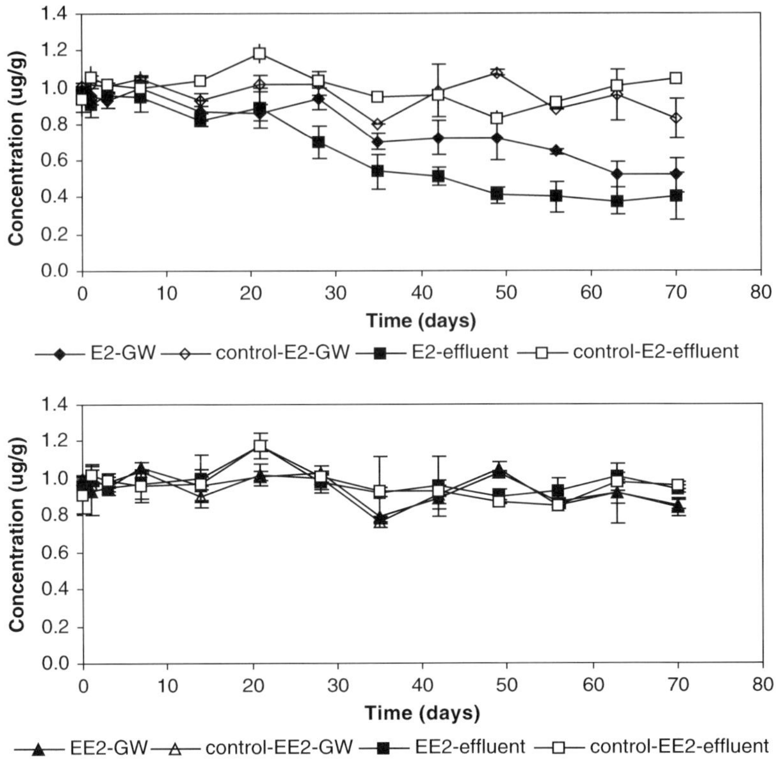

Fig. 4.3.2. Degradation of E2 and EE2 in a sediment–groundwater media (1:1, w/w) and a sediment–effluent media (1:1, w/w) under anaerobic conditions. GW indicates the sediment–groundwater media and "effluent" means the sediment–effluent media. Reprinted from Ref. [40] with permission of the National Ground Water Association. Copyright 2004.

anaerobic conditions the mobility and residence time of E2 and EE2 in aquifer would primarily depend on their sorption.

4.3.2.3 Removal of drug residues by soil-aquifer treatment (SAT)

Different authors [20,21,41] are describing the fate of pharmaceutical residues during SAT when treated municipal wastewater effluents are used for aquifer recharge. The knowledge on the fate of such refractory compounds is not only of scientific but also of public interest. Thus, in a few semi-arid locations in the US, waters processed by SAT are recovered and utilized to satisfy potable water needs [41]. Drewes et al. [20,21] and Sedlak and Pinkston [42] studied the fate of several drug residues during SAT of sewage effluents after secondary treatment of municipal wastewater. They reported a high efficacy for the removal of several pharmaceuticals such as the analgesics diclofenac, ibuprofen, fenoprofen, ketoprofen and naproxen or the beta-blocker metoprolol. However, SAT did not remove residues of other compounds such as the anti-epileptic drugs carbamazepine and primidone or the analgesic drug propyphenazone [20,21].

For the Sweetwater Recharge Facility (SRF) in Tucson, AZ, Quanrud et al. [43] and Wilson et al. [44] reported that percolation of treated wastewater through an unconsolidated sediment and underground storage for periods in the order of months reliably decreased levels of dissolved organic carbon (DOC) by an order of magnitude or more. However, it is also known that several specific trace organics survive SAT processes [45]. Quanrud et al. [41] investigated the estrogen activity in wastewater effluent before and after SAT using both a (hER-beta) competitive binding assay and a transcriptional activation (yeast estrogen screen, YES) assay. Besides several methodological difficulties, Quanrud et al. [41] concluded that "there are circumstances under which SAT does not completely remove estrogenic activity during municipal wastewater effluent polishing". Other, on a first view oppositional results were reported by Mansell and Drewes [46] who examined the fate of steroid hormones, represented by the androgen testosterone and the estrogens E2 and E3, during SAT. In this study enzyme-linked immunosorbent assays were applied to samples from controlled laboratory soil-column studies and from two water reuse field sites where treated wastewater is fed to groundwater recharge basins. In general, the mobility of the selected hormones in subsurface systems was low. In groundwater monitoring wells or shallow lysimeters representing water samples after 1.5 m of travel through porous media estriol and testosterone were not detected above their limits of detection of 0.6 ng/L [46]. In the same samples, E2 was consistently found at concentrations below 2 ng/L [46]. But no breakthrough was observed for all target compounds in groundwater samples collected downstream of the surface spreading facility operational for more than 13 years [46]. Adsorption to the porous media matrix were suggested as the primary removal mechanisms of E2, E3 and testosterone, during groundwater recharge via SAT [46,47]. Additional attenuation below the detection limit occurred in the presence of bioactivity that occurred regardless of the type of organic carbon matrix present (hydrophobic acids, hydrophilic carbon vs. colloidal carbon) or the dominating redox conditions (anoxic vs. aerobic) [46,47].

In combination, the results from the studies reported by Quanrud et al. [41] and by Mansell and Drewes [46] are both indicating that the majority of residues of estrogenic compounds found in municipal wastewater effluents are removed during SAT. However, it is not clear if SAT is always capable to completely remove such residues. The investigations by Quanrud et al. [41] indicate that under unappreciable conditions a break-through scenario is possible. However, their methodological approach does not allow an identification of the chemicals causing the biological response. Even though different studies [48,49] have shown that estrogenic disrupting activities in municipal sewage effluents are mainly caused by steroid hormones including E2, E3 and EE2, other synthetic chemicals, such as bisphenol A, 4-nonylphenol and 4-tert-octylphenol, are to a lesser extent also contributing to the total estrogenic potential of wastewater. Thus, these or other still unknown estrogens might be refractory during SAT causing the effects observed by Quanrud et al. [41].

Drewes et al. [50] investigated the behavior of triiodinated benzene derivatives in domestic effluents and their fate during subsequent groundwater recharge by SAT. Measurements were conducted by analyzing the content of the adsorbable organic iodine (AOI) as a surrogate for the triiodinated benzene derivatives that are mainly used in hospitals as X-ray contrast media. In the effluents of seven wastewater treatment facilities located in different southwestern states of the USA, organic iodine concentrations varied between 5 and 40 μg iodine/L [50]. In laboratory biodegradation experiments and in field studies, negligible removal was observed under aerobic redox conditions while anoxic conditions led to partial removal of organic iodine [50]. This result was also confirmed in a consecutive study [21] presented in more detail below. After travel times of 8–10 years, residues of iodinated X-ray contrast media were still detected at concentrations between 8 and 15 μg iodine/L in the groundwater recharge systems. Drewes et al. [50] assumed that beside appropriate redox conditions, biologically available organic carbon seems to be a key factor for the biodegradation organic iodine in the environment.

Drewes et al. [21] examined the fate of selected residues of pharmaceuticals at two water reuse sites located in the Southwestern USA where secondary and tertiary-treated wastewater is used for subsequent groundwater recharge by SAT. Wastewater effluents infiltrated at the field site shown schematically in Fig. 4.3.3 was received from a municipal STP with activated sludge treatment including nitrification/denitrification followed by disinfection and tertiary filtration. The

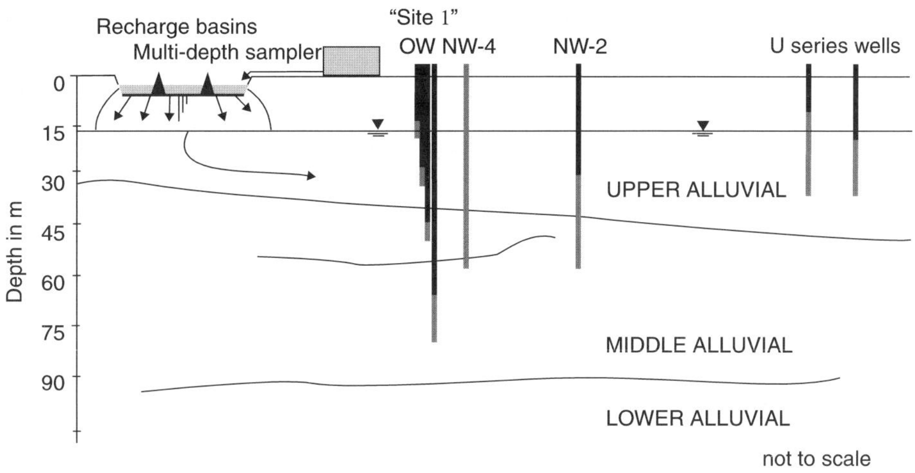

Fig. 4.3.3. Cross section of an SAT site in Southwestern USA indicating screening intervals of groundwater monitoring wells sampled for the analysis of pharmaceutical residues. Reprinted from Ref. [21] with permission of the National Ground Water Association. Copyright 2004.

effluent was applied to four surface spreading basins for groundwater recharge. Groundwater monitoring wells with screened intervals of 23–56 m below ground surface (NW2, NW4), and with screened intervals of 14–25 m below ground surface (e.g. 2U and 6U) were located down gradient of the percolation ponds. Additionally, two piezometer wells OW1 and OW2 located on the property of the water reclamation plant with four screened intervals of 6 m (from 15 m, 27 m, 40 m and 61 m below ground surface) were monitored during the study. Travel times were estimated and reclaimed water plume movements were defined based on the results obtained for sulfate used as a tracer compound. Retention times were ranging from a few months to more than eight years for the monitoring wells (Fig. 4.3.3) analyzed for pharmaceutical residues in this study. The results from these investigations are compiled in Table 4.3.2. Additionally, DOC and specific UV absorbance (SUVA) were used to assess changes in the bulk of organics. The total volume of water recharged since initiation of operations was ~37 million m^3 which could produce a plume of 3.9 km^2 in the upper alluvial unit downgradient from the recharge basins [21].

Drewes et al. [21] concluded that caffeine, diclofenac, ibuprofen, ketoprofen, naproxen, fenoprofen and gemfibrozil were efficiently removed to concentrations near or below the detection limits of the analytical method after retention times of less than six months during

TABLE 4.3.2

Residues of pharmaceuticals and bulk organics in tertiary effluent and groundwater samples from monitoring wells at an SAT site in Southwestern USA. Please refer to Fig. 4.3.3 for the assignment of the monitoring wells

		Tertiary effluent	OW2-90 (6–18 months)	NW-4 (6–18 months)	NW-2 (~16 months)	2U (~2 years)	6U (> 8 years)	Groundwater background
	Mean DOC (mg/L)	5.62 (±0.87)	1.43 (±0.35)	1.49 (±0.26)	1.16 (±0.18)	1.12 (±0.11)	0.9 (±0.12)	0.5
	SUVA (L/mg m)	1.54 (±0.36)	1.87 (±0.38)	1.79 (±0.22)	1.84 (±0.37)	2.46 (±0.61)	2.18 (±0.51)	N/A
Psycho-motor simulants (ng/L)	Caffeine	n.d.	n.d.	n.d.	n.d.	n.d.	n.d.	n.d.
X-ray contrast agents (μg/L)	Organic iodine (μg I/L)	19.3 (Sun)30.5 (Wed)	15.4	8	8.9	6.7	12.5 (±0.9)	0.6
Lipid regulators (ng/L)	Gemfibrozil	n.d.	n.d.	n.d.	n.d.	n.d.	n.d.	n.d.
	Clofibric acid	n.d.	n.d.	n.d.	n.d.	n.d.	n.d.	n.d.
	Fenofibrate	n.d.	n.d.	n.d.	n.d.	n.d.	n.d.	n.d.
Antiepileptics (ng/L)	Carbamazepine	175 (±28.3)	115	235	125	145	85	n.d.
	Primidone	202 (±24.75)	140	120	160	90	100	n.d.
Analgesics/anti-inflammatory drugs (ng/L)	Diclofenac	n.d.	n.d.	n.d.	n.d.	n.d.	n.d.	n.d.
	Ibuprofen	16(±0)	n.d.	n.d.	16	n.d.	n.d.	n.d.
	Ketoprofen	n.d.	n.d.	n.d.	n.d.	n.d.	n.d.	n.d.
	Naproxen	8 (±0)	n.d.	n.d.	n.d.	<10	<10	n.d.
	Fenoprofen	n.d.	n.d.	n.d.	n.d.	n.d.	n.d.	n.d.
	Propyphenazone	n.d.	n.d.	n.d.	n.d.	n.d.	n.d.	n.d.
	Meclofenamic acid	n.d.	n.d.	n.d.	n.d.	n.d.	n.d.	n.d.
	Tolfenamic acid	n.d.	n.d.	n.d.	n.d.	n.d.	n.d.	n.d.
Blood viscosity agent (ng/L)	Pentoxifylline	n.d.	n.d.	n.d.	n.d.	n.d.	n.d.	n.d.

n.d., Not detectable; N/A, not available.

Source: Reprinted from Ref. [21] with permission of the National Ground Water Association. Copyright 2003.

groundwater recharge. In contrast, the anti-epileptic drugs, carbamazepine and primidone, were not removed during SAT under either anoxic saturated or aerobic unsaturated flow conditions with travel times of up to eight years. Triiodinated contrast agents were present in both secondary and tertiary effluent in concentrations between 11 and 30 μg iodine per liter and were partially degradable during SAT under anoxic saturated flow conditions. Under aerobic flow conditions no decrease of the surrogate parameter AOI was observed.

4.3.2.4 Removal of drug residues from contaminated surface waters by bank filtration

Bank filtration has been recognized as important, effective and cheap technique for surface water treatment and removal of microbes, inorganic and some organic contaminants. Bank filtration is successfully and continuously used for more than 100 years in several European countries but it is also used in several other parts of the world including the USA. In the last years, bank filtration was again gaining increased attention by public water suppliers as it might be used as a natural and cheap alternative to extensive artificial drinking water purification processes. Bank filtration can be used as a first "pre-filtration" step in the purification train of drinking water production. A properly designed and operated bank filtration facility may also completely substitute extensive treatment procedures. In Berlin, Germany, more than 50% of the production relies on surface water treated by bank filtration only followed by aeration and sand filtration but without further disinfection procedures. But even if bank filtration is only used and needed for the recharge of the groundwater aquifers, it might additionally be used as a first clean-up step to remove bacteria and viruses but also some other anthropogenic chemicals. Reducing the burden of contaminants also means to save chemicals and operational costs in drinking water purification and to minimize the formation of disinfection by-products [6,51].

Bank filtration is an interesting option for major municipalities and/or semi-arid and arid areas which often have an increased demand for potable water that cannot be kept into sufficient supply by the naturally occurring groundwater resources. Downstream from agricultural areas, surface water used for the infiltration at bank filtration sites is often contaminated by nutrients, animal drugs and pesticides [6]. Large municipalities have been identified as the main sources for the occurrence of pharmaceutical residues from human medical applications in the aquatic environment [4]. In the surface water of such areas, drug residues are often present at concentrations higher than those of the "classical" chemicals, e.g. pesticides. At several sites, e.g. in Berlin, Germany, bank filtration is used in drinking water production even when these sites are located downstream from areas of discharge of purified municipal sewage. Thus, the behavior and fate of pharmaceutical residues during the infiltration of surface water under the influence of wastewater may also be relevant for public drinking-water supply.

The fate of pharmaceutical residues during bank filtration has been studied in several investigations [5–7,11,12,52–54]. Brauch et al. [11] studied the behavior of several selected polar organic contaminants, including two drug residues, at two waterworks near the lower Rhine River. The waterworks Wittlaer, run by the Stadtwerke Duisburg AG, produces water from wells also parallel to the banks of the Rhine River. In the above-mentioned study, raw water samples, containing known amounts of bank-filtered water mixed with groundwater, and groundwater samples from an additional sampling point (M1t), located between the Rhine River and the well galleries of the Wittlaer Waterworks, were analyzed. On average, the raw water produced by the waterworks in Wittlaer contains 60 percent of bank-filtered water from the Rhine River. Water from this monitoring well consists only of bank-filtered water and is not influenced by groundwater. In their investigations, Brauch et al. [11] detected the analgesic diclofenac and the anti-epileptic drug carbamazepine in samples from the Rhine River. As demonstrated by the results presented in Fig. 4.3.4, diclofenac was removed almost completely during the bank filtration process. Carbamazepine was found at average concentrations of 0.11 μg/L in the Rhine River and at concentrations of 0.11 and 0.067 μg/L in the samples collected from location M1t and in the raw water from the Wittlaer waterworks, respectively. Significant removal of carbamazepine was not observed during bank filtration (Fig. 4.3.5). The decrease in concentrations was assumed to be caused only by dilution with uncontaminated groundwater. However, ozonation and charcoal filtration were able to remove carbamazepine from the raw water.

Further information on the transport and removal of drug residues was derived from field studies carried out at groundwater recharge sites [5,52–60]. In Berlin, Germany, different drug residues including analgesics, antibiotics, anti-epileptic drugs, blood lipid regulators and estrogenic steroids were investigated between 2002 and 2005 in terms of an interdisciplinary research project entitled NASRI (Natural and Artificial Systems for Recharge and Infiltration) [52–54]. The transport

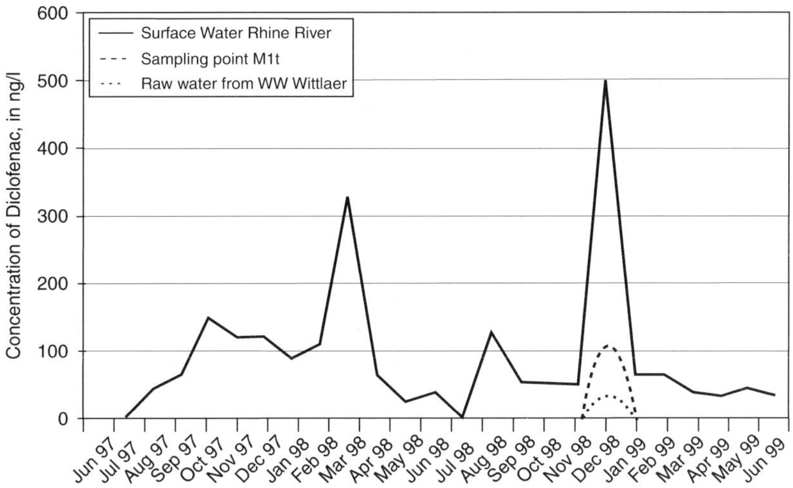

Fig. 4.3.4. Concentrations of diclofenac in the Rhine River (732.1 km), at sampling location M1t, and in the raw water from the Wittlaer Water works. Reproduced with permission from Ref. [11].

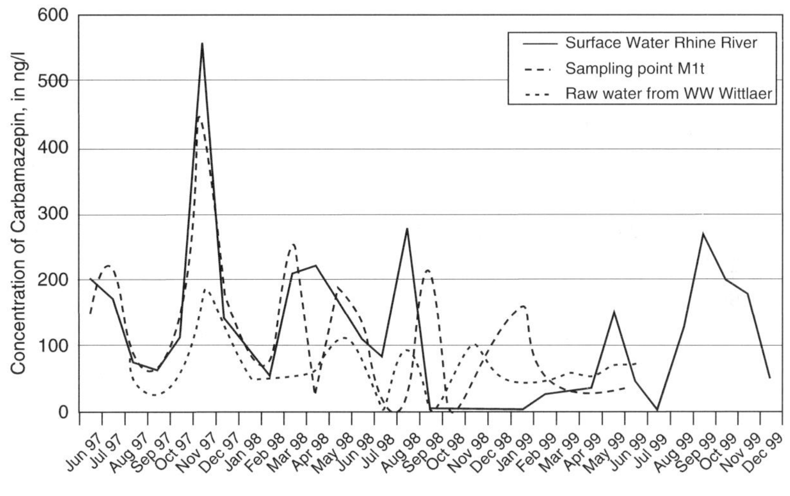

Fig. 4.3.5. Concentrations of carbamazepine in the Rhine River (732.1 km), at sampling location M1t, and in the raw water from the Wittlaer Waterworks. Reproduced with permission from Ref. [11].

and fate of drug residues was investigated across different bank filtration sites (Fig. 4.3.6) using surface water under the influence of municipal wastewater for the recharge of groundwater used for drinking-water supply.

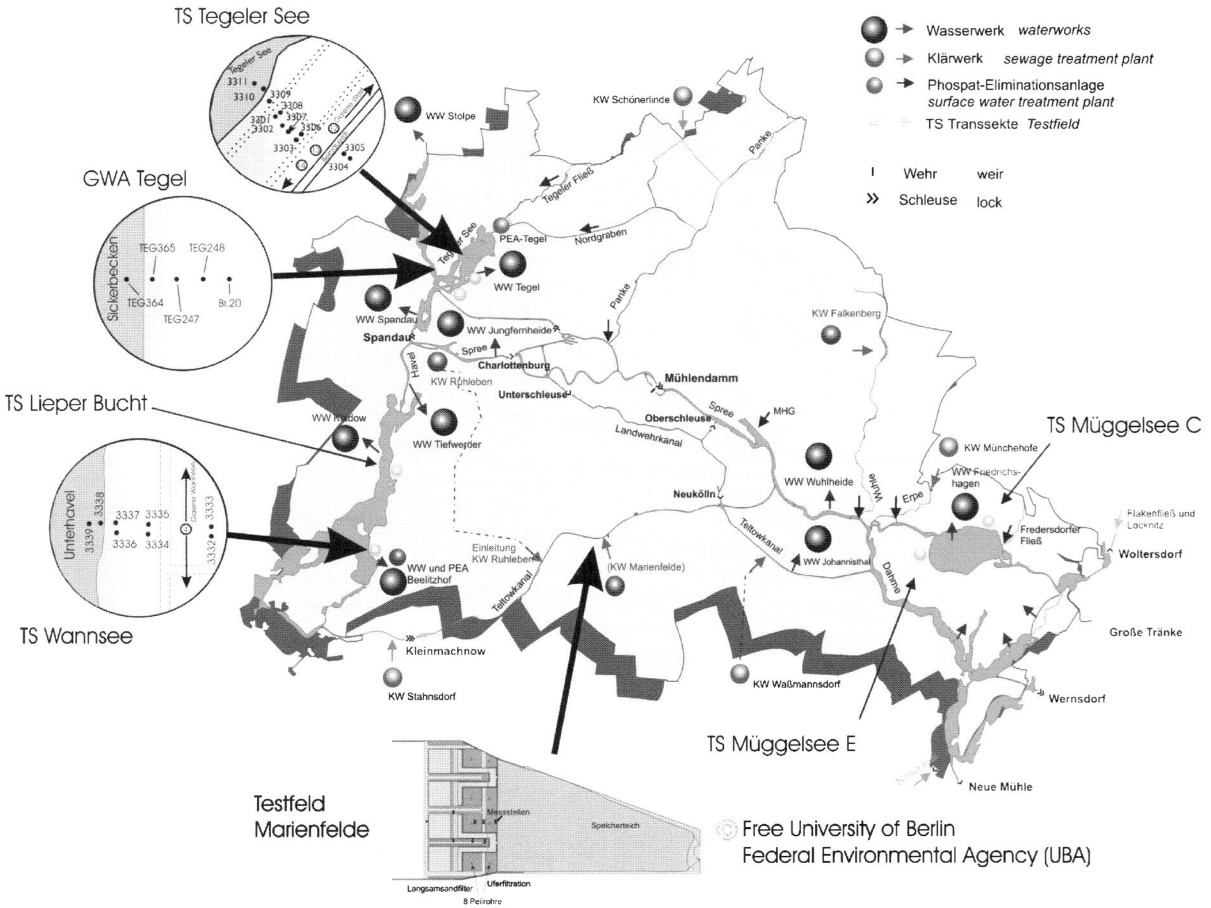

Fig. 4.3.6. Map showing the locations of the bank-filtration transects at Lake Wannsee (TS Wannsee) and Lake Tegel (TS Tegler See), the transect at the groundwater enrichment pond near lake Tegel (GWA Tegel), and the semi-technical facility (recharge pond-Testfeld Marienfelde) in Berlin, Germany. Reprinted from Ref. [53] with permission of the National Ground Water Association. Copyright 2004.

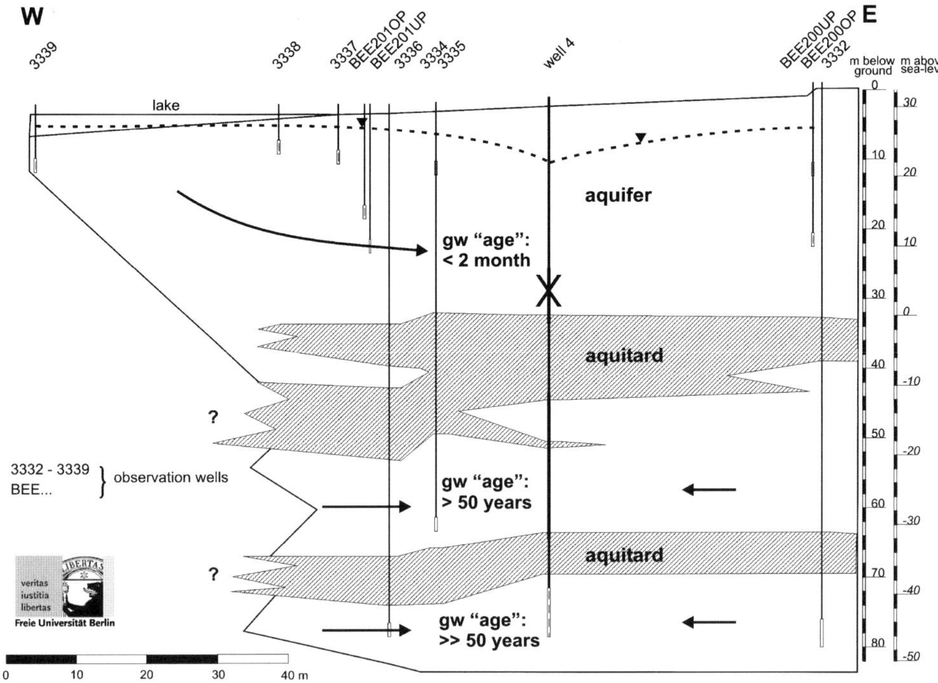

Fig. 4.3.7. Schematic hydrogeologic section of the transect at Lake Wannsee, location of monitoring wells (shallow and deep) and WSW 4. The darker areas on the bottom of the wells are the individual screens. Reprinted from Ref. [53] with permission of the National Ground Water Association. Copyright 2004.

Figure 4.3.7 shows the hydrogeological cross section from a bank-filtration transect located at lake Wannsee and Table 4.3.3 compiles the corresponding results measured for several polar drug residues and related trace organics found in the surface water of the lake and the groundwater of the bank filtration site. At the bank filtration site at lake Wannsee, residues of six pharmaceuticals including the analgesic drugs diclofenac and propyphenazone, the anti-epileptic drugs carbamazepine and primidone, and the drug metabolites clofibric acid and 1-acetyl-1-methyl-2-dimethyl-oxamoyl-2-phenylhydrazide (AMDOPH) were found to leach from the contaminated streams and lakes into the groundwater. Residues of these compounds were also detected at low concentrations in receiving public-supply wells. [53]

The water produced by water-supply wells represents a mixture of landward groundwater with bank-filtered water recharged recently (younger bank filtrate) or even several years ago (old bank filtrate). Thus, the results obtained for the water samples collected from the

TABLE 4.3.3

Organic compounds with positive findings and their mean concentrations (ng/L) in the monitoring wells 3335, 3337, 3338 and 3339 and the receiving water-supply wells 3, 4 and 5 at transect "Lake Wannsee" between May and October 2002 ($n = 6$)

Transect Wannsee (May–October 2002)	Affiliation	Surface water	3339	3338	3337	3335	Well 5	Well 4	Well 3
Diclofenac	Analgesic	25	30	30	15	20	5	<5	10
Clofibric acid	Metabolite of a BLR	60	25	20	15	5	125	115	95
Propyphenazone	Analgesic	145	85	100	75	55	105	30	160
AMDOPH	Metabolite of an analgesic	170	115	150	180	175	580	280	330
Carbamazepine	Anticonvulsant	330	215	365	325	360	30	15	70
Primidone	Anticonvulsant	60	70	60	50	60	35	10	40
Indometacine	Analgesic	15	<5	n.d.	n.d.	n.d.	n.d.	n.d.	n.d.
Bezafibrate	BLR	60	10	n.d.	n.d.	n.d.	n.d.	n.d.	n.d.
Bentazone	Herbicide	25	20	30	35	20	10	10	15
Mecoprop	Herbicide	15	15	20	10	20	10	10	15
p,p'-DDA	DDT—metabolite	20	20	25	25	10	15	10	20
o,p'-DDA	DDT—metabolite	5	5	5	5	<5	<5	<5	<5
NPS	Metabolite of a corrosion inhibitor	20	20	25	25	15	165	55	115
TCEP	Flame retardant	315	300	255	495	400	70	95	210
TCIPP	Flame retardant	2100	1130	1800	1345	1145	295	190	390

AMDOPH, 1-Acetyl-1-methyl-2-dimethyl-oxamyl-2-phenylhydrazide; BLR, blood lipid regulator; DDT, dichloro diphenyl trichloroethane; DDA, Bis(4-chlorophenyl)-acetic acid; NPS, *N*-(Phenylsulfonyl)-sarcosine; TCEP, Tris(2-chloroethyl)-phosphate; TCIPP, Tris(2-chloroisopropyl)-phosphate; n.d., Not detected.

Source: Reprinted from Ref. [53] with permission of the National Ground Water Association. Copyright 2004.

water-supply wells provide some information about the contamination of the raw water used for drinking water production but will not provide any information on the removal efficiency of the analytes if only compared to the results measured in the lake.

Estrogenic steroids are difficult to study in environmental samples because they only appear at very low but still environmentally important concentrations of between 0.1 and 10 ng/L in wastewater effluents [54–57]. Theoretically, steroid hormones should not be expected to occur in groundwater because of their hydrophobicity resulting in easy sorption and due to their potential for biotransformation. Nevertheless, Kuch and Ballschmiter [58] and Adler et al. [59] reported several detections of estrogens including EE2 in German ground- and drinking water samples. Other investigations of steroid hormones in sewage, surface and bank-filtered water do, however, not indicate the presence of such compounds for ground or drinking water [54]. Thus, E1 was the only compound found at concentrations above the limit of determination of 0.1 ng/L in surface water under the influence of municipal wastewater used for the infiltration. Zuehlke et al. [54] observed that even short distances between the river or lake banks and the monitoring wells led to dramatic decreases of E1 concentrations illustrating the potential of groundwater recharge systems for the retention of estrogenic steroids.

Fanck and Heberer [60] detected residues of five antibiotics, the sulfonamide sulfamethoxazole, the sulfonamide synergist trimethoprim, the macrolides clarithromycin and roxithromycin, and the lincosamide clindamycin in surface water under the influence of municipal wastewater used for bank filtration. Additionally, dehydro-erythromycin, the metabolite of the macrolide erythromycin and acetyl-sulfamethoxazole, the main human metabolite of sulfamethoxazole, were found. With the exception of sulfamethoxazole, antibiotic residues were not found in the water-supply wells. Sulfamethoxazole was detected at trace-levels in samples collected from monitoring and water-supply wells whereas most of the other compounds are readily removed close to the bank where the surface water is infiltrated. Most of the compounds were not or only found at trace levels in the first two monitoring wells located close to the bank [60]. Fanck and Heberer [60] assumed that for sulfamethoxazole and dehydro-erythromycin an improved degradation occurs under reduced conditions, whereas compounds such as clindamycin were preferably degraded under oxic conditions.

In conclusion, bank filtration either decreased the concentrations by dilution (e.g., for carbamazepine and primidone) and partial removal (e.g., for diclofenac and sulfamethoxazole), or totally removed such residues (e.g., bezafibrate, indomethacine, most antibiotics, and estrogens). However, a complete removal of all potential pharmaceutical residues by bank filtration cannot be guaranteed. Thus, in areas with surface waters under the influence of municipal sewage effluents polar drug residues have to be considered as being relevant for the quality of potable water when groundwater recharge is used in drinking water production. Nevertheless, bank filtration has proven as being an efficient method for the removal of pharmaceutical residues by natural attenuation and as a useful tool for the pre-treatment of surface water under the influence of sewage effluents used for drinking-water supply.

However, Heberer et al. [53] also concluded that drug residues such as carbamazepine or primidone that are readily transported during bank filtration might be used as indicators of sewage impacts. Thus, it might be possible to evaluate whether surface water is impacted by contaminations from municipal sewage effluents or whether contaminations associated with sewage effluent can be transported into groundwater at groundwater recharge sites.

4.3.2.5 Removal of drug residues from contaminated surface waters by groundwater replenishment (GWR)

Groundwater recharge of surface water is also conducted using groundwater replenishment (GWR) facilities operated similar to the SAT site in Fig. 4.3.3. Figure 4.3.8 shows the hydrogeological cross section of such a field site located near lake Tegel in Berlin, Germany. The behavior of various organic contaminants including some drug residues was also investigated in the course of the already mentioned NASRI project.

At this site, lake water under the influence of sewage effluents is applied to a recharge pond (RP3) and drawn into the aquifer by a gradient produced by a gallery of water-supply wells including water-supply well 20 (WSW 20) abstracting more than 80% of artificially recharged groundwater [61]. The water produced by WSW 20 is a mixture of younger and older water recharged only recently (travel time of 50 days) or even several years ago [61]. Age dating investigations also confirmed this assumption yielding an average T/He age of 2.3 years for the water produced by this well [61]. Thus, the results obtained for the water samples collected from WSW 20 provide some information about the contamination of the raw water used for drinking water production but

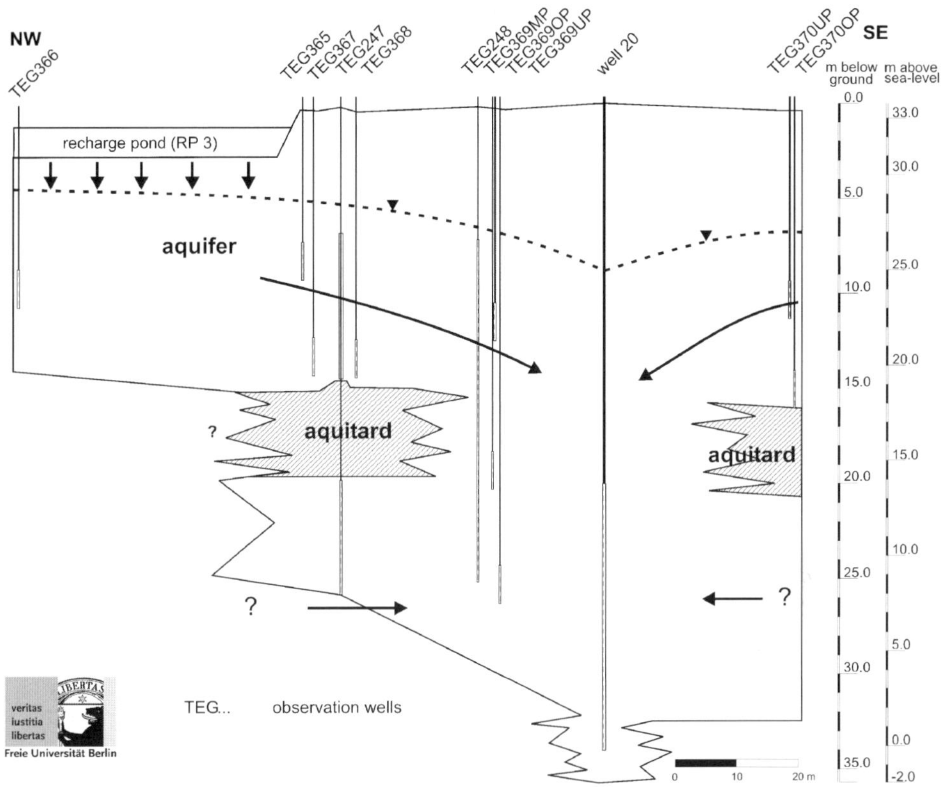

Fig. 4.3.8. Hydrogeological cross section of a groundwater recharge transect (recharge pond RP3, monitoring well field and water-supply well WSW 20) located at the groundwater replenishment site in Berlin-Tegel, Germany. Figure taken from Ref. [62], courtesy of CSIRO PUBLISHING.

it will not provide any information on the removal efficiency of the analytes if only compared to the results measured for the surface water applied to the recharge ponds. Table 4.3.4 compiles the results for several drug residues detected in samples collected from the recharge pond, the monitoring wells and from WSW 20 [62].

As also shown in Table 4.3.5, Heberer and Adam [62] categorized the residues according to their removal behavior. They concluded that drug residues with log P_{OW} values above 3 such as bezafibrate or indomethacin (group 3) are readily removed by bank filtration whereas compounds with low log P_{OW} values (groups 1 and 2) are not completely eliminated. However, the log P_{OW} values alone are not sufficient to predict the behavior of the individual compound because additional parameters such as other (polar) sorption mechanisms and especially

TABLE 4.3.4

Concentrations of pharmaceutical residues in ng/L detected in recharge pond RP3, in the monitoring wells (for well assignments please refer to Fig. 4.3.4) and in water-supply well WSW20. Arrows indicate groundwater flow direction toward WSW20. Samples were collected between July 2002 and June 2003 ($N = 6–12$)

	Pond RP3	Teg 366	Teg 365	Teg 247	Teg 368	Teg 248	Teg 369 OP	Teg 369 UP	**WSW 20**	Teg 370 OP	Teg 370 UP
N	12	12	12	8	6	8	6	6	11	6	6
Flow direction		→	→	→	→	→	→	→		←	←
AMDOPH	455	440	395	425	390	315	300	330	**1570**	1085	3915
Carbamazepine	470	545	430	385	460	430	220	230	**210**	20	20
Primidone	135	140	125	115	170	95	80	90	**100**	30	70
Propyphenazone	120	20	20	30	15	20	10	10	**40**	10	55
Clofibric acid	20	5	5	10	10	5	5	5	**5**	15	n.d.
Diclofenac	135	15	45	5	15	10	<5	<5	**10**	n.d.	n.d.
Indomethacin	20	<10	<10	n.d.	n.d.	n.d.	n.d.	n.d.	**n.d.**	n.d.	n.d.
Bezafibrate	30	n.d.	n.d.	n.d.	n.d.	n.d.	n.d.	n.d.	**n.d.**	n.d.	n.d.

n.d., Not detected; LOD, limit of detection (<1 ng/L for all analytes).
Source: Table taken from Ref. [62], courtesy of CSIRO PUBLISHING.

TABLE 4.3.5

Classification of target compounds according to their removal rates during groundwater recharge at a groundwater replenishment plant in Berlin-Tegel (Fig. 4.3.4)

Group	Compound	log P_{OW}	Mean removal at Teg 248 (%)	Mean decrease of concentrations at WSW20 (%)
Low removal rates (0–50%)	AMDOPH	Unknown	31	−245 (exceptional case[a])
	Carbamazepine	2.45	9	55
	Primidone	0.91	30	26
Medium removal rates (51–95%)	Propyphenazone	2.05	83	67 (exceptional case[a])
	Clofibric acid	3.1	75	75
	Diclofenac	1.13	93	93
High removal rates (>95%)	Indomethacin	4.27	>95	>95
	Bezafibrate	4.2	>97	>97

Source: Table taken and modified from Ref. [62], courtesy of CSIRO PUBLISHING.
[a]Increased concentrations caused by former contamination of the groundwater body.

their physico-chemical and microbial persistence are also important factors for the removal of such residues.

At the same field site, Massmann et al. [63] investigated the impact of variable temperatures on the redox conditions and the behavior of pharmaceutical residues during artificial recharge. They recognized that aerobic conditions prevailed during winter, whereas anaerobic conditions were reached below the pond when temperatures exceeded 14°C. As long as oxygen was present, residues of the analgesic phenazone were fully degraded before reaching the first groundwater well. In contrast, phenazone was not fully eliminated when conditions turned anaerobic. Massmann et al. [63] also confirmed that AMDOPH and the anti-epileptic drug carbamazepine are very persistent but there was also evidence that AMDOPH may be slightly degradable under aerobic conditions.

In general, the results from the GWR site were almost similar to those observed at the bank filtration sites. Again, the two anti-epileptic drugs, carbamazepine and primidone, and the drug metabolite AMDOPH were identified as highly persistent and very mobile compounds that easily leach through the sub-soil into groundwater and are thus also relevant to public-water supply. AMDOPH and propyphenazone are, however, also representing special cases because the increase of the concentrations of AMDOPH and the low decrease of the concentrations observed for propyphenazone can only be explained by mixtures of naturally recharged groundwater containing no residues and artificially recharged groundwater with equal to lower (recently infiltrated surface water) or even higher concentrations (earlier infiltrated water, high historical background concentrations) of these analytes.

4.3.3 REMOVAL IN WATERWORKS

4.3.3.1 Flocculation

According to Ternes et al. [64,65], flocculation with Fe(III)chloride and aluminium sulphate did not result in a significant elimination of the drug residue from raw water. Using flocculation in the lab-scale Jar test, the relative concentration levels (c/c_0) were $96 \pm 11\%$ for diclofenac, $87 \pm 10\%$ for clofibric acid, $111 \pm 15\%$ for bezafibrate, $87 \pm 12\%$ for carbamazepine and $110 \pm 14\%$ for primidone. Monitoring of upscaled flocculation processes in two waterworks yielded similar results and confirmed the transferability of the lab-scale results to waterworks conditions.

4.3.3.2 Aeration and rapid sand filtration

Most of the drug residues which have been identified as being relevant for drinking-water supply are not affected by aeration or rapid sand filtration. One exception are phenazone-type residues that have been detected in raw water samples from all Berlin water works located near the Havel river in Berlin, Germany [25,28]. The residues were most likely caused by spills from a former production plant located in a city upstream of Berlin [25].

In Berlin, production of drinking water is exclusively based on groundwater mainly (by approximately 70%) obtained from groundwater recharge. The groundwater abstracted by means of vertical and horizontal filter wells with submersible pumps is then only aerated and processed through open and closed rapid filters without any addition of chemicals. Nevertheless, phenazone-type residues such as phenazone, propyphenazone and dimethylaminophenazone detected at individual concentrations up to 4 μg/L in the raw water of several water works undergo a significant degradation in the receiving drinking-water treatment plants (Tables 4.3.6 and 4.3.7). Thus, phenazone and propyphenazone are removed by 90%, respectively, and dimethylaminophenazone was no longer detected after purification.

TABLE 4.3.6

Elimination of phenazone residues during drinking-water treatment at a public drinking waterworks in Berlin, Germany

Substance	Limit of quantification in ng/L	Aerated raw water		Drinking water		Removal	
		∅ concentration in ng/L	RSD (%)	∅ concentration in ng/L	RSD (%)	Rate (%)	RSD (%)
Phenazone[a]	0.05	3.95	26	0.4	69	90	7
Propyphenazone[a]	0.005	1.23	29	0.12	54	90	5
Dimethylaminophenazone	0.05	0.4	—	ND	—	>99	—
AMDOPH	0.01	1.2	—	0.9	—	25	—
AMPH	0.02	E 0.02–0.1	—	E 0.03	—	—	—
DMOAS	0.01	Traces	—	Traces	—	—	—

ND, Not detected; E, estimated values.
Source: Reproduced with permission from Ref. [25] © 2002 Elsevier.
Note: Concentration values are given in μg/L.
[a]$N = 6$, for all other compounds $N = 2$.

TABLE 4.3.7

Concentrations of DP, PDP and DMADP (4-(*N*,*N*-dimethyl)-amino-1,5-dimethyl-1,2-dehydro-3-pyrazolone) in raw- and drinking water samples from two public drinking water works in the north west of Berlin, Germany

Sample type	Phenazone	DP	Propyphenazone	PDP	Dimethylaminophenazone	DMADP
Raw water[a]	2.50	1.15	0.88	0.32	0.24	<LOD
Drinking water[a]	0.25	1.10	0.08	0.24	<LOD	<LOD
Raw water[b]	1.10	0.98	0.39	0.25	<LOD	<LOD
Drinking water[b]	0.05	0.29	<LOD	0.10	<LOD	<LOD

LOD, Limit of detection (0.001 μg/L for DP and PDP, 0.002 μg/L for DMADP and 0.05 μg/L for the parent compounds).

Source: Reproduced with permission from Ref. [28] © 2004 Elsevier.

Note: All values are given in μg/L.

[a]Sample collected in water works 1.

[b]Sample collected in water works 2.

Reddersen et al. [25] and Zuehlke et al. [28] identified several metabolites namely AMDOPH, 1-acetyl-1-methyl-2-phenylhydrazide (AMPH), dimethyloxalamide acid-(N'-methyl-N-phenyl)-hydrazide (DMOAS), 1,5-dimethyl-1,2-dehydro-3-pyrazolone (DP) and 4-(2-methylethyl)-1,5-dimethyl-1,2-dehydro-3-pyrazolone (PDP) (Fig. 4.3.9) which are formed during rapid sand filtration and/or already occur in the raw water. Among these metabolites, DP, PDP and AMDOPH were recognized as persistent residues occurring at concentrations up to the μg/L-level in purified drinking water (Tables 4.3.6 and 4.3.7).

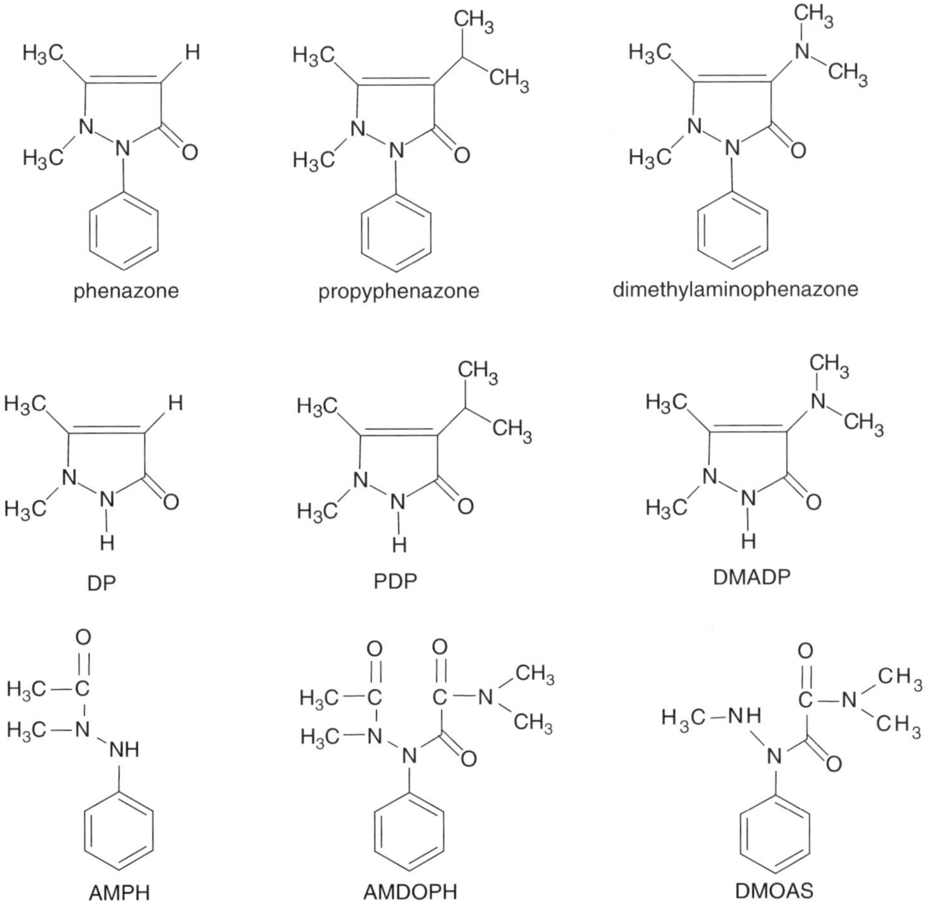

Fig. 4.3.9. Structural formulae of phenazone-type residues investigated and identified in raw and drinking water samples in Berlin, Germany.

4.3.3.3 Ozonation

Ozone is increasingly used for drinking-water treatment and for some municipal and industrial wastewater applications. In the United States, more than 200 drinking water plants were using ozonation for drinking water disinfection at the end of the 1990s [66]. Ozone is used as an almost odorless alternative to the treatment with chlorine that also produces more troublesome disinfection by-products. From industrial and municipal wastewater applications ozone is known to be very efficient in decreasing the loads of organic compounds. Thus, ozonation was also recognized as a promising technique for the removal of pharmaceutical residues either from municipal wastewater or drinking water.

Obviously, doses of ozone as they are currently used in waterworks for drinking water disinfection are usually not sufficient to transform significant amounts of persistent organic compounds. At increased ozone doses a significant decrease of the concentrations was observed for several of the drug residues identified as being relevant for public drinking-water supply [65,67,71]. But Hua et al. [18] also observed significant decreases of carbamazepine concentrations in a Canadian drinking-water treatment pilot plant applying ozone treatment as part of the water processing. Huber et al. [67] concluded that ozonation and advanced oxidation processes (AOPs) are promising for the efficient removal of pharmaceuticals in drinking waters. In bench-scale experiments, they determined second-order rate constants for the reactions of selected pharmaceutical compounds with ozone (k_{O_3}) and OH radicals (k_{OH}). High reactivities with ozone (k_{O_3}) were observed for diclofenac ($\sim 1 \times 10^6\,M^{-1}\,s^{-1}$) and carbamazepine ($\sim 3 \times 10^5\,M^{-1}\,s^{-1}$). The results indicate that these two substances are very rapidly transformed during ozonation. Transformation with lower reactivities were measured for bezafibrate ($590 \pm 50\,M^{-1}\,s^{-1}$), diazepam ($0.75 \pm 0.15\,M^{-1}\,s^{-1}$), ibuprofen ($9.6 \pm 1\,M^{-1}\,s^{-1}$) and iopromide ($< 0.8\,M^{-1}\,s^{-1}$). Ternes et al. [65] observed an appreciable decrease of the concentration levels of bezafibrate and primidone during ozonation. But ozone only exhibited limited efficiency in removing clofibric acid [65].

In general, the efficiency of the ozonation process for the removal of the drug residues turned out to be very compound specific. Thus, at small ozone doses of 0.5 mg/L the concentrations of diclofenac and carbamazepine could be decreased by more than 97% while clofibric acid was decreased by only 10–15%. Concentrations of primidone and bezafibrate were declined by 50% at doses of ozone of approximately 1.0 mg/L and 1.5 mg/L, respectively. For clofibric acid, even extremely high-ozone doses up to 2.5–3.0 mg/L resulted only in a decrease of the initial concentrations of $\leqslant 40\%$ [65]. Owing to the missing active sites that are susceptible to ozone attack for clofibric acid, reactions of ozone with this compound are expected to be very slow and OH-radical reactions should be predominant with $k_{OH} \sim 5 \times 10^9\,M^{-1}\,s^{-1}$ [67]. Andreozzi et al. [68] claimed that ozonation and hydrogen peroxide photolysis are capable of enabling a fast removal of clofibric acid in aqueous solution with an almost complete conversion of the organic chlorine content into chloride ions.

Quiang et al. [69] determined the absolute second-order rate constants of ozone with the two amine-based antibiotics lincomycin (containing a free amine group and a sulfur group) and spectinomycin (containing two free amine groups). For lincomycin, Quiang et al. [69] measured absolute rate constants of $2.76 \times 10^6\,M^{-1}\,s^{-1}$ and $3.26 \times 10^5\,M^{-1}\,s^{-1}$ for the attack with ozone at the neutral and the monoprotonated form, respectively. For spectinomycin, ozone attacks with absolute rate constants of $1.27 \times 10^6\,M^{-1}\,s^{-1}$ and $3.30 \times 10^5\,M^{-1}\,s^{-1}$ were determined for the neutral and the monoprotonated form, respectively. These results indicate that ozone reacts quickly with both compounds. The authors also state [69] that the protonated amine is non-reactive toward ozone and the reaction rate significantly depends on pH of the solution. They assumed that lincomycin and spectinomycin will be efficiently transformed by ozonation processes at neutral pH values.

Boyd et al. [70] reported that chlorination, ozonation and dual media filtration processes decreased the concentrations of naproxen found in Mississippi River and Detroit River waters below the detection limits and also reduced the concentrations of clofibric acid detected in Detroit River waters.

The above-mentioned studies have shown that the concentrations of drug residues may significantly be decreased by ozonation. This decrease is, however, only documented by the disappearance of the parent compounds analyzed by instrumental analysis mainly applying GC-MS or LC-MS/MS. The nature and the toxicity of the transformation products remain often unknown. In general, the oxidation processes should increase the polarity of the formed products which is often associated with a reduced biological toxicity and an enhanced microbial degradability. For diclofenac, the main oxidation product showed an increase of the molecular weight of 16 mass units, which evidently results from a substitution of a hydrogen atom by a hydroxy moiety [67].

McDowell et al. [71] conducted a study to investigate the formation of products from the reaction of carbamazepine with ozone and OH radicals. They identified the three oxidation products BQM (1-(2-benzaldehyde)-4-hydro-(1*H*,3*H*)-quinazoline-2-one), BQD (1-(2-benzaldehyde)-(1*H*,3*H*)-quinazoline-2,4-dione) and BaQD (1-(2-benzoic acid)-(1*H*,3*H*)-quinazoline-2,4-dione) shown in Fig. 4.3.10. Further kinetic studies with the ozonation products showed only very slow, subsequent oxidation kinetics for the reaction with ozone (second-order rate constants, $k_{O_3} = \sim 7\,M^{-1}\,s^{-1}$ and $\sim 1\,M^{-1}\,s^{-1}$ at pH = 6 for BQM and BQD, respectively). Rate constants for reactions with OH radicals (k_{OH}) of $\sim 7 \times 10^9\,M^{-1}\,s^{-1}$ and $\sim 5 \times 10^9\,M^{-1}\,s^{-1}$ were determined for BQM and BQD, respectively. Consequently, further oxidation of the primary oxidation products was mainly achieved by reactions with OH radicals. BQM and BQD were also identified in ozonated water from German waterworks with residues of carbamazepine occurring in its raw water [71].

Huber et al. [70] investigated the oxidation of EE2 during ozonation of fortified aqueous solutions for future applications in drinking water or wastewater purification. In an earlier publication, Huber et al. [67]

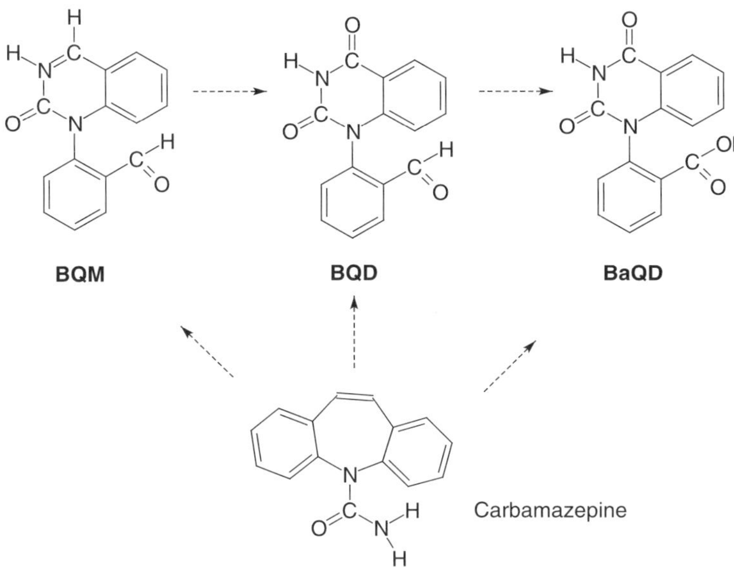

Fig. 4.3.10. Formulae of oxidation products BQM (1-(2-benzaldehyde)-4-hydro-(1H,3H)-quinazoline-2-one), BQD (1-(2-benzaldehyde)-(1H,3H)-quinazoline-2,4-di-one) and BaQD (1-(2-benzoic acid)-(1H,3H)-quinazoline-2,4-dione) formed during ozonation of carbamazepine.

already reported an extremely high second-order rate constant (k_{O_3}) $3\times10^6\,M^{-1}\,s^{-1}$ (at pH 7) for the reaction of EE2 with ozone resulting in a half-life of approximately 10 ms for an O_3 concentration of 1 mg/L. Thus, doses of ozone typically applied for the disinfection of drinking waters were found to be sufficient to reduce the estrogenicity by a factor of more than 200, measured with a recombinant yeast estrogen screen (YES) assay [72]. Nevertheless, it was not possible to completely remove the estrogenic activity with between 0.1 to 0.2% of the initial EE2 concentration remaining after ozonation. Additionally, Huber et al. [72] also identified several of oxidation products formed during ozonation of EE2, E2 and E1. The chemical structures of the oxidation products shown in Fig. 4.3.11 and 4.3.12 were significantly altered as compared to the parent [72]. More generally, the study by Huber et al. [72] has shown that the selective oxidation of the

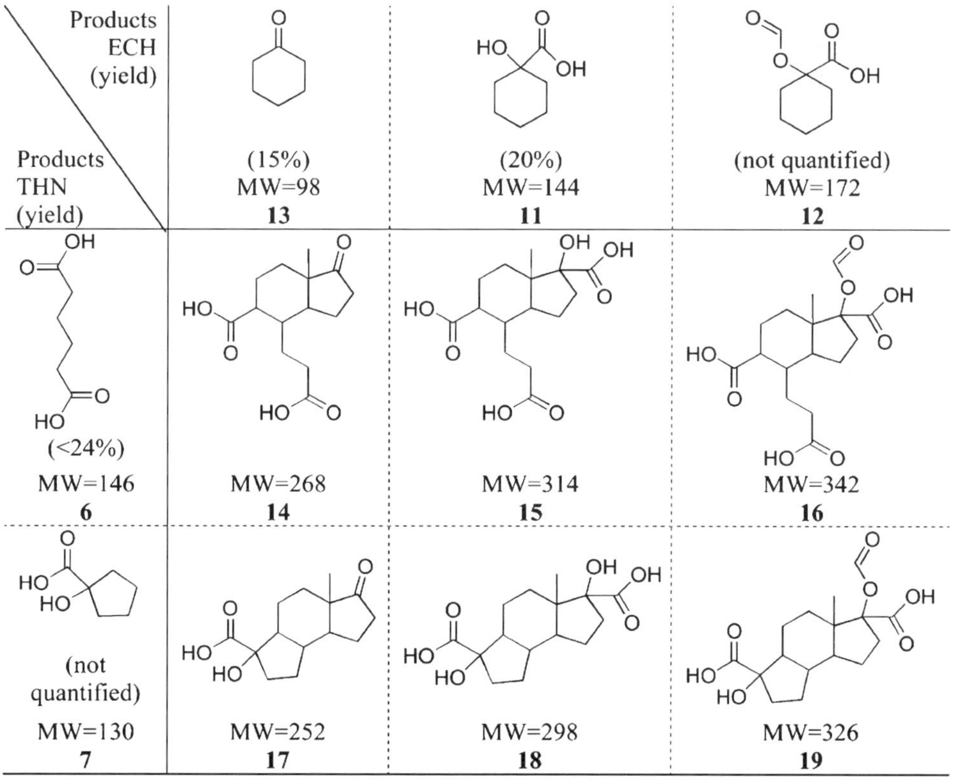

Fig. 4.3.11. Structures of products formed by the reaction of O_3 with EE2. Reproduced with permission from Ref. [72]. Copyright 2004 American Chemical Society.

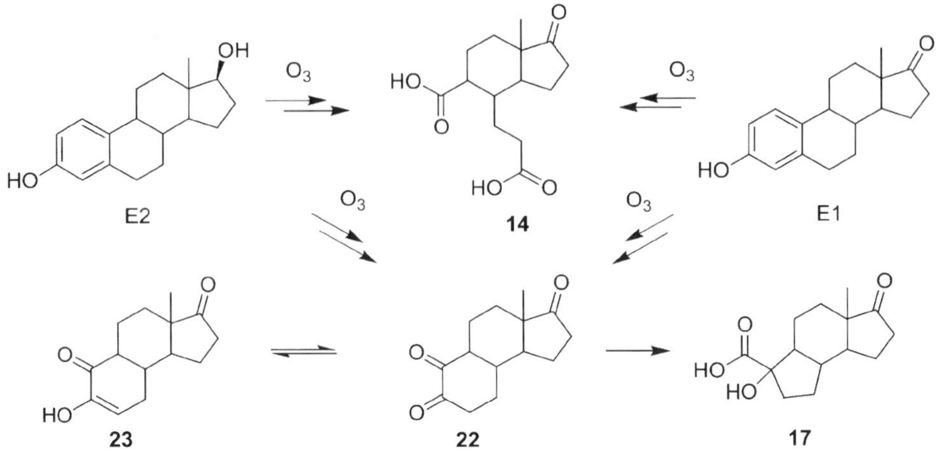

Fig. 4.3.12. Structures of products formed by the reaction of O_3 with E1 and E2. Reproduced with permission from Ref. [72]. Copyright 2004 American Chemical Society.

phenolic moiety efficiently reduces the estrogenicity of EE2-containing solutions.

4.3.3.4 Treatment with chlorine dioxide

Huber et al. [73] also investigated the several pharmaceuticals during water treatment with chlorine dioxide. Only four out of nine investigated compounds exhibited an appreciable reactivity. For these compounds the following second-order rate constants (at a pH of 7 and 20°C) were determined: $6.7 \times 10^3\,M^{-1}\,s^{-1}$ for sulfamethoxazole, $2.2 \times 10^2\,M^{-1}\,s^{-1}$ for roxithromycin, $2 \times 10^5\,M^{-1}\,s^{-1}$ for EE2 and $1.05 \times 10^4\,M^{-1}\,s^{-1}$ for diclofenac. Additional experiments with natural water showed that chlorine dioxide also reacted rapidly with other sulfonamides, with macrolide antibiotics, with E1 and E2, and with pyrazolone derivatives including phenazone, propyphenazone and dimethylaminophenazone. Nevertheless, many compounds investigated in this study were found to be refractive to the treatment with chlorine dioxide.

The authors concluded [73] that compared to ozone, chlorine dioxide reacts more slowly and with a fewer number of compounds. Thus, this treatment technique will only be efficient enough to be used for the oxidation of a broad range of potential drug residues.

4.3.3.5 GAC filtration

Treatment of contaminated raw water by filtration with granular activated carbon (GAC) has been proven as being a very efficient removal process in drinking water production [64,74]. With the exception of clofibric acid, even relatively high concentrations of carbamazepine, diclofenac and bezafibrate could almost completely be removed by GAC at specific throughputs over 70 m^3/kg [74]. Carbamazepine showed the highest adsorption capacity of the compounds investigated by Ternes et al. [74]. It was removable at a specific throughput of about 50 m^3/kg in a carbon layer of 80 cm and at a specific throughput of 70 m^3/kg in a layer of 160 cm. Clofibric acid is less prone to adsorption but could be removed completely at a specific throughput of 15–20 m^3/kg [74]. This assumption is also supported by results reported for clofibric acid from studies with powdered activated carbon conducted by Boyd et al. [70] and Boyd and Grimm [75].

Ternes et al. [74] investigated the behavior of various drug residues occurring in the raw water of a full-scale waterwork using flocculation with Fe(III) chloride and GAC filtration. Pharmaceutical residues were not affected by flocculation but significant removal was observed after GAC filtration. Thus, diclofenac and bezafibrate were no longer detected above the limit of quantitation. The concentrations of carbamazepine and primidone were decreased by more than 75% and for clofibric acid a removal of only 20% was observed.

Another refractory compound was identified by Boyd et al. [70]. Naproxen, found in Louisiana and Ontario surface waters at 22–107 ng/L, was not removed from Mississippi River waters when applying conventional drinking-water treatment processes (coagulation, flocculation and sedimentation) and continuous addition of powdered activated carbon at a dosage of 2 mg/L [70].

4.3.3.6 Membrane filtration

Membrane filtration is one of the most promising techniques for the removal of drug residues and other sewage-borne contaminants. High pressure-driven membranes such as nanofiltration (NF) or reverse osmosis (RO) membranes might be applicable to remove pharmaceutical residues from contaminated raw waters to be used for drinking-water production [76].

Despite higher operational costs an increasing number of sewage or drinking water facilities is now using membrane filtration as final purification method. RO is frequently used for desalination but, like NF, it is also applied for the treatment of waste- and drinking water. NF distinguishes itself from RO by only retaining multivalent ions making it a very economic alternative whenever the retention of monovalent salts is not required [77]. One of the key objectives for the use NF or RO filtration in water and wastewater treatment is the removal of trace contaminants. Nevertheless, the retention of such pollutants has until recently not sufficiently been understood [77]. Now, several new studies have investigated the rejection capacities of membrane filtration for the removal of pharmaceutical residues and also provided an insight into mechanisms influencing the retention behavior of drug residues using different filtration techniques and materials.

Drewes et al. [20] studied different treatment technologies used for municipal wastewater purification including activated sludge, trickling filter, SAT, NF and RO. The aim of this study was to evaluate the capability of the different techniques to remove residues at full-scale facilities in Arizona and California used for indirect potable reuse. In contrast to SAT, none of the investigated drug residues were detected in tertiary effluents after treatment with NF or RO.

The performance of two membrane-based mobile drinking water purification units (MDWPU) for the removal of pharmaceutical residues and other sewage-borne contaminants was investigated in two field-trials [78,79]. Both MDWPUs were applying different pre-filtration techniques such as bag, slit or ultrafiltration and final purification by RO. In both field-trials, potable water was generated from highly contaminated raw-water sources (contaminated surface water and municipal sewage effluents, respectively) simulating "worst-case" conditions representing reasonable scenarios in civil disaster operations or in military out of area missions. Even under these "worst-case" scenarios all of the investigated drug residues were efficiently removed. The residues occurring in the raw water at concentrations up to the μg/L-level were decreased below their analytical limits of detection (<1–10 ng/L) [78,79].

Adams et al. [80] evaluated conventional drinking-water treatment processes including RO under typical water treatment plant conditions to determine their efficacy for the removal of seven common antibiotics. All experiments were conducted with synthetic solutions of both distilled/deionized water and Missouri River water fortified with the studied compounds. In these experiments, RO has been shown to be successful in removing all studied antibiotic compounds.

The removal of estrogenic steroids by NF or RO was investigated in several research studies [77,81–84]. Schaefer et al. [77] investigated the retention behavior of E1 at RO and NF membranes and concluded that size exclusion dominates retention with the tighter membranes. In the case of NF membranes exhibiting low retention of ions, both size exclusion and adsorptive effects appeared to be instrumental in maintaining high retention. The authors assumed that these effects may be driven by hydrogen bonding between E1 and the membrane. Deprotonation of E1 or high concentrations of sodium chloride led to a significant decrease in the retention of E1 at "open" NF membranes but did not affect retention by RO [77].

Xu et al. [85] studied the rejection of drug residues by a variety of commercially available RO, NF and ultra-low-pressure RO (ULPRO) membranes by simulating operational conditions for drinking-water treatment and wastewater reclamation. The rejection rates obtained for the ionic drug residues exceeded 95% using NF-90, XLE and TFC-HR membranes, and was above 89% for the NF-200 membrane. Xu et al. [85] observed that the presence of effluent organic matter improved the rejection of ionic organics by tight NF and RO membranes most likely as a result of a decreased negatively charged membrane surface. However, the presence of effluent organic matter could suspend the effect of hydrodynamic operating condition on rejection performance.

Kimura et al. [76,86] investigated the rejection efficiency of drug residues by polyamide NF/RO membranes. They observed that, regardless of other physical/ chemical properties, negatively charged compounds (e.g. the analgesic drug diclofenac) can be rejected to a great extent (i.e. $>90\%$) by electrostatic repulsion [76,87]. On the other hand, the rejection of non-charged compounds was mainly influenced by the size of the compounds. However, solute affinity for the membrane also influenced the rejection efficiency. NF was in several cases not able to remove non-charged compounds such as phenacetin whereas primidone, another non-charged compound, was always rejected by more than 70%. Thus, the authors concluded [86] that additional processes might be responsible for a good rejection of such compounds. Nghiem et al. [88] pointed out that speciation of drug residues may lead to a dramatic change in retention being a function of pH resulting. Generally, much greater retention was observed for ionized, negatively charged compounds but ibuprofen also adsorbs considerably to the membrane in its neutral form because of its relatively high hydrophobicity. Nghiem et al. [88] also identified a high dipole moment as another important intrinsic physicochemical property of polar organic compounds that can substantially affect their retention. In a more general statement they concluded "that retention of pharmaceuticals by a tight NF membrane is dominated by steric (size) exclusion, whereas both electrostatic repulsion and steric exclusion govern the retention of ionizable pharmaceuticals by a loose NF membrane".

Kimura et al. [76] also examined the retention behavior of drug residues by RO membranes using two different materials (polyamide and cellulose acetate). The polyamide membrane exhibited a better performance in terms of rejection but often retention was not complete (57–91%). Kimura et al. [76] concluded that the molecular weight of the test compounds can generally indicate the tendency of rejection for the polyamide membranes (size exclusion dominated the retention by the polyamide membrane) while polarity may be used to describe the retention trend of the tested compounds by the cellulose acetate membrane. In contrast, salt rejection or molecular weight cut-off (MWCO) often used to characterize membrane rejection properties do not provide quantitative information on the rejection of pharmaceutical residues by NF/RO membranes [76]. Kimura et al. [86] also observed another interesting phenomenon. Experiments conducted at feed water concentrations of 100 ng/L resulted in lower rejection efficiency as compared to experiments conducted at 100 μg/L [86].

4.3.4 OVERALL CONCLUSIONS

Some recent investigations have shown that residues of pharmaceuticals occurring at concentrations up to the μg/L-level in municipal sewage effluents and in surface waters under the influence of such discharges can also be relevant for groundwater and public drinking-water supply.

The efficacy of natural attenuation using recharge techniques such as SAT, bank filtration and artificial GWR for the removal of pharmaceutical residues was investigated in several research studies. Treatment of sewage effluents by SAT and treatment of surface water under the influence of municipal sewage effluents by bank filtration or artificial GWR were able to remove several of the drug residues that were detected in the raw water. Thus, properly designed and operated bank filtration or artificial GWR facilities were recognized as efficient, low-cost technologies for the pre-treatment of contaminated surface waters used for drinking-water supply. However, a complete removal of all potential drug residues cannot be guaranteed. Especially, the anti-epileptics primidone and carbamazepine and several drug metabolites such as AMDOPH or clofibric acid were identified as refractory/ persistent compounds which are not or only slightly attenuated during groundwater recharge.

Investigations have also shown that waterworks applying only flocculation and sand filtration will not be able to remove substantial amounts of drug residues from contaminated raw water (ground- or surface water). However, investigations of phenazone-type residues in several Berlin waterworks have shown that even rapid sand filtration was able to decrease the loads of the parent drug compounds by more than 90%. This decrease was caused by microbial degradation of these compounds by bacteria forming an active biofilm in the filter units. Unfortunately, the metabolites AMDOPH, DP and PDP formed by microbial degradation from DMAA, phenazone and propyphenazone, respectively, were not only detected in the aerated raw water but also identified as persistent compounds in purified drinking water.

From today's knowledge, contamination of raw waters by polar drug residues can only sufficiently be removed by using more advanced techniques such as ozonation or activated carbon filtration/ treatment. An even more efficient alternative to ozonation and GAC filtration are membrane filtration techniques such as NF, RO or ULPRO that are able to remove such residues completely often independent from the physico-chemical properties of the molecules (this is especially true for RO and ULPRO). Operational (energy) costs may, however, be seen as a disadvantage of pressure-driven membrane filtration techniques.

Verstraeten and Heberer [51] proposed a multi-barrier approach including bank filtration for an efficient pre-treatment of contaminated surface water that also reduces the total carbon loads of feed water applied to a subsequent treatment by membrane filtration. The fundamental question is, however, if a complete removal of traces of pharmaceuticals from drinking water is really necessary and if it justifies the application of expensive treatment technologies such as ozonation, GAC filtration or membrane filtration? An extensive treatment would also only be necessary at those facilities identified as being susceptible for sewage borne contaminations. Thus, polar (non-toxic) residues of pharmaceuticals may also be used as excellent indicators enabling to evaluate if and how surface water is impacted by contaminations from municipal sewage effluents and to check if other contaminations associated with sewage effluents might also be present in ground- or drinking water.

From a toxicological point of view, traces of drug residues as they have been reported to occur in drinking water are, from today's scientific knowledge, not harmful for lifetime human consumption [25,28,89]. Residues of pharmaceuticals that have not been identified as being harmful to humans are currently not explicitly regulated in any drinking water directive. In Europe, the precautionary principle urges the public water suppliers, however, to minimize the concentrations of anthropogenic compounds in drinking water as far as technicality possible and economically reasonable.

REFERENCES

1 Th. Heberer, *Toxicol. Lett.*, 131 (2002) 5.

2 A.B.A. Boxall, L.A. Fogg, P.A. Blackwell, P. Kay, E.J. Pemberton and A. Croxford, *Rev. Environ. Contam. Toxicol.*, 180 (2004) 1.

3 Th. Heberer and M. Adam, Occurrence, fate, removal of pharmaceutical residues in the aquatic environment: A review of recent research data. In: D.R. Dietrich, T. Petry and S. Webb (Eds.), *Advances in Pharmacology: Hot Spot Pollutants*, Vol. 52-Pharmaceuticals in the Environment, Academic Press, Elsevier Publishers, Amsterdam, 2005, pp. 11–36.

4 Th. Heberer, *J. Hydrol.*, 266 (2002) 175.

5 Th. Heberer, K. Reddersen and A. Mechlinski, *Water Sci. Technol.*, 46 (2002) 81.

6 I.M. Verstraeten, Th. Heberer and T. Scheytt. In: C. Ray, G. Melin and R.B. Linsky (Eds.), *Riverbank Filtration: Improving Source-Water Quality*, Dordrecht, Kluwer Academic Publishers, 2002, p. 175.

7 Th. Heberer and A. Mechlinski, *Hydroplus*, 137 (2003) 53.

8 P.E. Stackelberg, E.T. Furlong, M.T. Meyer, S.D. Zaugg, A.K. Henderson and D.B. Reissman, *Sci. Total Environ.*, 329 (2004) 99.

9 Th. Heberer and H.J. Stan, *Vom Wasser*, 86 (1996) 19.

10 Th. Heberer and H.J. Stan, *Int. J. Environ. Anal. Chem.*, 67 (1997) 113.

11 H.J. Brauch, F. Sacher, E. Denecke and T. Tacke, *gwf Wasser Abwasser*, 14 (2000) 226.

12 W. Kuehn and U. Mueller, *J. AWWA*, 12 (2000) 60.

13 Th. Heberer, U. Duennbier, Ch. Reilich and H.J. Stan, *Fresenius' Environ. Bull.*, 6 (1997) 438.

14 F. Sacher, F. Th. Lange, H.-J. Brauch and I. Blankenhorn, *J. Chromatogr. A*, 938 (2001) 199.

15 T.A. Ternes, *Pharmaceuticals and Personal Care Products in the Environment: Scientific and Regulatory Issues*. In: C.G. Daughton and T. Jones-Lepp (Eds.), Symposium Series 791, American Chemical Society, Washington, DC, 2001, p. 39.

16 Th. Heberer, B. Fuhrmann, K. Schmidt-Bäumler, D. Tsipi, V. Koutsouba and A. Hiskia, *Pharmaceuticals and Personal Care Products in the Environment: Scientific and Regulatory Issues*. In: C.G. Daughton and T. Jones-Lepp (Eds.), Symposium Series 791, American Chemical Society, Washington, DC, 2001, p. 70.
17 M. Rabiet, A. Togola, F. Brissaud, J.L. Seidel, H. Budzinski and F. Elbaz-Poulichet, *Environ. Sci. Technol.*, 40 (2006) 5282.
18 W.Y. Hua, E.R. Bennett and R.J. Letcher, *Water Res.*, 40 (2006) 2259.
19 G.A. Loraine and M.E. Pettigrove, *Environ. Sci. Technol.*, 40 (2006) 687.
20 J. Drewes, Th. Heberer and K. Reddersen, *Water Sci. Technol.*, 46 (2002) 73.
21 J. Drewes, Th. Heberer, T. Rauch and K. Reddersen, *Ground Water Monit. R.*, 23 (2003) 64.
22 W.P. Eckel, B. Ross and R.K. Isensee, *Ground Water*, 31 (1993) 801.
23 J.V. Holm, K. Ruegge, P.L. Bjerg and T.H. Christensen, *Environ. Sci. Technol.*, 29 (1995) 1415.
24 K.K. Barnes, M.J. Focazio, S.C. Christenson, D.W. Kolpin, E.T. Furlong, S.D. Zaugg, M.T. Meyer and L.B. Barber, *Ground Water Monit. R.*, 24 (2004) 119.
25 K. Reddersen, Th. Heberer and U. Duennbier, *Chemosphere*, 49 (2002) 539.
26 C.G. Daughton and T.A. Ternes, *Environ. Health Persp.*, 107 (1999) 907.
27 M. Ahel and I. Jelicic, *Pharmaceuticals and Personal Care Products in the Environment: Scientific and Regulatory Issues*. In: C.G. Daughton and T. Jones-Lepp (Eds.), Symposium Series 791, American Chemical Society, Washington, DC, 2001, p. 100.
28 S. Zuehlke, U. Duennbier and Th. Heberer, *J. Chromatogr. A*, 1050 (2004) 201.
29 H.-J. Stan, Th. Heberer and M. Linkerhägner, *Vom Wasser*, 83 (1994) 57.
30 G.E. Cordy, N.L. Duran, H. Bouwer, R.C. Rice, E.T. Furlong, S.D. Zaugg, M.T. Meyer, L.B. Barber and D.W. Kolpin, *Ground Water Monit. R.*, 24 (2004) 58.
31 T. Scheytt, P. Mersmann, R. Lindstädt and Th. Heberer, *Chemosphere*, 60 (2005) 245.
32 T. Scheytt, P. Mersmann, R. Lindstädt and Th. Heberer, *Water Air Soil Poll.*, 165 (2005) 3.
33 G. Preuss, U. Willme and N. Zullei-Seibert, *Acta Hydroch. Hydrob.*, 29 (2001) 269.
34 P. Mersmann, T. Scheytt and Th. Heberer, *Acta Hydroch. Hydrob.*, 30 (2002) 275.
35 T. Scheytt, P. Mersmann, M. Leidig, A. Pekdeger and Th. Heberer, *Ground Water*, 42 (2004) 767.
36 T.J. Scheytt, P. Mersmann and Th. Heberer, *J. Contam. Hydrol.*, 83 (2006) 53.

37 A.W. Garrison, J.D. Pope and F.R. Allen, GC/MS analysis of organic compounds in domestic wastewaters. In: C.H. Keith (Ed.), *Identification and Analysis of Organic Pollutants in Water*, Chapter 30, Ann Arbor Science Publishers, Ann Arbor, 1976, p. 517.
38 C. Hignite and D.L. Azarnoff, *Life Sci.*, 20 (1977) 337.
39 K.M. Lai, K.L. Johnson, M.D. Scrimshaw and J.N. Lester, *Environ. Sci. Technol.*, 34 (2000) 3890.
40 G.G. Ying, R.S. Kookana and P. Dillon, *Ground Water Monit. R.*, 24 (2004) 102.
41 O. Conroy, D.M. Quanrud, W.P. Ela, D. Wicke, K.E. Lansey and R.G. Arnold, *Environ. Sci. Technol.*, 39 (2005) 2287.
42 D.L. Sedlak and K.E. Pinkston, Water Resour. Update (2001) 56–64.
43 D.M. Quanrud, J. Hafer, M.M. Karpiscak, J. Zhang, K.E. Lansey and R.G. Arnold, *Water Res.*, 37 (2003) 3401.
44 L.G. Wilson, G.L. Amy, C.P. Gerba, H. Gordon, B. Johnson and J. Miller, *Water Environ. Res.*, 67 (1995) 371.
45 Y. Fujita, W.H. Ding and M. Reinhard, *Water Environ. Res.*, 68 (1996) 867.
46 B.L. Mansell and J.E. Drewes, *Ground Water Monit. R.*, 24 (2004) 94.
47 J. Mansell, J.E. Drewes and T. Rauch, *Water Sci. Technol.*, 50 (2004) 229.
48 J.E. Drewes, J. Hemming, S.J. Ladenburger, J. Schauer and W. Sonzogni, *Water Environ. Res.*, 77 (2005) 12.
49 S.A. Snyder, D.L. Villeneuve, E.M. Snyder and J.P. Giesy, *Environ. Sci. Technol.*, 35 (2001) 3620.
50 J.E. Drewes, P. Fox and M. Jekel, *J. Environ. Sci. Health A*, 36 (2001) 1633.
51 I.M. Verstraeten and Th. Heberer. In: C. Ray, G. Melin and R.B. Linsky (Eds.), *Riverbank Filtration: Improving Source-Water Quality*, Dordrecht, Kluwer Academic Publishers, 2002, p. 321.
52 Th. Heberer and A. Mechlinkski, Hydroplus 2003, 137—Hydrosciences—October (2003) 53.
53 Th. Heberer, A. Mechlinkski, B. Fanck, A. Knappe, G. Massmann, A. Pekdeger and B. Fritz, *J. Ground Water Monit. R.*, 24 (2004) 70.
54 S. Zuehlke, U. Duennbier, Th. Heberer and B. Fritz, *J. Ground Water Monit. R.*, 24 (2004) 78.
55 T.A. Ternes, M. Stumpf, J. Mueller, K. Haberer, R.D. Wilken and M. Servos, *Sci. Total Environ.*, 225 (1999) 81.
56 C.H. Huang and D.L. Sedlak, *Environ. Toxicol. Chem.*, 20 (2001) 133.
57 C. Baronti, R. Curini, G. D'Ascenzo, A. Di Corcia, A. Gentili and R. Saperi, *Environ. Sci. Technol.*, 34 (2000) 5059.
58 H.M. Kuch and K. Ballschmiter, *Environ. Sci. Technol.*, 35 (2001) 3201.
59 P. Adler, T. Steger-Hartmann and W. Kalbfus, *Acta Hydroch. Hydrob.*, 29 (2001) 227.

60 B. Fanck and Th. Heberer, Transport and attenuation of antibiotic residues during river bank filtration in Berlin, Germany. Poster at the 5th International Symposium on Management of Aquifer Recharge (ISMAR 2005), 11–16 June 2005, Berlin, Germany.
61 A. Pekdeger, personal communication.
62 Th. Heberer and M. Adam, *Environ. Chem.*, 1 (2004) 22.
63 G. Massmann, J. Greskowiak, U. Duennbier, S. Zuehlke, A. Knappe and A. Pekdeger, *J. Hydrol.*, 328 (2006) 141.
64 T.A. Ternes, M.L. Janex-Habibi, T. Knacker, N. Kreuzinger and H.S. Siegrist, Final report of the EU project POSEIDON, Contract No. EVK1-CT-2000-00047, www.eu-poseidon.com
65 T.A. Ternes, M. Meisenheimer, D. McDowell, F. Sacher, H.J. Brauch, B. Haist-Gulde, G. Preuss, U. Wilme and N. Zulei-Seibert, *Environ. Sci. Technol.*, 36 (2002) 3855.
66 R.G. Rice, *Ozone-Sci. Eng.*, 21 (1999) 99.
67 M.M. Huber, S. Canonica, G.Y. Park and U. von Gunten, *Environ. Sci. Technol.*, 37 (2003) 1016.
68 R. Andreozzi, V. Caprio, R. Marotta and A. Radovnikovic, *J. Hazard. Mater.*, 103 (2003) 233.
69 Z. Qiang, C. Adams and R. Surampalli, *Ozone-Sci. Eng.*, 26 (2004) 525.
70 G.R. Boyd, H. Reemtsma, D.A. Grimm and S. Mitra, *Sci. Total Environ.*, 311 (2003) 135.
71 D.C. McDowell, M.M. Huber, M. Wagner, U. von Gunten and T.A. Ternes, *Environ. Sci. Technol.*, 39 (2005) 8014.
72 M.M. Huber, T.A. Ternes and U. von Gunten, *Environ. Sci. Technol.*, 38 (2004) 5177.
73 M.M. Huber, S. Korhonen, T.A. Ternes and U. von Gunten, *Water Res.*, 39 (2005) 3607.
74 T.A. Ternes, M. Meisenheimer, D. McDowell, F. Sacher, H.J. Brauch, B. Haist-Gulde, G. Preuss, U. Wilme and N. Zulei-Seibert, *Environ. Sci. Technol.*, 36 (2002) 3855–3863.
75 G.R. Boyd and D.A. Grimm, *Ann. NY Acad. Sci.*, 948 (2001) 80.
76 K. Kimura, S. Toshima, G. Amy and Y. Watanabe, *J. Membr. Sci.*, 245 (2004) 71.
77 A.I. Schäfer, L.D. Nghiem and T.D. Waite, *Environ. Sci. Technol.*, 37 (2003) 182.
78 T. Heberer and D. Feldmann, Removal of pharmaceutical residues from contaminated raw water sources by membrane filtration. Chapter 29. In: K. Kuemmerer (Ed.), *Pharmaceuticals in the Environment*, 2nd ed, Springer-Verlag, Berlin, 2004, pp. 391–410.
79 Th. Heberer, D. Feldmann, K. Reddersen, H.J. Altmann and T. Zimmermann, *Acta Hydroch. Hydrob.*, 30 (2002) 24.
80 C. Adams, Y. Wang, K. Loftin and M.T. Meyer, *J. Environ. Eng. ASCE*, 128 (2002) 253.

81 L.D. Nghiem and A.I. Schäfer, *Environ. Eng. Sci.*, 19 (2002) 441.
82 L.D. Nghiem, A.I. Schäfer and T.D. Waite, *Water Sci. Technol.*, 46 (2002) 265.
83 L.D. Nghiem, A.I. Schäfer and T.D. Waite, *Desalination*, 147 (2002) 269.
84 Th. Wintgens, M. Gallenkemper and Th. Melin, *Desalination*, 146 (2002) 387.
85 P. Xu, J.E. Drewes, C. Bellona, G. Amy, T.U. Kim, M. Adam and Th. Heberer, *Water Environ. Res.*, 77 (2005) 40.
86 K. Kimura, G. Amy, J. Drewes, Th. Heberer, T.U. Kim and Y. Watanabe, *J. Membr. Sci.*, 227 (2003) 113.
87 K. Kimura, G. Amy, J. Drewes and Y. Watanabe, *J. Membr. Sci.*, 221 (2003) 89.
88 L.D. Nghiem, A.I. Schafer and M. Elimelech, *Environ. Sci. Technol.*, 39 (2005) 7698.
89 B.W. Schwab, E.P. Hayes, J.M. Fiori, F.J. Mastrocco, N.M. Roden, D. Cragin, R.D. Meyerhoff, V.J. D'Aco and P.D. Anderson, *Regul. Toxicol. Pharmacol.*, 42 (2005) 296.

Chapter 5

Conclusions and future research needs

Damia Barceló and Mira Petrović

5.1 GENERAL REMARKS

The objective of this overview and final chapter is to identify some of the priority areas and, the way forward for scientific research in pharmaceutical residues in the water cycle. Pharmaceuticals are an extraordinarily diverse group of chemicals used in veterinary medicine, agricultural practices and human health. Many pharmaceuticals are highly bioactive, most are polar, many are optically active and all, when present in the environment, occur usually at not more than trace concentrations. They are a class of new, the so-called "emerging" contaminants that have raised great concern in the last years.

Although the problems related to pharmaceutical residues in the environment can be considered new they bear many similarities with endocrine disruption. In this respect we should mention the article by J. Sumpter [1] entitled "Lessons from endocrine disruption and their application to other issues concerning trace organics in the aquatic environment", where he specifically mentioned that hopefully we have learned our lessons from endocrines and we will be able to apply them to this difficult issue of how to best approach future concerns about the potential impact of other new and emerging contaminants, e.g. pharmaceuticals in the water cycle. Although there are examples that fully support this statement, for instance, the problems associated with the incomplete removal of pharmaceuticals and endocrine disruptors after passage through wastewater treatment plants (WWTPs), from our point of view we believe that still quite a lot of information is missing as regards to the fate and behaviour of pharmaceuticals in the environment. We will indicate in these conclusions some of the gaps that certainly need to be thoroughly researched into the near future.

Comprehensive Analytical Chemistry 50
Petrović and Barceló (Eds)
Volume 50 ISSN: 0166-526X DOI: 10.1016/S0166-526X(07)50016-4

5.2 LEGISLATION

Pharmaceuticals are not yet included in any priority list either in US or in Europe. Yet US EPA has made some progress on the lists of potential new drinking water contaminants by considering herbicide degradates, e.g. atrazine-desethyl, alachlor ESA and other acetanilide degradation products [2].

Regulatory agencies have issued detailed guidelines on how pharmaceuticals should be assessed for possible unwanted effects in the environment. The first requirement was introduced by the European Union in 1995 with 92/18 EEC directive and the corresponding note for guidance [3] for veterinary pharmaceuticals. The European Commission released a draft guideline (Directive 2001/83/EC) specifying that an authorization for a medicinal product for human use must be accompanied by an experimental risk assessment. Evaluation of the potential environmental risk posed by pharmaceuticals should be evaluated on a case-by-case basis. Any arrangements to limit environmental impact should be considered. In any event, the impact should not constitute criteria for refusal or authorization [4].

The US Food and Drug Administration (FDA) published guidance for the assessment of veterinary drugs already in 1980 whereas the guidance for human drugs was published later [5]. Current FDA regulation indicates that applicants in the US are required to provide an environmental assessment report when the expected introduction concentration of the active ingredient of the pharmaceutical in the aquatic environment is equal or higher than 1 μg/L that corresponds to about 40 tonnes as trigger value. This trigger value of 40 tonnes per year is controversial. Indeed, under current legislation a company can obtain a "categorial exclusion" and do not have to perform an environmental assessment if they manufacture less than 40 tonnes. This figure assumes that the pharmaceutical is spread uniformly across the US. Such categorial exclusion does not take into account the input from multiple companies that might all be making the same active pharmaceutical ingredient. For instance, if 10 companies are manufacturing a drug at 30 tonnes per year each, for a total of 300 tonnes per year, there is no trigger to perform an environmental assessment, or, if one company surpasses 40 tonnes per year that company's environmental assessment would not account for production from other companies. Pharmaceuticals like carbamazepine are manufactured in the US by 12 companies with different names, so maybe this is the time to question if the FDA rule concerning environmental impact needs to be revised and if the EU requires a similar action.

5.3 CHEMICAL ANALYSIS

One of the reasons for the increasing concern on pharmaceuticals has certainly been the improvement on analytical techniques during the last few years [6]. After sampling and conventional solid phase extraction of a surface river water sample followed by LC-tandem MS it is possible to easily detect nanogram per litre level of common pharmaceutical residues in natural water samples.

There are, however, few questions that still need to be solved. The main drawback of the conventional analytical approach is to target compound monitoring, which is often insufficient to assess the environmental relevance of emerging contaminants. Are we sure that other pharmaceutical compounds are in the water sample? What about their transformation products? Most of the literature reviewed on the methods used for pharmaceutical residue analysis uses either only a class of compounds, e.g. β-blockers, anti-inflammatory drugs or, in the case of multi-residue methods, includes at the maximum 30–40 target analytes using triple quadrupole instruments. Generally, very few transformation products are included in the monitoring programmes of pharmaceuticals in the environment. There are several reasons for that: not all the transformation products are commercially available or they are too expensive. The alternative is to use a second analyser, like time of flight (TOF) or ion trap (IT) to look for "known unknowns", such as possible transformation products of the pharmaceuticals degraded by microbial action and/or UV light that can be present in the samples. This obviously requires a second analysis; besides there is often a lack of sensitivity and perhaps, also of experience of the laboratory to identify the tentative list of the breakdown products formed. In addition to that, the problems related to conjugated metabolites, like glucuronide and sulfate conjugates that can be deconjugated by microbial action during wastewater treatment processes needs also further and careful investigation. Indeed, since many pharmaceuticals are excreted as conjugates by humans, they may actually increase in concentration after passage through the WWTP. Thus, the matrix of pharmaceuticals compounds exiting the plant may be very different from those entering.

Although most of the work published on analytical techniques has been performed with the common pharmaceutical compounds, in the last few years attention has also been paid to illicit drugs. After the publication of the paper by Zuccato et al. [7], the issue of cocaine and other illicit drugs in water has been raised. Recently a paper published by the same group [8] indicated that in this specific case of illicit drugs the metabolites of cocaine, morphine and methadone are the key target analytes to be followed by LC-tandem MS. This type of work can be used for different purposes: in one way, as an environmental study to look for possible effect of illicit drugs in the aquatic environment, similarly as conventional pharmaceutical residues, and secondly as a tool to combat drug abuse by identifying hot spots through monitoring of influents and effluents of WWTP.

Although water analysis is no longer a problem, the extraction and further analysis of pharmaceuticals in solid samples is still a challenge and methods are only available for a limited number of compounds in soil, sediment and sludge samples and it has been restricted to several groups of antibiotics. In the last few years more studies have taken place like sorption studies on sediment for compounds like ivermectin, diazepam, carbamezapim and for several β-blockers, like atenolool, metopralol, sotalol and propanolol [9,10].

Of the various solid samples, sewage sludge is one of the more complex one. Some of the biggest analytical challenge is that a "complete" analysis of sewage sludge include overcoming the large negative surface charges and intersticial spaces that provide multiple active sites for charged compounds and the clean-up step for removing the bulk material (e.g. fats, proteins and surfactants) that are co-extracted with the pharmaceuticals. Research efforts involving novel extraction devices like pressuried liquid extraction, as well as highly selective clean-up procedures using molecularly imprinted polymers should be explored to isolate new pharmaceutical residues and their transformation products from complex sludge samples [11].

Another relevant issue is the quality of data and the performance of inter-laboratory tests combined with the use of reference materials. Although the first European Union inter-laboratory study in anti-inflammatory drug residues in water has been recently organized [12], there is still a lack of such studies. We should be able to compare monitoring data from Europe and US and for this reason an international inter-laboratory study needs to be organized for the most common pharmaceuticals that are being measured in surface waters and wastewaters in Europe and US. In addition and complementary to that, the availability of reference materials to be used by the laboratories performing the monitoring studies should be also recommended.

To summarize this section, we should indicate that good analytical chemistry is needed to perform any environmental work on fate and behaviour and on the impact of pharmaceuticals in the environment. Integrated studies of effect direct assays (EDA) combining analytical chemistry (applying techniques that are able to identify unknown compounds/metabolites/transformation products) and biology (effect monitoring) seems to be an appropriate way to tackle the complex problem of environmental contamination by pharmaceuticals.

5.4 OCCURRENCE, FATE AND BEHAVIOUR AND MODELLING

Most of the current literature publications on pharmaceutical residues in the water cycle have addressed the contamination of surface waters and wastewaters. Water cycle and pharmaceutical residue analysis should also include all the compartments especially groundwater and the leaching of emerging contaminants through the soil and to groundwater. Attention should be paid to the distribution of emerging contaminants between groundwater and surface water in certain parts of the river like alluvial plains, and to the quality and environmental impact of such waters for their possible use as drinking water, since many aquifers are used as a source of water supply. A recent example of pharmaceuticals in artificial recharge of groundwater indicated that the fate of phenazone was shown to be indirectly controlled by the infiltration water temperature through its effect on the aquifer redox conditions [13].

While metabolism of pharmaceuticals in the human body and in other mammals has been extensively studied, the kinetics and mechanisms of microbial degradation of these compounds in the environment are still largely unknown. The degradation in natural environments or in WWTP can be attributed to biotic and abiotic processes (photolysis and hydrolysis). Partial or total elimination of pharmaceuticals from the environment by microorganisms capable of using organic compounds as an energy source lead to the formation of degradation products and ultimately mineralization. Many organic compounds are biodegraded by organisms that utilize the compounds for growth. Another important biodegradation process is co-metabolism, in which an organic compound is modified but not utilized for growth. A recent review reporting the microbial degradation of frequently occurring pharmaceuticals in the environment has been published and indicated that research is still needed in that area [14].

Direct and indirect photolyses are important depending on the chemical structure of the pharmaceutical compound, its quantum yield and the presence of sensitizing agents like humic acid and nitrate. From the various studies reported, it seems that photosensitizing effects at the different river waters need to be undertaken to understand the persistence of pharmaceuticals in the environment as indicated in Chapter 3.2 and other studies [15].

An even more important question that should be addressed is whether these pharmaceutical residues are bioavailable and, if so, what the environmental impact will be. Treated wastewater used for irrigation, as well as sewage sludge-derived soil amendments and animal manures, should be further investigated as potential sources of human and veterinary pharmaceuticals. In this respect it is expected that water-soluble contaminants will be detected in high water-content plants whereas the more hydrophobic compounds will be found in the plant lipids. The bioavailability of pharmaceuticals is the key parameter to elucidate the routes and pathways of contaminants from source (suspended particles, sediment) to targets (organisms, populations and ecosystems), which implies highly complex processes with a multitude of interactions between abiotic environment and the different parts of the biocenosis (different organisms from bacteria to fish). There are a very limited number of papers reporting levels of pharmaceuticals in biota samples. One of the few examples on the levels of pharmaceuticals in fish detected residues of fluoxetine, carbamezapime and sulfametoxazole in fish muscle and liver at levels as high as 80 ng/g [16].

There is also a lack of studies concerning the formation of transformation products in the environment following natural degradation or water treatment. An interesting paper by Bedner and colleague William MacCrehean [17] investigated acetaminophen present in water and its interaction with hypochlorite, a chlorinating agent used to treat wastewater and drinking water. The concentrations of reactants used were those frequently present at wastewater treatment plants. In both clean water and wastewater backgrounds, eleven new compounds were formed within one hour, the time the reactants would likely to be in contact at any plant. Two of these were identified as the toxicants 1, 4-benzoquinone (a genotoxic and mutagenic agent) and *N*-acetyl-*p*-benzoquinone imine (a hepatotoxin produced during acetaminophen metabolism that is responsible for overdose deaths). Together, these compounds represented the fate of nearly 27% of the original drug concentration. Fortunately, these are unstable compounds, especially in the presence of sulfite, which is sometimes used to dechlorinate-treated water, so they are unlikely to persist in the environment for long. However, they could accumulate where treated wastewater is returned to rivers and the effects of re-supply over long periods are unknown. They might also be formed when drinking water is chlorinated, but they would also break down quickly so it is unlikely they should be troublesome. However, the results obtained under conditions that simulate wastewater disinfection raise the question of what other drug-derived toxins are out there and what is happening under real-world conditions.

Which models can we use to predict the behaviour of pharmaceuticals in the environment? Going back to the introduction of this chapter, we should be able to learn other polar pollutants in the environment. Pharmaceuticals are constituted by a broader group of chemical structures, much larger than commonly studied endocrine disruptors. The modelling of emerging contaminants (e.g. pharmaceuticals) should learn from other polar pollutants that have been studied considerably in river basins including groundwater studies like the case of polar pesticides. Modelling should address all relevant scales starting from micro-scales watershed interactions, the transport of dissolved species of pollutants as well as suspended matter in soil and groundwater systems at the catchment scale and river basin. The use of integrative modelling will be of help to the improvement of river basin management concerning emerging contaminants.

5.5 REMOVAL FROM WWTP

WWTP using secondary biological sewage treatment plants have brought enormous benefits to society and the environment. Considering the short hydraulic residence time (few hours), the large reduction in the amount of natural and xenobiotic compounds is remarkable. Four key factors are critical in predicting the impact of each WWTP plant: (1) the size of the human population connected to the WWTP, (2) the flow through the works, (3) the type of treatment employed and (4) the available dilution in the receiving water.

Certainly the flow into the works will change during the day, so taking into account sewer transit and typical activated sludge treatment, the 8–9 am peak flush would not probably emerge in the effluent until 8–9 pm. Given these variations in "human discharge" and flow, the most valuable way to make measurements of pharmaceuticals clearly associated with human health in effluents is from 24 h composite samples. Another key issue is the hydraulic retention time (HRT). Activated sludge is the most intensive biological treatment in which bacteria are suspended in a tank and vigorously aerated, with HRT varying from 5 to 20 h. It has recently been reported [18] that larger HRT provide a better elimination of pharmaceuticals. The tanks have, in certain occasions, a first anoxic or anaerobic stage to encourage denitrification, which may also play a role in removing trace organics from WWTP.

It can be noticed in the literature that large differences are observed when comparing elimination rates for certain pharmaceuticals, for instance diclofenac, from various plants. Pharmaceuticals can be eliminated by sorption onto the sludge or through microbial degradation. In many cases, the metabolites formed during biodegradation are more polar than the parent compound. The high polarity combined with the low biodegradability that some pharmaceutical compounds exhibit results in inefficient elimination. The efficiency of contaminant removal is strongly dependent on the type of treatment technology (e.g. physicochemical vs. biological treatment) as well as on the operational parameters of the plant. The factors indicated above can contribute to these differences and another conclusion is that there is a need for an increased understanding of the mechanisms of degradation and elimination of pharmaceuticals in WWTP at environmentally relevant concentrations.

To understand the process taking place in the WWTP and to increase the knowledge on biodegradation of contaminants in WWTP, biodegradation studies of pharmaceuticals under laboratory controlled conditions simulating WWTPs should be conducted. A few studies have investigated biodegradation pathways in various environmental compartments and reported identities of biotransformation products during primary biodegradation. The identification of degradates in environmental samples is a challenging task because not only are they present in very low concentrations but they are also mixed with complex matrices that interfere with detection. There is a need to increase our knowledge about the fate of pharmaceuticals during sewage treatment for the implementation of better removal technologies. Future work on WWTP will show to what extend pharmaceuticals can be removed from wastewater and to what extend the implementation of an improved technology is feasible taking into account other macro- and micropollutants as well as the broad variety of complex matrices.

One of the technologies that looks most promising is the use of membrane bioreactors (MBR). The increased use of a membrane bioreactor with a similar process as the one taking place in secondary treatment seems to be an excellent alternative to improve the biodegradation of pharmaceuticals in the environment to increase their removal rates. MBR provides three basic aspects: (i) adsorption, improved physical sludge characteristics, with higher biomass concentration and more effective surface; (ii) biodegradation, cultivation of metabolic speciation, with high-sludge age, low mass organic load favouring biological synthesis of broader substrate spectrum and (iii) direct and complete separation through membrane with entire removal of all contaminants bound to colloids and particulate matter. Apart from the technical aspects, one of the key aspects of the success of MBR during the last few years has been its cost with a price drop from 2001 to 2004 estimated to be from \$0.8 m^{-3} to \$0.5 m^{-3} [19].

A recent paper reported a detailed study on the removal of more than 30 pharmaceutical compounds of different chemicals groups [20]. Several pharmaceuticals (e.g. ibuprofen, naproxen, acetaminophen, ketoprofen, diclofenac, bezafibrate, gemfibrozil, ranitidine, ofloxacin, hydrochlorothiazide and paroxetine) with high-attenuation rates can be expected to be completely removed from wastewater during membrane treatments by sorption, degradation or the combination of both. For most of the investigated compounds, MBR effluent concentrations were significantly lower than in the effluent of a conventionally activated sludge (CAS) treatment. Hydrochlorothiazide and paroxetine had slightly higher elimination percentages in CAS. Some substances were removed neither in MBR nor in CAS process (e.g. carbamazepine). However, no relationship was found between the structures of target compounds and their removal during wastewater treatments. Furthermore, the range of variation of the removal rates of the MBR system was small for most of the compounds, while in the conventional treatment stronger fluctuations were observed and the MBR system turned out to be a lot less sensitive to changes in operational parameters (temperature, flow rate, etc.). Membrane treatment processes should be optimized by a modification of the membranes (variation of materials and reduction of molecular mass cut-off limits) and/or by modification of the treatment process (inoculation of special microorganisms). The efficiencies of diverse microbial populations in the elimination of selected pharmaceuticals and the optimization of design and operating parameters of a laboratory-scale MBR should be considered as a future research needed in this area. Scale up from pilot MBR to real-world WWTP should also be investigated to assess if the processes and elimination in the pilot pant are still valid in a large-scale plant.

Other methods of removal of pharmaceuticals are reported in Chapter 4.3 on the removal of pharmaceuticals during drinking-water treatment. Methodologies involving natural attenuation methods such as bank filtration or artificial groundwater recharge, the use of graphitised activated carbon and various types of membrane filtration techniques can be of help to remove the pharmaceuticals residues from water matrices. WWTPs and drinking-water suppliers are deeply interested in such technological developments to improve the quality of the water. The limiting factors are the costs of all these technologies when they need to be implemented at real scale, since they will have a direct cost for the consumer, therefore increasing water prices. So, compromises will always be needed in selecting the most appropriate technology that is cost-effective. It is also clear that more efforts should also be directed towards reducing the contaminants loads to WWTPs, so, for instance, by not throwing away unused pharmaceutical into the waste or into the toilet. 33% and 25% of the unused drugs in Germany and Austria, respectively, are disposed with the household waste or down the drain. Such compounds will enter the environment intact. Disposal habits of the American public indicate that only 1.4% of Americans returned unused medication to the pharmacy, whereas 54% threw them away and 35.4% disposed of them in the sink/toilet [21]. With regard to veterinary medicines, the issue is difficult to predict and to solve since in many EU countries the number of pigs and other animal farms is still growing. A positive measure since January 2006 is the one recently implemented by the European Union indicating that all growth promoters, mainly antibiotics, in pigs are banned.

5.6 TOXICITY

The growing occurrence of human and veterinary pharmaceuticals in the environment is becoming of increasing concern and improving their ecological and human risk assessment constitute a challenge for the scientific community. Historically, each therapeutic class has been designed for humans, mammals and birds used in agriculture to target specific organs, metabolic pathways and receptors resulting in the modulation of the physiological functions of the organism so that a disease or infection can be treated and a healthy state restored.

The scientific community was certainly shocked by the recent example of the direct correlation between diclofenac residues and renal failure on three species of vultures in India and Pakistan, which was attributed to the diclofenac-treated livestock through feeding [22]. Diclofenac is being commonly used without any prescription and it can be detected in most of the river waters that are being analysed due to its poor removal by secondary wastewater treatment plants.

Owing to the fact that most of pharmaceuticals currently monitored were not designed for aquatic species clearly a lot knowledge is still lacking concerning the effects of pharmaceuticals on aquatic organisms. The recent review article of K. Fent [23] is a very comprehensive one and certainly reflects the state of the art on the issue of the toxic effects caused by pharmaceuticals in the aquatic environment. In this review, the authors indicate that certainly there are quite a few studies on the acute effects of pharmaceuticals, but not much has been done on chronic toxicity data, especially on fish. Key questions are: What acute and chronic ecotoxicological effects may be elicited by pharmaceuticals and by mixtures? What are the effect concentrations of pharmaceuticals and how do they relate to environmental levels? Looking at the various papers published in the literature, the concentration detected in both waste and surface waters are in general from 100 to 1000 times higher than the levels reported to cause acute toxicity but concerning chronic effects the margin of safety is narrow for some of the most ubiquitous compounds. As a wide spectrum of pharmaceuticals has been detected in natural waters, the effects of mixtures should also be taken into account; hence, the overall toxicity could be the result of the sum of individual concentrations or the interaction of different compounds, occurring effects at the NOEC of individual substances [18].

There is generally a lack of chronic toxicity data on pharmaceuticals, especially on fish. Current data on acute and chronic toxicity of pharmaceuticals support the conclusion that more target, or biomolecule-oriented or mode of action-based investigations will allow more relevant insights into effects on survival, growth and reproduction than traditional standard ecotoxicity testing. Unless more is known about the possible chronic effects of individual pharmaceuticals and mixtures, conclusions on the hazards or risk of pharmaceuticals to the aquatic ecosystem are premature yet. It can be still maintained that a gap exists between our current knowledge of the toxic effects of emerging contaminants on organisms derived from laboratory studies (both in vitro and in vivo) and exposure routes and the real effects occurring in nature on different structural and functional levels (organisms, populations and ecosystems).

Of the various groups of pharmaceuticals, antibiotics have traditionally been studied, due to the problems associated with resistant bacteria. The continued land application of manure containing tetracyclines and other antibiotics can exert selective pressure on soil microbial populations and promote the selection of resistant microbes. Once in the environment, resistant genes are capable of being transferred from bacteria of gastrointestinal origin to native soil bacteria. The main issues of concern are veterinary antibiotics, which have been addressed by

Boxall et al. [24] and which are summarized here: (1) research considering that antibiotics have a significant role in developing antibiotic resistance and multiple antibiotic resistant bacterial populations; (2) if there is a direct relationship between antibiotic residues and antibiotic resistant bacteria in the environment and (3) whether continuous exposure to low levels of complex mixtures of antibiotics has negative effects on the quality of water and ecosystem health. The third point is obviously quite applicable to all pharmaceuticals. It should be added that every month new papers are published in the scientific literature covering some aspects of the toxic effects of a certain pharmaceutical in the environment on a different organism, vertebrate or invertebrate or aquatic plant.

To summarize, we need further information about the proportion of contaminants really causing adverse effects and we need to look especially into the combined effects of pharmaceutical mixtures. Of a complex nature but of high relevance is the assessment of ecological risk in relation to pharmaceutical mixtures and in this case, the risk assessment of environmentally relevant mixtures would benefit from the use of quantitative metrics based on understanding of the processes of the potential interactions between the components at the toxicokinetic and toxicodynamic levels [25]. New technologies have emerged to assist scientists in dissecting these mechanisms such as ecotoxicogenomics, probabilistic methods and quantitative structure–activity relationships and these may be of great value to both the ecological risk assessment and to the human risk assessment of pharmaceuticals in drinking water.

REFERENCES

1 J. Sumpter and A.C. Johnson, *Environ. Sci. Technol.*, 39 (2006) 4321.
2 S.D. Richardson and T. Ternes, *Anal. Chem.*, 77 (2005) 3807.
3 EMEA, Note for guidance: Environmental risk assessment for veterinary medicinal products other than GMO-containing and immunological products, 1998, EMEA, London (EMEA/CVMP/055/96).
4 EMEA, Note for guidance: Environmental risk assessment of medicinal products for human use, CMP/SWP/444/draft, The European agency for the evaluation of medicinal products, 2005, EMEA, London .
5 FDA-CDER, Guidance for industry environmental assessment of human drugs and biologics applications, Revision 1, FDA, 1998, Centre for drug evaluation and research, Rockville.
6 D. Barceló and M. Petrović, *TRAC-Trends Anal. Chem.*, 26 (2007) 2.

7 E. Zuccato, Ch. Chiabrando, S. Castiglioni, D. Calamari, R. Bagnati, S. Schiarea and R. Fanelli, *Environ. Health*, 4 (2005) 14.
8 S. Castiglioni, E. Zuccato, E. Crisci, Ch. Chiabrando, R. Fanelli and R. Bagnati, *Anal. Chem.*, 78 (2006) 8421.
9 D. Loffler, J. Römbke, M. Meller and T. Ternes, *Environ. Sci. Technol.*, 39 (2005) 5209.
10 X.-S. Miao, J.-J. Yang and Ch.D. Metcalfe, *Environ. Sci. Technol.*, 39 (2005) 7469.
11 F. Chapuis, J.U. Mullot, V. Pichon, G. Tuffal and M.C. Hennion, *J. Chromatogr. A*, 1135 (2006) 127.
12 Network of reference laboratories and related organisations for monitoring and bio-monitoring of emerging pollutants, NORMAN EU project, 2005–2008.
13 J. Greskowiak, H. Prommer, G. Massmann and G. Nutzmann, *Environ. Sci. Technol.* (in press).
14 S. Perez, D.S. Aga and D. Barceló, Biodegradation of pharmaceuticals in the environment. In: L. Verschaeve (Ed.), *Applied Microbiology and Biotechnology*, Research Singpost, India, 2006, pp. 2–32.
15 A. Yu-Chen Lin and M. Reinhard, *Environ. Toxicol. Chem.*, 24 (2005) 1303.
16 A.J. Ramirez, M.A. MottaLeb, B.W. Brooks and C. Chambliss, LC-tandem MS screening analysis of selected pharmaceuticals in aquatic organisms collected form from an effluent-dominated stream, Abstract 81 presented at the SETAC 27 Annual Meeting in North America held in Montreal, Canada, November 2006.
17 M. Bedner and W.A. Maccrehean, *Environ. Sci. Tehcnol.*, 40 (2006) 516.
18 M. Gros, M. Petrović and D. Barceló, *Environ. Toxicol. Chem.* (in press).
19 G.T. Daigger, B.C. Rittmann, A. Adham and G. Andreottola, *Environ. Sci. Technol.*, 39 (2005) 399A.
20 J. Radjenovic, M. Petrović and D. Barceló, *Anal. Bioanal. Chem.*, 387 (2007) 1365.
21 J.P. Bound and N. Voulvoulis, *Environ. Health Persp.*, 113 (2005) 1705.
22 R. Risebrough, *Nature*, 427 (2004) 596.
23 K. Fent, A.A. Weston and D. Caminada, *Aquat. Toxicol.*, 76 (2006) 122.
24 A.K. Sarmah, M.T. Meyer and A.B.A. Boxall, *Chemosphere*, 65 (2006) 725.
25 J.L.C.M. Dorne, L. Skinner, G.K. Frampton, D.J. Spurgeon and A.M.J. Ragas, *Anal. Bioanal. Chem.*, 387 (2007) 1259.

Appendix: List of pharmaceutical compounds

Compound	Therapeutical group	CAS number	Structure	Elemental formula	Molecular weight	pK_a	log K_{ow}
Acebutolol	β-blocker	34381-68-5	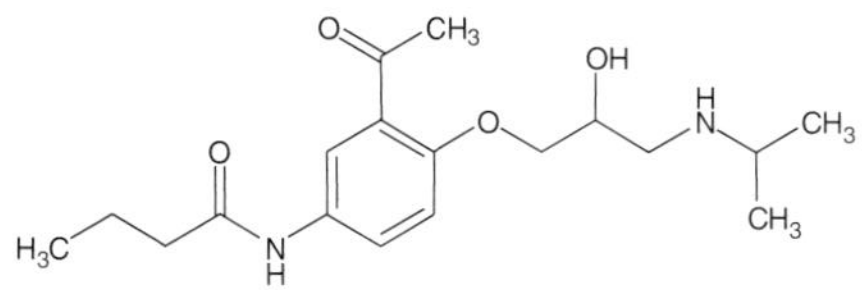	$C_{18}H_{28}N_2O_4 \cdot HCl$	372.89	9.2	1.71
Acecarbromal	Sedative, hypnotic	77-66-7	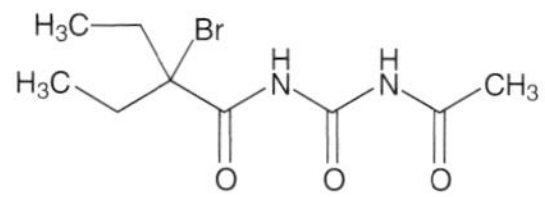	$C_9H_{15}BrN_2O_3$	279.13	—	1.88
Aceclofenac	Anti-inflammatory; analgesic	89796-99-6	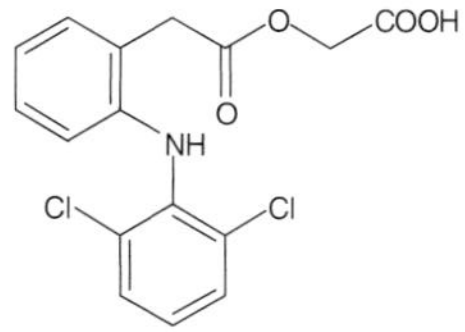	$C_{16}H_{13}Cl_2NO_4$	354.19	—	—
Acemetacin	Anti-inflammatory	53164-05-9	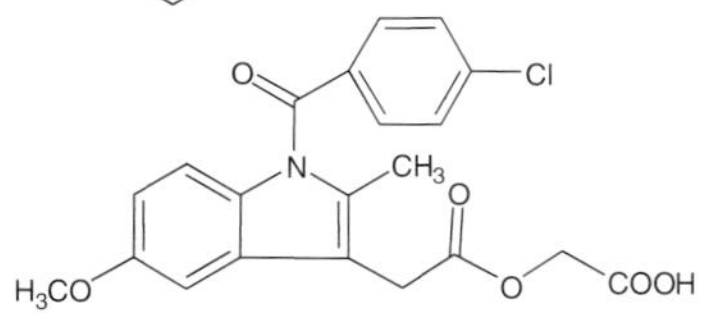	$C_{21}H_{18}ClNO_6$	415.83	—	4.13

Appendix (Continued)

Compound	Therapeutical group	CAS number	Structure	Elemental formula	Molecular weight	pK_a	log K_{ow}
Acetaminophen (paracetamol)	Analgesic; antipyretic	103-90-2	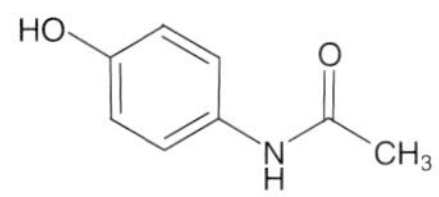	$C_8H_9NO_2$	151.16	9.38	0.46
Acetylsalicylic acid	Analgesic; antipyretic; anti-inflammatory	50-78-2	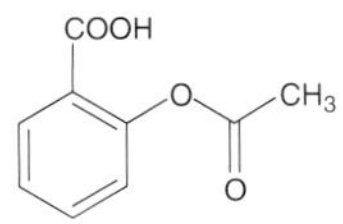	$C_9H_8O_4$	180.16	3.49	1.19
Acyclovir	Antiviral	59277-89-3	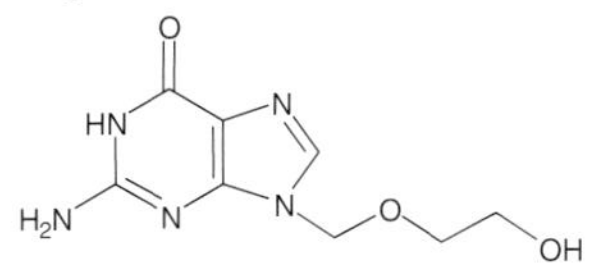	$C_8H_{11}N_5O_3$	225.20	—	−1.56
Albuterol	Bronchodilator; tocolytic	18559-94-9 51022-70-9 (sulfate)	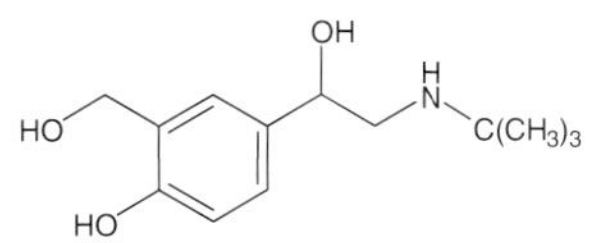	$C_{13}H_{21}NO_3$	239.31	10.3	0.64
Alclofenac	Analgesic; antipyretic; anti-inflammatory	22131-79-9	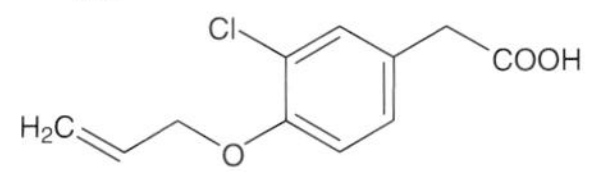	$C_{11}H_{11}ClO_3$	226.66	4.5	2.48
Allobarbital	Sedative, hypnotic	52-43-7	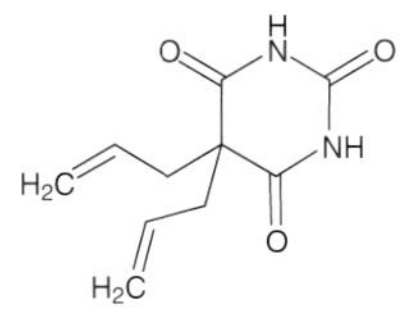	$C_{10}H_{12}N_2O_3$	208.21	7.77	1.15

Amlodipine	Antianginal; antihypertensive	88150-42-9 111470-99-6 (benzenesulfonate)	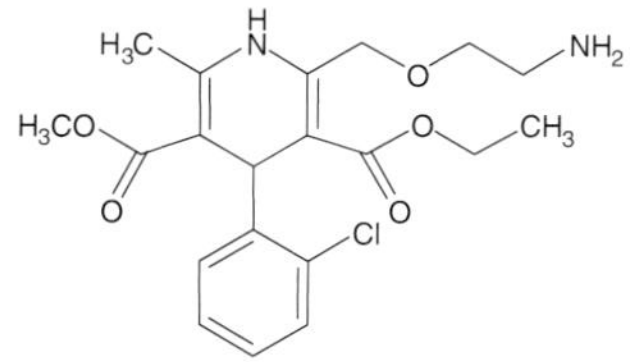	$C_{20}H_{25}ClN_2O_5$ $C_{20}H_{25}ClN_2O_5 \cdot C_6H_5SO_3H$ (Benzenesulfonate)	408.88 567.06	—	—
Amobarbital	Sedative, hypnotic	57-43-2	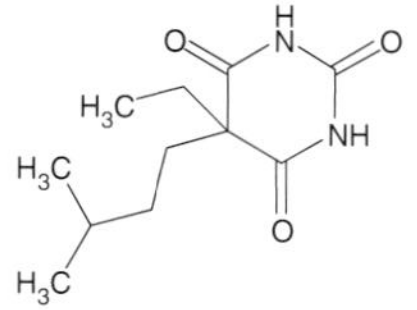	$C_{11}H_{18}N_2O_3$	226.27	7.84	2.07
Amoxicillin	Antibacterial	26787-78-0	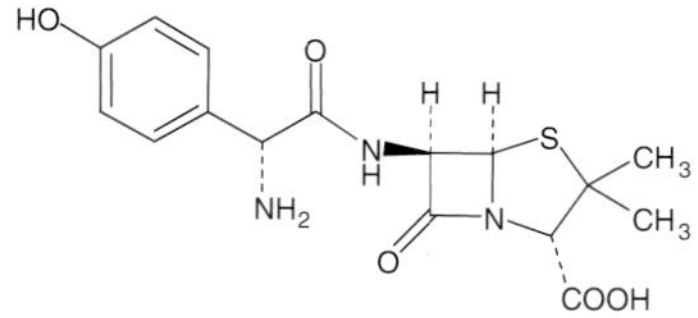	$C_{16}H_{19}N_3O_5S$	365.41	—	0.87
Ampicillin	Antibacterial	69-53-4	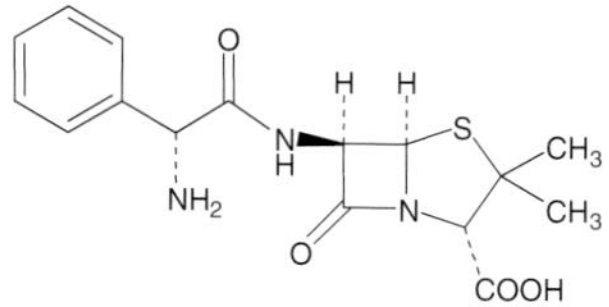	$C_{16}H_{19}N_3O_4S$	349.41	—	1.35
Androstenedione	Estrogen	63-05-8	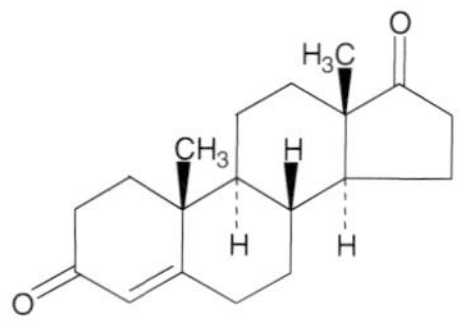	$C_{19}H_{26}O_2$	286.41	—	2.75
Aprobarbital	Sedative, hypnotic	77-02-1	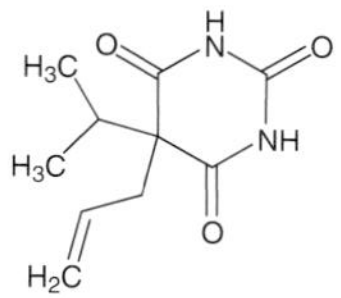	$C_{10}H_{14}N_2O_3$	210.23	7.99	1.15

Appendix (Continued)

Compound	Therapeutical group	CAS number	Structure	Elemental formula	Molecular weight	pK_a	log K_{ow}
Atenolol	β-blocker	29122-68-7	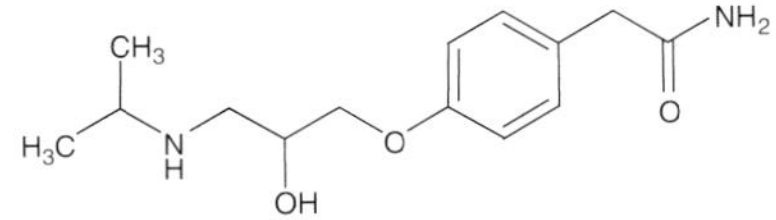	$C_{14}H_{22}N_2O_3$	266.34	9.6	0.16
Azithromycin	Antibacterial	83905-01-5	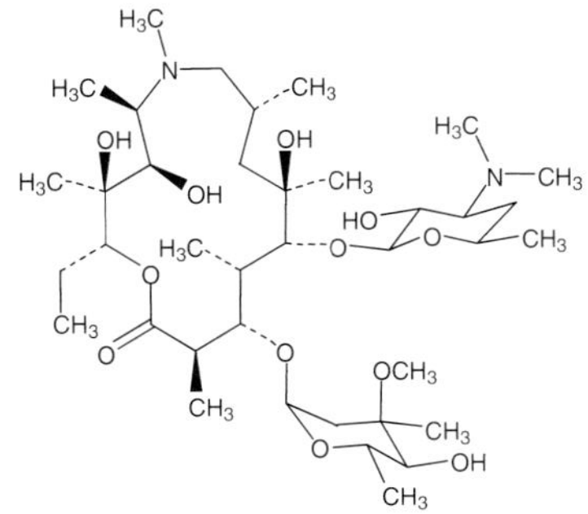	$C_{38}H_{72}N_2O_{12}$	748.98	pK_1 8.7 pK_2 9.5	4.02
Betaxolol	β-blocker	63659-18-7	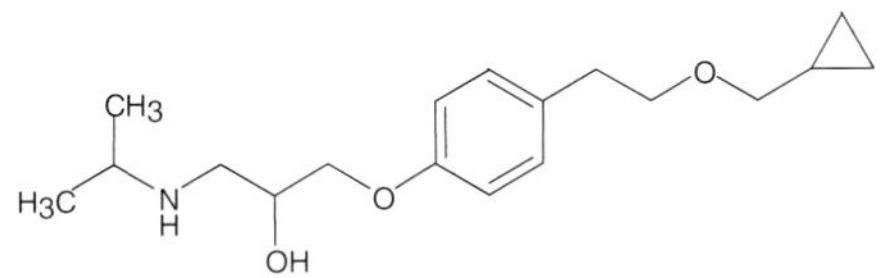	$C_{18}H_{29}NO_3$	307.43	—	2.81
Bezafibrate	Lipid regulator	41859-67-0	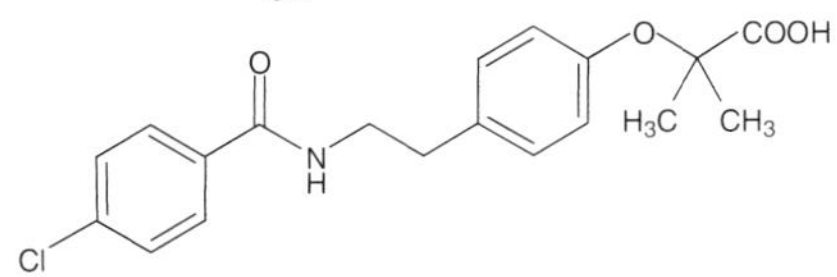	$C_{19}H_{20}ClNO_4$	361.83	3.60	4.25
Bisoprolol	β-blocker	66722-44-9	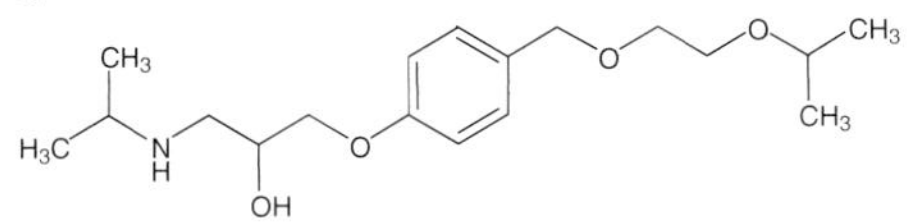	$C_{18}H_{31}NO_4$	325.44	—	1.87

Appendix: List of pharmaceutical compounds

Bromazepam	Anxiolytic	1812-30-2	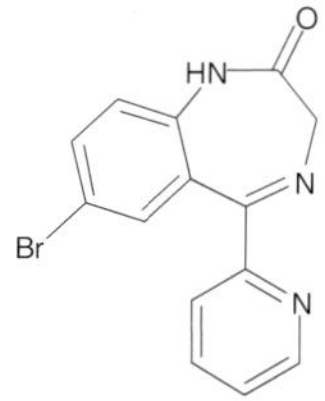	$C_{14}H_{10}BrN_3O$	316.16	—	2.05
Butalbital	Sedative, hypnotic	77-26-9	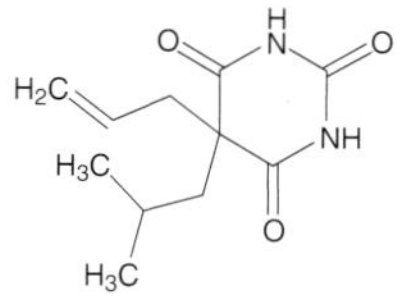	$C_{11}H_{16}N_2O_3$	224.26	—	1.87
Caffeine	Cardiac and respiratory stimulant; diuretic	58-08-2	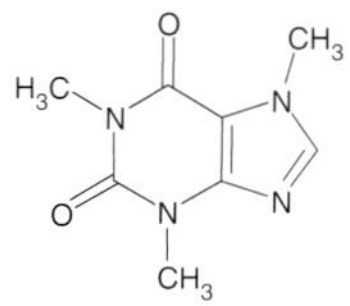	$C_8H_{10}N_4O_2$	194.19	10.4	−0.07
Carazolol	β-blocker	57775-29-8	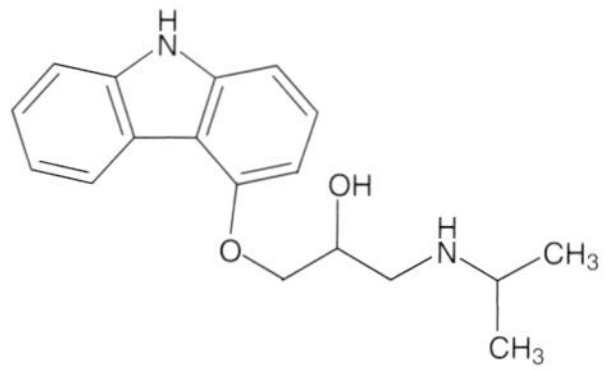	$C_{18}H_{22}N_2O_2$	298.38	—	3.59
Carbamazepine	Antiepileptic	298-46-4	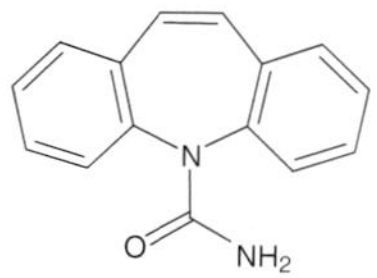	$C_{15}H_{12}N_2O$	236.27	13.9	2.45
Chloramphenicol	Antibacterial	56-75-7	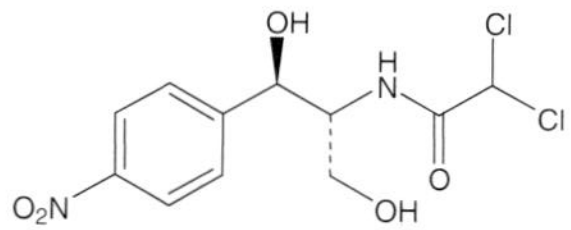	$C_{11}H_{12}Cl_2N_2O_5$	323.13	—	1.14

Appendix (Continued)

Compound	Therapeutical group	CAS number	Structure	Elemental formula	Molecular weight	pK_a	log K_{ow}
Chlortetracycline (Aureomycin)	Antibacterial; antiamebic	57-62-5	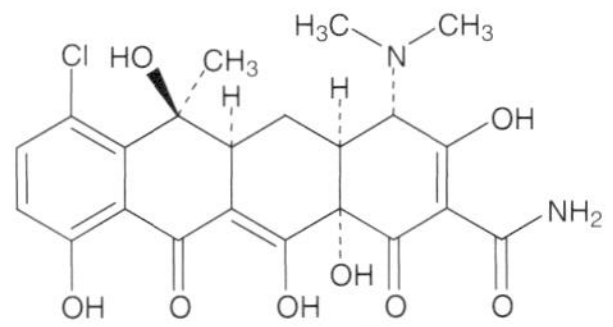	$C_{22}H_{23}ClN_2O_8$	478.89	—	−0.62
Ciprofloxacin	Antibacterial	85721-33-1	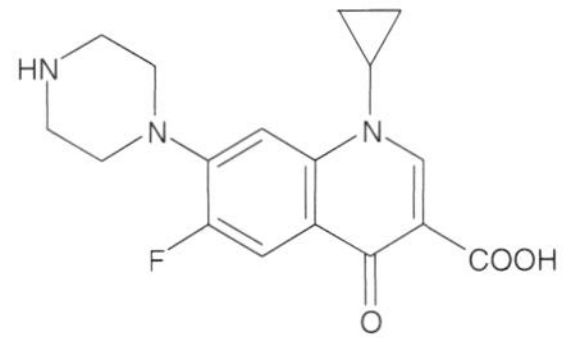	$C_{17}H_{18}FN_3O_3$	331.34	6.09	0.28
Clarithromycin	Antibacterial	81103-11-9	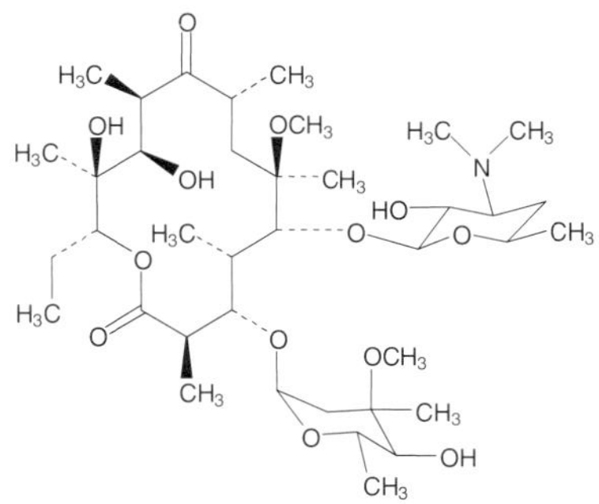	$C_{38}H_{69}NO_{13}$	747.95	8.99	3.16
Clenbuterol	Bronchodilator	37148-27-9	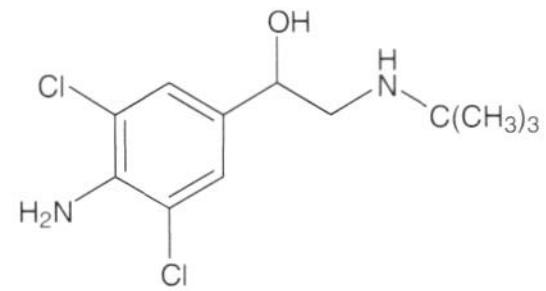	$C_{12}H_{18}Cl_2N_2O$	277.20		2.00

Appendix: List of pharmaceutical compounds

Clofibric acid	Lipid regulator and cholesterol lowering drugs	882-09-7	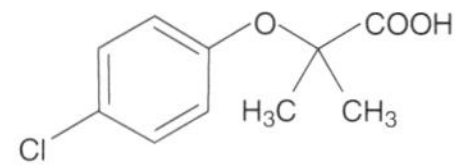	$C_{10}H_{11}ClO_3$	214.65		2.57
Cloxacillin	Antibacterial	61-72-3	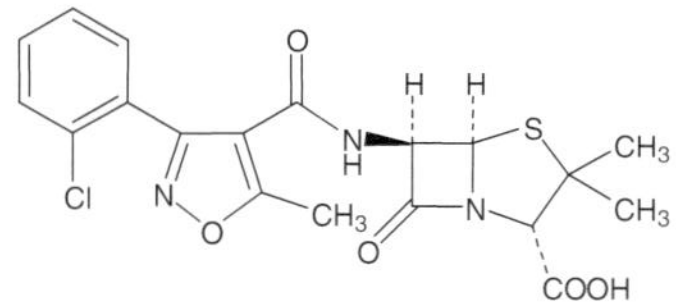	$C_{19}H_{18}ClN_3O_5S$	435.89	2.78	2.48
Codeine	Analgesic (narcotic)	76-57-3	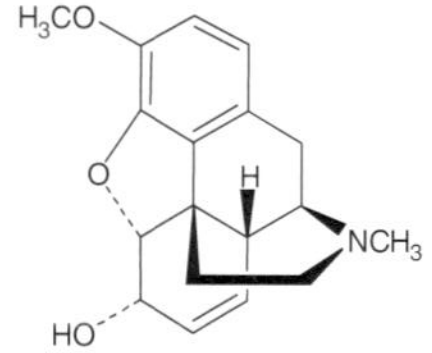	$C_{18}H_{21}NO_3$	299.36	8.21	1.19
Cyclophosphamide	Antineoplastic	6055-19-2; 50-18-0 (anhydrous form)	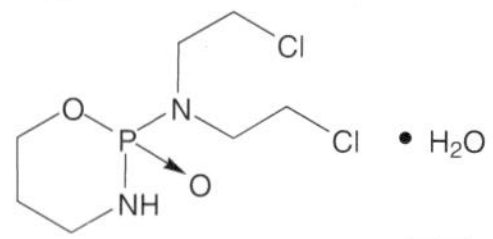	$C_7H_{15}Cl_2N_2O_2P \cdot H_2O$	279.10		0.63
Danofloxacin	Antibacterial	112398-08-0	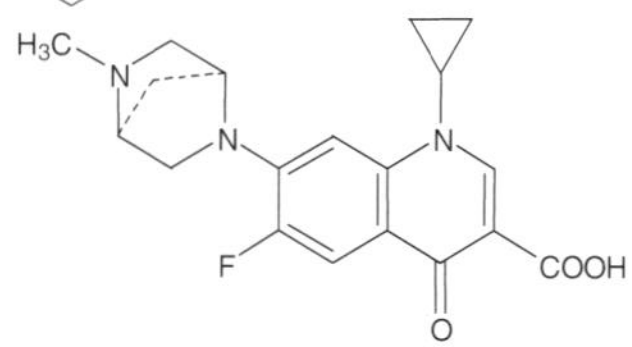	$C_{19}H_{20}FN_3O_3$	357.38	—	—
Daunorubicin	Antineoplastic	20830-81-3	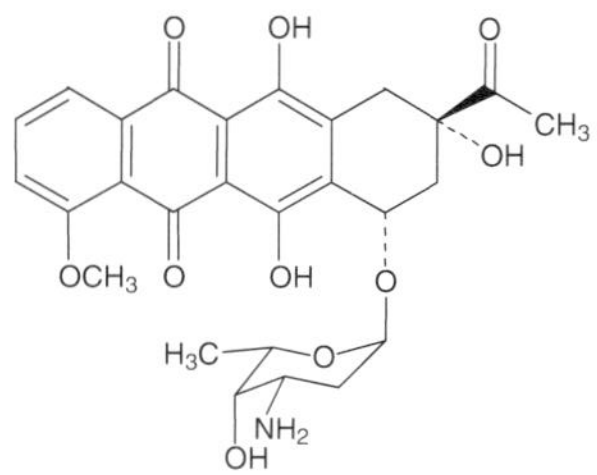 Daunomycinone Daunosamine	$C_{27}H_{29}NO_{10}$	527.52		1.83

Appendix (Continued)

Compound	Therapeutical group	CAS number	Structure	Elemental formula	Molecular weight	pK_a	log K_{ow}
Dextropropoxyphene (propoxyphene)	Analgesic (narcotic)	469-62-5	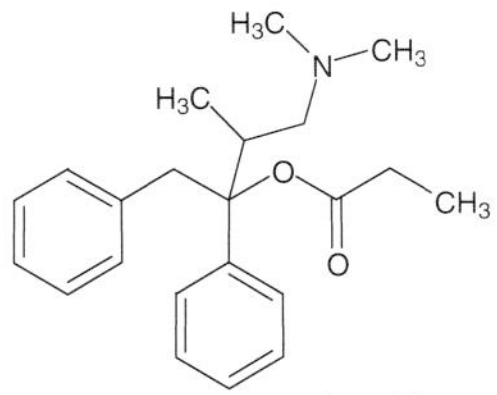	$C_{22}H_{29}NO_2$	339.47	—	4.18
Diatrizoate	X-ray contrast media	737-31-5	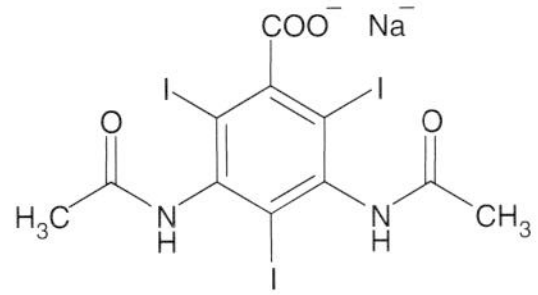	$C_{11}H_8I_3N_2NaO_4$	635.90		−1.28
Diazepam	Anxiolytic; muscle relaxant	439-14-5	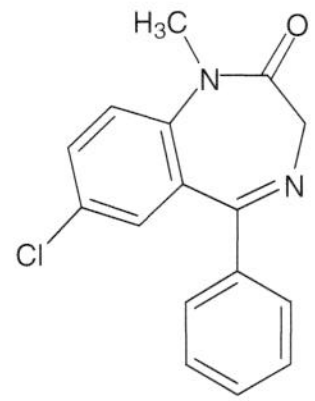	$C_{16}H_{13}ClN_2O$	284.75	3.4	2.82
Diclofenac	Anti-inflammatory	15307-86-5	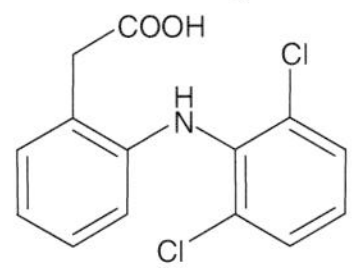	$C_{14}H_{11}Cl_2NO_2$	296.15	4.15	4.51
Dicloxacillin	Antibacterial	3116-76-5	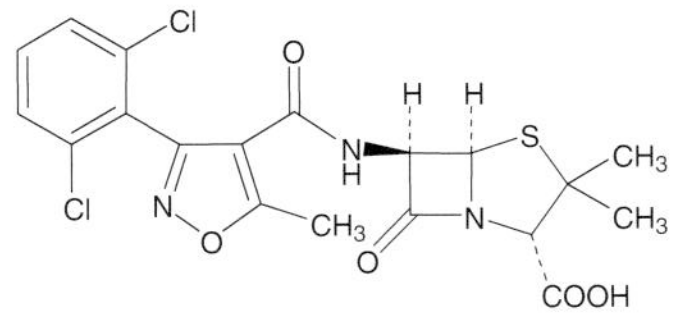	$C_{19}H_{17}Cl_2N_3O_5S$	470.33		2.91

Diphenhydramine	Antiemetic	523-87-5	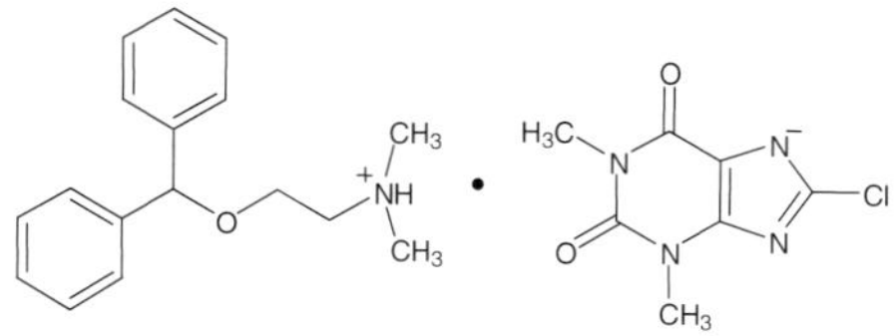	$C_{24}H_{28}ClN_5O_3$	469.97	9.0	−0.39
Doxorubicin	Antineoplastic	23214-92-8	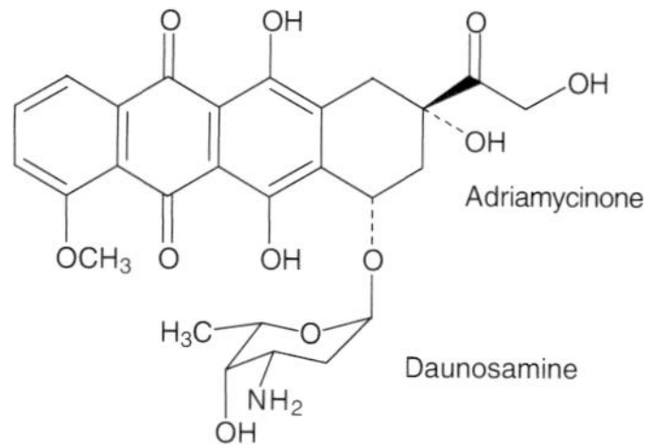	$C_{27}H_{29}NO_{11}$	543.52		1.27
Doxycycline	Antibacterial	17086-28-1 (monohydrate)564-25-0 (anhydrous)	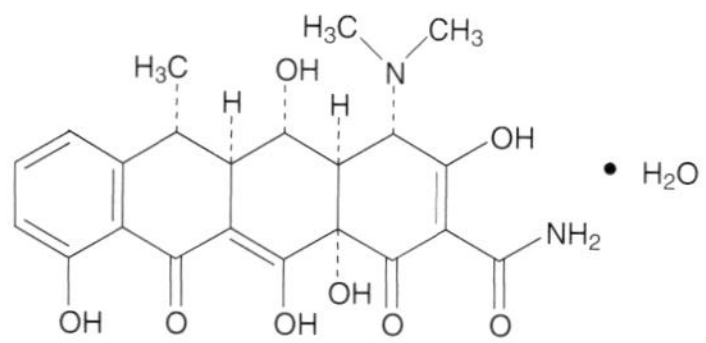	$C_{22}H_{24}N_2O_8 \cdot H_2O$	462.45		−0.02
Enalapril	Antihypertensive ACE inhibitor	75847-73-3	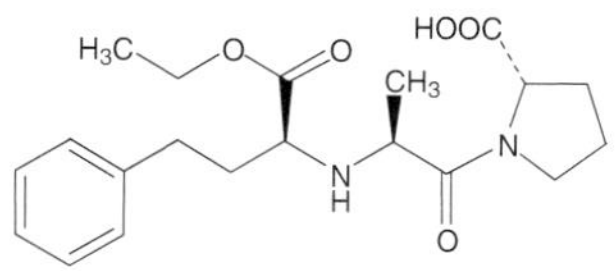	$C_{20}H_{28}N_2O_5$	376.45		0.07
Enoxacin	Antibacterial	74011-58-8	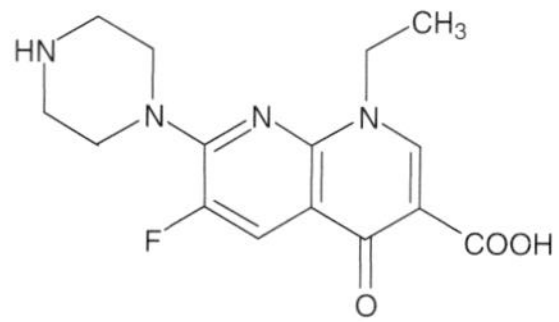	$C_{15}H_{17}FN_4O_3$	320.32		−0.20

Appendix (Continued)

Compound	Therapeutical group	CAS number	Structure	Elemental formula	Molecular weight	pK_a	log K_{ow}
Enrofloxacin	Antibacterial	93106-60-6	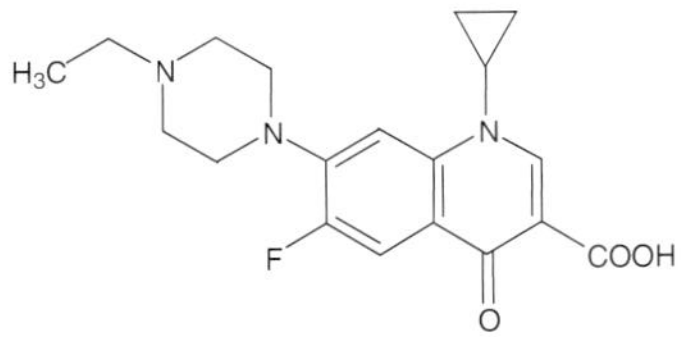	$C_{19}H_{22}FN_3O_3$	359.39		0.70
Epirubicin	Antineoplastic	56420-45-2	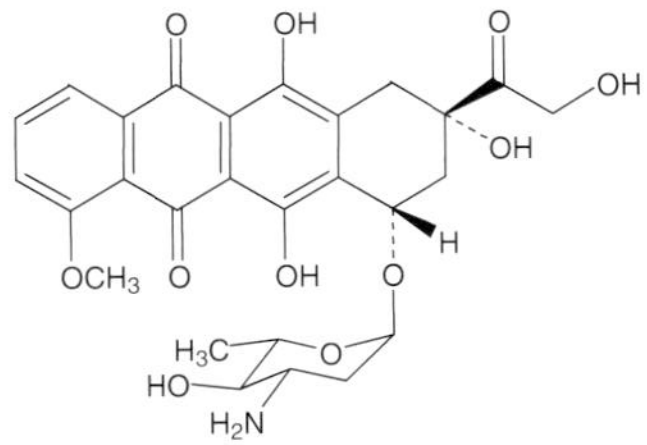	$C_{27}H_{29}NO_{11}$	543.52		1.85
Erythromycin	Antibacterial	114-07-8	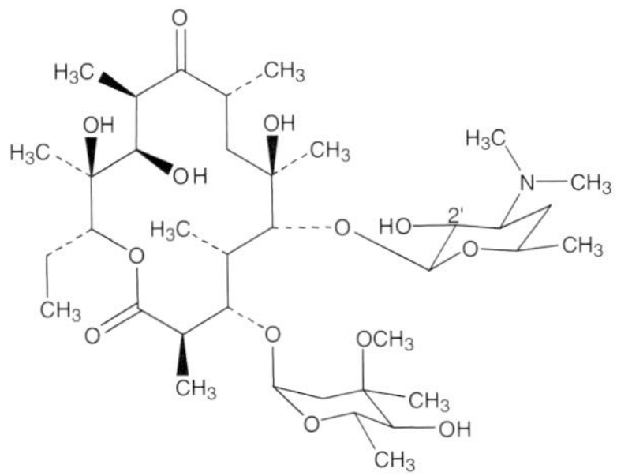	$C_{37}H_{67}NO_{13}$	733.92	8.88	3.06

β-estradiol	Estrogen	50-28-2	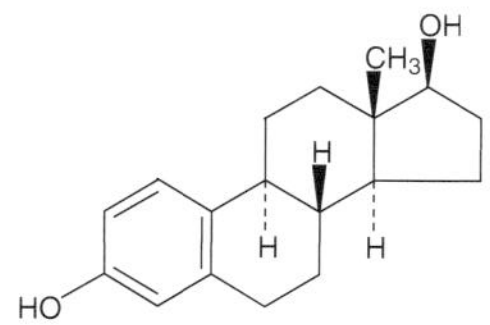	$C_{18}H_{24}O_2$	272.38	10.4	4.01
Estriol	Estrogen	50-27-1	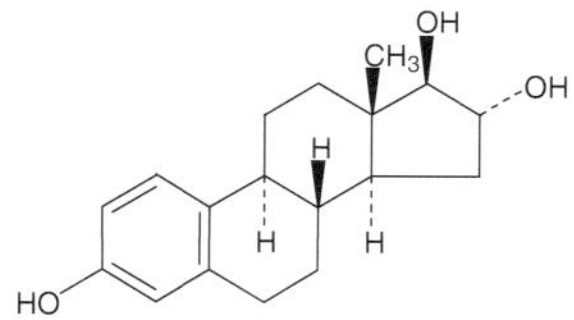	$C_{18}H_{24}O_3$	288.38	10.4	2.45
Estrone	Estrogen	53-16-7	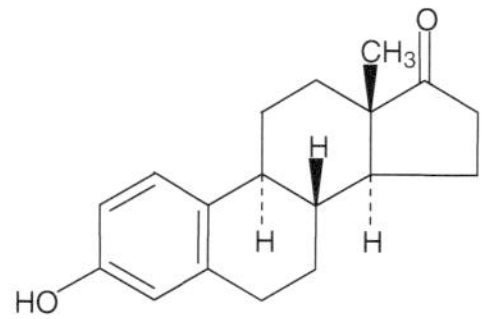	$C_{18}H_{22}O_2$	270.37	10.4	3.13
Ethinyl estradiol	Estrogen. In combination with progestogen as oral contraceptive	57-63-6	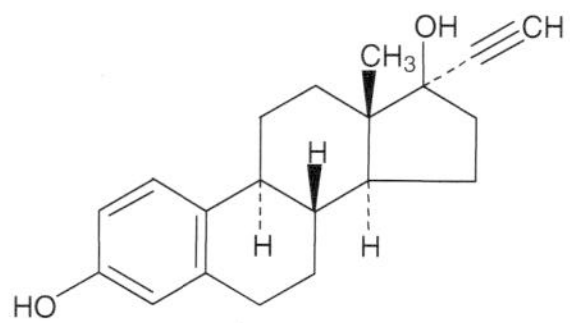	$C_{20}H_{24}O_2$	296.40	10.4	3.67
Etofibrate	Lipid regulator	31637-97-5	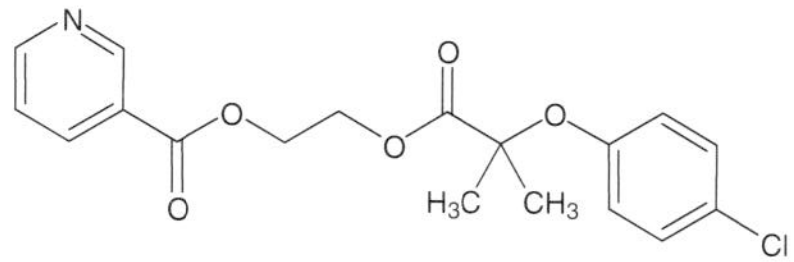	$C_{18}H_{18}ClNO_5$	363.80	—	3.43
Famotidine	Antiulcerative	76824-35-6	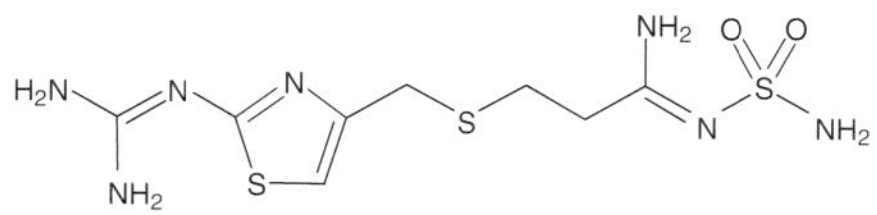	$C_8H_{15}N_7O_2S_3$	337.45	—	−0.64

Appendix (Continued)

Compound	Therapeutical group	CAS number	Structure	Elemental formula	Molecular weight	pK_a	log K_{ow}
Fenfluramine	Anorexic	458-24-2	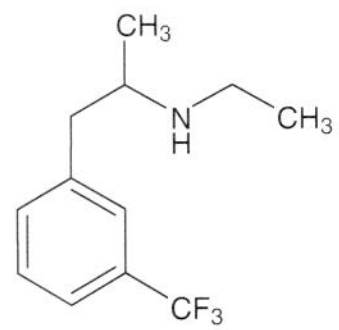	$C_{12}H_{16}F_3N$	231.26	—	3.36
Fenofibrate	Lipid regulator	49562-28-9	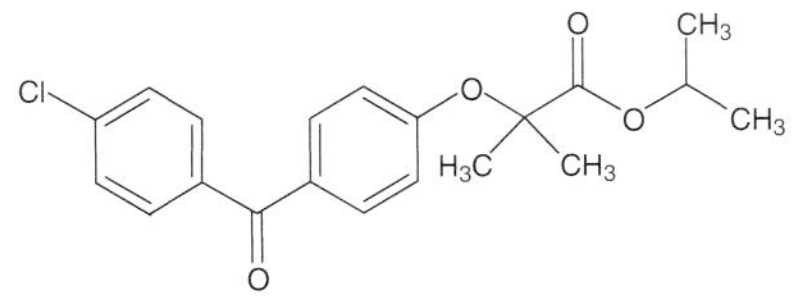	$C_{20}H_{21}ClO_4$	360.84	—	5.19
Fenoprofen	Anti-inflammatory; analgesic	31879-05-7 53746-45-5 (calcium salt dihydrate)	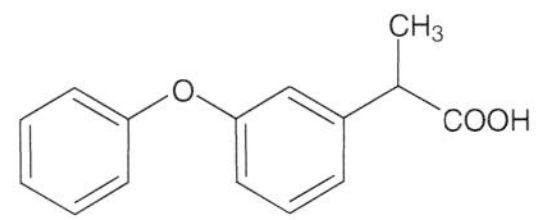	$C_{15}H_{14}O_3$	242.27	7.3	3.90
Fenoterol	Bronchodilator; tocolytic	13392-18-2	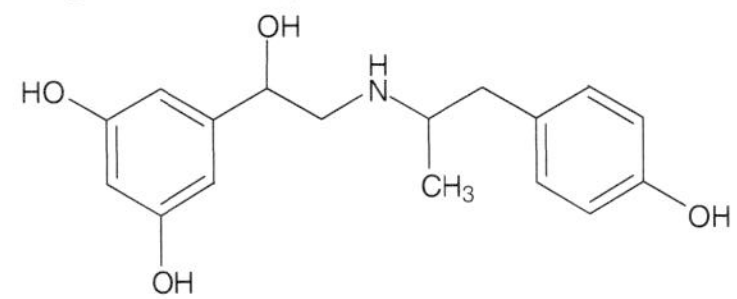	$C_{17}H_{21}NO_4$	303.35	—	1.22
Fluoxetine	Antidepressant	54910-89-3 59333-67-4 (hydrochloride)	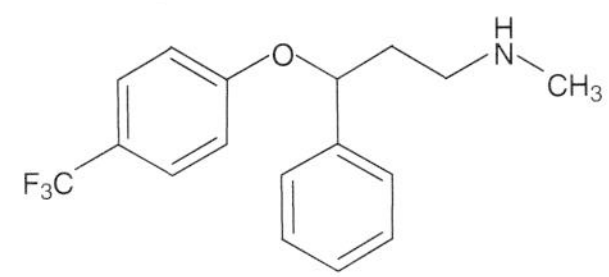	$C_{17}H_{18}F_3NO$	309.33	9.5	4.05

Flumequine	Antibacterial	42835-25-6	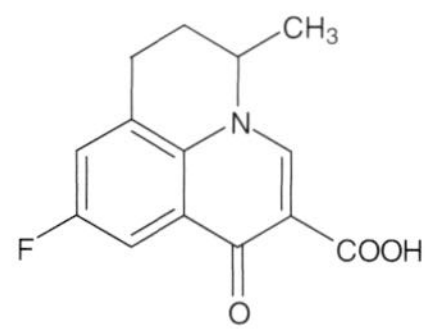	$C_{14}H_{12}FNO_3$	261.25	—	1.60
Fluorouracil (5-FU)	Antineoplastic	51-21-8	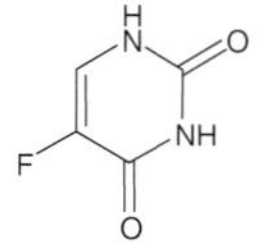	$C_4H_3FN_2O_2$	130.08	8.02	−0.89
Fluvoxamine	Antidepressant	54739-18-3	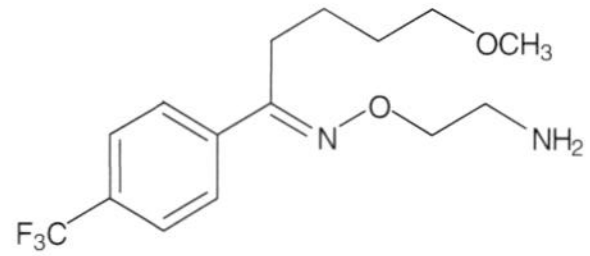	$C_{15}H_{21}F_3N_2O_2$	318.33	—	—
Furosemide	Diuretic; antihypertensive	54-31-9	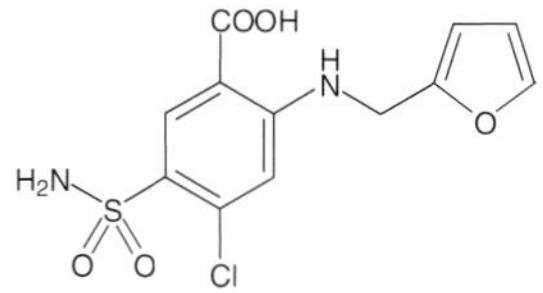	$C_{12}H_{11}ClN_2O_5S$	330.75	3.9	2.03
Gemfibrozil	Lipid regulator	25812-30-0	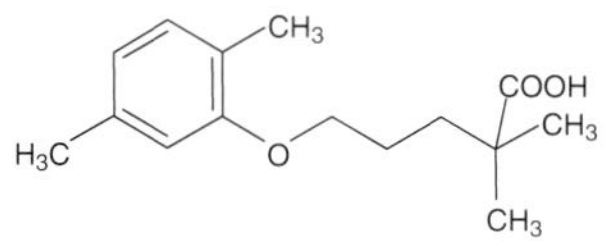	$C_{15}H_{22}O_3$	250.33	—	4.77
Glyburide (glibenclamid, glybenzcyclamide)	Antidiabetic	10238-21-8	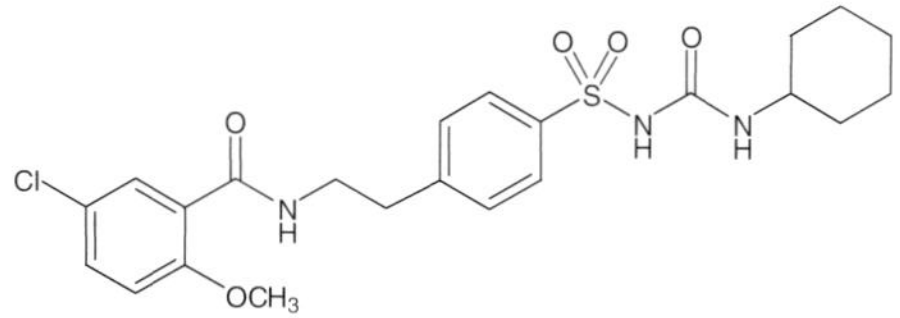	$C_{23}H_{28}ClN_3O_5S$	494.01	—	4.79

Appendix (Continued)

Compound	Therapeutical group	CAS number	Structure	Elemental formula	Molecular weight	pK_a	log K_{ow}
Hexobarbital	Sedative, hypnotic	56-29-1	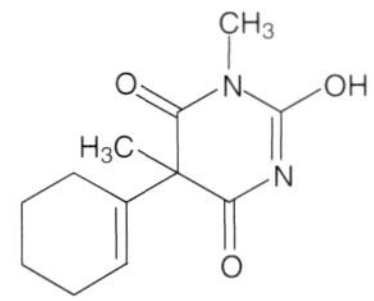	$C_{12}H_{16}N_2O_3$	236.27	8.2	1.98
Hydrochlorothiazide	Diuretic	58-93-5	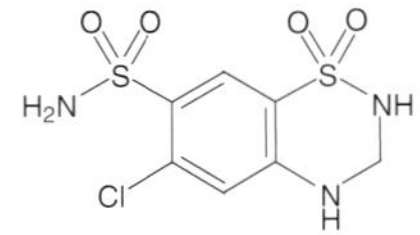	$C_7H_8ClN_3O_4S_2$	297.74	7.9	−0.07
Hydrocodone	Analgesic (narcotic)	125-29-1	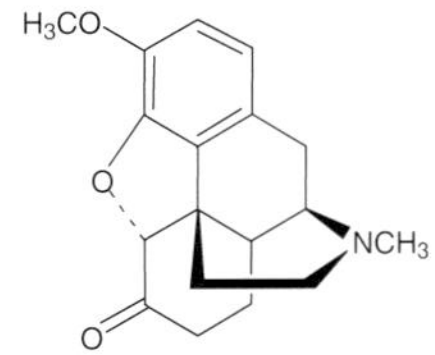	$C_{18}H_{21}NO_3$	299.36	8.48	2.16
Ibuprofen	Anti-inflammatory; analgesic; antipyretic	15687-27-1	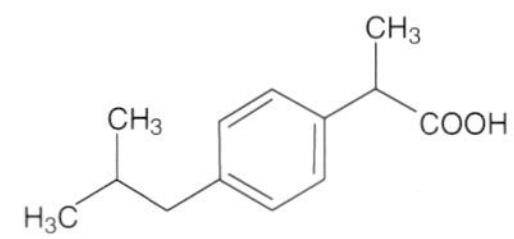	$C_{13}H_{18}O_2$	206.28	4.91	3.97
Ifosfamide	Antineoplastic	3778-73-2	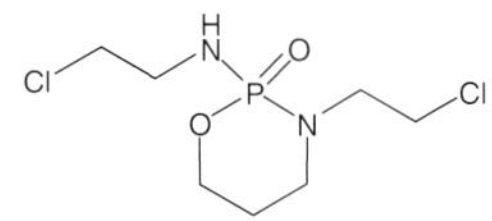	$C_7H_{15}Cl_2N_2O_2P$	261.09	—	0.86

Indomethacin	Anti-inflammatory, antipyretic, analgesic	53-86-1	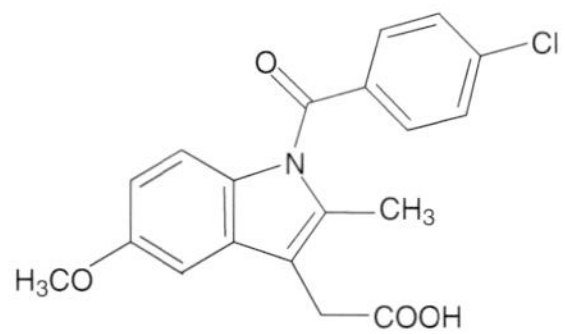	$C_{19}H_{16}ClNO_4$	357.79	4.5	4.27
Iohexol	X-ray contrast media	66108-95-0	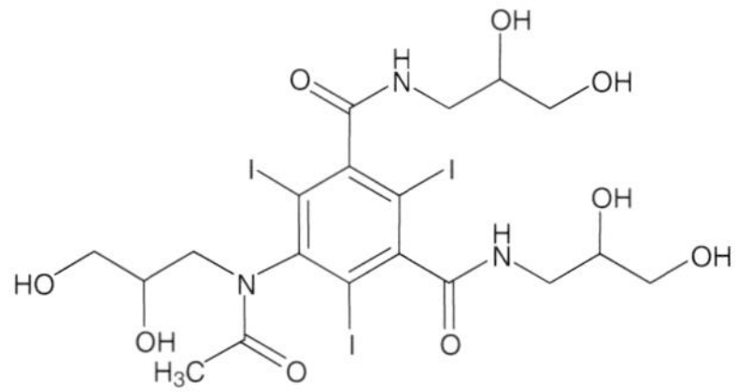	$C_{19}H_{26}I_3N_3O_9$	821.14	—	−3.05
Iomeprol	X-ray contrast media	78649-41-9	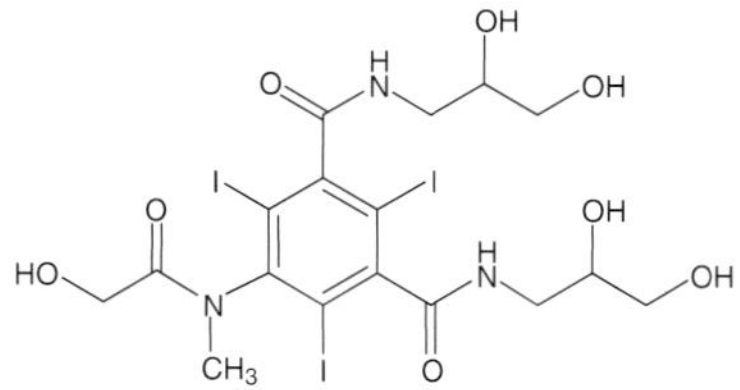	$C_{17}H_{22}I_3N_3O_8$	777.08	—	—
Iopamidol	X-ray contrast media	60166-93-0	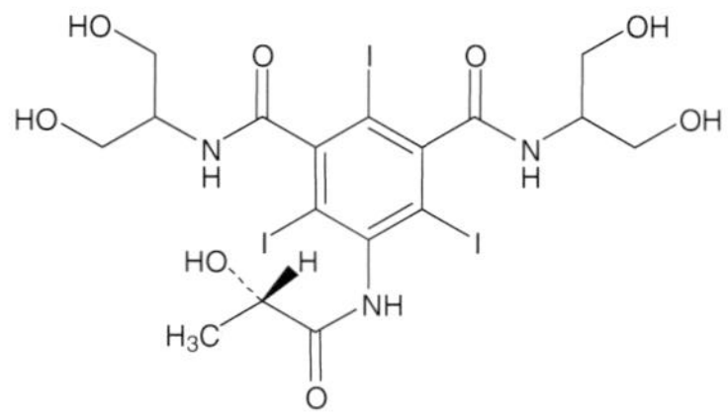	$C_{17}H_{22}I_3N_3O_8$	777.08	—	−2.42
Iopromide	X-ray contrast media	73334-07-3	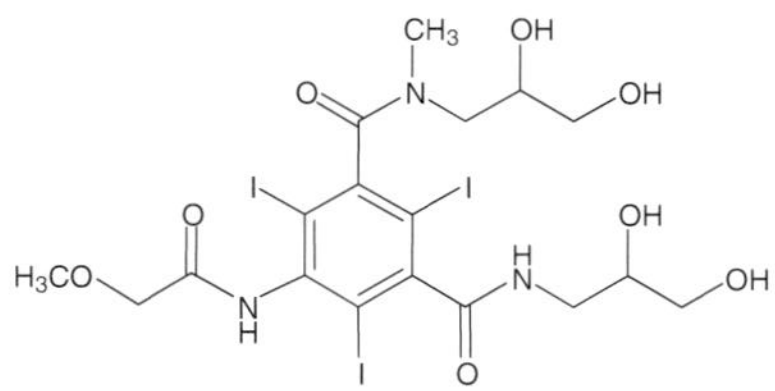	$C_{18}H_{24}I_3N_3O_8$	791.11	—	−2.05

Appendix (Continued)

Compound	Therapeutical group	CAS number	Structure	Elemental formula	Molecular weight	pK_a	log K_{ow}
Ivermectin	Anthelmintic	70288-86-7	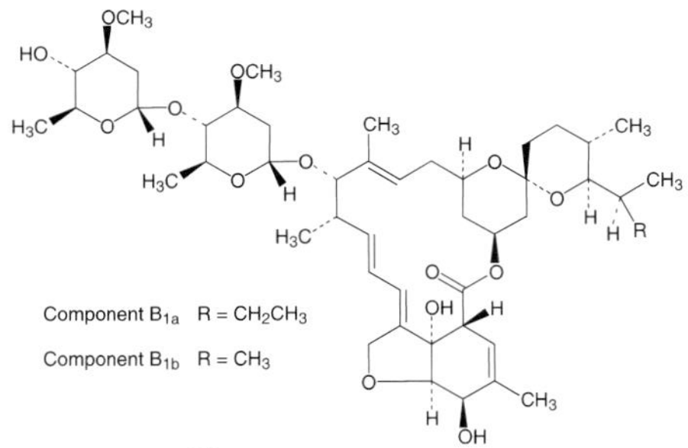	$C_{48}H_{74}O_{14}$ Component B_{1a} $C_{47}H_{72}O_{14}$ Component B_{1b}	875.09 861.06	—	—
Josamycin	Antibacterial	16846-24-5	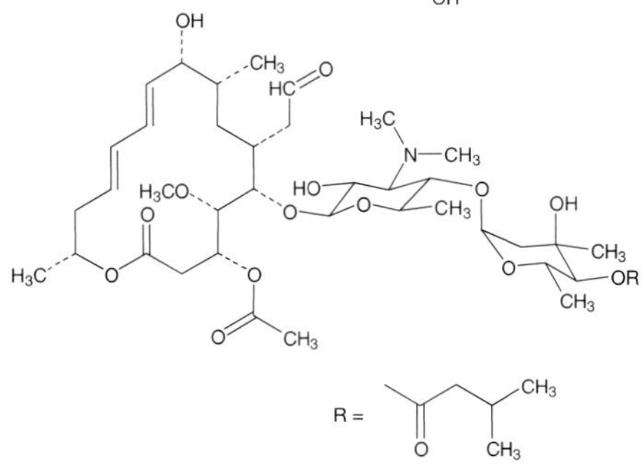	$C_{42}H_{69}NO_{15}$	827.99	—	3.16
Ketoprofen	Anti-inflammatory; analgesic	22071-15-4	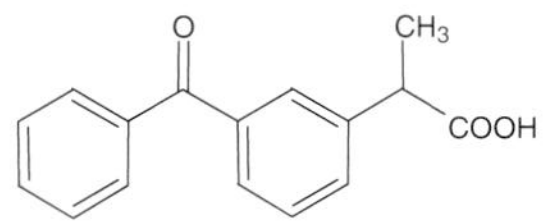	$C_{16}H_{14}O_3$	254.28	4.45	3.12

Lansoprazole	Antiulcerative	103577-45-3	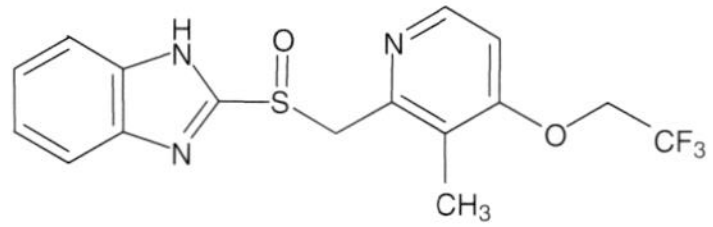	$C_{16}H_{14}F_3N_3O_2S$	369.37	—	—
Levonorgestrel	Progestogen; oral contraceptive; as contraceptive implant	797-63-7	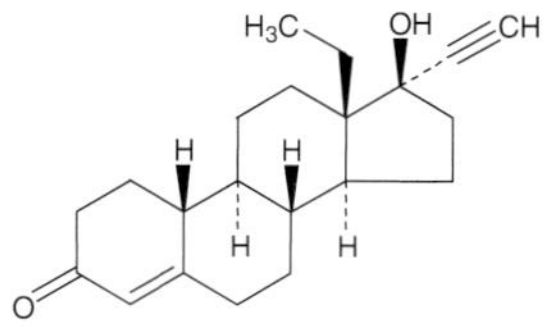 (−)-form	$C_{21}H_{28}O_2$	312.44	—	3.48
Lincomycin	Antibacterial	154-21-2	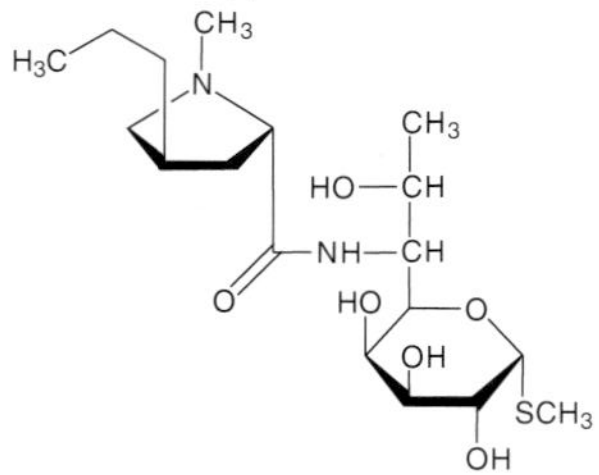	$C_{18}H_{34}N_2O_6S$	406.54	—	0.56
Lisinopril	Antihypertensive	83915-83-7	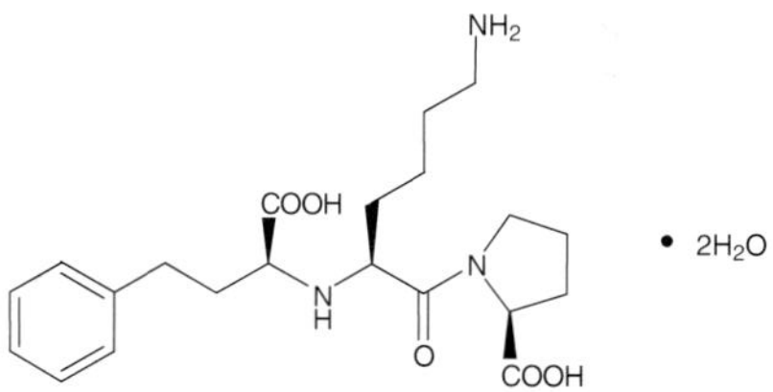	$C_{21}H_{31}N_3O_5.2H_2O$	441.52	2.5	−1.01
Loratadine	Antihistaminic	79794-75-5	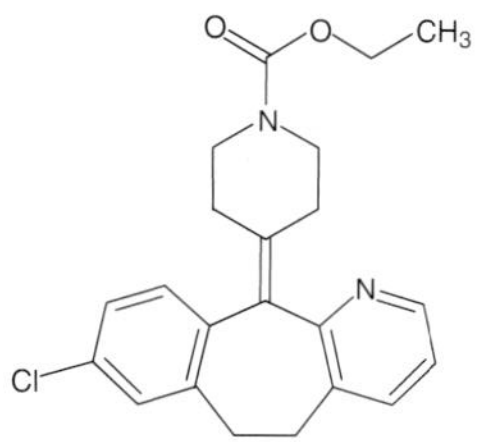	$C_{22}H_{23}ClN_2O_2$	382.89	—	5.20

Appendix (Continued)

Compound	Therapeutical group	CAS number	Structure	Elemental formula	Molecular weight	pK_a	log K_{ow}
Lorazepam	Anxiolytic	846-49-1	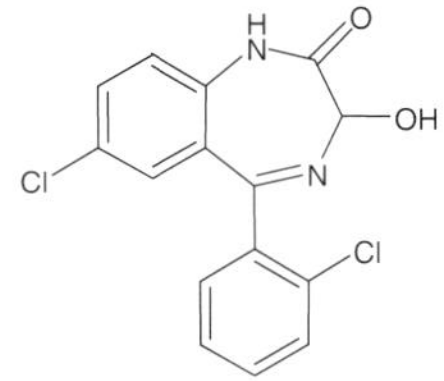	$C_{15}H_{10}Cl_2N_2O_2$	321.16	pK_1 13 pK_2 11.5	2.39
Lovastatin	Lipid regulator	75330-75-5	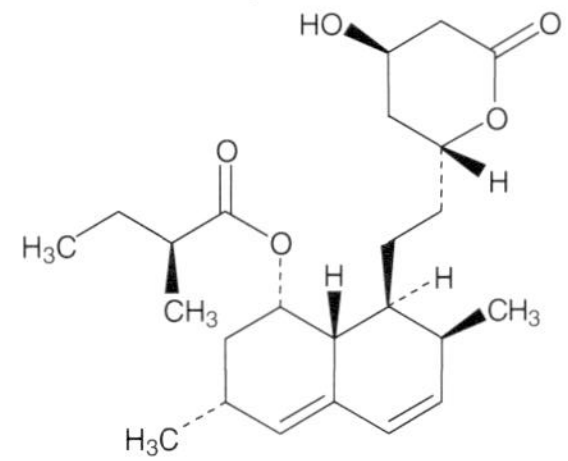	$C_{24}H_{36}O_5$	404.54	—	4.26
Meclofenamic acid	Anti-inflammatory; antipyretic	644-62-2	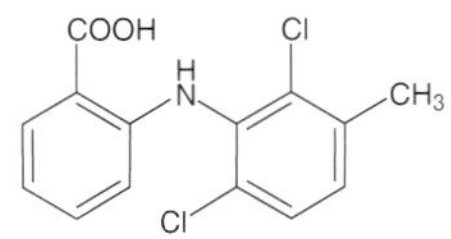	$C_{14}H_{11}Cl_2NO_2$	296.15	—	6.02
Mefenamic acid	Anti-inflammatory; analgesic	61-68-7	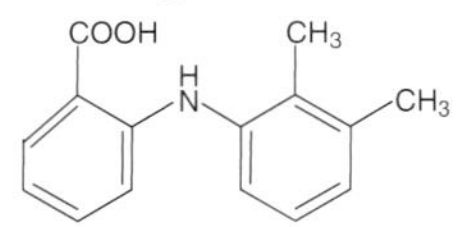	$C_{15}H_{15}NO_2$	241.28.	4.2	5.12
Meprobamate	Anxiolytic	57-53-4	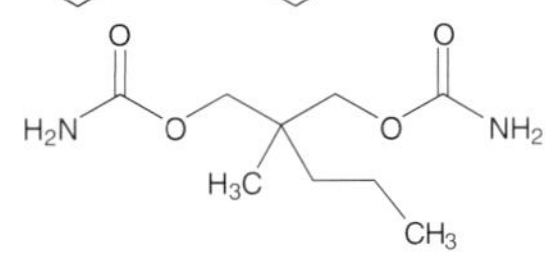	$C_9H_{18}N_2O_4$	218.25	—	0.70

Mestranol	Estrogen; in combination with progestogen as oral contraceptive	72-33-3	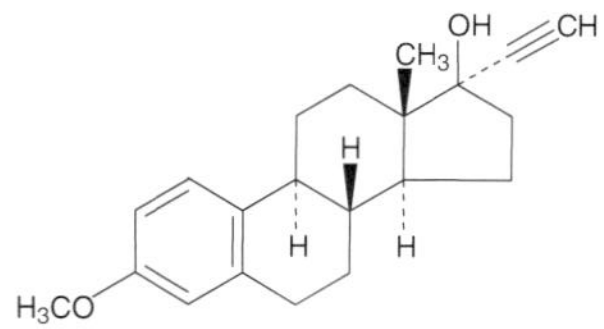	$C_{21}H_{26}O_2$	310.43	—	4.68
Metformin	Antidiabetic	657-24-9	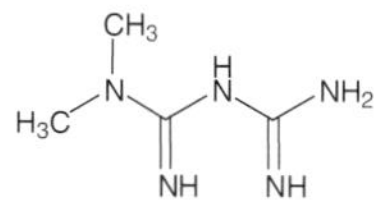	$C_4H_{11}N_5$	129.16	—	−2.64
Methicillin	Antibacterial	132-92-3	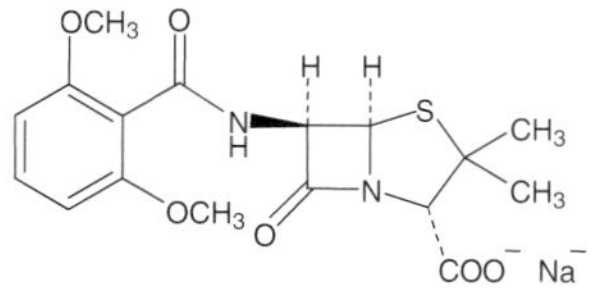	$C_{17}H_{19}N_2NaO_6S$	402.40	—	−2.66
Metoprolol	β-blocker	37350-58-6	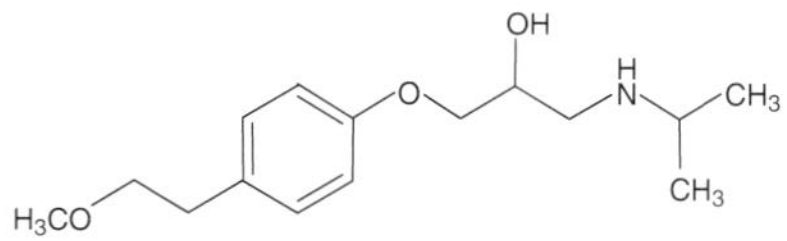	$C_{15}H_{25}NO_3$	267.36	9.6	1.88
Mevastatin	Lipid regulator	73573-88-3	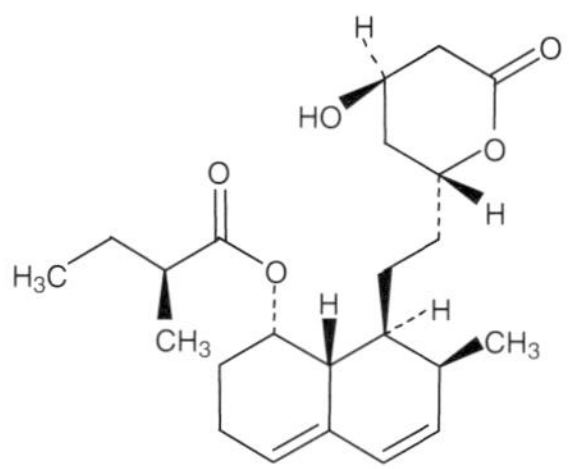	$C_{23}H_{34}O_5$	390.51	—	3.95
Minocycline	Antibacterial	10118-90-8	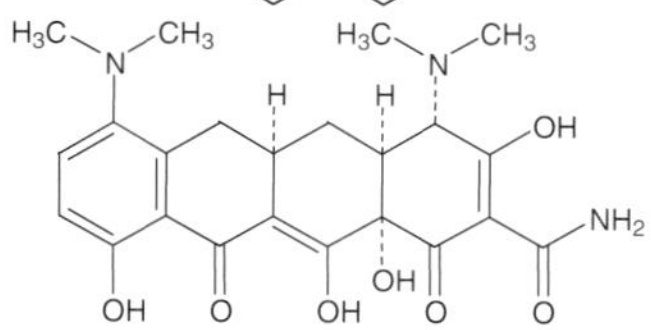	$C_{23}H_{27}N_3O_7$	457.48	—	0.05

Appendix (Continued)

Compound	Therapeutical group	CAS number	Structure	Elemental formula	Molecular weight	pK_a	log K_{ow}
Nadolol	Antihypertensive; antianginal	42200-33-9	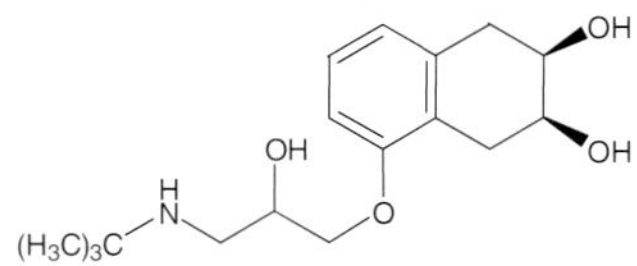	$C_{17}H_{27}NO_4$	309.40	9.67	0.81
Nafcillin	Antibacterial	147-52-4	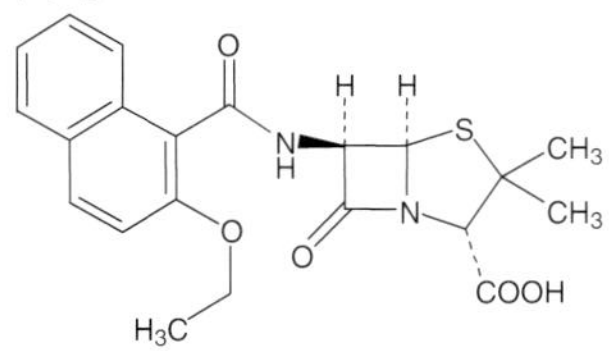	$C_{21}H_{22}N_2O_5S$	414.48	—	3.79
Naproxen	Anti-inflammatory; analgesic; antipyretic	22204-53-1	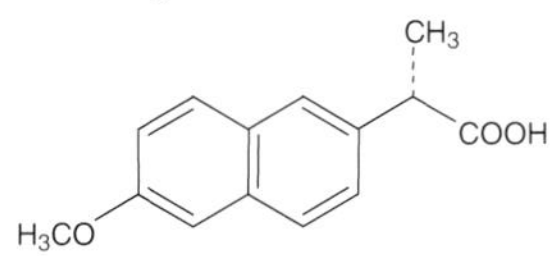	$C_{14}H_{14}O_3$	230.26	4.15	3.18
Norethindrone	Progestogen	68-22-4	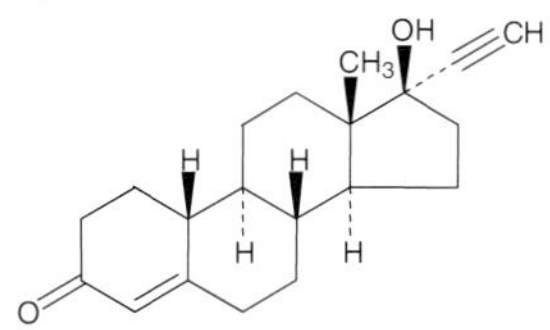	$C_{20}H_{26}O_2$	298.42	—	2.97
Norfloxacin	Antibacterial	70458-96-7	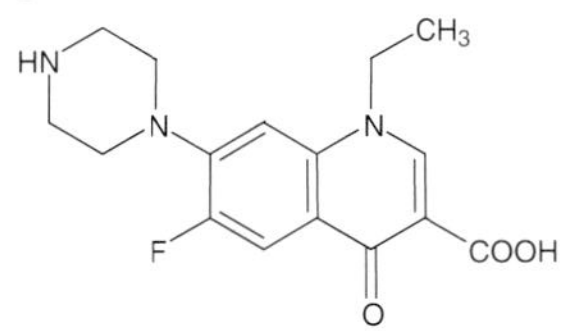	$C_{16}H_{18}FN_3O_3$	319.33	6.26	−1.03

Novobiocin	Antibacterial	303-81-1	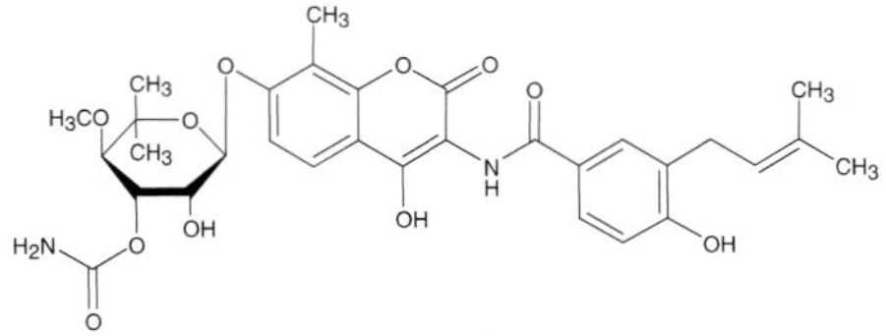	$C_{31}H_{36}N_2O_{11}$	612.62	4.3	2.45
Ofloxacin	Antibacterial	82419-36-1	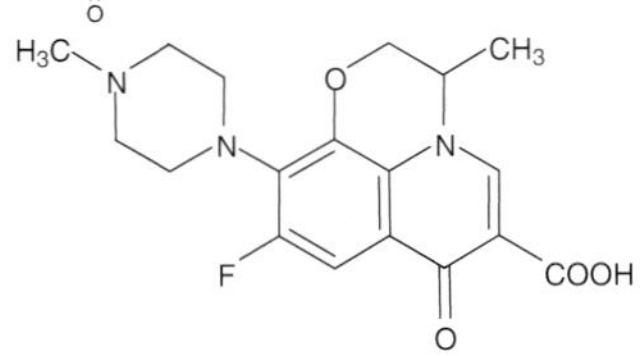	$C_{18}H_{20}FN_3O_4$	361.37	5.97	−0.39
Olanzapine	Antipsychotic	132539-06-1	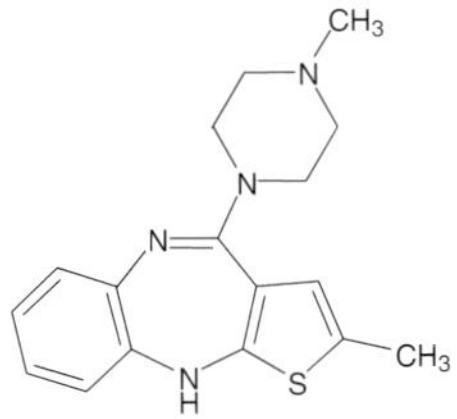	$C_{17}H_{20}N_4S$	312.44	—	—
Oleandomycin	Antibacterial	3922-90-5	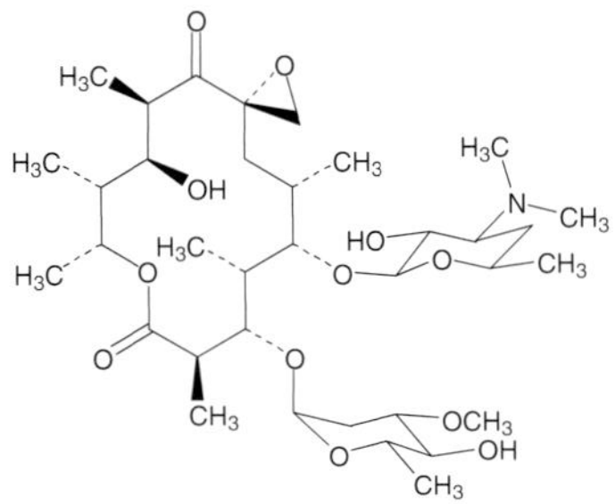	$C_{35}H_{61}NO_{12}$	687.86	8.84	1.69
Omeprazole	Antiulcerative	73590-58-6	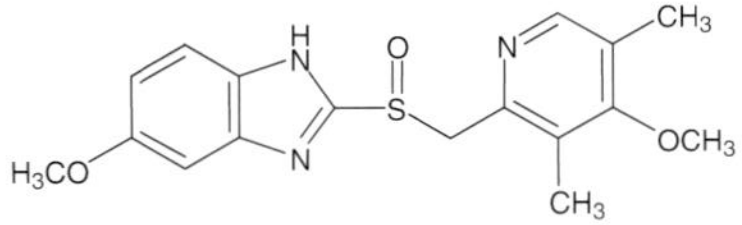	$C_{17}H_{19}N_3O_3S$	345.42	—	2.23

Appendix (Continued)

Compound	Therapeutical group	CAS number	Structure	Elemental formula	Molecular weight	pK_a	log K_{ow}
Oxacillin	Antibacterial	66-79-5	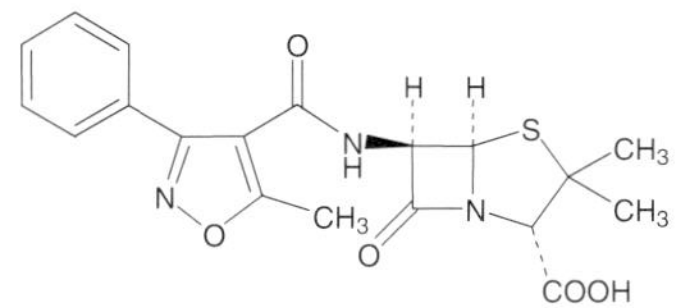	$C_{19}H_{19}N_3O_5S$	401.44	2.72	2.38
Oxazepam	Anxiolytic	604-75-1	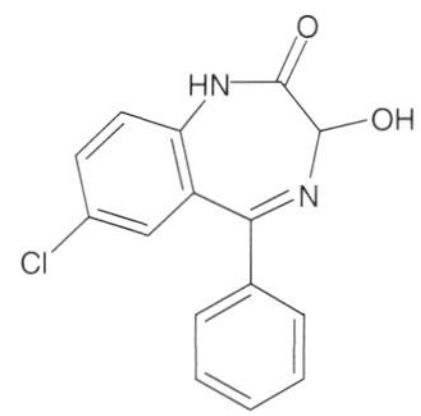	$C_{15}H_{11}ClN_2O_2$	286.72	—	2.24
Oxytetracycline	Antibacterial	79-57-2	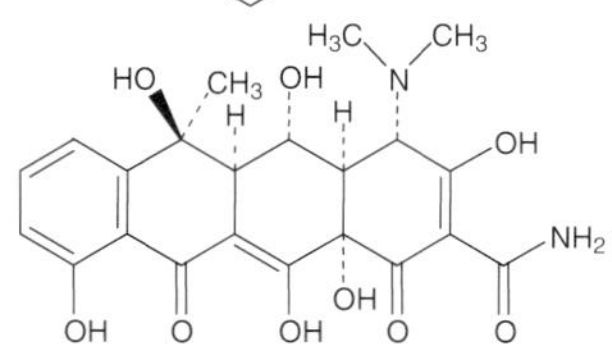	$C_{22}H_{24}N_2O_9$	460.43	3.27	−0.90
Oxprenolol	β-blocker	6452-71-7	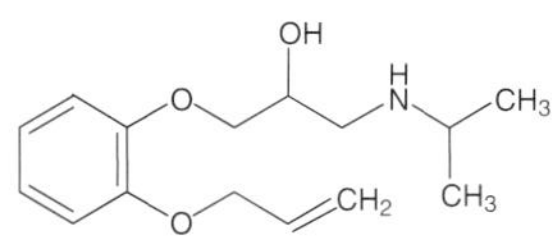	$C_{15}H_{23}NO_3$	265.35	—	2.10

Paroxetine	Antidepressant	61869-08-7	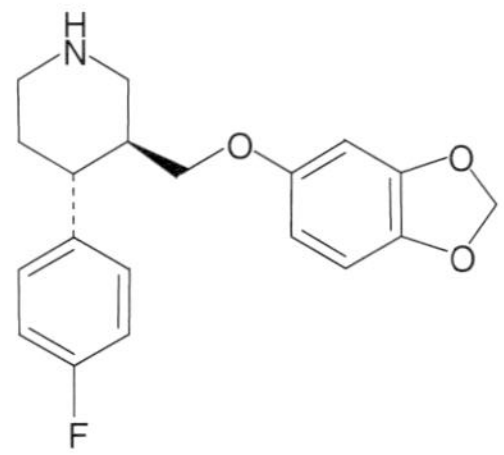	$C_{19}H_{20}FNO_3$	329.36	9.0	3.95
Penicillin G	Antibacterial	61-33-6	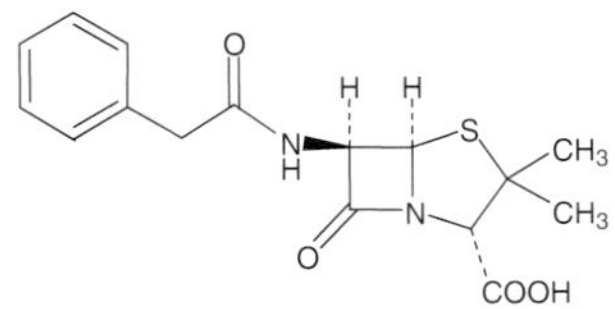	$C_{16}H_{18}N_2O_4S$	334.40	2.74	1.83
Penicillin V	Antibacterial	87-08-1	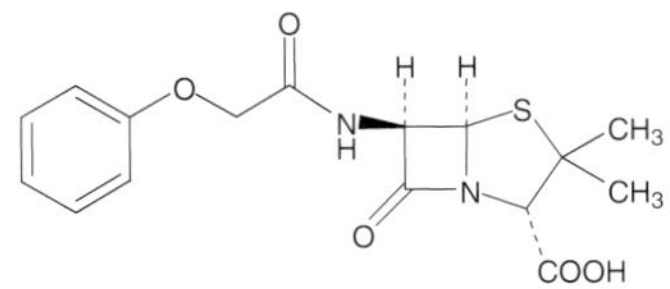	$C_{16}H_{18}N_2O_5S$	350.39	2.79	2.09
Pentobarbital	Sedative, hypnotic	76-74-4	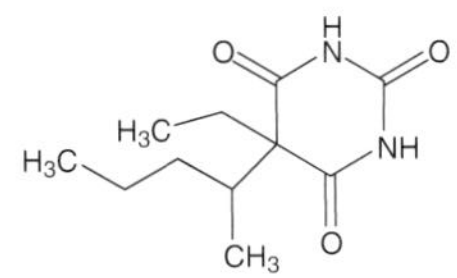	$C_{11}H_{18}N_2O$	226.27	8.11	2.10
Pentoxifylline	Blood viscosity agent	6493-05-6	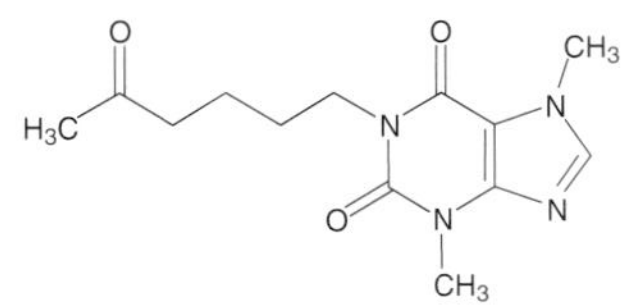	$C_{13}H_{18}N_4O_3$	278.31	—	0.29
Phenobarbital	Anticonvulsant; sedative; hypnotic	50-06-6	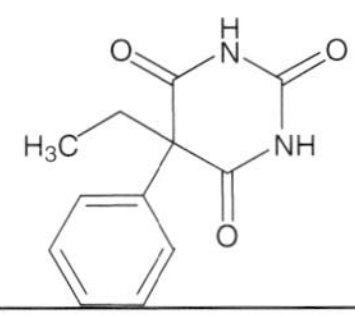	$C_{12}H_{12}N_2O_3$	232.23	7.3	1.47

Appendix (Continued)

Compound	Therapeutical group	CAS number	Structure	Elemental formula	Molecular weight	pK_a	log K_{ow}
Phenazone	Analgesic	60-80-0	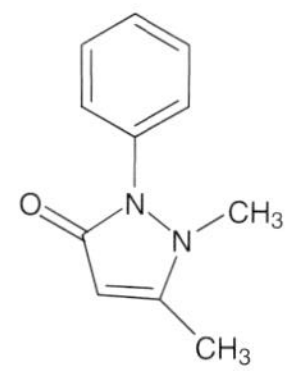	$C_{11}H_{12}N_2O$	188.23	1.4	0.38
Phenylbutazone	Anti-inflammatory	50-33-9	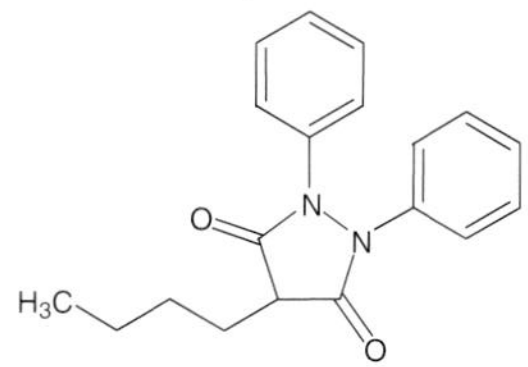	$C_{19}H_{20}N_2O_2$	308.37	4.5	3.16
Pindolol	β-blocker	13523-86-9	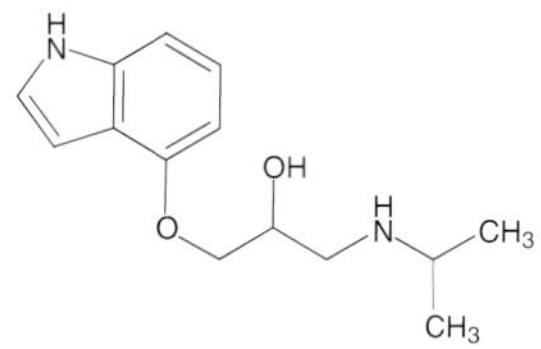	$C_{14}H_{20}N_2O_2$	248.32	9.25	1.75
Pravastatin	Lipid regulator	81131-70-6	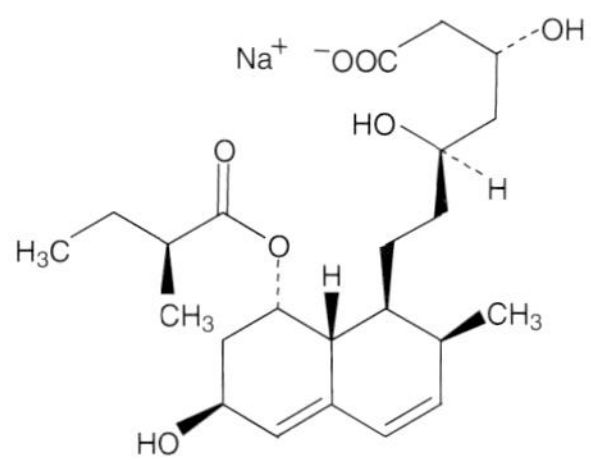	$C_{23}H_{35}NaO_7$	446.51	—	−0.23

Primidone	Anticonvulsant	125-33-7	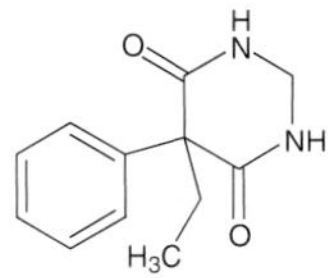	$C_{12}H_{14}N_2O_2$	218.25	—	0.91
Pregnenolone	Estrogen	145-13-1	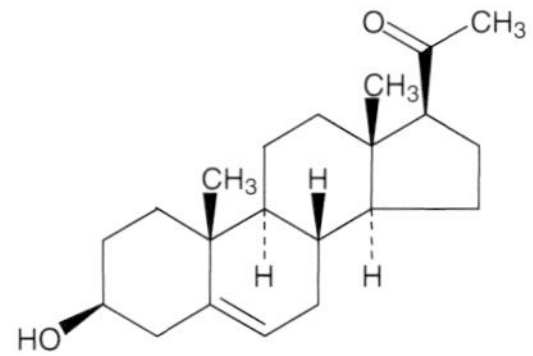	$C_{21}H_{32}O_2$	316.48	—	4.22
Progesterone	Progestogen	57-83-0	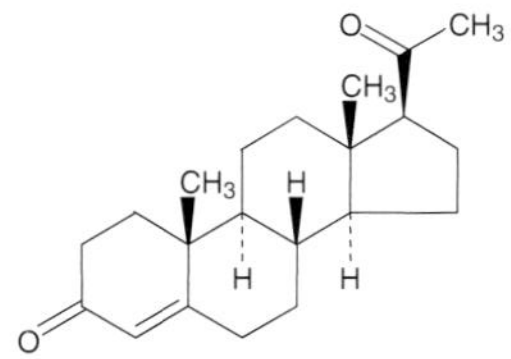	$C_{21}H_{30}O_2$	314.46	—	3.87
Propranolol	β-blocker	525-66-6	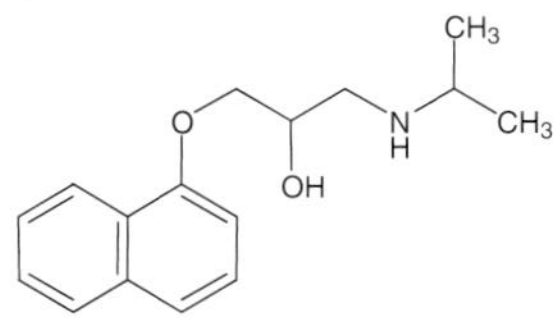	$C_{16}H_{21}NO_2$	259.34	9.42	3.48
Propyphenazone	Analgesic; antipyretic; anti-inflammatory	479-92-5	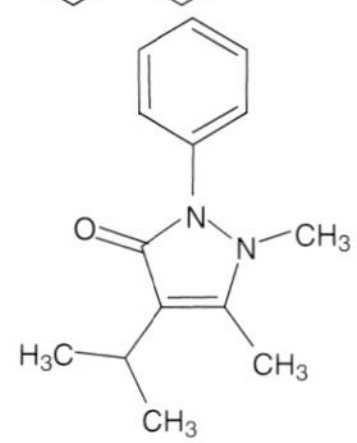	$C_{14}H_{18}N_2O$	230.30	—	1.94
Ranitidine	Antiulcerative	66357-35-5	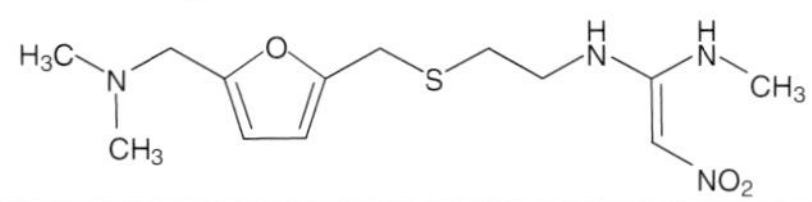	$C_{13}H_{22}N_4O_3S$	314.41	2.4	0.27

Appendix (Continued)

Compound	Therapeutical group	CAS number	Structure	Elemental formula	Molecular weight	pK_a	log K_{ow}
Roxithromycin	Antibacterial	80214-83-1	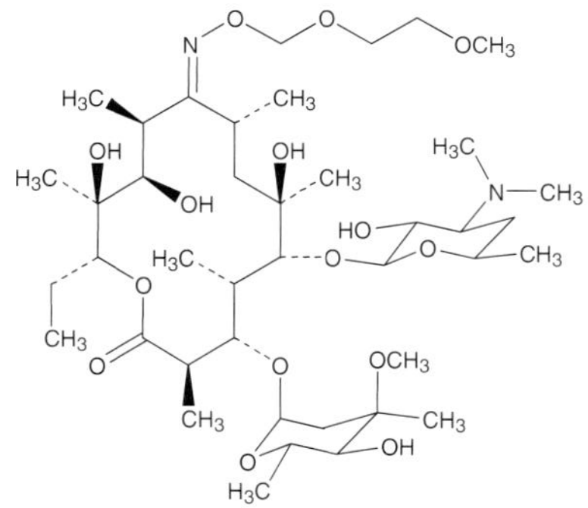	$C_{41}H_{76}N_2O_{15}$	837.04	8.8	2.75
Secobarbital sodium	Sedative; hypnotic	309-43-3	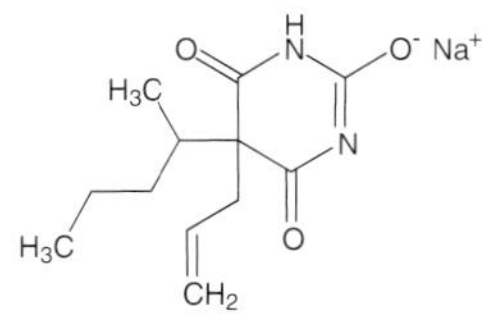	$C_{12}H_{17}N_2NaO_3$	260.26	—	−1.26
Sotalol	β-blocker	3930-20-9	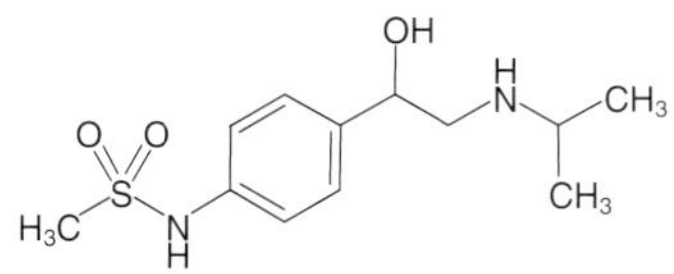	$C_{12}H_{20}N_2O_3S$	272.37	pK_1 8.2 pK_2 9.8	0.24

Spiramycin	Antibacterial	8025-81-8	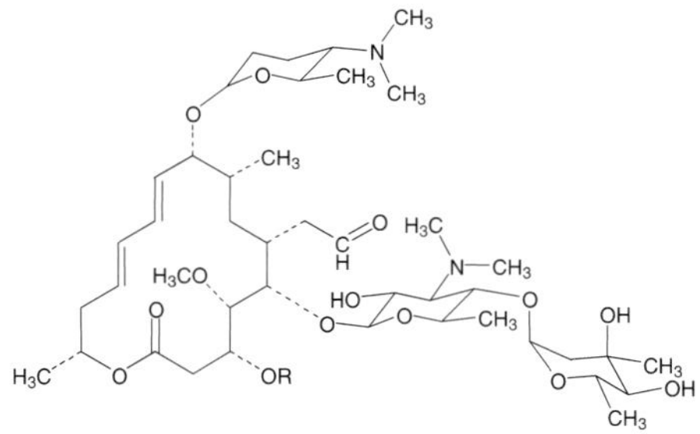 Spiramycinl R=H Spiramycinll R=COCH3 Spiramycinlll R=COCH2CH3	$C_{43}H_{74}N_2O_{14}$	843.05	—	—
Sulfadiazine	Antibacterial	68-35-9	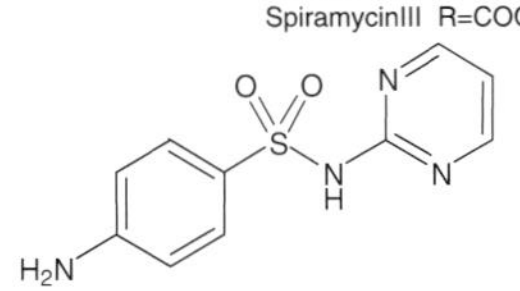	$C_{10}H_{10}N_4O_2S$	250.28	pK_1 6.36 pK_2 2.1	−0.09
Sulfamerazine	Antibacterial	127-79-7	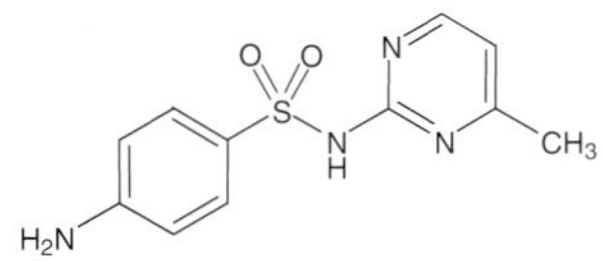	$C_{11}H_{12}N_4O_2S$	264.31	—	0.14
Sulfamethazine	Antibacterial	57-68-1	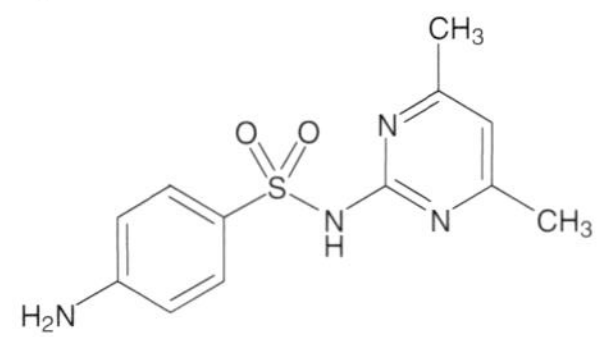	$C_{12}H_{14}N_4O_2S$	278.34	pK_1 7.59 pK_2 2.3	0.89

Appendix (Continued)

Compound	Therapeutical group	CAS number	Structure	Elemental formula	Molecular weight	pK_a	log K_{ow}
Sulfamethoxazole	Antibacterial	723-46-6	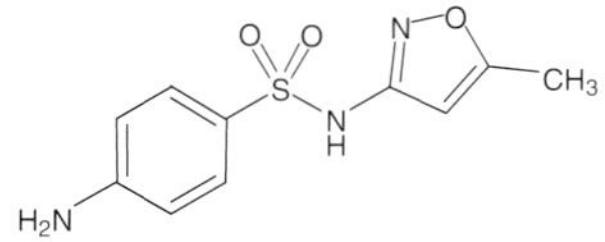	$C_{10}H_{11}N_3O_3S$	253.28	pK_1 5.7 pK_2 1.8	0.89
Sulfapyridine	Antibacterial	144-83-2	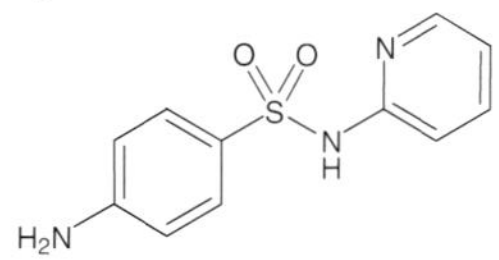	$C_{11}H_{11}N_3O_2S$	249.29	pK_1 8.43 pK_2 2.3	0.35
Terbutalin (bambuterol)	Bronchodilator	81732-65-2	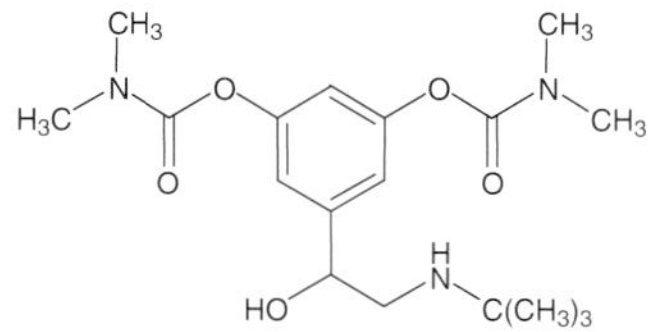	$C_{18}H_{29}N_3O_5$	367.44	—	-
Tetracycline	Antibacterial	992-21-2	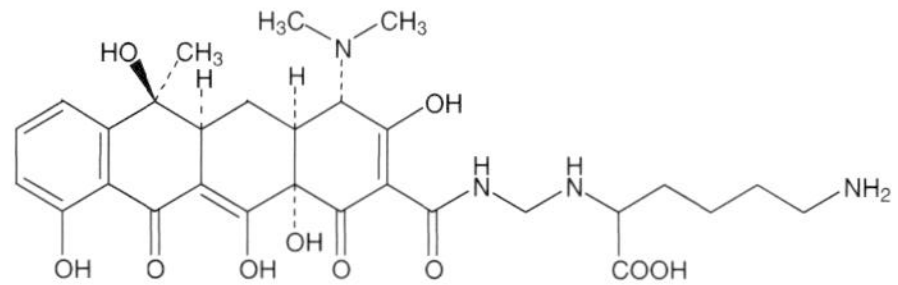	$C_{29}H_{38}N_4O_{10}$	602.63	—	−3.22
Tiamulin	Antibacterial	55297-95-5	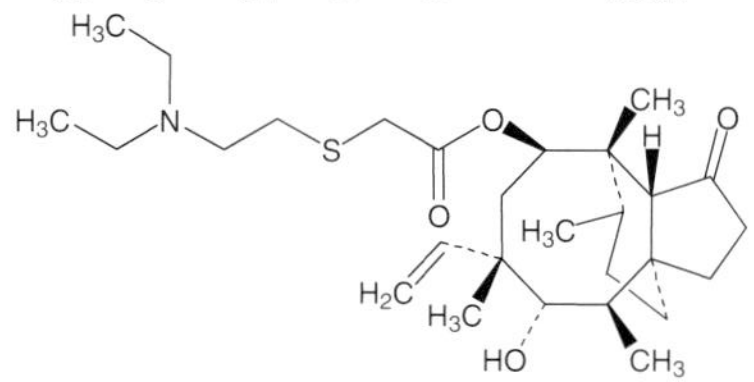	$C_{28}H_{47}NO_4S$	493.75	—	4.75

Tilmicosin	Antibacterial	108050-54-0	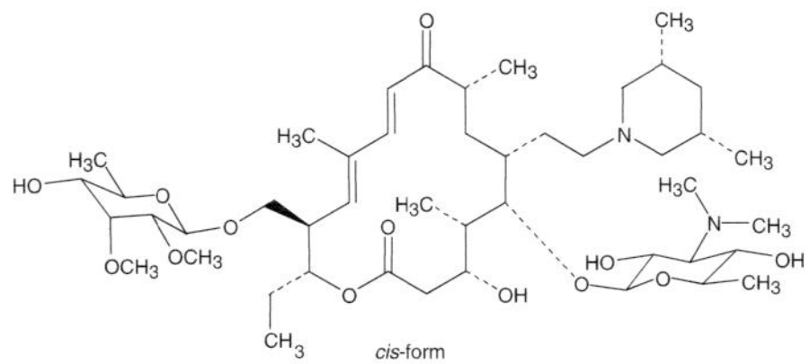	$C_{46}H_{80}N_2O_{13}$	869.13	8.18	3.80
Timolol	β-blocker	26839-75-8	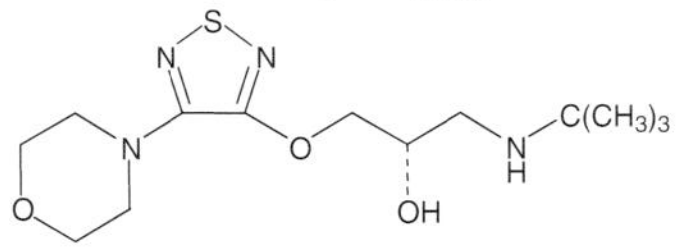	$C_{13}H_{24}N_4O_3S$	316.43	9.21	1.83
Tolfenamic acid	Anti-inflammatory; analgesic	13710-19-5	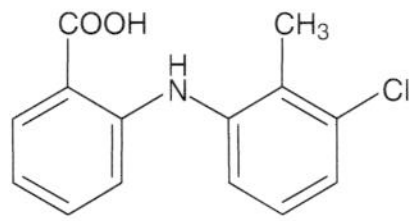	$C_{14}H_{12}ClNO_2$	261.71	—	5.17
Trimethoprim	Antibacterial	738-70-5	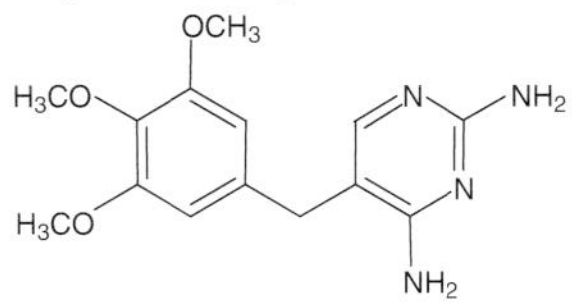	$C_{14}H_{18}N_4O_3$	290.32	7.12	0.91
Tylosin	Antibacterial	1401-69-0	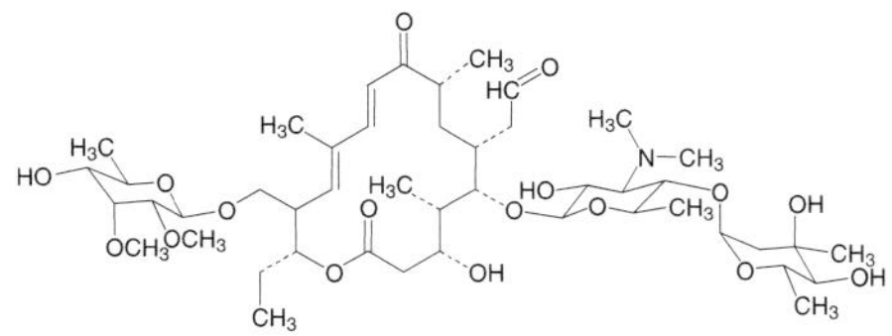	$C_{46}H_{77}NO_{17}$	916.10	7.73	1.63
Verapamil	Antihypertensive; antianginal; antiarrhythmic	52-53-9	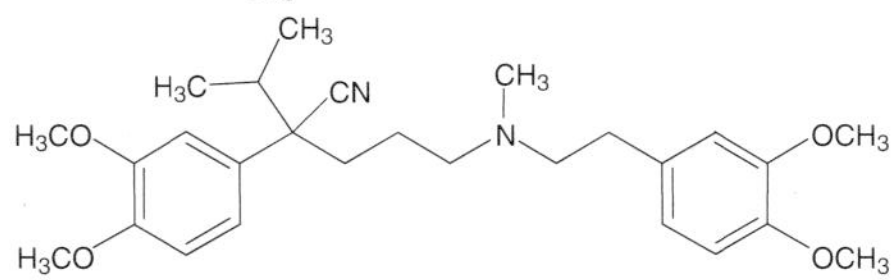	$C_{27}H_{38}N_2O_4$	454.60	8.92	3.79

Appendix (*Continued*)

Compound	Therapeutical group	CAS number	Structure	Elemental formula	Molecular weight	pK_a	log K_{ow}
Warfarin	Anticoagulant	81-81-2	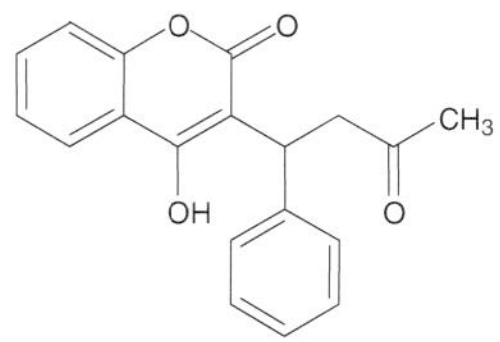	$C_{19}H_{16}O_4$	308.33	—	2.70

Subject Index